AF615146

Proceedings of the

INTERNATIONAL CONFERENCE ON NUCLEAR PHYSICS

Proceedings of the

INTERNATIONAL CONFERENCE ON NUCLEAR PHYSICS

Organised on the occasion of

The Golden Jubilee of the Indian National Science Academy

Bhabha Atomic Research Centre, Bombay, India
December 27-31, 1984

Editors: B. K. Jain
Nuclear Physics Division
Bhabha Atomic Research Centre
Bombay 40085

B. C. Sinha
Variable Energy Cyclotron Centre
Bhabha Atomic Research Centre
Calcutta 700064

Sponsored by

Board of Research in Nuclear Sciences
Department of Atomic Energy, India
Indian National Science Academy

World Scientific
Singapore • Philadelphia

Published by

World Scientific Publishing Co. Pte. Ltd.
P. O. Box 128, Farrer Road, Singapore 9128
242, Cherry Street, Philadelphia PA 19106-1906, USA

Library of Congress Cataloging in Publication Data
Main entry under title:

International Conference on Nuclear Physics

(1984 + Bhabha Atomic Research Centre)
Nuclear Physics.
1. Nuclear Physics – Congresses. I. Jain, B. K. (Bimal Kumar),
1935- II. Sinha, B. C. III. Indian National Science Academy. IV. India.
Board of Research in Nuclear Sciences. V. Title.
QC770.1515 1984 539.7 85-10805
ISBN 9971-978-28-8

Printed in Singapore by Kyodo-Shing Loong Printing Industries Pte Ltd.

INTERNATIONAL ADVISORY COMMITTEE

R. Ramanna (Chairman)	India
A. Arima	Japan
C.V.K. Baba	India
D. Brink	UK
D.A. Bromley	USA
T.E.O. Ericson	CERN
J. Hüfner	FRG
J. Huizenga	USA
P.K. Iyengar	India
S.S. Kapoor	India
M.K. Mehta	India
A.N. Mitra	India
J.O. Newton	Australia
P. Radvanyi	France
J. Rasmussen	USA
J.P. Schiffer	USA
D.K. Scott	USA
B. Sinha	India
H. Specht	FRG
V.M. Strutinsky	USSR
A. Winther	Denmark

ORGANISING COMMITTEE

P.K. Iyengar (Chairman)	Bombay
C.V.K. Baba	Bombay
A.S. Divatia	Calcutta
B.K. Jain	Bombay
S.S. Kapoor	Bombay
G.K. Mehta	Kanpur
A.N. Mitra	Delhi
M.K. Pal	Calcutta
J. Parikh	Ahmedabad
L. Satpathy	Bhubaneswar
B. Sinha	Calcutta
C. Warke	Bombay

PREFACE

The International Conference on Nuclear Physics was held at Bhabha Atomic Research Centre, Bombay, during December 27-31, 1984, on the occasion of the golden jubilee of the Indian National Science Academy. The Conference was aimed at providing an overview of the recent advances in our understanding of nucleons and nuclei. Nuclear Physics Research is passing through an exciting phase in recent years. While conventional studies relating to nuclear structure and low energy reactions are becoming more and more detailed, the current experimental and theoretical investigations over a wide range of energies are bringing out many new features such as the sub-nucleonic degrees of freedom in nuclei, new phases of nuclear matter etc. These frontier areas were reviewed during the Conference by about 40 invited speakers from different countries. In addition, about 120 contributions were also received for presentation during the Conference. While, for reasons of time, only 22 of these could be orally presented in plenary sessions, the remaining were presented in poster sessions.

Nuclear Physics Research in India owes a great deal to Dr. Raja Ramanna, who completed 60 years of age on 28th January 1985. Apart from his personal contributions in the area of nuclear fission, he played a pivotal role in planning and nurturing Nuclear Research in the country. A special session on fission and related phenomena was, therefore, organised during the Conference to commemorate the 60th birthday of Dr. Raja Ramanna.

The Conference was attended by about 350 participants from India and abroad. A booklet containing the abstracts of all the invited talks and the contributed papers was distributed to the participants during the Conference. The present volume is a collection of the detailed write-ups of the invited talks.

An evening programme of Indian Classical music and dance arranged during the Conference added novelty and a very welcome departure from the seriousness of the Conference. The programme consisted of a sitar recital and a colourful display of Bharata Natyam by eminent artists.

A large number of individuals and organisations have contributed to the overall success of the Conference. The Organising Committee, working in consultation with the International Advisory Committee, was responsible for the selection and organisation of the programme. The personal interest and supervision of Dr. P.K. Iyengar, Director, Bhabha Atomic Research Centre added to make the Conference a success.

Bombay
March 8, 1985

B.K. Jain
B.C. Sinha

TABLE OF CONTENTS

---

*Manuscript not received

Proceedings of the

INTERNATIONAL CONFERENCE ON NUCLEAR PHYSICS

FRAGMENTATION OF EXPANDING DROPS

V. R. Pandharipande

Department of Physics
University of Illinois at Urbana-Champaign
1110 W. Green Street, Urbana, Illinois 61801

A. Vicentini and G. Jacucci

Dipartimento di Fisica
Universita di Trento, 38050 Povo, Italy
and
Department of Physics
University of Illinois at Urbana-Champaign
1110 W. Green Street, Urbana, Illinois 61801

ABSTRACT

We have studied time evolution of hot drops of matter containing ~ 230 or ~ 130 particles with the molecular dynamics method. The initial states have uniform density and sharp surface. A wide range of initial temperatures is chosen to study the phenomenon of evaporation, fragmentation and total vaporization in a unified fashion. Evaporation and vaporization are respectively low and high energy phenomenon whereas fragmentation occurs at intermediate energies. We find that the liquid-gas phase transition plays a crucial role in determining the initial conditions that lead to fragmentation. In classical systems fragmentation seems to occur at densities ~ 1/4th the liquid density, in the region of adiabatic instabilities. This region has a maximum temperature of ~ $T_c/2$, where T_c is the critical temperature for the phase transition. It thus appears unlikely that fragmentation of small drops can be used to study the isothermal critical region of the phase transition. The properties of monomers emitted in the expansion of the drop are discussed.

1. INTRODUCTION

It is hoped that we may observe phase transitions in nuclear matter via heavy ion reactions. At lower energies the liquid-gas transition is of interest, while at higher energies we hope to see the effects of a phase transition to quark-gluon plasma. During the course of this meeting many speakers will discuss these possibilities (see papers by Hüfner, Sinha, Satpathy and Satpathy). In this paper we discuss the effect of liquid-gas phase transition on fragmentation of hot classical drops in a very elementary fashion.

The phase diagram of extended nuclear matter, obtained by switching off the coulomb interaction, is shown in Fig. 1. The curves in Fig. 1 have been calculated by Ravenhall[1)] by fitting a Skyrme type equation of state to results of microscopic calculations.[2)] CE denotes the coexistence curve, while ITS and AS are the isothermal and adiabatic spinodal curves. The spinodals are locus of points at which $\partial P/\partial \rho$ along an isotherm or adiabat is zero. Matter is dynamically unstable towards isothermal (adiabatic) density fluctuations in the regions enclosed by ITS (AS) curves. We note that the ITS meets the CE curve at the critical temperature T_c, while the AS has a maximum temperature of $\sim T_c/2$.

The interest in studies of fragmentation of drops containing ~ 200 particles was generated by observation of nuclear fragmentation in high energy proton-nucleus[3)] and heavy-ion collisions.[4)] Many researchers[5-10)] have attempted to understand fragmentation as a consequence of the liquid-gas phase transition. These studies have raised interesting questions, the most fundamental being "do such small systems show effects of phase transitions expected in extended systems containing billions of particles?". The simplest theoretical attempts[5,6)] assume that fragmentation occurs at well defined values of temperature and density, and relate the fragment mass yield to the equilibrium thermodynamic composition at that temperature and density. In these attempts the exponential $A^{-\lambda}$ yield of fragments of mass A, observed in some experiments, is attributed to fragmentation at the critical temperature T_c. On the

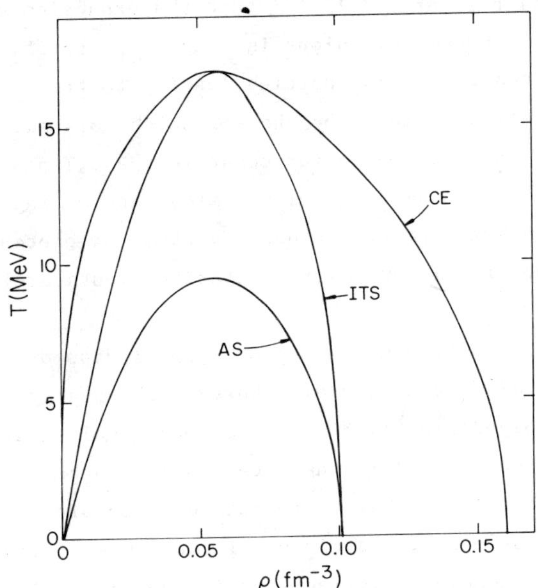

Fig. 1: Phase diagram of nuclear matter.

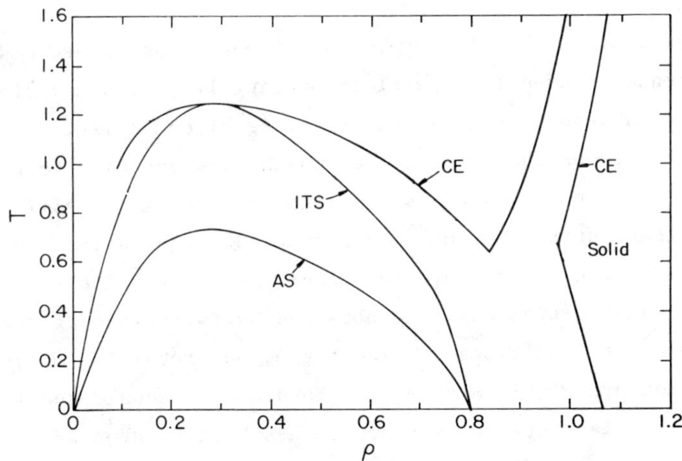

Fig. 2: Phase diagram of argon, as obtained with the truncated Lennard-Jones potential, in reduced units.

other hand it has been argued[7,9,10] that the expansion of hot nuclear matter produced in these collisions is almost adiabatic. In this case we expect density inhomogeneities leading to fragmentation to develop in the region enclosed by the AS, which does not reach up to T_c. Hence where does fragmentation occur in the ρ,T plane, and how is it related to the phase diagram, are also interesting questions. The properties (energy spectra, abundance etc.) or protons, neutrons and the light nuclei emitted in fragmentation events are also of interest.

The expansion and fragmentation of a small drop of hot quantum liquid is a very difficult problem. However the analogous classical problem can be solved exactly by computer bases molecular dynamics. Many of the arguments given in the literature on nuclear fragmentation do not specifically relate to quantum mechanics. Hence we can hope that the solutions of the classical problem will give us more insight into the phenomena of fragmentation and its relation to evaporation and total vaporization.

2. CALCULATIONS

We have studied the time evolution of hot drops of argon with a truncated Lennard-Jones interparticle potential. The phase diagram of argon, calculated with this potential, is given in Fig. 2. But for the existence of the solid phase, which does not seem to play a significant role in the dynamics of fragmentation, it is similar to that of nuclear matter. A cube containing 432 argon atoms is equilibrated at desired temperature and density with periodic boundary conditions. The initial state is obtained by switching off the periodic boundary conditions at time t = 0, and retaining only those particles that are within a sphere whose diameter equals the length of the cube. This state corresponds to a spherical drop of matter containing ~ 230 particles at the desired values of ρ and T. However, the surface of this drop is artifically sharp. The expansion of this drop is studied by calculating the motion of each of the particles by solving the coupled Newton's equations with the molecular dynamics method. We calculate the positions and velocities of all the particles as a function of time. From these we estimate the variation of the average density and temperature of the matter inside the expanding drop with time. The composition of the final state is also calculated.

The expansion from a given initial condition is called an event. The events are denoted by X(e), where the letter X gives the density and e the energy (per atom) of the initial state in reduced units. The reduced unit for energy if 119.8 K, and the binding energy of argon at T = P = 0 is ~ 6 units. X = B,C and D respectively denote events with initial density .82, .58 and 1.2, in reduced units. The reduced unit of length is 3.405 Å, and the density of argon liquid at low temperatures is ~ .82 in these units. D-events with $e < 2$ have solid initial states, but we do not find noticeable effects of it. The e is a constant of motion, and is found to be a very important parameter.

The details of the calculation can be found in ref. 11. Here we merely show how a typical fragmentation event looks on a computer. Figures 3-6 show the X,Y coordinates of the particles at times t = 4, 7, 16 and 60 in units of 10^{-12} sec in the event B(1.62). The velocity of sound in argon liquid at $\rho = .84$ is 860 m/sec. Thus the time required for sound to travel a distance equal to the radius of the initial drop is 1.6×10^{-12} sec. At t = 0 the particles were uniformly distributed in a sphere of radius four units at density = 0.82. At t = 4 the average central density ρ_{ac} has dropped down to ~ 0.2, and density inhomogeneities are noticable, though the system has not fragmented. By t = 16 the fragmentation is almost over, and at t = 60 the fragments are well separated. Detailed studies of ~ 25 events with ~ 230 particles, and 9 events with ~ 130 particles are reported in ref. 11. The events with ~ 130 particles are denoted by b(e); they have initial density of .82. Here we merely summarize the conclusions drawn from this study.

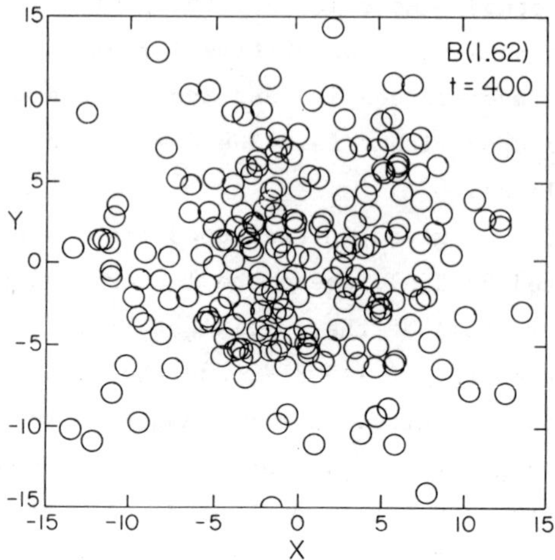

Fig. 3: The X and Y coordinates of particles in event B(1.62) at $t = 4 \times 10^{-12}$ sec in reduced units. The circles representing the particles have diameter of ~ 1, while the range of interaction is 3.

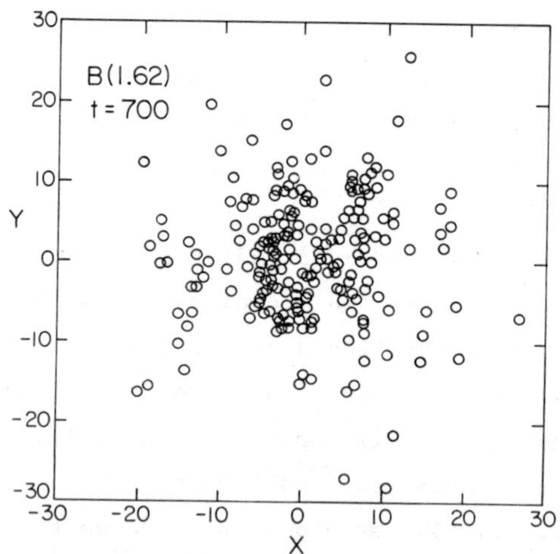

Fig. 4: The X and Y coordinates of particles in event B(1.62) at $t = 7 \times 10^{-12}$ sec. See caption of Fig. 3 for more details.

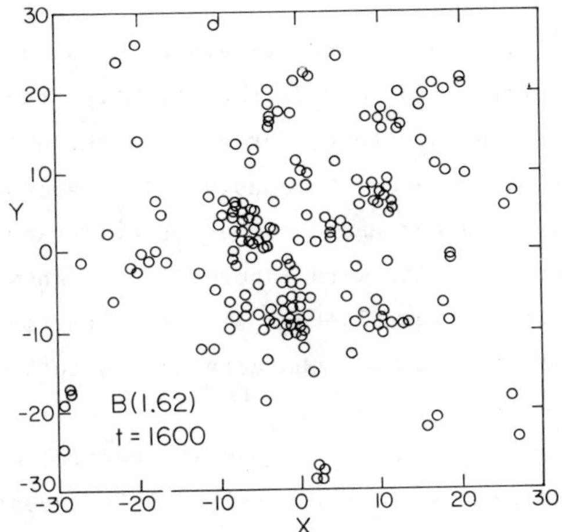

Fig. 5: The X and Y coordinates of particles in event B(1.62) at $t = 16 \times 10^{-12}$ sec. See caption of Fig. 3 for more details.

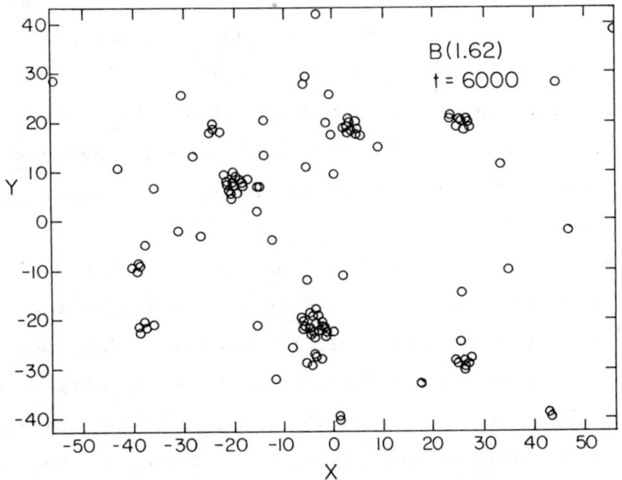

Fig. 6: The X, Y coordinates of particles in event B(1.62) at $t = 60 \times 10^{-12}$ sec. See caption of Fig. 3 for more details.

3. CONCLUSIONS

1. The compositions at t = 60 are given in Tables I and II. The final ($t \rightarrow \infty$) compositions may differ slightly from these as discussed in ref. 11. We see that, for all the three initial densities, events with low e can be considered as evaporation; they have a large residual cluster and a number of evaporated monomers and few-body clusters. Events with intermediate e show fragmentation; the final state has few clusters with many particles. At large e we see total vaporization characterized with a lack of any large cluster in the final state.

2. The calculated values of the average central density ρ_{ac} and average central temperature T_{ac} for a selected few events are plotted in Figs. 7, 8 and 9. The trajectories obtained by joining the ρ_{ac} and T_{ac} points at successive times map the evolution of the expansion in the ρ,T plane. We see that outside the region of adiabatic instabilities the expansion is almost adiabatic (adiabats are shown in these figures by dashed lines). Total vaporization results when the entropy of the initial state is so large that the adiabat passing through the initial state goes above the region of adabatic instabilities (events B(23.5), B(8.9) and C(2.8) in Fig. 7). Such trajectories may pass through the region enclosed by isothermal and adiabatic spinodals, but this does not result into fragmentation. Thus at least classical liquid drops do not fragment at or near the critical temperature T_c.

3. When matter enters the region of adiabatic instabilities density inhomogeneities start to build up (as in Fig. 3 for example). Entropy is created at this stage. If the energy is low enough the expansion stops and matter heats up due to phase separation (see trajectories B(0.72) and D(-1.6) in Figs. 8 and 9 respectively). Fragmentation occurs only if the expansion continues to densities lower than ~ 1/4th the liquid density. If the system phase separates at $\rho_{ac} > \rho_{liquid}/4$, the liquid occupies > 1/4th the volume. Under such conditions we can expect the liquid to form a single connected blob from percolation theory. This blob can have a rather crooked

Table I. Composition of small clusters.

Event	N_i	T_i	N_1	E_1	N_2	E_2	N_3	E_3	N_4
B(-2.71)	230	.89	22	1.23	0		0		0
D(-1.6)	230	1.29	33	1.58	5	1.23	0		1
B(-1.5)	234	1.38	42	1.44	4	1.23	2		0
B(-1.0)	231	1.60	69	1.76	5	0.75	1		0
B(-.72)	231	1.95	74	1.76	4	0.92	2		1
C(-.55)	224	1.41	77	1.40	6	1.28	1		2
D(-.12)	225	1.78	56	1.74	6	3.33	0		0
C(0.0)	226	1.79	95	1.70	4	1.07	2		1
B(0.1)	222	2.20	83	1.79	9	4.21	0		3
C(.38)	217	1.99	95	1.88	9	2.93	6	1.52	0
D(.42)	220	2.23	67	2.24	7	2.48	3		0
B(.63)	233	2.54	85	2.34	15	3.14	6	2.54	1
B(1.62)	225	2.76	98	3.39	12	6.81	3		2
D(1.78)	223	2.60	73	4.05	12	4.34	3		1
C(2.0)	225	2.68	119	3.60	12	3.45	5	4.13	3
B(2.2)	220	2.98	103	4.10	13	4.17	5	11.5	3
C(2.8)	225	3.36	112	4.23	20	5.40	11	4.37	5
B(3.0)	223	3.70	108	4.58	19	7.01	5	5.72	3
B(3.9)	222	4.25	120	5.87	17	9.92	7	5.72	2
D(4.6)	230	3.69	91	6.89	14	11.6	9	13.2	2
D(6.6)	232	4.07	110	8.69	17	15.3	10	16.1	2
B(6.9)	225	5.58	140	8.64	22	9.65	6	14.9	2
C(7.1)	222	5.46	153	8.51	27	8.63	2		1
B(8.9)	230	6.51	152	12.1	22	11.5	3		3
B(24)	231	14.1	208	24.6	10	31.4	1		0

Columns N_i and T_i give the temperature and the number of particles in the initial state. The number of monomers, dimers, 3- and 4-body clusters are listed under columns N_{1-4}, and their mean energies are given in columns E_{1-3}.

Table II. Composition of Large Clusters

Event	N_5	N_6	N_7	N_8	Clusters with $N_c>8$	E_1
B(-2.71)	0	0	0	0	208	1.23
D(-1.6)	0	0	0	0	183	1.58
B(-1.5)	0	0	0	0	178	1.44
B(-1.0)	1	0	0	0	144	1.76
B(-.72)	1	0	0	0	134	1.76
C(-.55)	1	1	0	0	113	1.40
D(-.12)	1	0	1	1	18,18,26,31,44	1.74
C(0.0)	2	1	0	0	97	1.70
B(.10)	0	0	0	0	9,18,21,61	1.79
C(.38)	1	0	0	1	9,23,41	1.88
D(.42)	2	0	2	0	11,11,12,16,16,24	2.24
B(.63)	0	1	1	0	9,10,14,19,31	2.34
B(1.62)	1	2	1	0	9,10,21,21	3.39
D(1.78)	4	2	1	1	10,11,11,16,18	4.05
C(2.0)	2	2	1	0	12,14	3.60
B(2.2)	3	2	0	0	10,27	4.10
C(2.8)	0	1	2	0		4.23
B(3.0)	2	2	1	1	13	4.58
B(3.9)	3	1	0	1	10	5.87
D(4.6)	3	4	0	2	10,11	6.89
D(6.6)	3	2	1	2		8.69
B(6.9)	3	0	0	0		8.64
C(7.1)	1	0	0	0		8.51
B(8.9)	1	0	0	0		12.1
B(24.)	0	0	0	0		24.6

The number of 5 to 8-body clusters is listed under columns N_{5-8}, and clusters with more than 8-particles are listed separately as in Table I.

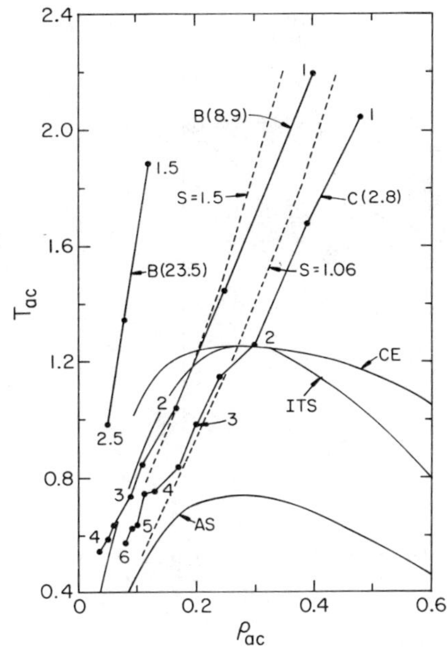

Fig. 7: The evolution trajectories of events are shown together with the phase diagram. The numbered dots, circles and triangles indicate time in units of 10^{-12} sec. The dots are at intervals of $.5 \times 10^{-12}$ sec, the circles at intervals of 3×10^{-12} sec, and the triangles as marked. Adiabats are shown by dashed lines and labeled with entropy/particle.

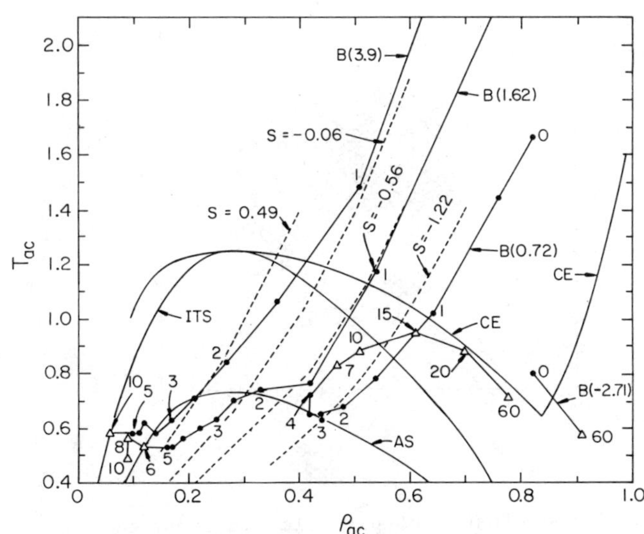

Fig. 8: Evolution trajectories of events (see caption of Fig. 7).

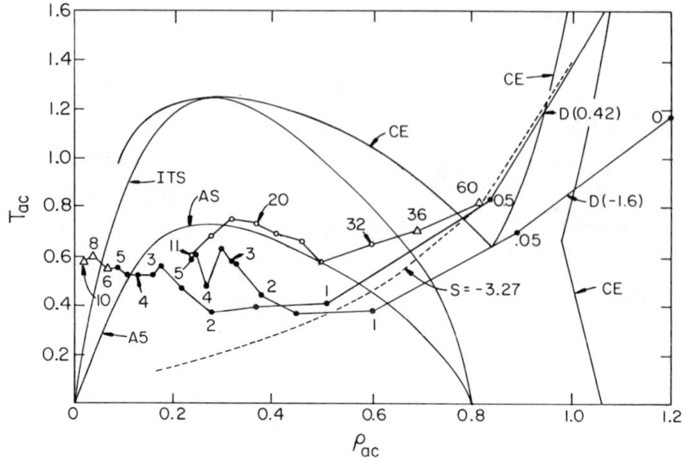

Fig. 9: Evolution trajectors of events (see caption of Fig. 7).

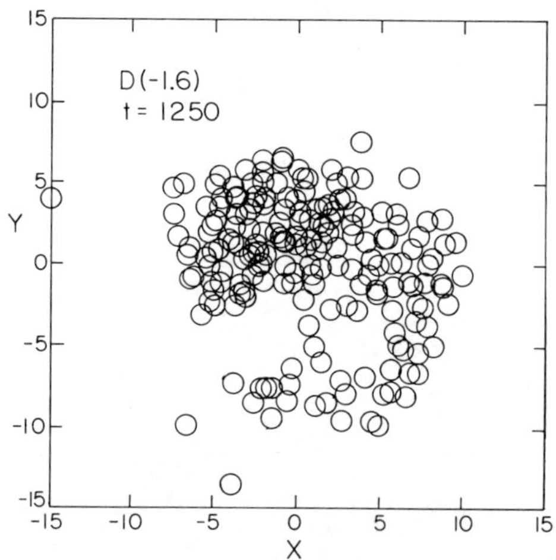

Fig. 10: The X, Y coordinates of particles in event D(-1.6) at t = 12.5×10^{-12} sec. See caption of Fig. 3 for more details.

surface (see for example the X,Y coordinates of particles at t = 12.5 in event D(-1.6) shown in Fig. 10). This surface relaxes to spherical shape almost isothermally, the released surface energy is mostly carried away by evaporation.

4. When the energy is sufficient to drive the expansion in the region of adiabatic instability to $\rho < \rho_{liquid}/4$ the system fragments. Thus fragmentation seems to have an energy threshold. The mass of the fragments appears to be largest near the threshold (see Table II). The fragments are created with rather crooked surface (Fig. 5), and they relax to spherical shape almost isothermally at a temperature of ~ .6 in reduced units. We do not have a quantitative explanation of this temperature. It may be determined by a balance of the rates of heat generation by relaxation of the surface, and heat removal by evaporation.

5. The energy distribution of the monomers emitted in fragmentation events is not Maxwellian. The monomers emitted before fragmentation have higher mean enegies than those emitted after. The monomer spectrum in fragmentation events needs better understanding.

6. Some of the gross properties of the emitted monomers seem to be interesting. Fig. 11 shows E_1 the mean energy of the emitted monomers as a function of e. The function $E_1(e)$ does not seem to depend much on the initial density in the range studied. However it has a sharp bend at the fragmentation threshold value of e. The ratio of the number of monomers N_1 to the total number of particles N_i in the initial state seems to depend upon both e and the initial density (Fig. 12).

4. OUTLOOK

This study has shown the expansion of hot drops of classical liquids containing ~ 200 particles is very much influenced by the phase diagram of extended matter. In particular the maximum entropy below which fragmentation can occur is simply the critical (or maximum) entropy of the adiabatic spinodal curve. The need for expansion to low average densities provides a threshold energy for

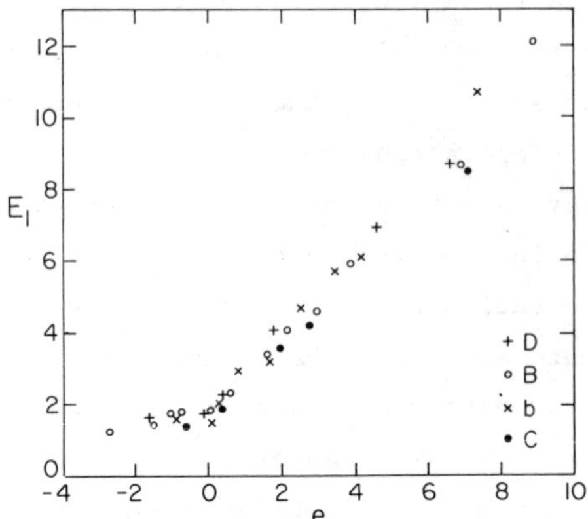

Fig. 11: The mean energy of monomers E_1 as a function of e in B, C and D events with ~ 230 and b-events with ~ 130 particles.

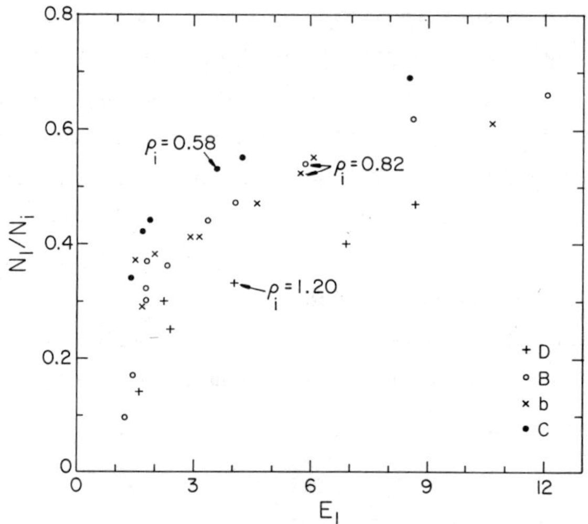

Fig. 12: The fraction of monomers in B, C and D events with ~ 230 particles and b-events with ~ 130 particles, as a function of E_1 the mean energy of monomers.

fragmentation. Thus it appears that a unified study of the phenomenon of vaporization, fragmentation and evaporation can provide information on the equation of state in the region of the liquid-gas phase transition. In so far as fragmentation occurs in the region of adiabatic instabilities, it is a consequence of the liquid-gas phase transition.

Molecular dynamic calculations of the fragmentation of charged liquid drops are being carried out. Calculations in which the initial condition corresponds to the collision between two cold drops have been carried out in the past[12,13], though without an emphasis on the liquid-gas phase transition. It would be interesting to see how the phase transition influences the dynamics of the collision. The treatment of quantum effects however is much more difficult, and so is that of realistic nuclear forces with their exchange character.

5. ACKNOWLEDGEMENTS

We thank Drs. W. G. Hoover, D. G. Ravenhall and P. J. Siemens for discussions and suggestions. This work was supported by the National Science Foundation Grant PHY-81-21399.

REFERENCES

1. Ravenhall, D.G., private communication (1984).
2. Friedman, B. and Pandharipande, V.R., Nucl. Phys. A361, 502 (1981).
3. Finn, J.E., Agarwal, S., Bujak, A., Chuang, J., Gutay, L.J., Hirsch, A.S., Minich, R.W., Porile, N.T., Scharenberg, R.P., Stringfellow, B.C. and Turkot, F., Phys. Rev. Lett. 49, 1321 (1982); Minisch, R.W., Agarwal, S., Bujak, A., Chuang, J., Finn, J.E., Gutay, L.J., Hirsch, A.S., Porile, N.T., Scharenberg, R.P., Stringfellow, B.C. and Turkot, F., Phys. Lett. 118B, 458 (1982).
4. Chitwood, C.B., Fields, D.J., Gelbke, C.K., Lynch, W.G., Panagiotou, A.D., Tsand, M.B., Utsunomiya, H. and Friedman, W.A., Phys. Lett. 131B, 289 (1983); Warwic, A.I., Wieman,

H.H., Gutbrod, H.H., Maier, M.R., Peter, J., Ritter, H.G., Stelzer, H., Weik, F., Freedman, M., Henderson, D.J., Kufman, S.B., Steinberg, E.P. and Wilkins, B.D., Phys. Rev. C27, 1083 (1983).

5. Panagiotou, A.D., Curtin, M.W., Toki, H., Scott, D.K. and Siemens, P.J., Phys. Rev. Lett. 52, 496 (1984).
6. Goodman, A.L., Kapusta, J.I. and Mekjian, A.Z., Phys. Rev. C30, 851 (1984).
7. Lopez, J.A. and Siemens, P.J., Nucl. Phys. A (1984) in press.
8. Schulz, H., Kampfer, B., Barz, H.W., Ropke, G. and Bondorf, J., Phys. Lett. 147B, 17 (1984).
9. Bertsch, G. and Siemens, P.J., Phys. Lett. 126B, 9 (1983).
10. Bertsch, G., Nucl. Phys. A400, 221 (1983).
11. Vicentini, A., Jacucci, G. and Pandharipande, V.R., submitted to Phys. Rev. C (1984).
12. Bodmer, A.R. and Panos, C.N., Phys. Rev. C15, 1342 (1977).
13. Molitoris, J.J., Hoffer, J.B., Kruse, H. and Stocker, H., Phys. Rev. Lett. 53, 899 (1984).

Experimental Search for New Forms of Nuclear Matter

Shoji Nagamiya

Department of Physics, Faculty of Science
University of Tokyo, Hongo, Bunkyo-ku, Tokyo, 113, Japan

In this talk I would like to discuss two subjects that I am currently interested in. The first one is the search for quark-gluon plasma and the second one is the creation and usage of unstable nuclei far from the nuclear stability line. Both subjects are related to the study of new forms of nuclear matter, since in the density-temperature plane the quark-gluon plasma is expected far beyond the region of normal nuclei, and in the isospin space unstable nuclei are expected again far beyond the region of normal nuclei.

In Sec. 1, current hopes and expectations for nuclear matter at high density and high temperature are discussed. Then, in Sec. 2 our experimental program using 15 GeV/nucleon S beams at AGS at Brookhaven National Laboratory is described. Finally, in Sec. 3 our recent results at the Bevalac using unstable nuclear beams are discussed.

1. Nuclear Matter at High Density and High Temperature

Normal nuclei on the earth have the following properties: Density (ρ) is nearly equal to $\rho_0 \sim 0.17$ nucleons/fm^3 and temperature (T) is almost zero. Recently, both nuclear and particle physicists started to study nuclear matter under extreme conditions. Shown in Fig. 1 is a familiar phase diagram of nuclear matter in the plane of density (ρ) and temperature (T). Various new phases are expected in this plane in the region far beyond the normal nuclear matter, and among all, the most attractive one is the quark-gluon plasma at extremely high density and high temperature[1)]. First, I would like to explain the mechanism why this new phase shows up.

If nuclear matter is heated up along the line (B) in Fig. 1 without changing baryon density, then, pions start to be created, as shown in Fig. 2. Nuclear matter that contains nucleons, pions and other excited baryons is called the hadronic plasma. At sufficiently high temperatures, pions and nucleons would be overlapped significantly, as

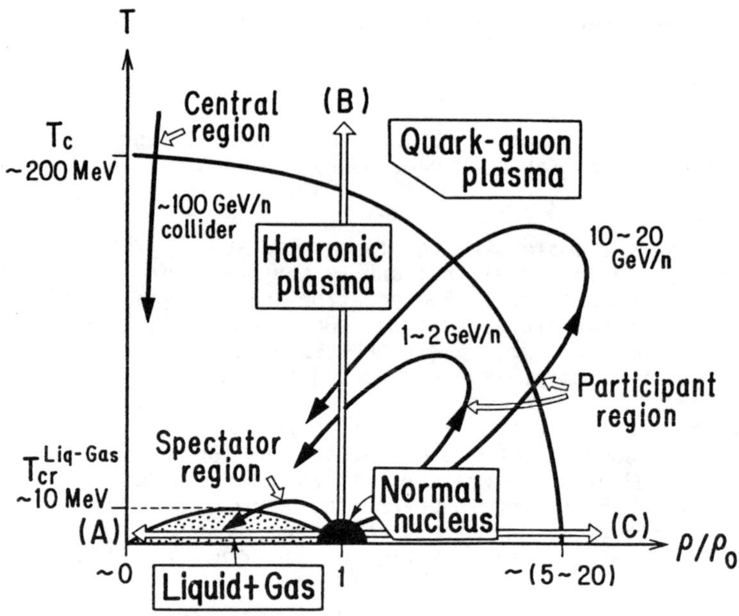

Fig. 1 Expected new forms of nuclear matter in the (ρ,T) plane. Possible collisions paths are indicated by solid curves.

shown in the middle of Fig. 2; the interquark distance is smaller there than the size of nucleon or pion itself ($r \sim 0.8$ fm). In this case it might be better to describe the system as an assembly of quarks and gluons, since a pion is made of two quarks ($q\bar{q}$) and a nucleon of three quarks (qqq). This system is called the quark matter or the quark-gluon plasma.

Why do we expect a sharp phase transition from hadronic plasma to quark-gluon plasma? In Fig. 2 the quantity ε/T^4 is plotted as a function of T for both hadronic plasma and quark-gluon plasma, where ε is the energy density per unit volume. In hadronic plasma both pions and nucleons are regarded as elementary particles, and the system would have a limiting temperature T_C (which is called the Hagedorn limiting temperature[2]) because of the following reason. We know that the temperature of water never exceeds 100° C. At around 100° C, even heat is

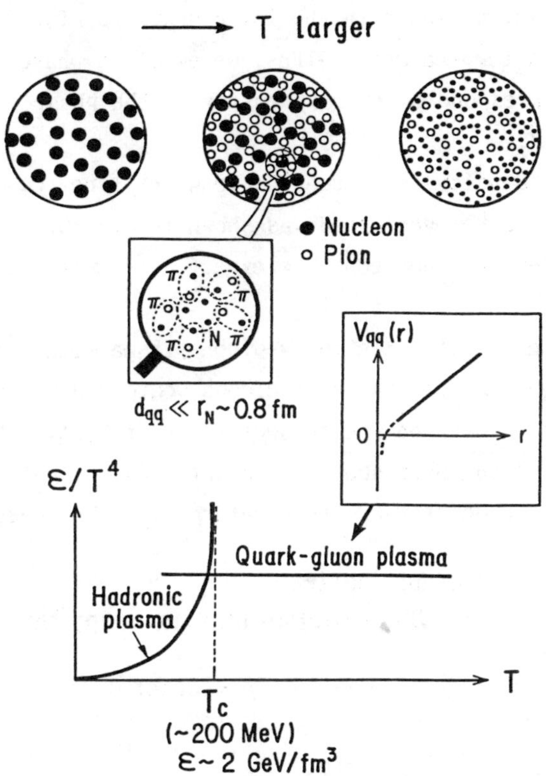

Fig. 2 Behavior of nuclear matter when T is increased along the line (B) in Fig. 1.

supplied more, most of the heat energies are used to forming bubbles and not to increasing kinetic energies of H_2O molecules. Similarly, at $T \sim T_C$ in hadronic plasma most of the energies are used to forming pion bubbles, and the boiling temperature (T_C) in this case is on the order of pion mass of 140 MeV.

On the other hand, it is well known that quark-quark interactions become weaker as interquark distance becomes shorter (asymptotic freedom). At sufficiently high quark density, therefore, the system would behave like a free gas of quarks and gluons (if we describe the system in terms of quarks). There, the Steffan-Boltzmann law ($\varepsilon \propto T^4$) is expected to hold, as shown in the lower graph in Fig. 2; that is, there are no limiting temperatures. Thus, we expect a phase transition at T $\sim T_C$. The corresponding energy density at the phase transition point is $\sim$ 2 GeV/fm^3.

If baryon density is increased along the line (C) in Fig. 1, then nucleons start to be overlapped, as shown in Fig. 3. Again, at sufficiently high density, nucleons are expected to melt into quarks and gluons.

Can we then create this quark-gluon plasma with high-energy heavy-ion beams? As shown in Fig. 4, nucleons would crush each other in the overlapped region between the projectile and target, like we see in a Bombay street or in Tokyo subway. In this region, called the participant region, both density and temperature would increase first. After

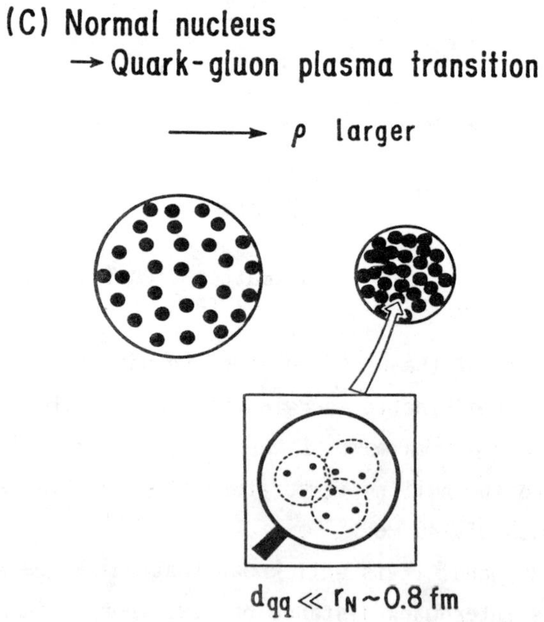

(C) Normal nucleus
→ Quark-gluon plasma transition

⟶ ρ larger

$d_{qq} \ll r_N \sim 0.8$ fm

Fig. 3 Behavior of nuclear matter when ρ is increased along the line (C) in Fig. 1.

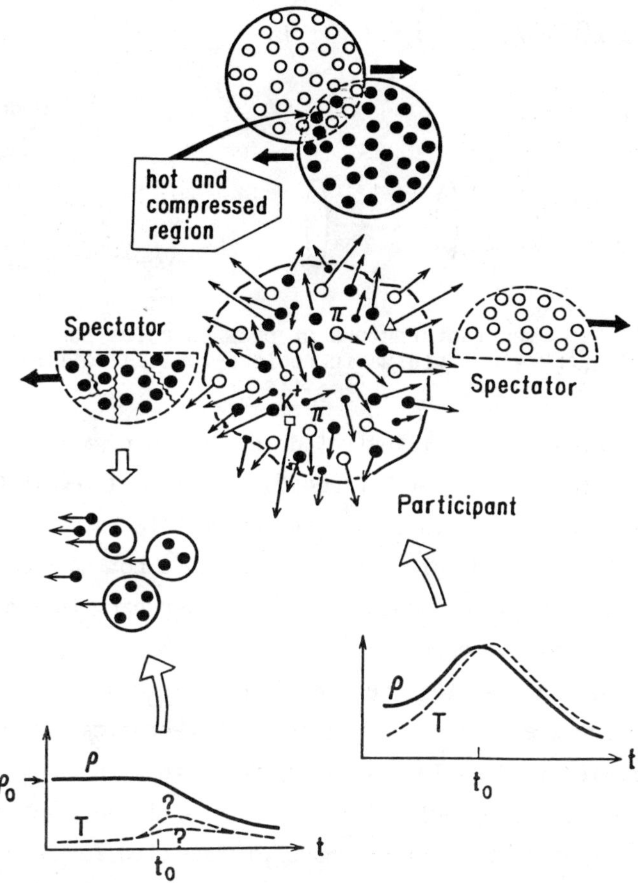

Fig. 4 Participant-spectator separation in nuclear collision, and the expected time dependences of ρ and T.

the violent collision the system would then starts to expand and to be cooled down. The corresponding dynamical path in the (ρ, T) plane is drawn by solid curve(s) in Fig. 1.

Currently, theorists believe that there are two important energy regions for the creation of quark-gluon plasma. The first one is 10-20 GeV/nucleon (lab) at which colliding nuclei stop each other so that both density and temperature of nuclear matter would increase sharply

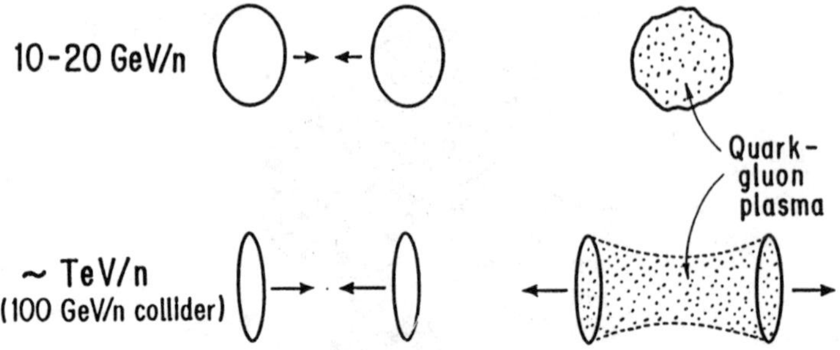

Fig. 5 Two important regions of beam energies for the creation of quark-gluon plasma in nucleus-nucleus collisions.

during the collision, as illustrated in the upper part in Fig. 5. The other energy region is ~ 1 TeV/nucleon (lab) or a few 10 GeV/nucleon in a collider mode. In this case the projectile and target would pass through each other, and along their passages a hot baryon-free region is likely to be created, as shown in the lower part in Fig. 5. At ~ 1 TeV/nucleon (lab) the energy density of this baryon-free region may go up high enough so that this region can melt into quarks and gluons. Of course, these expectations are mainly the theoretical hope based on the current knowledge of nuclear stopping power and thus have large ambiguities. Also, since nuclear collision is a time-dependent dynamical process, whereas a quark-gluon phase is predicted for a static nuclear matter, the creation of this new phase is not at all guaranteed. Nevertheless, as experimentalists, let us have a hope and plan to search for this new phase with high-energy nuclear beams, since this is one of the most exciting subjects in nuclear physics.

2. Actual Experimental Plan

Based on the hope described in the previous section, both Brookhaven National Laboratory (BNL) and CERN have proposed ambitious plans to accelerate heavy ions at higher beam energies than the Bevalac. At BNL the present Tandem is used as an injector to AGS. The transfer line from Tandem to AGS is now under construction so that ^{32}S beams at 15 GeV/nucleon can be obtained in the summer of 1986. If a proposed 1 GeV booster ring is approved, beams up to ^{197}Au would be available in 1988. In addition, RHIC (Relativistic Heavy Ion Collider) is currently proposed with which to provide 100 GeV/nucleon U + 100 GeV/nucleon U beams in a collider mode. At CERN ^{16}O beams at much higher energies (225 and 50 GeV/nucleon) will be available at SPS in the fall of 1986. This CERN program will be discussed more in detail by Dr. Specht at this Conference[3)].

The advantage of BNL is its anticipated plenty amount of beam time. At CERN only a few weeks of running time per year are allocated for heavy ions for 1986 and 1987 (but not for 1988). We therefore decided to perform an experiment at BNL. About 40-50 physicists[4)] from BNL, MIT, LBL, and Japan gathered together last summer to make an experimental proposal (Spokesmen: O. Hansen and myself). In October, 1984, 1000 hours of beam time were approved.

We have three goals in this experiment. The first one is to study a question of how nuclear matter is heated up during the course of the collision. Shown in Fig. 6 are proton and pion energy distributions, obtained at the Bevalac[5)], which were measured at c.m. 90° as a function of the c.m. energies E^* of emitted particles. Ne + NaF collisions are selected here at three bombarding energies, 0.4, 0.8, and 2.1 GeV/nucleon. The spectrum shape is nearly exponential for all the cases. I call this slope parameter E_0 the "temperature" with quotation marks attached, since it is difficult to believe that thermal equilibrium is reached in such a collisions. No matter how this E_0 is called, however, this quantity E_0 is an important one, since it is closely related to the average kinetic energy carried by emitted particles, and therefore, would indicate how nuclear matter is heated up at the stage when these particles are emitted.

If we plot the observed values of E_0 as a function of the c.m. beam energy per nucleon, then they increase monotonically as the beam energy

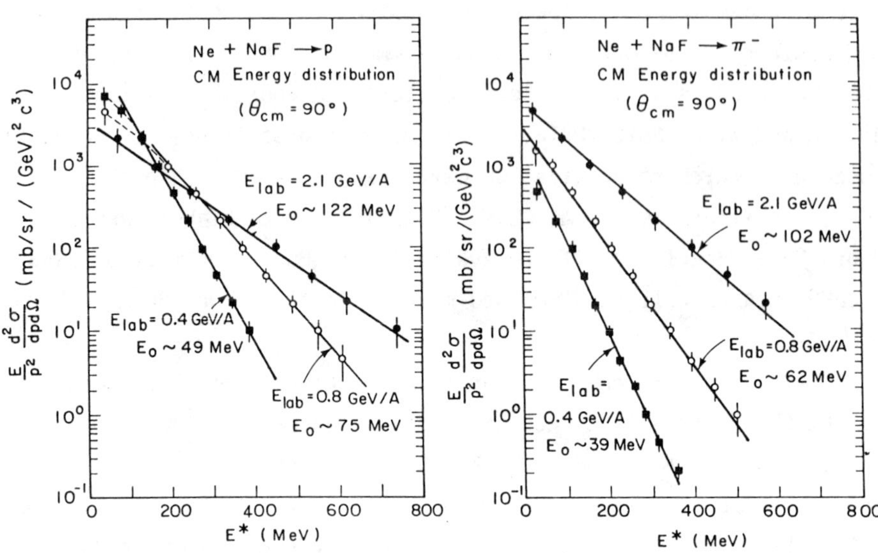

Fig. 6 Proton and pion energy distributions measured at the Bevalac.

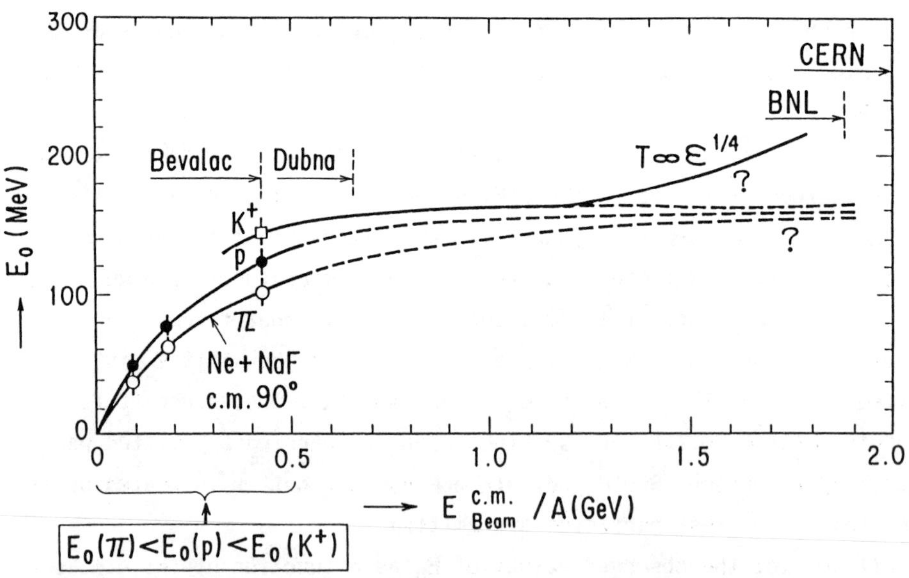

Fig. 7 Values of E_0 for p, π, and K^+ vs. beam energy per nucleon.

increases, as shown in Fig. 7. This fact indicates that nuclear matter is heated up more at higher beam energies. An important and interesting question to ask here is whether or not these E_0 values converge to a certain value at much higher beam energies than 2 GeV/nucleon (lab).

There is a good reason to expect that these values of E_0 may converge. In pp data at CERN[6] the average transverse momentum approaches a constant at high energies (Fig. 8). For production of both pions and kaons the corresponding temperature values are close to the Hagedorn limiting value of 140 MeV. On the other hand, there is another good reason to expect that these values of E_0 may not converge. As mentioned in Fig. 2, if the system stays in a hadronic phase, the temperature value should saturate at a value of T_C even if the beam energy is increased. However, if a quark-gluon plasma is created, the temperature value can increase up with beam energies because of the

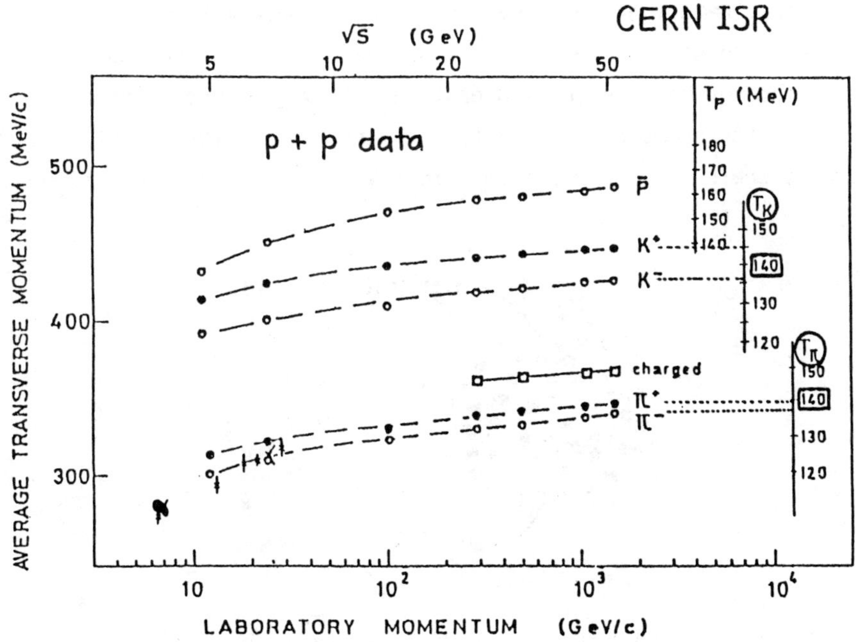

Fig. 8 Average p_T values in p+p collisions[6].

relation $T \propto \varepsilon^{1/4}$. Perhaps, such an increase may be seen only for K^+, because K^+ has a long mean free path in hadronic matter, whereas $\pi^{\pm}$ has a short mean free path so that it may reflect only a cold stage of the collision at which the system returned back to the original hadronic phase. Anyway, it is interesting to test it.

The second goal is to measure production of strange mesons. In hadronic phase it has been known experimentally that the production rate of $s\bar{s}$ is about 1/10 of the production rate of $u\bar{u}$ (or $d\bar{d}$)[7]. If a quark-gluon plasma is created, however, the production of $u\bar{u}$ or $d\bar{d}$ may be suppressed[8], because nucleons are made of u and d quarks, and thus, the Pauli blocking effect prohibits production of such pairs (see Fig. 9). Currently, theorists predict that the production rate of $s\bar{s}$ is nearly equal to that of $u\bar{u}$ (or $d\bar{d}$) in the quark-gluon phase[8]. We therefore plan to measure $\phi(s\bar{s})$ and $K^+(u\bar{s})$ mesons to test if this prediction is correct. These two mesons are particularly useful, since both of them have long mean free paths in hadronic matter.

The third goal is to study how nuclear collision evolves in time in the plane of density and temperature. If we measure identical particle interferometry, called the Hanbury-Brown/Twiss effect[9], then we can determine the interaction volume V. At the same time, if we

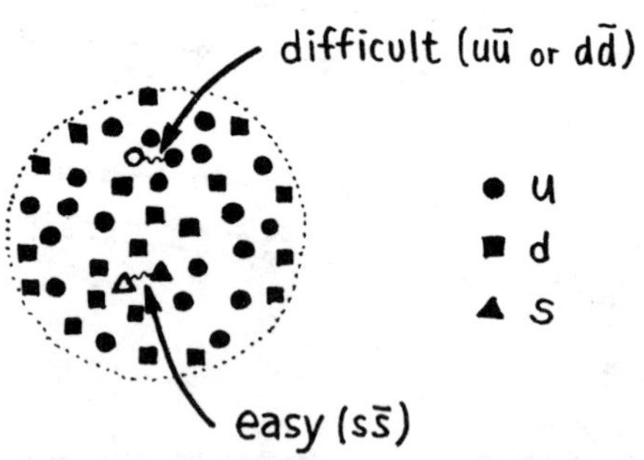

Fig. 9 Suppression mechanism of $u\bar{u}$ or $d\bar{d}$ creation in quark-gluon plasma.

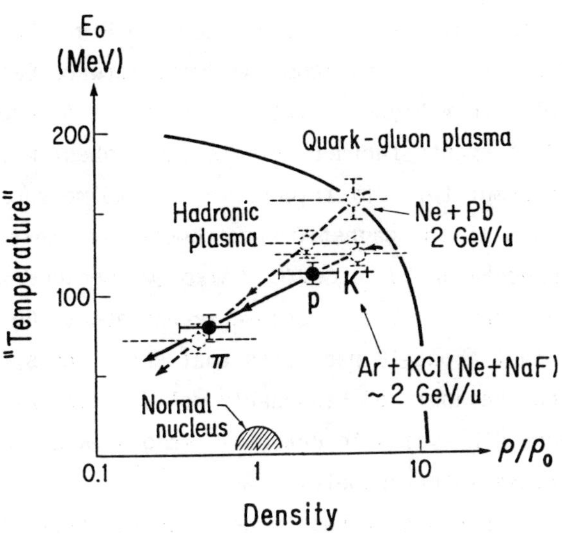

Fig. 10 Time evolution of nuclear collision estimated from the Bevalac data.

measure nucleon multiplicity m_N, then we can experimentally determine the "effective density" from relation $\rho = m_N/V$. Pion and K^+ may probe different density values, because K^+ is likely to sample an earlier hot and compressed stage of the collision, whereas pion to sample a later expanded stage.

Since the value of "temperature" E_0 can be determined from the spectrum shape, we would also have a hope to plot experimental points in the (ρ, T) plane for each type of emitted particles. By connecting the experimental points we may further learn how nuclear collision evolves in time. Fig. 10 shows the experimental points obtained in this manner from the known Bevalac data[10]. For K^+, only the value of E_0 is known but ρ is not, primarily because K^+K^+ interferometry is not feasible at the Bevalac. At BNL energies the K^+K^+ interferometry is possible. We thus may be able to plot many experimental points in the (ρ, T) plane for K^+, K^-, π^+ and π^-.

In order to achieve these three goals we prepare a magnetic spectrometer with a solid angle of 25 msr, as shown in Fig. 11. Forward veto counters (of heavy fragments) combined with forward Cerenkov counters select relatively high-multiplicity events. A streamer-tube array of about 5000 elements further selects the highest multiplicity events. With such a multiplicity trigger we plan to measure $p^{\pm}$, $\pi^{\pm}$, $K^{\pm}$, $d^{\pm}$, $\alpha^{\pm}$, etc. with the magnetic spectrometer at angles larger than 5° and with momenta up to 20 GeV/c. Also, we measure π^0 with 300 elements of Pb glass array. The spectrometer has about 100 segmentation (or more) so that about 10 particles that enter the spectrometer at the same time can be analyzed reasonably well. With this segmentation we also measure K^+K^- pairs to detect ϕ mesons, K^+K^+ pairs or $\pi\pi$ pairs for two-particle interferometry.

Each element in this set up has to be designed carefully. For example, the timing is crucial. Recently we have achieved ≦ 100 ps time resolution for 1 meter scintillation counter for minimum ionizing particles. Anyway, we are moving ahead so that we can take beams in the summer of 1986.

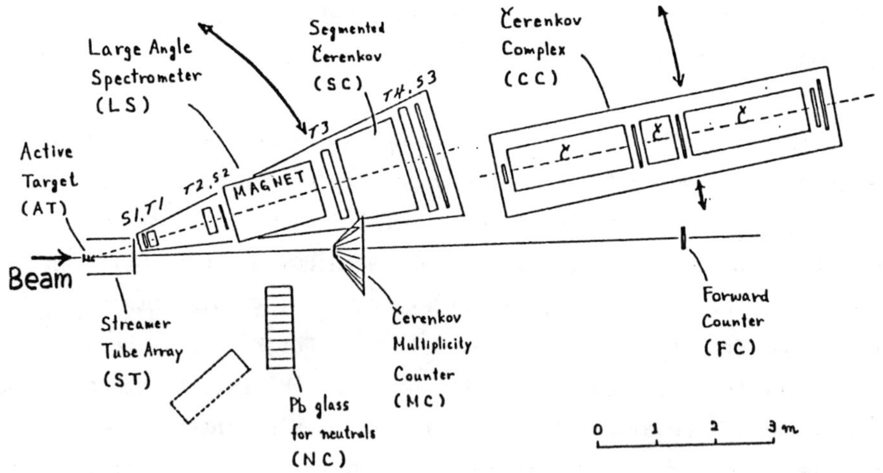

Fig. 11 Layout of experimental set-ups for our AGS experiment.

3. Experiment with Unstable Nuclear Beams

One of early-day measurements at the Bevalac was the study of projectile fragmentation. Projectile fragments are created from the projectile spectator (see Fig. 4), which is a part of the projectile nucleus that has not been scraped out by the collision. Therefore, they keep various memories that the original projectile had before the collision. For example, they move at the same speed with the original projectile.

An interesting application of projectile fragments is their usage as primary beams, because they are emitted sharply at forward angles with a unique velocity which is nearly equal to the beam velocity. Therefore, our group[11)] has recently prepared a Be target at the exit of the Bevalac, and unstable isotopes created there were transported along the beam line to the experimental area, as shown in Fig. 12. In this way we could use beams of unstable nuclei that cannot be obtained on the earth.

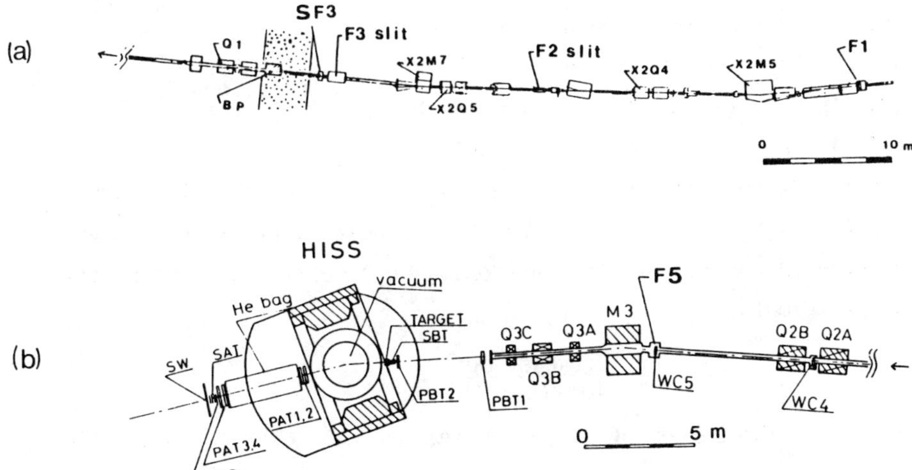

Fig. 12 A secondary beam line at the Bevalac for experiment with unstable nuclear beams.

The beam spot at the target was about 1 cmϕ at $\Delta p/p \sim 3$ %. Typical intensities were 10^{6-7}/s for ^{6}He and 10^3/s for ^{8}He. In the first experiment we concentrated on He-isotope beams, because they are the easiest case. The main contributors to this experiment were I. Tanihata and K. Sugimoto.

When we tuned the beam line to accept $A/Z \sim 3$ particles, we had t, ^{6}He, and ^{9}Li, as shown in Fig. 13. Beams of ^{6}He were then selected by the pulse-height cut in the scintillation counter.

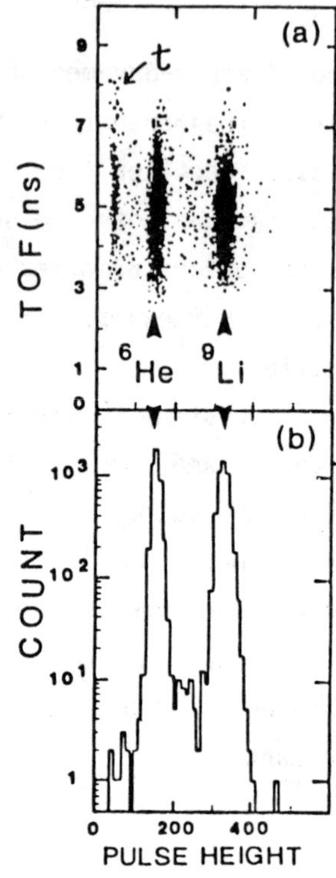

Fig. 13 Pulse-height distributions for A/Z = 3 particles.

The first experiment was the measurements of total reaction cross sections for these He-isotope beams on three targets, Be, C, and Al. The experimental set-up was very simple. Two sets of detector systems were prepared, one before and the other after the target, as shown in Fig. 14. Beam particles (such as ^{6}He) were counted by these detector systems; N_0 (before the target) and N_1 (after the target). Then, take a ratio $R = N_1/N_0$ which is a fraction of beam particles that have not experienced any nuclear interactions. We determine the cross section by the formula:

$$\sigma_I = \frac{1}{n} \cdot \frac{R(\text{target-out})}{R(\text{target-in})},$$

where n is the number of target nuclei per cm^2. The ratio of R(target-out) is used for the purpose of correcting for background contributions from experimental areas other than the target. Practically, however, this values of R(target-out) was close to one. As shown in Fig. 14,

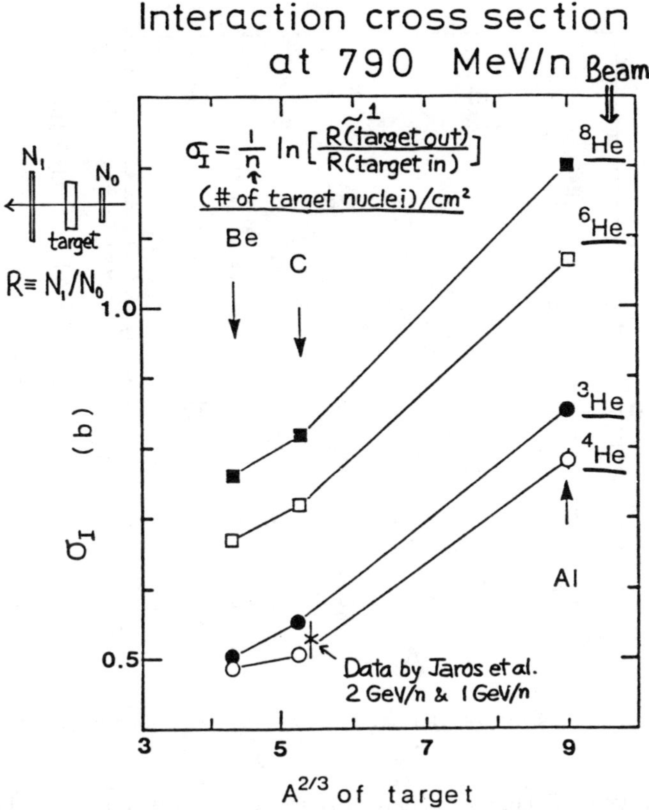

Fig. 14 Interaction cross sections with He-isotope beams on Be, C, and Al targets.

our results of σ_I for ^{4}He + ^{12}C agree with the previous results by Jaros et al.[12] (marked by ×).

Now, we operationally define radii of the projectile (R_P) and target (R_T) by

$$\sigma_I = \pi(R_P+R_T)^2.$$

With this definition, differences of $R(^n\text{He}) - R(^3\text{He})$ can be determined from the data, as plotted in Fig. 15. ^{4}He is smaller than ^{3}He. Also, ^{6}He and ^{8}He are larger than those extrapolated by the $A^{1/3}$ systematics.

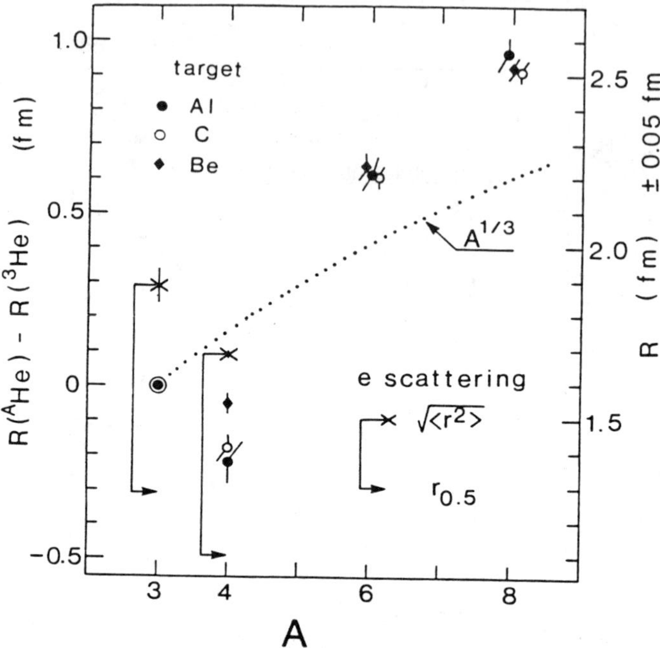

Fig. 15 Radii of He isotopes deduced from the data shown in Fig. 14.

I think that we can think of many experiments with unstable beams, e.g., measurements of magnetic moments, lifetimes, masses, etc. of these isotopes. Part of our Japanese group will thus continue to stay at Lawrence Berkeley Laboratory to measure these quantities.

About 2-3 years ago I proposed In Tokyo to construct an accelerator with two rings; one to accelerate heavy ions and the other to accumulate and cool unstable isotope beams so that high-quality unstable nuclear beams can be obtained. Such a machine will be constructed at GSI, Germany, in a few years, although I am still pushing it in Tokyo.

I would like to thank my colleagues for stimulating daily discussions; especially O. Hansen, S. Steadman, T. Ludlam, and Y. Miake for the BNL part and I. Tanihata and K. Sugimoto for the LBL part. This work was supported in part by INS-LBL Collaboration Program and in part by the Japan-U.S. Collaboration Program in High Energy Physics.

References

1) For example, see Quark Matter Formation and Heavy Ion Collisions, M. Jacob and H. Satz (editors), World Scientific Pub., Singapore, 1982. Or, see Proceedings of Quark Matter '83, T. W. Ludlam and H. E. Wegener (editors), Nucl. Phys. **A418** (1984).
2) R. Hagedorn, in Cargese Lectures in Physics, Vol. 6, Gordon and Breach (1973); R. Hagedorn, in Proceedings of Quark Matter '84 (to be published).
3) H. Specht, Talk presented at this Conference.
4) At the time when the proposal was submitted the participants on the list were:
BNL • • • D. Alberger, P. D. Bond, C. Chasman, Y. Y. Chu, J. B. Cumming, E. Duek, Ole Hansen, P. Haustein, S. Katkoff, M. J. Le-Vine, T. Ludlam, J. Olness, A. Pfoh, L. P. Remsberg, A. Shor, A. Sunyar, M. Tanaka, M. J. Tannenbaum, P. Thieberger, J. M. van Dijk, P. Vincent, and H. Wagner.
Hiroshima • • • T. Sugitate.
LBL • • • D. Greiner, T. Mulera, and V. Perez-Mendez.
MIT • • • H. A. Enge, L. Grodzins, R. J. Ledoux, S. G. Steadman, and D. Woodruff.
Tokyo • • • Y. Akiba, H. Hamagaki, O. Hashimoto, S. Homma, Y. Miake, S. Nagamiya, I. Tanihata, and M. Torikoshi.
Waseda • • • T. Doke and J. Kikuchi.
5) S. Nagamiya, M.-C. Lemaire, E. Moeller, S. Schnetzer, G. Shapiro, H. Steiner, and I. Tanihata, Phys. Rev. **C24** (1981) 971.
6) G. Giacomelli and M. Jacob, Phys. Reports **C55** (1979) 1.
7) M. Bourquin et al., Z. Physik **C5** (1980) 275.
8) J. Rafelski, Nucl. Phys. **A418** (1984) 215c, and references therein.
9) R. Hanbury-Brown and R. Q. Twiss, Nature **178** (1956) 1046; also see F. B. Yano and S. E. Koonin, Phys. Lett. **75B** (1978) 556.
10) S. Nagamiya, Nucl. Phys. **A416** (1984) 239c.
11) I. Tanihata, H. Hamagaki, O. Hashimoto, S. Nagamiya, Y. Shida, N. Yoshikawa, O. Yamakawa, K. Sugimoto, T. Kobayashi, D. E. Greiner, N. Takahashi, and Y. Nojiri, submitted to Phys. Rev. Lett. (LBL Report LBL-18821, 1984).
12) J. Jaros et al., Phys. Rev. **C18** (1978) 2273.

TIME-DEPENDENT CALCULATIONS ON HEAVY-ION COLLISION AND FISSION

M. K. PAL
Saha Institute of Nuclear Physics
Calcutta 700 009
INDIA

ABSTRACT

Time-dependent calculations on heavy-ion collision and an adiabatic version for the computation of fission are described. A semi-classical modification using the phase-space distribution function is also discussed.

1. INTRODUCTION

Unlike my predecessors in this session, who described ultra high-energy nuclear phenomena, I shall deal with the theoretical treatment of fairly cold nuclear processes in the typical energy regime of a few MeV per nucleon. The theory has been applied by several groups to heavy ion collisions, specially to deep inelastic processes and fusion of the colliding nuclei. An adiabatic version of the theory suitable for application to fission and sub-barrier fusion will also be discussed.

The theory treats the time-evolution of the many-body wave function for a given nuclear Hamiltonian. The mean-field approximation is used which renders the many-body wave function to a determinantal form comprising single-particle wave functions of the individual nucleons in the mean field. Although approximate, the theory has the merit of being fully microscopic. Furthermore, it allows various semi-classical variants some of which will also be discussed. Extensions beyond the mean field approximation have been worked out but their applications to real nuclear collisions are still scarce. Several reviews[1)] of the mean-field theory, its applications, and extension to include residual nucleon-nucleon collision effects exist in the literature which may be consulted for details. A few pertinent aspects of the theory is presented in the next section. My own work contained in this report has been done in collaboration with A. K. Mukherjee.

2. TIME-DEPENDENT MEAN FIELD THEORY

In this theory, as already remarked, the time-dependent many-body Schrodinger equation

$$i\hbar \frac{\partial \Psi}{\partial t} = H\Psi \tag{1}$$

is solved with a determinantal form for the wave function Ψ. The nuclear Hamiltonian H consists of the kinetic energy of the nucleons, and the sum of effective two-body interaction amongst all pairs of nucleons. For the effective interaction a simplified density-dependent Skyrme potential is most often used, together with a two-body Yukawa potential and the Coulomb potential for the protons. The Yukawa and Coulomb parts are treated approximately by neglecting exchange effects.

Eq. (1) leads to the single-particle time-dependent Hartee-Fock (TDHF) equation

$$i\hbar \frac{\partial \phi}{\partial t} = (T + \mathcal{V})\phi \tag{2}$$

where ϕ is a single-nucleon wave function in the mean field $\mathcal{V}$, computed from the effective two-body potential in the usual way, and T is the kinetic energy operator of the nucleon. This equation represents an initial value problem wherein the insertion of the initial wave function enables one to numerically compute the wave function at all subsequent times. Plots of the local single-particle density

$$\rho(\underset{\sim}{r}) = \sum_{\alpha}^{occ} |\phi_\alpha(\underset{\sim}{r})|^2, \tag{3}$$

Where the summation runs over all the occupied single particle states in the determinant, are made as a function of time. Various approximations are used to feed in information on angular momentum into the computation, and in reducing the dependence on spatial co-ordinates of $\rho(\underset{\sim}{r})$ to two co-ordinates instead of three. In one variant, for example, a moving co-ordinate frame is used in which the line joining the centres of the two heavy ions is treated as a symmetry axis, and the additional energy due to the rotation of this axis is fed into the Hamiltonian with a moment of inertia computed from the density. In another variant the single particle wave function perpendicular to the reaction plane is taken to be frozen in time, thereby reducing the time-dependent calculation to that of the wave function in the moving reaction plane.

Fusion cross section, and fusion window in angular momentum have been computed for many collision processes. Deep inelastic reactions of heavy ions have also been investigated extensively.

The TDHF theory is not suited for application to fission and sub-barrier fusion. There is no natural way in this theory to compute the fission barrier, and restricting the path followed by the nucleus to the bottom of the potential energy valley. Fission is a slow nuclear process where primarily a few collective shape variables of the nucleus change adiabatically with time. As a result the nucleus, in the given configuration in the space of the collective variables, tends to minimise its energy and thus stay at the bottom of the potential energy valley leading from the equilibrium shape to the saddle point shape. The TDHF wave function behaves like a classical wave packet and once it enters the nucleus it cannot come out of the barrier by barrier penetration which is essentially a quantum mechanical phenomenon.

To circumevent this shortcoming an adiabatic version (ATDHF) of the TDHF theory has been developed by several authors[2)] and it has been shown that in this version of the theory the path can be restricted to the bottom of the potential energy valley, and a collective Hamiltonian can be set up for the motion of the collective variables. A quantum mechanical treatment of the collective Hamiltonian then leads in a natural way to barrier penetration. Major features of the ATDHF is presented in section 3.

3. ATDHF THEORY

This theory is cast in terms of a set of collective co-ordinate variables q and the corresponding canonically conjugate momenta p. For simplicity only one such co-ordinate and momentum variable will be kept in the following work based on the formulation by Villars. It is now well-known that other alternative formulations of the theory are essentially identical in content.

Starting with a determinantal wave function $\Phi(q)$, dependent on the collective variable q and even under time-reversal, another determinantal wave function $\Psi(q,p)$ is generated according to

$$\Psi(q,p) = e^{ipQ(q)} \Phi(q) \tag{4}$$

where Q is a time-even one-body operator dependent on the collective variable q, and p is a real parameter to play the role of the momentum variable. Under adiabatic conditions p is small and hence an expansion in powers of p to a few low orders is used.

It is stipulated that the solution Ψ of eq. (1) is in the form of (4), and the time-dependence of Ψ is implicity contained in the time-dependence of the parameters q and p. Thus, we shall use

$$i\frac{\partial}{\partial t} = i\dot{q}\frac{\partial}{\partial q} + i\dot{p}\frac{\partial}{\partial p} \tag{5}$$

where the overhead dot denotes time-derivative.

Requiring Ψ to be determined by the variational equation

$$\delta \int_{t_1}^{t_2} dt \; \langle \Psi(q,p) | \left(i\frac{\partial}{\partial t} - H \right) | \Psi(q,p) \rangle = 0 \tag{6}$$

which is equivalent to eq. (1), one obtains the classical equation

$$\delta \int_{t_1}^{t_2} dt \; \{ \dot{q} p - \mathcal{H}(q,p) \} = 0 \tag{7}$$

where

$$\mathcal{H}(q,p) = \langle \Psi(q,p) | H | \Psi(q,p) \rangle. \tag{8}$$

In writing (7) from (6) we have used

$$\langle \Phi(q) | [Q, P] | \Phi(q) \rangle = i \tag{9}$$

as a normalisation condition for the operator Q, and P occurring (9) has been defined by

$$P \Phi(q) = i \frac{\partial \Phi}{\partial q}. \tag{10}$$

Eq. (7), as in classical mechanics, leads to the canonical equation of motion for q and p

$$\dot{q} = \frac{\partial \mathcal{H}}{\partial p}, \qquad \dot{p} = -\frac{\partial \mathcal{H}}{\partial q} \tag{11}$$

where $\mathcal{H}$ plays the role of the Hamiltonian. Retaining terms up to second order in p, the collective Hamiltonian (8) reduces to

$$\mathcal{H}(q,p) = \frac{p^2}{2\mathcal{M}(q)} + \mathcal{V}(q) \tag{12}$$

with

$$\mathcal{M}^{-1}(q) = \langle \Phi(q) | [[H, iQ], iQ] | \Phi(q) \rangle \tag{13a}$$

and

$$\mathcal{V}(q) = \langle \Phi(q) | H | \Phi(q) \rangle. \tag{13b}$$

Obviously, $\mathcal{M}(q)$ is the collective mass, and $\mathcal{V}$ the collective potential energy.

Equations determining the operator Q and the wave function Φ can also be obtained by expanding in powers of p. In the lowest two orders, one obtains the following equations which are known as Villars equations :

$$\langle \delta\Phi | (H - \frac{\partial V}{\partial q} Q) | \Phi(q) \rangle = 0 \tag{14a}$$

$$\langle \delta\Phi | [H, iQ] - m^{-1} P | \Phi(q) \rangle = 0. \tag{14b}$$

While solving these equations, it is necessary and sufficient to treat $\delta\Phi$ as a complete set of particle-hole excited states with respect to the determinant Φ(q).

Since the first Villars equation (14a) has to be valid at all points q, the following equation obtained by differentiating (14a) with respect to q should also hold :

$$\langle \delta\Phi | [H, iP] + \frac{\partial^2 V}{\partial q^2} Q + \frac{\partial V}{\partial q} \frac{\partial Q}{\partial q} | \Phi(q) \rangle = 0. \tag{15a}$$

At the minima, maxima and saddle points of V(q) where $\partial V / \partial q$ is zero, this equation reduces to

$$\langle \delta\Phi | [H, iP] + \frac{\partial^2 V}{\partial q^2} Q | \Phi(q) \rangle = 0. \tag{15b}$$

Equations (14b) and (15b) can together be rewritten as

$$[R(q)]^2 \begin{pmatrix} P \\ P^* \end{pmatrix} = [m^{-1}(q) \frac{\partial^2 V}{\partial q^2}] \begin{pmatrix} P \\ P^* \end{pmatrix} \tag{16}$$

where R(q) is the local RPA matrix at the point q. Thus the displacement operator P of the wave function Φ(q) at the minimax points is determined by the eigen vectors of the corresponding RPA matrix. At all other points the third term in (15a) persits and hence P is <u>not</u> in the direction of an RPA-mode. A local harmonic approximation, considered by some authors, in which the wave function Φ(q) is displaced everywhere in the direction of the lowest local RPA-mode to generate the path, is therefore a wilful prescription, and the path so obtained is obviously not in conformity with the original time-evolution equation (1).

A difficulty encountered in solving the two Villars equations will now be discussed. Eq. (14a) clearly determines the operator Q (except for a normalisation factor) in terms of H at all points where the collective force $\partial \mathcal{V}/\partial q$ is not zero. With Q so determined, the displacement operator P is obtained from (14b), simaltaneously imposing the normalisation condition (9). Thus starting at any point q in the collective parameter space, where $\partial \mathcal{V}/\partial q$ is not zero, it is possible to generate a path by giving small displacements through δq with P determined as above. In fact it can be shown in general that the solutions of (14a) and (14b) are the lines of force corresponding to the potential field $\mathcal{V}(q)$. At any point the path leads in the direction of grad $\mathcal{V}$. At the minimum Hartree-Fock point where grad $\mathcal{V}$ is zero, considerable ambiguity is encountered if the solution of the two equations are to be generated starting from such a point. In an analytical derivation[3] with a three-level Lipkin model containing two local RPA-modes it has been demonstrated that indeed all the infinite number of solutions, except one, at the minimum point start out in the direction of the lowest RPA-mode, whereas one solution starts out in the direction of the higher mode. Energetically the solution starting along the lower RPA mode is the relevant one for computing fission, and the ambiguity of obtaining such solutions at the minimum point is thus to be resolved. It has been shown[4] that the second-order equation in p, along with zeroth and first-order equations (14a,b) serves to resolve the ambiguity as discussed below.

The second-order equation, above referred to, is given by

$$\langle \delta\Phi | [[H, iQ], iQ] + \frac{\partial m^{-1}}{\partial q} Q - 2m^{-1} \frac{\partial Q}{\partial q} | \Phi \rangle = 0. \qquad (17a)$$

Together with eq. (15a) which also contains $\partial Q/\partial q$, this equation leads to a consistency condition

$$\langle \delta\Phi | \tfrac{1}{2}\lambda [[H,iQ],iQ] + i m^{-1}[H,P] + \omega Q | \Phi \rangle = 0 \qquad (17b)$$

where

$$\lambda(q) = \frac{\partial V}{\partial q}, \qquad \omega = \frac{1}{2\lambda}\frac{\partial}{\partial q}(\lambda^2 m^{-1}). \qquad (17c)$$

An interesting consequence of the consistency condition (17b) has been proved analytically, which asserts that the path satisfying this condition obeys

$$\delta\,|\mathrm{grad}\,V|^2 - \omega\,\delta V = 0. \qquad (18)$$

That is to say, the path, everywhere on a patch of the equipotential surface, moves in the direction of the least gradient of V, and hence is the valley path. That the ATDHF theory delineates such a path has also been discussed by Klein et al[5].

Summarising the results in this section, we now state that for computation of the fission path eqs. (14a,b) have to be solved subject to the consistency condition (17b). The mass-parameter $m(q)$ and the collective potential energy $V(q)$ are to be computed at each point q, and finally the collective Hamiltonian $\mathcal{H}(q,p)$ is to be treated quantum mechanically for the barrier penetration probability.

4. HYDRODYNAMIC APPROXIMATION

For economy in computing it would be desirable to obtain a hydrodynamic approximation to the time-dependent problem and check if it works for given nuclear processes. The standard procedure is well-known[6] and proceeds via the Wigner transform $f(\mathbf{R},\mathbf{k})$ of the non-local density operator $\rho(\mathbf{r},\mathbf{r}')$:

$$f(\underset{\sim}{R},\underset{\sim}{k}) = \int d^3s\, e^{-\frac{i}{\hbar}\underset{\sim}{k}\cdot\underset{\sim}{s}}\ \rho(\underset{\sim}{R}+\tfrac{1}{2}\underset{\sim}{s}\,,\ \underset{\sim}{R}-\tfrac{1}{2}\underset{\sim}{s}) \tag{19a}$$

where

$$\underset{\sim}{R} = \tfrac{1}{2}(\underset{\sim}{r}+\underset{\sim}{r}'), \qquad \underset{\sim}{s} = \underset{\sim}{r} - \underset{\sim}{r}'. \tag{19b}$$

The time-dependent equation

$$i\hbar\frac{\partial\rho}{\partial t} = [h, \rho] \tag{20a}$$

$$h = T + v \tag{20b}$$

which is the equivalent of the TDHF equation (2), then yields in a straightforward manner

$$\frac{\partial f}{\partial t} + \frac{2}{\hbar} f \sin[\tfrac{\hbar}{2}(\overleftarrow{\nabla}_R\cdot\overrightarrow{\nabla}_k - \overleftarrow{\nabla}_k\cdot\overrightarrow{\nabla}_R)]\,\tilde{h} = 0. \tag{21}$$

Here $\tilde{h}$ is the Wigner transform of the HF Hamiltonian (20b), and the arrows indicate the direction in which the corresponding gradient operators act.

To the lowest order in $\hbar$, eq. (21) reduces to the classical Vlasov equation, that is, the collisionless Boltzmann equation,

$$\frac{\partial f}{\partial t} + [\tfrac{\underset{\sim}{k}}{m}\cdot\nabla_R - (\nabla_R v)\cdot\nabla_k]\,f = 0. \tag{22}$$

It is possible to extend this equation by introducing collision terms and some applications[7)] with collision terms have been discussed for the collision of two nuclear slabs.

Multiplying eq. (22) with various powers of k and integrating, we obtain the hierarchy of hydrodynamic equations :

$$\frac{\partial\rho}{\partial t} + \mathrm{div}(\rho\underset{\sim}{u}) = 0 \tag{23a}$$

$$m\rho\left(\frac{\partial u_i}{\partial t} + \underset{\sim}{u}\cdot\mathrm{grad}\,u_i\right) = -\rho\,\mathrm{grad}_i v - \sum_{j=1}^{3}\frac{\partial P_{ij}}{\partial R_j} \tag{23b}$$

where

$$\underset{\sim}{u} = \text{average velocity} = \overline{\underset{\sim}{k}}/m, \; \underset{\sim}{U} = \frac{\underset{\sim}{k}}{m} - \underset{\sim}{u}$$
$$P_{ij} = \text{pressure tensor} = m\rho\, \overline{U_i U_j} \tag{23c}$$

and the overhead bar indicates average. To close the system of equations, one has to approximate the pressure tensor from, say, the Thomas-Fermi model.

In the adiabatic version of the theory[8)], the density ρ is expanded in various powers of p. Retaining terms up to the second order, we write

$$\rho = \rho^{(0)} + \frac{p}{\lambda}\rho^{(1)} + \frac{1}{2}\frac{p^2}{\lambda^2}\rho^{(2)} \tag{24a}$$

$$f = f^{(0)} + \frac{p}{\lambda} f^{(1)} + \frac{1}{2}\frac{p^2}{\lambda^2} f^{(2)} \tag{24b}$$

where λ is as defined in eq. (17c).

The two Villars equations then can be written as

$$f^{(1)} = -\{\tilde{h}^{(0)}, f^{(0)}\} \tag{25a}$$

$$\frac{\lambda}{\mathcal{M}(q)}\frac{\partial f^{(0)}}{\partial q} = \{\tilde{h}^{(0)}, f^{(1)}\} + \{\tilde{h}^{(1)}, f^{(0)}\} \tag{25b}$$

where the curly brackets denote Poisson brackets. The equation corresponding to the second-order consistency condition can also be written and is not being displayed here in details.

Taking the required k-moments of these equations, we obtain

$$j_i^{(1)}(\underset{\sim}{R}, q) = -\sum_{j=1}^{3} \tau_{ij}^{(0)}(\underset{\sim}{R}, q) + \rho^{(0)}(\underset{\sim}{R}, q)\frac{\partial v^{(0)}}{\partial x_i} \tag{26a}$$

$$\lambda \frac{\partial \rho^{(0)}}{\partial q} + \nabla_R \cdot \underset{\sim}{j}^{(1)} = 0. \qquad (26b)$$

The various quantities occuring in these equations are defined as

$$\rho^{(0)}(\underset{\sim}{R},q) = \frac{1}{(2\pi\hbar)^3} \int d^3k \; f^{(0)}(\underset{\sim}{R},\underset{\sim}{k},q)$$

$$j^{(1)}(\underset{\sim}{R},q) = \frac{1}{(2\pi\hbar)^3} \int d^3k \; \underset{\sim}{k} f^{(1)}(\underset{\sim}{R},\underset{\sim}{k},q) \qquad (27)$$

$$\tau_{ij}^{(0)}(\underset{\sim}{R},q) = \frac{1}{(2\pi\hbar)^3} \int d^3k \; m^{-1} k_i k_j \; f^{(0)}(\underset{\sim}{R},\underset{\sim}{k},q).$$

Once again the hierarchy is to be closed by approximating the kinetic energy tensor τ. Details will be discussed in a forthcoming publication.

1) Negele, J. W., Revs. Mod. Phys. 54, 913 (1982).

2) Villars, F., Nucl. Phys. A285, 269 (1985).
Baranger, M. and Veneroni, M., Annals of Phys. 114, 123 (1978).
Brink, D.M., Giannoni, M. J. and Veneroni, M., Nucl. Phys. A258, 237 (1976).
Rowe, D.J. and Basserman, R., Can. J. Phys. 54, 1941 (1976).
Holzwarth, G. and Yukawa, T., Nucl. Phys. A219, 125 (1974).
Marumori, T., Prog. Theor. Phys. 57, 112 (1977).
Goeke, K., Cusson, R.Y., Grummer, F., Reinhard, P.G. and Reinhardt, H., Suppl. of Prog. Theor. Phys. Nos. 74 and 75, 33 (1983).

3) Mukherjee, A.K. and Pal, M.K., Phys. Lett. 100B, 457 (1981).

4) Mukherjee, A.K. and Pal, M.K., Nucl. Phys. A373, 289 (1982) ; Lecture Notes in Physics, Springer-Verlag, 171, 358 (1982).

5) Klein, A., Marumori, T. and Une, T., Phys. Rev. C29, 240 (1984)
Klein, A. and Tanabe, K., Phys. Lett. 135B, 255 (1983).

6) Ring, P. and Schuck, P., The Nuclear Many-Body Problem, Springer-Verlag, p.553 (1980).

7) Grange, P., Richert, J., Wolschin, G. and Weidenmuller, H.A., Nucl. Phys. A356,260 (1981).

8) Brink, D.M. and Di Toro, M., Nucl. Phys. A372, 151(1981).

PION-INDUCED PION AND ETA MESON PRODUCTION ON A FREE NUCLEON AND IN NUCLEI - A REVIEW

R.S. Bhalerao

Theoretical Physics Group
Tata Institute of Fundamental Research
Colaba, Bombay 400 005
INDIA

ABSTRACT

The status of the (π,η) and $(\pi,2\pi)$ reactions on a free nucleon and in nuclei is reviewed. The relevance of these reactions to other areas of medium energy physics is pointed out.

1. INTRODUCTION AND MOTIVATION

The pionic production of an eta meson in a nucleus belongs to a domain of medium energy physics that has just started being explored. The study of pionic production of a single pion in a nucleus is also in a very preliminary stage though it has a somewhat longer history. In this talk I will review the experimental and theoretical work that has been done on the $(\pi,2\pi)$ and (π,η) reactions on a free nucleon and in nuclei. I will restrict myself to incident π laboratory kinetic energies T_π below 1 GeV. In this energy region other πN reactions such as $\pi N \rightarrow K\Lambda$ and $\pi N \rightarrow K\Sigma$ are also possible. These reactions, however, have much smaller cross sections compared to $\pi N \rightarrow \pi\pi N$ and $\pi N \rightarrow \eta N$ reactions. The thresholds for the reactions $\pi^- p \rightarrow \pi^+\pi^- n$, $\pi^- p \rightarrow \eta n$, $\pi^- p \rightarrow K^o\Lambda$, and $\pi^- p \rightarrow K^+\Sigma^-$ are at T_π = 172, 561, 768, and 905 MeV, or at c.m. energies $\sqrt{s}$ = 1219,1488,1613, and 1691 MeV, respectively. (For simple kinematic reasons, the thresholds for the

---

(Invited talk at the International Conference on Nuclear Physics, Bhabha Atomic Research Centre, Bombay, India, December 1984.)

same reactions in nuclei are expected to be much lower.) We now discuss the motivations for studying the $(\pi,2\pi)$ and (π,η) reactions.

1.1 $\pi N \to \pi\pi N$

In terms of the overall magnitude of the cross section, this is the most important πN reaction in the energy region of our interest. This reaction was studied near threshold due to its ability to discriminate among various chiral-dynamics models. Specifically, the chiral-symmetry-breaking parameter ξ appearing in the model Lagrangian can be determined by comparing near-threshold experimental cross sections for the reaction $\pi N \to \pi\pi N$ with those obtained in the soft-pion theory[1]). One of the most recent attempts in this direction has been described in ref. [2]). This reaction also serves as a probe of low energy $\pi\pi$ scattering which otherwise is difficult to study experimentally[1,3]). Finally, knowledge of this reaction is essential in understanding $(\pi,2\pi)$ reaction in nuclei.

1.2 $\pi d \to \pi\pi NN$

The deuteron provides the simplest nuclear environment in which pion production involving more than one nucleon can be studied. The deuteron being a loosely bound system, the medium modification of the basic $\pi N \to \pi\pi N$ amplitude is expected to be minimal. The internal structure of the deuteron has been studied extensively and is sufficiently well understood. As the π double-charge-exchange reaction is not possible on the deuteron, detection of an outgoing pion of charge opposite to that of the incoming pion necessarily implies that the π production has taken place. This allows single-arm detection devices to be used to achieve higher counting rates. Finally, the $\pi d \to \pi\pi NN$ reaction is perhaps the ideal place to study the socalled double - Δ mechanism[4]) for π production and to look for the quasibound $\pi^- nn$ state predicted to exist[5]).

1.3 $A(\pi,2\pi)B$

The $(\pi,2\pi)$ reaction in a complex nucleus is also of great interest, since it may provide a new avenue for studying the pion field inside a nucleus. Eisenberg[6,7]) has proposed this reaction as a means to selectively excite pionlike levels ($J^{\pi} = 0^-, 1^+, 2^-, \ldots;\ T = 1$), to study the spin-isospin oscillations in closed-shell nuclei, and to look for possible indications of pion condensation precursor phenomena. Since the basic $\pi N \rightarrow \pi\pi N$ cross section becomes comparable to the πN elastic scattering cross section at pion kinetic energies $T_{\pi} \sim 600$ MeV and exceeds the elastic scattering cross section at $T_{\pi} > 1$ GeV, we believe that the $(\pi,2\pi)$ reaction in complex nuclei represents one of the most powerful tools to study nuclei at pion energies above the much studied (3,3) resonance region.

1.4 $\pi N \rightarrow \eta N$

This is an important πN reaction in the energy region $T_{\pi} = 0.6$ - 1.0 GeV. In terms of the cross-section magnitude, it is the second most important πN inelastic channel, second only to π-induced π production. Knowledge of this reaction is essential in understanding (π,η) reaction in nuclei.

1.5 $A(\pi,\eta)B$

Nuclear (π,η) reactions are also of great interest. In the SU(6) model, η differs from π^o by having an additional $s\bar{s}$ quark - antiquark pair in its wave function. The systematics of ηN and $\pi^o N$ scattering could yield information on the dynamical role of this $s\bar{s}$ pair in meson - baryon interactions. Like (γ,η) and other modes of nuclear η production, (π,η) reactions will enable one to study ηN scattering. Since hadronic and electromagnetic η production processes are different, details of ηN scattering extracted from different types of nuclear reactions will complement each other. Because π^o and η do not belong to the same isospin multiplet, comparative studies of $A(\pi^{\pm},\eta)B$ and $A(\pi^{\pm},\pi^o)B$ reactions are also of interest.

2. EXPERIMENTAL SITUATION

2.1 $\pi N \rightarrow \pi\pi N$

Total cross sections for the reactions

$$\pi^- p \to \pi^+ \pi^- n,$$
$$\pi^- p \to \pi^o \pi^o n,$$
$$\pi^+ p \to \pi^+ \pi^+ n, \qquad (1)$$
$$\pi^- p \to \pi^o \pi^- p,$$

and

$$\pi^+ p \to \pi^o \pi^+ p,$$

are available over a wide energy range, starting at T_π between 200 and 300 MeV and going up to $T_\pi \sim$ a few GeV. An extensive list of experimental papers can be found in ref.[2]). Thus cross sections very near threshold ($T_\pi < 200$ MeV) are not available. These data show that among the five π-production channels (eq. (1)), cross sections for $\pi^- p \to \pi^+ \pi^- n$ are the largest for $T_\pi < 1$ GeV [8]). $\sigma(\pi^- p \to \pi^+ \pi^- n)$ reaches a maximum (~12 mb) at $T_\pi \sim 900$ MeV. Some differential cross sections are also available [9]).

2.2 $\pi d \to \pi\pi NN$

Single-and double-differential and total cross sections for the reactions $\pi^- d \to \pi^+ \pi^- nn$ and $\pi^+ d \to \pi^- \pi^+ pp$ at $T_\pi = 256$, 331, and 450 MeV have been measured very recently [10,11]). The total cross sections were found to be very close to those for the $\pi^- p \to \pi^+ \pi^- n$ reaction. In refs.[10,11]) upper limits for formation of bound $\pi^- nn$ and $\pi^+ pp$ systems have also been given.

2.3 $A(\pi,2\pi)B$

For the $(\pi,2\pi)$ reaction in a complex nucleus the only published data are from an old emulsion experiment [12]). In this experiment, 322 events of the type $\pi^- A \to \pi^+ \pi^- A'$ were detected, and the total cross section as a function of T_π was deduced. Recently, Dropesky et al. [13]) have used the activation technique to measure $(\pi,2\pi)$ cross sections leading to all particle - stable states for the reactions

$^{27}Al(\pi^+,2\pi^+)^{27}Mg$, $^{51}V(\pi^+,2\pi^+)^{51}Ti$, $^{45}Sc(\pi^-,2\pi^-)^{45}Ti$, and $^{63}Cu(\pi^-,2\pi^-)^{63}Zn$ at T_π = 350 MeV. More $A(\pi,2\pi)B$ data based on in-beam counter techniques are expected to be available in the next few years [14].

2.4 $\pi N \to \eta N$

The η meson was discovered in 1961 and since then total and differential cross sections for the $\pi N \to \eta N$ reaction have been measured at a large number of energies [15]. Above the threshold the total production cross section rises rapidly with pion energy and reaches a maximum ($\sim$2.5 mb) at $T_\pi \sim$ 661 MeV. Polarization data are available at c.m. energies between 1764 and 2272 MeV [16] but not in the region where σ_{tot} peaks.

2.5 $A(\pi,\eta)B$

To my knowledge there are no published data for the (π,η) reaction in a complex nucleus. Preliminary data for the reaction $^3He(\pi^-,t)\eta$ have recently been obtained at LAMPF by J.C. Peng et al. [17]. Future plans include measurements of total and differential $(\pi^\pm,\eta)$ cross sections on several targets at many energies [17].

3. THEORETICAL SITUATION

3.1 $\pi N \to \pi\pi N$

3.1.1 Near threshold. The $\pi N \to \pi\pi N$ reaction at low energies has often been treated in the framework of the "chiral dynamics" method of Weinberg [18,1]. In this method an effective Lagrangian is constructed which when used in the lowest-order perturbation theory gives precisely the results of the current algebra method but in a much simpler way. The invariant amplitude $T(\pi N \to \pi\pi N)$ calculated in this method has terms corresponding to the "tree" diagrams shown in fig.1. With the exceptions of the

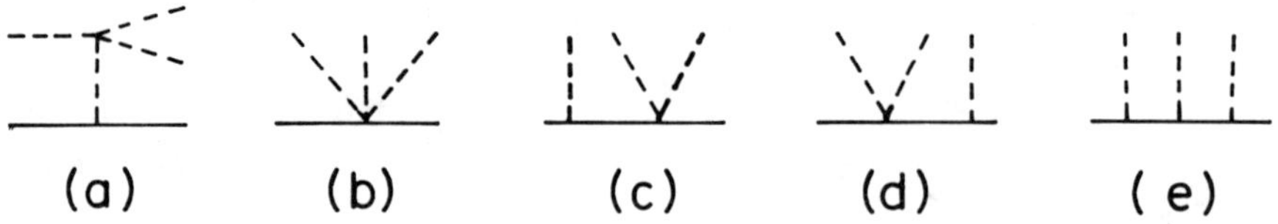

Fig.1 (a) and (b) one-point, (c) and (d) two-point, and (e) three-point tree diagrams for the reaction $\pi N \to \pi\pi N$.

reactions $\pi^+ p \to \pi^o \pi^+ p$ and $\pi^- p \to \pi^o \pi^- p$, total cross sections for the reactions in eq. (1), calculated using the chiral dynamics method do not agree [19]) with the lowest-energy data that are available. Hence $\pi N \to \pi\pi N$ amplitudes based on this method are not universally suitable for use in nuclear calculations.

Recently, Kälbermann and Eisenberg [20]) have treated the $\pi N \to \pi\pi N$ reaction in the framework of the chiral bag models with pion penetration of the bag.

3.1.2 Resonance region. Taking $\pi^- p \to \pi^+ \pi^- n$ reaction as an example, we illustrate below some of the simple diagrams that contribute to the $\pi N \to \pi\pi N$ reaction in the resonance region. Here α and β are

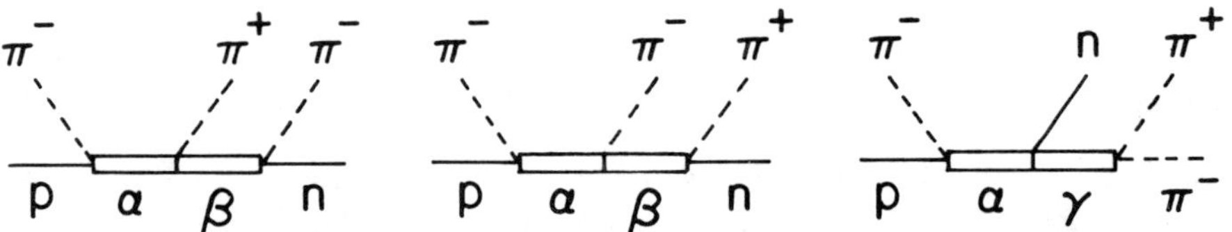

Fig.2 $\pi N \to \pi\pi N$ in the resonance region.

N^* or Δ baryon resonances and $\gamma = \rho, \varepsilon \cdots$. Several phenomenological isobar models based on subsets of these diagrams have been proposed in the literature [21-27]). They generally contain many free parameters which are fitted to the experimental $\pi N \to \pi\pi N$ data. Aitchison and Brehm [28]) have improved the phenomenology by presenting an isobar model which incorporates unitarity and analyticity in each of the two-body subenergy channels.

3.2 $\pi d \to \pi\pi NN$

Some of the simple diagrams which contribute to this reaction are shown in fig.3. Important

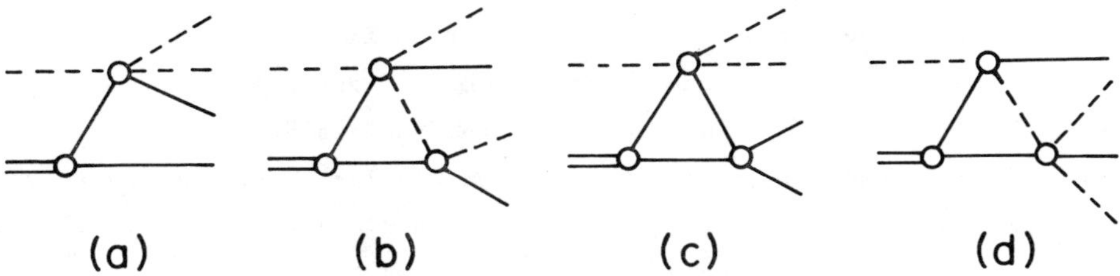

Fig.3 Diagrams for $\pi d \to \pi\pi NN$.

ingredients of these diagrams are an off-shell model for the π-production amplitude, deuteron wave functions and off-shell $\pi N \to \pi N$ and $NN \to NN$ t-matrices. We have studied the $\pi^- d \to \pi^+\pi^- nn$ reaction in the quasifree approximation (fig. 3a) and found that this mechanism dominates [29], see section 4 for details. A similar conclusion was drawn recently by Rockmore [30].

3.3 $A(\pi,2\pi)B$

In recent years, several authors have predicted total cross sections for the $A(\pi,2\pi)$ reactions as a function of the target nucleus mass and/or incident pion energy [7, 31-35]. All of these authors use as a basic input the $\pi N \to \pi\pi N$ amplitude derived from Weinberg's effective Lagrangian [18,1] in some approximation. In ref. [19] we showed that these approximate schemes contain only minimal dynamical information in that their predictions differ from phase space prediction by only a multiplicative constant. Also the $\pi^- p \to \pi^+\pi^- n$ cross sections based on the approximate theories [7,31-34] are generally an order of magnitude smaller than the experimental data. Predicted cross sections for $A(\pi,2\pi)$ reactions suffer from these drawbacks. In particular, if the existence of pion condensation precursor phenomena is to be tested in $A(\pi,2\pi)$ reactions, then the calculation of the basic $\pi N \rightarrow \pi\pi N$ amplitude will have to be done more accurately than has been reported. Otherwise, the importance of precursor phenomena, if they exist at all, would be overestimated to compensate for the low theoretical cross sections arising from the use of the approximate theories [7,31-34]. Angular correlations of the outgoing pions and the energy spectrum of the outgoing neutron calculated in the exact theory and with the approximate schemes of refs. [7,31-35] differ significantly in magnitude and shape, indicating the additional inadequacies of the latter theories. In particular, analysis of the measured $\pi\pi$ angular correlation in $A(\pi,2\pi)$ reactions with an approximate theory may lead to erroneous conclusions about pion propagation in a nucleus. (See ref. [19] for more details).

Hence it is important to have $A(\pi,2\pi)$ calculations performed with a more realistic $\pi N \rightarrow \pi\pi N$ amplitude.

3.4 $\pi N \rightarrow \eta N$

Existing theoretical models [36] for the $\pi N \rightarrow \eta N$ amplitude are based either on the K-matrix approach or on a Breit-Wigner type parameterization with the masses and the partial widths of various resonances determined by fitting the experimental η production cross sections. However, these models are not suitable for calculating (π,η) reactions in a nucleus because they do not contain form factors to allow a meaningful off-shell extrapolation of the amplitude. They also cannot provide an ηN elastic scattering amplitude, a quantity indispensible for calculating the final-state interaction in $A(\pi,\eta)B$ reactions. In ref. [37] we have presented a model which overcomes these shortcomings. It is discussed in detail in section 5.

3.5 $A(\pi,\eta)B$

These calculations are in progress and will be reported later[38].

4. QUASIFREE CALCULATION OF $\pi d \rightarrow \pi\pi NN$

We have calculated the double differential cross sections $d^2\sigma/d\Omega dT$ for the reaction $\pi^- d \rightarrow \pi^+\pi^- nn$ as a function of T [29]. Here Ω

and T refer to the solid angle and kinetic energy of the outgoing π^+ in the c.m. frame. Calculations were performed in the plane-wave quasifree approximation (fig.3a). The experimental double differential cross sections for $\pi N \to \pi\pi N$ were used to parameterize the π-production vertex. The calculation involves no other free parameters. Calculated $d^2\sigma/d\Omega dT$ are in surprisingly good agreement with the general features of the experimental data [10,11]), although some discrepancies do exist. At T_π = 256 MeV our calculation reproduces the shape of the spectra very well, but the absolute value of the cross section is 20% below the measured value. At 331 MeV both the shape and the integrated cross section obtained from the quasifree calculation agree with the measurement within the experimental uncertainty. At 450 MeV there are some slight discrepancies between the calculated and measured values at backward angles. These small discrepancies may be due to the quasifree approximation made in this work or due to the simple parameterization used for the basic production amplitude [29,11]).

The above observations along with the fact that $\sigma(\pi d \to \pi\pi NN) \simeq \sigma(\pi N \to \pi\pi N)$ suggest that two-nucleon effects contribute little to pion production on the deuteron. This makes the study of $A(\pi,2\pi)$ more interesting.

5. AN OFF-SHELL MODEL FOR $\pi N \to \eta N$ AND FOR ηN SCATTERING

In this section, I shall describe briefly an off-shell isobar model for the $\pi N \to \pi N$, $\pi N \to \eta N$, and $\eta N \to \eta N$ t-matrices, that we presented recently [37]). We believe it will play a useful role in the development of "η-nuclear physics".

In the energy region $\sqrt{s}$ = 1488 to 1600 MeV that we are interested in, there are only three important channels in πN collision: $\pi N \to \pi N$, $\pi N \to \pi\pi N$, and $\pi N \to \eta N$. We incorporate all of them in our model. The diagrams for the interaction matrix elements are shown in figs. 4a to 4c, where α denotes $N^*(I = 1/2)$ or

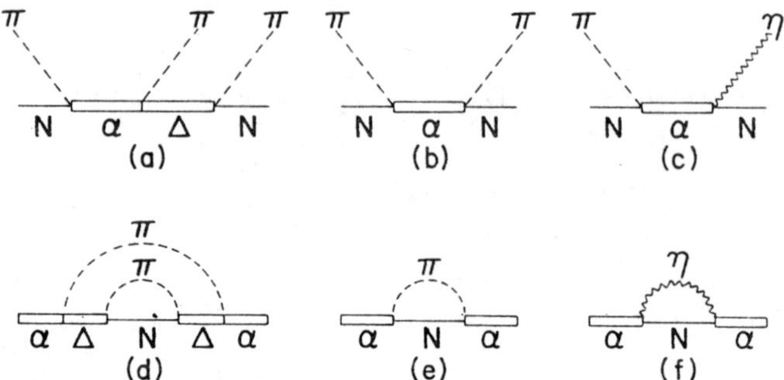

Fig.4 (a-c) Schematic representation of the separable interaction matrix elements, (d-f) various contributions to the self energy of a resonance α.

Δ(1232) (denoted henceforth Δ) having a bare mass m_α. Because the ηN system has I = 1/2, it can only couple to N^*. In figs.4a to 4c each vertex is described by a vertex function $h^{\ell}_{a\alpha}(k)$ to be specified later. Here a denotes the channels πN, πΔ, or ηN. The ℓ and k denote, respectively, the relative orbital angular momentum and the magnitude of the c.m. momentum in channel a. One can show that, upon formal elimination of the ππN channel, one obtains in the reduced model space a complex and energy-dependent effective interaction. The radial part of this interaction in a given partial wave ℓ is

$$\langle p'|V^{\alpha\ell}_{ij}(\sqrt{s})|p\rangle = \frac{h^{\ell}_{i\alpha}(p')\, h^{\ell}_{\alpha j}(p)}{\sqrt{s} - m_\alpha - \Sigma^{\alpha}_{2\pi}(\sqrt{s})} \,. \tag{2}$$

Here i = 1,2 and j = 1,2 with 1 and 2 labeling, respectively, the πN and ηN channels. The appearance of $\Sigma^{\alpha}_{2\pi}(\sqrt{s})$ in eq.(2) results from the formal elimination of the ππN channel from our model space. $\Sigma^{\alpha}_{2\pi}(\sqrt{s})$ is the self-energy of α, associated with a 2π intermediate state (fig.4d). The full solution of the coupled-channel equation $T = V + VG_oT$ with V of eq.(2) is:

$$\langle p'|T^{\alpha\ell}_{ij}(\sqrt{s})|p\rangle = \frac{h^{\ell}_{i\alpha}(p')\, h^{\ell}_{\alpha j}(p)}{\sqrt{s} - m_{\alpha} - \Sigma^{\alpha}_{\pi}(\sqrt{s}) - \Sigma^{\alpha}_{\eta}(\sqrt{s}) - \Sigma^{\alpha}_{2\pi}(\sqrt{s})} \quad . \qquad (3)$$

In eq.(3) $\sum^{\alpha}_{\pi}$ and $\sum^{\alpha}_{\eta}$ are the self-energies associated, respectively, with πN and ηN intermediate states (figs. 4e and 4f).

We have parameterized the vertex functions as

$$h^{\ell}(q) = \frac{g}{\sqrt{2\sqrt{s}}}\,[v_{\ell}(q/\Lambda)]^{\frac{1}{2}}\,\frac{\Lambda^2}{\Lambda^2 + q^2} \quad ,$$

where $v_{\ell}(\rho) \equiv [\rho^2\, n^2_{\ell}(\rho) + \rho^2\, j^2_{\ell}(\rho)]^{-1}$ is the "penetration factor" [39]). The parameters of our model are g, Λ, and the bare mass m_{α}. We included $\pi N\Delta$, $\pi\Delta\Delta$, πNN^*, $\pi\Delta N^*$, and ηNN^* vertices, with N^* being S11, P11 and D13 πN resonances.

We first determined the $\pi N\Delta$ and $\pi\Delta\Delta$ vertex functions and the bare mass of Δ by fitting the πN P33 phase shifts [40-42]). The resulting $g_{\pi N\Delta}$, $\Lambda_{\pi N\Delta}$, and m_{Δ} were then treated as fixed input in the determination of the parameters in the N^* channels. In general, we found two solutions for each of these channels. However, one of them could be readily eliminated because it gave unphysical $N^* \to \pi N$ and $N^* \to \eta N$ branching ratios. Although the πN scattering data alone cannot determine the signs of the coupling constants, g, the η production data helped us fix these signs. Satisfactory fits were obtained in all cases; as an example, we display in fig.5 our fits for the P33 and S11 channels. We summarize in Table I the parameters determined in this work.

The predicted $\pi N \to \eta N$ differential cross sections are compared with the experimental data [15]) in fig. 6. The agreement is excellent. We have calculated ηN scattering phase shifts. The ηN interaction was found to be attractive in the S11 channel.

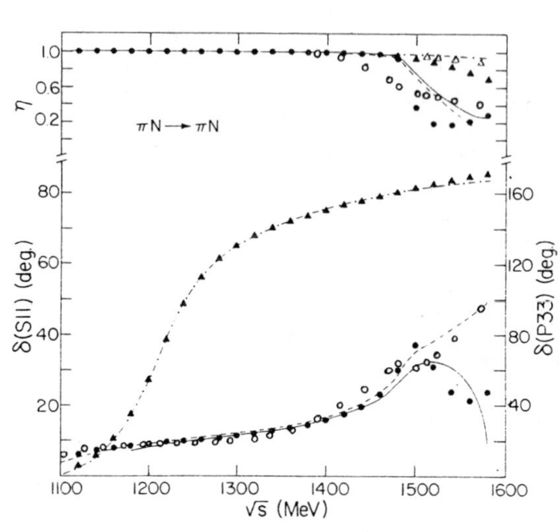

Fig.5. πN scattering phase shifts δ(in deg.) and inelasticity parameters η vs $\sqrt{s}$. The S11 data of ref. [42] (•), the S11 data of ref. [40] (o), and the P33 data of ref. [42] (▲) and of ref. [40] (Δ) are compared with the calculations shown, respectively, as the solid, dashed, and dot-dashed curves.

Table I. Coupling constants g, range parameters Λ (in MeV/c) and bare masses m_α (in MeV). Numbers with and without parentheses are due, respectively, to πN phase shifts of refs. [42] and [40] (see ref. [43]).

Coupling to →		πN	πΔ	ηN
P33	g	(2.347)	(1.247)	----
m_α = (1323.5)	Λ	(331.0)	(264.9)	----
S11	g	1.384 (1.301)	3.509 (7.080)	0.616 (0.769)
m_α = 1608.1 (2088.0)	Λ	344.3 (435.2)	203.3 (376.6)	465.1 (1503.0)
P11	g	1.814 (1.988)	5.019 (4.989)	0.267 (0.305)
m_α = 1588.6 (1664.9)	Λ	568.9 (573.5)	110.4 (159.8)	207.1 (187.9)
D13	g	3.073 (3.037)	1.411 (1.404)	0.334 (0.327)
m_α = 1579.3 (1567.8)	Λ	268.0 (289.5)	267.6 (250.1)	115.3 (156.2)

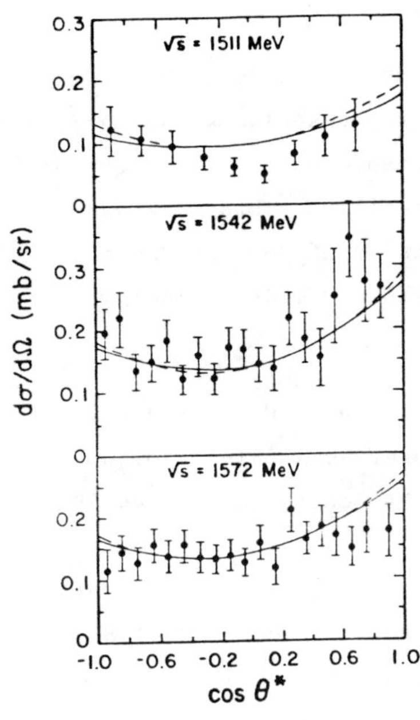

Fig. 6. Differential cross section for $\pi^- p \to \eta n$ vs cosine of the production angle in the cm frame. Solid (dashed) curves are obtained with parameters of Table I, based on πN phase shifts of ref. [42]), (ref. [40])). Data are from ref. [15]).

We conclude that our model is capable of describing the $\pi N \to \eta N$ reaction near threshold. We have noted that different πN phase-shift solutions can give very different g and Λ but very similar $\pi N \to \eta N$ cross sections. As different g and Λ imply different off-shell behavior of the interaction, we believe future studies of nuclear (π,η) reactions with our model will enable us to discriminate among various sets of πN phase shifts available in the literature.

6. CONCLUSION

I have briefly reviewed the status of the (π,η) and (π,2π) reactions and have pointed out areas where more experimental or theoretical work is needed. The study of these reactions has relevance to many other areas of medium energy physics. For example, many of the diagrams representing the π-nucleus double charge exchange have the $\pi N \to \pi\pi N$ vertex as an important ingredient [19]). The $\pi N \to \eta N$ model described here helps discriminate among various πN phase-shift solutions. The meson-baryon-baryon vertex functions deduced from the study of $\pi N \to \pi\pi N$ and $\pi N \to \eta N$ reactions can clearly be used for the modeling of many other nuclear reactions.

7. ACKNOWLEDGMENTS

The original work described here was done in collaboration with Dr. L.C. Liu and/or Dr. E. Piasetzky at the Los Alamos National Laboratory and the author thanks the Isotope and Nuclear Chemistry Division of the LANL for making his visit possible. This work was performed under the auspices of the Division of Nuclear Physics, Office of High Energy and Nuclear Physics, U.S. Department of Energy.

8. REFERENCES

1. Olsson, M.G. and Turner, L., "Use of Single-Pion Production to Remove Ambiguities in Partially Conserved Axial-Vector Current and Current-Algebra Predictions of $\pi\pi$ Scattering", Phys. Rev. Lett. 20, 1127-1130 (1968).
2. Manley, D.M., "Isospin Analysis of Low-Energy $\pi N \to \pi\pi N$ Data and Chiral-Symmetry Breaking", Phys. Rev. D 30, 536-540 (1984).
3. Martin, B.R., Morgan, D., and Shaw, G., "Pion-Pion Interactions in Particle Physics", Academic, New York, 1976.
4. Brown, G.E., Toki, H., Weise, W., and Wirzba, A., "Double-Δ(1232) Formation in Pion-Nucleon Absorption", Phys. Lett. 118, 39-44 (1982).
5. Garcilazo, H., "Possible Existence of a $\pi^- nn$ Bound State", Phys. Rev. C 26, 2685-2687 (1982); "Bound States of Negative Pions and Neutrons", Phys. Rev. Lett. 50, 1567-1570 (1983); "The πNN Bound-State Problem", Nucl. Phys. A408, 559-572 (1983).
6. Eisenberg, J.M., "Pion-Production in Pion-Nucleus Collisions", Nucl. Phys. A148, 135-144 (1970).
7. Eisenberg, J.M., "The $(\pi,2\pi)$ Process as a Possible Probe of Pion Condensation Precursor Phenomena", Phys. Lett. 93B, 12-15 (1980).
8. Olsson, M.G. and Yodh, G.B., "Analysis of Single-Pion Production Reactions $\pi + N \to \pi_1 + \pi_2 + N'$ below 1 BeV", Phys. Rev. 145, 1309-1326 (1966).
9. For the most recent data, see Manley, D.M., "Measurement and Isobar-Model Analysis of the Doubly Differential Cross Section for the π^+ Produced in $\pi^- p \to \pi^+\pi^- n$", Ph.D. Thesis, Univ. of Wyoming, Los Alamos National Laboratory Report No.LA-9101-T, 1981;

Bjork, C.W. et al., "Measurement of $\pi^- p \to \pi^- \pi^+ n$ near Threshold and Chiral-Symmetry Breaking", Phys. Rev. Lett. 44, 62-65 (1980).

10. Piasetzky, E. et al.,"Pion-Induced Pion Production on the Deuteron", Phys. Rev. Lett. 53, 540-543 (1984).

11. Lichtenstadt, J. et al., "Pion-Induced Pion Production on the Deuteron", Los Alamos National Laboratory Preprint.

12. Batusov, Yu.A. et al., "Production of Mesons by Mesons and Double-Charge Exchange of Negative Pions on Emulsion Nuclei in the Energy Interval 210-375 MeV", Sov. J. Nucl. Phys. 9, 221-222 (1969).

13. Dropesky, B.J. et al., unpublished.

14. Piasetzky, E., private communication.

15. Brown, R.M. et al., "Differential Cross Sections for the Reaction $\pi^- p \to \eta n$ between 724 and 2723 MeV/c", Nucl. Phys. B153, 89-111 (1979), and references therein.

16. Baker, R.D. et al., "Polarized Cross Section Measurements for the Reaction $\pi^- p \to \eta n$ from 1171 to 2267 MeV/c and a Partial-Wave Analysis of this Reaction in the Resonance Region", Nucl. Phys. B156, 93-110 (1979).

17. Peng, J.C., private communication.

18. Weinberg, S., "Dynamical Approach to Current Algebra", Phys. Rev. Lett. 18, 188-191 (1967).

19. Bhalerao, R.S. and Liu, L.C., "Comparison of Approximate Chiral-Dynamical $\pi N \to \pi\pi N$ Models Used in $A(\pi,2\pi)$ Calculations", Phys. Rev. C 30, 224-231 (1984).

20. Kälbermann, G. and Eisenberg, J.M., "Chiral Bag Models and the $\pi N \to \pi\pi N$ Reaction", Phys. Rev. D 28, 66-70 (1983).

21. Yodh, G.B. and Olsson, M., "Partial-Wave Analysis of the Reactions $\pi^+ p \to \pi^+ \pi^o p$ and $\pi^+ \pi^+ n$ below 1 BeV", Phys. Rev. 145, 1327-1331 (1966) and references therein.

22. Morgan, D., "Phenomenological Analysis of $I = \frac{1}{2}$ Single-Pion Production Processes in the Energy Range 500 to 700 MeV", Phys. Rev. 166, 1731-1759 (1968).

23. Herndon, D.J., Söding, P, and Cashmore, R.J., "Generalized

Isobar Model Formalism", Phys. Rev. D 11, 3165-3182 (1975); Herndon, D.J. et al., "Partial-Wave Analysis of the Reaction $\pi N \to \pi\pi N$ in the C.M. Energy Range 1300-2000 MeV", Phys. Rev. D 11, 3183-3213 (1975).

24. Longacre, R.S. et al., "K-Matrix Fits to $\pi N \to N\pi$ and $\pi N \to N\pi\pi$ in the Resonance Region $\sqrt{s}$ = 1.3 to 2.0 GeV", Phys. Rev. D 17, 1795-1825 (1978) and references therein.

25. Novoseller, D.E., "Partial-Wave Analysis Including π Exchange for $\pi N \to N\pi\pi$ in the C.M. Energy Range 1.65 - 1.97 GeV", Nucl. Phys. B137, 445-508 (1978); "Analysis of Decay of πN Resonances into $N\pi\pi$ Channels", Nucl. Phys. B137, 509-520 (1978).

26. Barnham, K.W.J. et al., "An Isobar Model Partial-Wave Analysis of Three-Body Final States in $\pi^+ p$ Interactions from Threshold to 1700 MeV c.m. Energy", Nucl. Phys. B168, 243-271 (1980).

27. Arndt, R.A. et al., "Isobar Production in $\pi^- p \to \pi^+ \pi^- n$ near Threshold", Phys. Rev. D 20, 651-682 (1979); Manley, D.M. et al., "Isobar-Model Partial-Wave Analysis of $\pi N \to \pi\pi N$ in the C.M. Energy Range 1320-1930 MeV", Phys. Rev. D 30, 904-936 (1984).

28. Aitchison, I.J.R. and Brehm, J.J., "Are there Important Unitary Corrections to the Isobar Model", Phys. Lett. 84B, 349-353 (1979), and references therein.

29. Bhalerao, R.S., Liu, L.C., and Piasetzky, E., "Contributions of Double-Δ and other Two-Nucleon Mechanisms to Pionic Pion Production on the Deuteron", Proc. Sympo. on Δ-Nucleus Dynamics, Argonne, U.S.A., 1983, p. 465-471. See also refs. [10,11]).

30. Rockmore, R., "Comparison of a Two-Body Threshold $(\pi,2\pi)$ Reaction Mechanism with the Usual One-Body Mechanism in the Deuteron", Phys. Rev. C 29, 1534-1536 (1984).

31. Cohen, J. and Eisenberg, J.M., "The $(\pi,2\pi)$ Reaction and Spin-Isospin Strength Distribution in Nuclei", Nucl. Phys. A395, 389-412 (1983).

32. Cohen, J., "The A Dependence and Threshold Behavior of the Inclusive $(\pi,2\pi)$ Reaction for T = 0 Nuclei", J. Phys. G 9, 621-629 (1983).

33. Cohen, J. and Eisenberg, J.M., "The $(\pi,2\pi)$ Reaction and Spin-Isospin Strength Distribution in Nuclei", Nuovo Cimento 76, 488-493 (1983).

34. Rockmore, R., "Threshold Pion Production in Pion-Nucleus Collisions: A Simple Estimate", Phys. Rev. C 11, 1953-1958 (1975).

35. Rockmore, R., "Threshold Estimates of $(\pi,2\pi)$ on Nuclei in the Fermi Gas Model: One-Body Mechanism", Phys. Rev. C 27, 2150-2157 (1983).

36. Carreras, B. and Donnachie, A., "Eta Production by Pions", Nucl. Phys. B16, 35-45 (1970) and references therein; Dobson, P.N., "Zero-Effective-Range Analysis of Pion-Nucleon Reactions near the η^o Production Threshold", Phys. Rev. 146, 1022-1025 (1966); Tuan, S.F., "Dynamical Basis for η-Baryon Interactions", Phys. Rev. 139, B1393-B1400 (1965).

37. Bhalerao, R.S. and Liu, L.C., "An Off-Shell Model for Threshold Pionic η production on a Nucleon and for ηN Scattering", Los Alamos National Laboratory Preprint No.LA-UR 84-3384.

38. Bhalerao, R.S. and Liu, L.C., unpublished.

39. Blatt, J.M. and Weisskopf, V.F., "Theoretical Nuclear Physics", John Wiley and Sons, New York, 1962.

40. Herndon, D.J. et al., "πN Partial-Wave Amplitudes", Lawrence Berkeley Laboratory Report No.UCRL-20030 πN.

41. Rowe, G., Salomon, M., and Landau, R.H., "Energy-Dependent Phase-Shift Analysis of Pion-Nucleon Scattering below 400 MeV", Phys. Rev. C 18, 584-589 (1978).

42. Arndt, R.A., private communication. We used his phase shift solution FP84.

43. The use of P33 phase shifts of ref. [41]) at T_π below 350 MeV, which give a better P33 resonance energy than ref. [40]), and the use of P33 phase shifts of ref. [40]) at higher energies led to parameter values very close to those given in Table I.

SOME TOPICS IN QUARK PHYSICS OF INTEREST TO NUCLEAR PHYSICISTS

Jishnu Dey
Laboratoire de Physique nucléaire-Université de Montréal
Case postale 6128, Succursale "A"
Montréal, P.Q. H3C 3J7-Canada

ABSTRACT

We discuss the possibility of collective motion of quarks in baryons. Since our ideas on vibration and rotational degrees of freedom are already published emphasis is placed on the new work on the relativistic Hartree Fock calculation.

1. MOTIVATION

Much of nuclear physics is concerned about energy levels of nuclei. It is found that many of the energy levels are due to the collective motion of the nucleons - either of the vibrational type (the size oscillation or breathing mode being one example) or the kind which forms a rotational band. Looking at the excitation spectrum of the nucleon, N, (iso-spin $T = 1/2$) and its $T = 3/2$ partner, the Δ-isobar - one is tempted to conclude that these are also based on collective motion of the quark mean field [1-6]. To summarise our findings on the vibrational mode of excitation one can stress the following three points:

1) The compression modulus (which is the coefficient of force of restitution against breathing mode vibration) turns out to be equal to the nucleon (or isobar) mass scale in the non-relativistic quark model (NRQM), the MIT bag model and the Skyrme model.
2) The inertia parameter, on the other hand, is very different

in these models - it is too small for the NRQM or bag model and too large for the Skyrmion. This results in a low-lying even parity excitation which is too high for the first two models and too low for Skyrmions.

3) It seems from the above discussion that one should allow for mean field degrees of freedom as in the Skyrme model but should also allow quark degrees of freedom which is suppressed in this model in favour of (σ,π) fields only. This brings us to the next topic of interest, the relativistic Hartree Fock.

2. INTRODUCTION

It is now generally believed that Quantum Chromodynamics (QCD) is the correct theory for strong interactions. In spite of great progress in the understanding of such a theory it is as yet difficult to compute the mass spectrum of the observed hadrons from QCD. There is then the need to make approximations and an attempt is made in this paper to calculate baryon properties keeping as close to QCD as present day techniques permit.*

To do this we recall that QCD is a theory of coloured particle with number of coloms $N_c = 3$. The existence of mesons as an extremely large number of narrow resonances and properties like the Zweig's rule can be understood, however, only when one is willing to think about the whole class of QCD like theories of arbitraty N_c. This was first suggested by 't Hooft[7] almost as soon as QCD was formulated and became an extremely powerful tool for understanding baryons after the work of Witten.[8] QCD is equivalent to a classical field theory in lowest order with coupling $1/N_c$ for any N_c, including $N_c = 3$, but it is virtually impossible to discover or prove this equivalence if one is willing to think only about the case $N_c = 3$. This $1/N_c$ expansion to leading order does give most of the qualitative features

* We have in mind only continuum calculations. It is conceivable that the steadily improving lattice calculations will give the correct baryon properties.

expected of low energy QCD such as confinement[7] and chiral symmetry breaking[9].

For mesons it leads to a Bethe-Salpeter equation with a quark-antiquark ($q\bar{q}$) potential and for baryons it leads to a Hartree-Fock equation. The alternative for baryons is to solve meson classical Lagrangians, like the Skyrme model, for example, but these meson field models are difficult to deduce from QCD without encountering terms which are sixth and higher order in the isospin current of the pions and it is difficult to justify the truncation at fourth order as in the Skyrme model. The other difficulty with the Skyrme model is that important low mass mesons like the ρ-meson are difficult to incorporate into the model. At this stage, therefore, relativistic HF with a potential seems a very promising alternative for describing baryons - with the potential chosen such that it fits the meson sector properties. If all the planar gluon diagrams of QCD could be summed this would yield this potential unambiguously but unfortunately this is a task beyond the present day capabilities so that a potential borrowed from the meson sector calculations is the best alternative. As it turns out the situation is quite satisfactory here, for one has the Richardson potential, which in spite of its simplicity is able to provide many of the major features of the QCD interaction with only one free parameter. It incorporates the correct renormalization group behaviour at high q^2. At small q^2 it has the correct $\frac{1}{q^4}$ behaviour predicted by the Schwinger-Dyson equs. of QCD - solved in the work of Mandelstam and Baker, Ball and Zachariasen[11], also pointed out from different considerations by Nair and Rosenzweig[12].* It fits the charmonium data and additional cross-checks are possible on its

* Note that although this $\frac{1}{q^4}$ behaviour is not directly based on $\frac{1}{N}$ expansion, the fact that we have confinement to leading order in $1/N_c$ implies that the $q\bar{q}$ potential in the $\frac{1}{N}$ expansion also should have a $\frac{1}{q^4}$ low q^2 behaviour

only free parameter Λ, which is related to string tension and hence to other parameters like critical temperature T_c through lattice gauge calculations - as discussed recently by Bailin. Cleymens and Scadron[13].

In the spirit of remaining close to the first principles of QCD, we have not, in our calculations, allowed the quark mass, m_q, to be a free parameter but have fixed it at the current algebra values. With the strange quark mass $m_s = 150$ MeV we find a very good agreement with the experimental Ω^- energy for a reasonable value for Λ, the Richardson parameter. There is some uncertainty in the current quark mass values. We find the remarkable result that the kinetic energy in the relativistic Hartree-Fock decreases as the quark mass increases so that the result in fact is not very sensitive to small changes in the mass chosen.

3. RESULTS AND DISCUSSION

We omit here the formal derivation of the Hartree Fock equations through the path integral method as it will be given elsewhere[14]. For number of colours N_c, we get the equation

$$N_c \int \phi^\dagger_{jm}(r)\ t\ \phi_{jm}(r)\ d^3r$$

$$- \frac{N_c^2-1}{4} \int d^3r_1 d^3r_2\ \phi^\dagger_{jm}(r_1)\ \phi^\dagger_{jm}(r_2)\ V(r_{12})\phi_{jm}(r_1)\ \phi_{jm}(r_2)$$

$$= N_c\ \varepsilon \int d^3r\ \phi^\dagger_{jm}(r)\ \phi_{jm}(r) \tag{1}$$

which upon variation with respect to $\phi^\dagger$ yields

$$[t - W_N(r_1)]\ \phi_{jm}(r_1) = \varepsilon\ \phi_{jm}(r_1) \tag{2}$$

In the above

$$t = \hat{\alpha}\cdot\vec{p} + \hat{\beta}\ m \tag{3}$$

is the Dirac kinetic energy and the self consistent potential is given by

$$W_n(r_1) = \frac{N_c^2-1}{2N_c} \int d^3r_2\ \phi^\dagger_{jm}(r_2)\ V(r_{12})\ \phi_{jm}(r_2) \tag{4}$$

Choosing the lowest single particle orbitals

$$\phi_{jm}(r) = \begin{pmatrix} i\,G(r) \\ \hat{\sigma}.r\,F(r) \end{pmatrix} \mathcal{Y}^{\frac{1}{2}}_{j\ell}(r) \tag{5}$$

where G and F satisfy

$$\frac{dG}{dr} - (m - W_N - \varepsilon)\,F = 0 \tag{6}$$

$$\frac{dF}{dr} - \frac{2}{r}\,F + (\varepsilon - W_N - m)\,G = 0 \tag{7}$$

with

$$W_N = -\frac{N_c^{\,2}-1}{2N_c} \int (G^2+F^2)\,V_o(r, r')\,r'^2 dr' \tag{8}$$

and

$$V_o(r, r') = \tfrac{1}{2} \int_o^{\pi} V(r_{12})\,\text{Sin}\,\theta\,d\,\theta \tag{9}$$

We start with trial solutions for quarks in the soliton bag model made available to us by Prof. L. Wilets and his group and put them in equation 8. Equation 6 and 7 are then solved for G and F and the results put back in 8. After seven to ten cycles consistency is achieved.

To eliminate the centre of mass effect in the energy the Peierls and Yoccoz method is satisfactory. We use the version described in detail by C.W. Wong[15]. Again we omit the details, which will be described in ref. 14. Our results are summarized in Table 1, where in terms of the only parameter Λ of the Richardson potential,[16] we show the variation in the energy. It is remarkable that the energy comes out right for the Ω^-(1670 MeV) in the range Λ = 300 to 325 MeV.

Table 1

Hartree Fock Energy, E_{HF} for Three Strange Quarks of Mass m_s = 150 MeV. All entries are in MeV.

Λ	E_{HF}	Centre of Mass Correction	Corrected E_{HF}
400	2247	151	2096
350	1951	112	1839
325	1831	110	1721
300	1690	108	1582

We have also tried to see the sensitivity of our results on the strange quark mass, m_s. In Table II we find that as the quark mass decreases, the kinetic energy increases to compensate, so that the energy is not a very sensitive function of m_s.

Table II

Variation of E_{HF} with quark mass m_s. All quantities in MeV.

Scale parameter Λ	Quark mass m_s	E_{HF}
400	500	2605
"	300	2320
"	150	2247
300	500	2198
"	300	1833
"	150	1690

Finally, we checked our program using the Martin potential[17)] and m_s = 518 MeV when we essentially reproduce the non-relativistic results of Richards[18)]. This is shown in Table III.

Table III

Ω^- energy with Martin potential and m_s = 518 MeV. All quantities are in MeV.

E_{HF}	Centre of Mass Correction	Corrected E_{HF}	K Harmonics result[18)]
1821	237	1584	1620

The difference between our result and the K-harmonics method may be due to relativistic effects included in our calculation.

In conclusion I wish to thank my collaborators for providing me with some insight into the physics of the baryon. I would also like to thank Professor Larry Wilets, Dr. V.P. Nair who suggested and collaborated with us on the Relativistic Hartree Fock Problem, and the Lewes Center for Physics where we first thought of the latter problem.

4. REFERENCES

1) Bhaduri, R.K.,Dey, J. and Preston, M.A., Phys. Lett.B 136, 289 (1984).

2) Dey, J., Dey, M. and Le Tourneux, J. in Solitons in Nuclear and Elementary Particle Physics, World Scientific, ed. A. Chodas et al. (1984) and to be published.

3) Toki, H., Dey, J. and Dey, M., Phys. Lett. B 133, 20 (1983).

4) Bhaduri, R.K., Jennings, B.K. and Waddington, J.C., Phys. Rev. D29 2051 (1984).

5) Murthy, M.V.N., Dey, M., Dey, J. and Bhaduri, R.K., Phys. Rev. D30, 152 (1984).

6) Bhaduri, R.K., Invited talk, International Conference on Nuclear Physics, Dec. 27-31, 1984. Proceeding to be published by World Scientific.

7) 't Hooft, G., Nucl. Phys. B72, 461 (1974), B75, 461 (1974).

8) Witten, E., Nucl. Phys. B160, 57 (1979).

9) Coleman, S. and Witten, E., Phys. Rev. Lett. 45, 100 (1980).

10) Skyrme, T.H.R., Proc. Roy. Soc. A260, 127 (1961).

11) Mandelstam, S., Phys. Rev. D20, 3223 (1979)
Baker, M., Ball, J and Zachariasen, F., Nucl. Phys. B186,531, 560 (1980).

12) Nair, V.P. and Rosenzweig, C., Phys. Lett B135, 450 (1984).

13) Bailin, D., Cleymans, J. and Scadron, M.D., Phys. Rev. D31, 164 (1985).

14) Dey, M., Dey, J. and Le Tourneux, J., Montréal preprint (in preparation).

15) Wong, C.W., Phys. Rev. D24, 1416 (1981)

16) Richardson, J.L., Phys. Lett. B82, 272 (1979).

17) Martin, A., Phys. Lett. B100, 511 (1981).

18) Richards, J.M., Phys. Lett. B100, 515 (1981).

PION PRODUCTION, SIDE-SPLASH AND THE NUCLEAR EQUATION OF STATE

S. Das Gupta
Physics Department, McGill University,
Montreal, Canada

ABSTRACT

We investigate the effect of the nuclear equation of state on pion production in heavy ion collisions and the sideways flow seen in Nb on Nb collision.

Heavy ion collisions at Bevalac started about a decade ago. During this time much data have been accumulated[1)]. Most data are of inclusive type. Here one measures the cross-section for the production of a particular type of particle irrespective of whatever else is produced. Extensive data exist on inclusive proton, pion and deuteron cross-sections. These are the most abundant charged species seen in heavy ion collisions. More exclusive experiments have also been done. We will refer to one such experiment[2)].

The picture one associates with such collisions is the following: There is initially a compressional stage; nuclear matter may acquire two to three times the normal density. This is followed by a disassembly stage; the matter expands and when the density has fallen below a certain value (often called the freezeout density), the matter is non-interacting. The detectors record the situation pertaining to this stage.

An interesting question is: what experiments can be done to

extract as much information as possible of this history of the collision. Do any of the inclusive measurements tell us anything about the compressional stage?

It was already pointed out by Siemens and Kapusta[3)] that the ratio of deuterons to protons contain valuable information about the entropy generation in the compressional phase. More recently it has been suggested[4)] that the pion production cross-sections give us information about the energy stored in compression. This conjecture was made using the results of a theoretical model. Let us describe this model, called the cascade model[5)].

The cascade model is a model of hard collisions. In the preliminary version of the model, nucleons are assigned positions inside the nuclear volume by a Monte Carlo simulation of constant density up to radius R. No Fermi momenta are given. The two nuclei are made to approach each other in the cm. with momenta appropriate for the experimental beam energy. In each sequential small time interval the computer checks how many collisions will take place. Two nucleons will collide if in the given time interval they pass the point of closest approach and this distance of closest approach is below a certain value determined by the total N-N cross-section. If there is collision, the following channels are considered:

$N+N \to N+N$; $N+N \to N+\Delta$; $N+\Delta \to N+N$; $N+\Delta \to N+\Delta$ and $\Delta+\Delta \to \Delta+\Delta$.
Where available, the differential and total cross-sections are taken from experiment. The choice of channels and the angle of scattering is made through Monte-Carlo samplings. In the simpler version of the code the Δ life-time is considered larger than the collision time. In some versions the decay of the Δ is also included.

$$\Delta \to \pi+N \; ; \quad \pi + N \to \Delta$$

Since these are Monte Carlo calculations, cascade calculations for each impact parameter have to be repeated many times to generate enough statistics.

The cascade model calculations show clear evidence of compression and subsequent expansion. It also overpredicts pion production cross-sections by about a factor of two.

Stock et. al[4] have experimentally determined the number of pions for central collisions of Ar on KCℓ at various incident energies. They also compare their data with cascade model calculations. The situation is depicted in Fig. 1. The calculation clearly overproduces pions and Stock et. al. conjecture that the reason for this overproduction is the following. It requires energy to compress a nucleus. The cascade model, on the other hand, is a hard collision model, has no field effects and will fail to account for the energy stored in compression. Let us assume that the cascade model is correct in all other aspects; it gives the correct compression and has the correct pion production mechanism. Then the comparison of the experimental data and the cascade model results give us information about energy stored in compression for each value of ρ/ρ_0. Then one can draw an equation of state curve as a function of ρ/ρ_0 (Fig. 1).

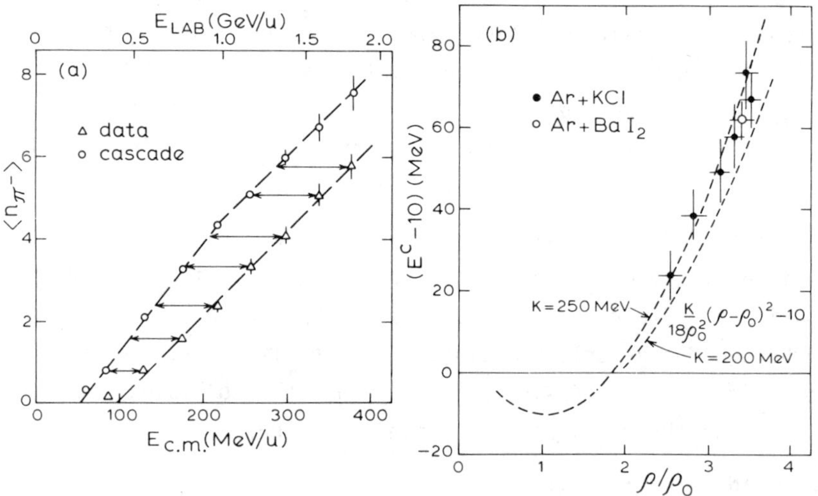

Fig. 1 (a) The mean π^- multiplicity as a function of bombarding energy for near central collisions of Ar + KCℓ. Horizontal arrows are the values of compressional energy per nucleon determined at each experimental point. (b) The values of compressional energy as a function of the calculated mean baryon density.

The argument is very appealing but some questions remain regarding the procedure. First of all one might expect to extract an equation of state at some high finite temperature. The rise of E/A against ρ/ρ_0 gets a major contribution from kinetic energy. In heavy ion collisions the two nuclei which collide are disjoint in momentum space and this effect would be absent.

Clearly what is needed is a cascade calculation which incorporates field effects. A calculational procedure which includes this effect was worked out in a recent paper[6)]. The method is based on transport equations and amounts to solving the Boltzmann equation using Monte Carlo methods. Let us summarise the main ingredients of this calculation.

(1) In a cascade model nucleons are assigned initial positions. This, in turn, implies a density. If we assume Skyrme type parametrisation then $u = u(\rho)$. In between collisions, test particles propagate according to $\dot{\vec{x}} = \vec{v}$; $\dot{\vec{p}} = - \vec{\nabla} u$.

(2) Nucleons are given initial Fermi momenta. In the case of an isolated nucleus the field term confines the nucleons to within the nuclear volume.

(3) Pauli blocking is checked for each collision. Let us assume that as a result of collision $(r_1 p_1)(r_2 p_2) \to (r_1 p_1')(r_2 p_2')$. The local environments of $(r_1 p_1')$ and $(r_2 p_2')$ are checked by drawing small six dimensional spheres about $(r_1 p_1')$ and (r_2, p_2'). The number of quantum states in this sphere is $N = \text{volume}/h^3$. If the actual number of nucleons is N' then based on the ratio of N'/N a Monte Carlo decision is made as to whether the scattering takes place or not. For example one finds that for Nb on Nb at 200 MeV/n incident energy >50% of collisions are Pauli blocked at the stage of highest compression. The number drops to 33% at 400 MeV/n.

(4) The calculations are done for 30 or 60 cascades running simultaneously. The average field that each test particle feels is the average field of these 30 ensembles.

In essence we have solved by Monte Carlo methods the transport equation

$$\frac{\partial f}{\partial t} + \vec{v}.\ \vec{\nabla}_r f - \vec{\nabla} u.\ \vec{\nabla}_p f = \left(\frac{\partial f}{\partial t}\right)_{coll.} \quad (1)$$

The left hand side, set equal to zero is the Vlasov equation. For $u(\rho)$ we use

$$\text{Stiff: } u(\rho) = (-124\rho/\rho_0 + 70.5\ (\rho/\rho_0)^2)\ \text{MeV} \quad (2)$$

$$\text{Soft : } u(\rho) = (-356\rho/\rho_0 + 303\ (\rho/\rho_0)^2)\ \text{MeV} \quad (3)$$

The first potential produces a compressibility coefficient of k = 375 MeV; the second potential has k = 200 MeV.

The results of our calculation can be stated very simply. We find that pion production is insensitive to the equation of state employed. Similar calculations but using different numerical techniques were also performed by Stöcker et. al.[7]. They also find that a hard or a soft equation of state gives about the same results. Similar conclusions were reached in a completely different model. Kapusta and Strottman using the hydrodynamical model also came to the conclusion that pion production is insensitive to the equation of state employed.

We now turn our attention to another recent experiment[2]. This is an exclusive experiment employing 4π detectors. Here the momentum and mass of each charged particle is measured. The number of charged particles depends upon the impact parameter but by restricting the observation to high multiplicity events one can select on nearly central collisions. For each event one measures the six elements of the symmetric matrix

$$F_{ij} = \sum_{\nu} p_i(\nu) p_j(\nu)/2m(\nu) \quad (4)$$

Here i,j are the three Cartesian coordinates, ν is the particle index and $m(\nu)$, the particle mass. Let θ be the angle (with respect to the original beam direction) of the eigenvector corresponding to the highest eigenvalue; θ is called the flow angle. For high multiplicity events in Nb on Nb collisions at 400 MeV/n lab energy the distribution of flow angles $dN/d(\cos\theta)$ as a function of θ shows a maximum at 30^0. (Fig. 2).

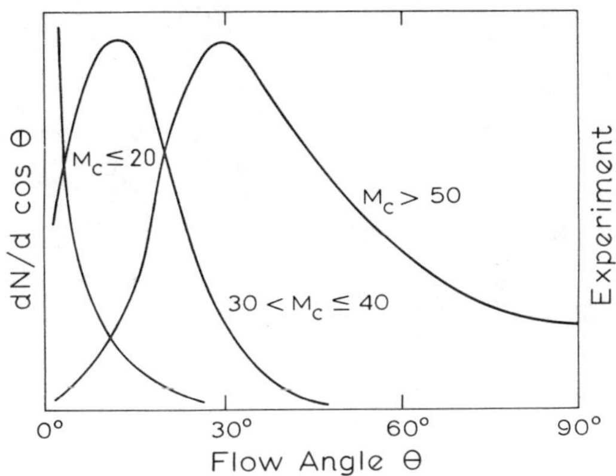

Fig. 2. Smooth curves through histograms of data points for $dN/d(\cos\theta)$ for different multiplicity tags. The data are for Nb on Nb at 400 MeV/n lab energy.

The need for performing an energy flow tensor analysis was pointed out by various authors in the past[9,10,11]. In one Fireball model the distribution of flow angles would be uniform between 0 and $\pi/2$; hydrodynamics predicts a side-splash for near central collisions. It is imperative to see what the Boltzman equation gives.

In the Boltzmann eqn. approach we can study the field effect alone by switching off the collision term; similarly we can study the effects of collision only by switching off the field term. This would be a standard cascade model prediction. These can be contrasted with a full calculation when both the effects are present (Fig. 3). It is interesting to note that neither the field effects nor the collisions themselves produce a sidewards flow; the combination however produces the desired effect. To compare with experimental data we have to integrate over impact parameter. Integration up to b = 3 fm. produces the results[12] shown in fig. 4. We find that the data are adequately reproduced by using a stiff equation of state.

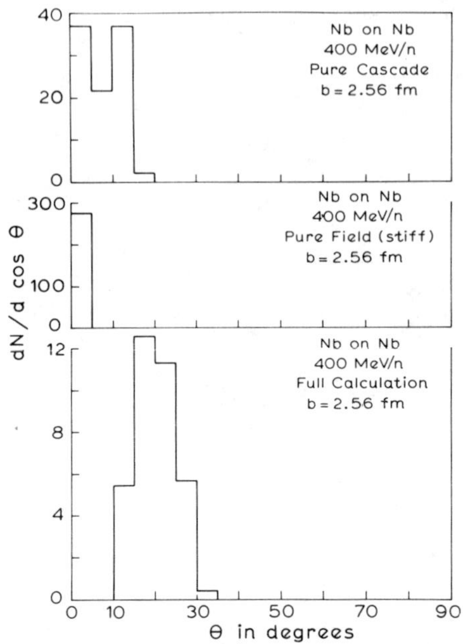

Fig. 3. For selected impact parameter, the distribution of flow angles for pure collisions, pure field and the full calculation.

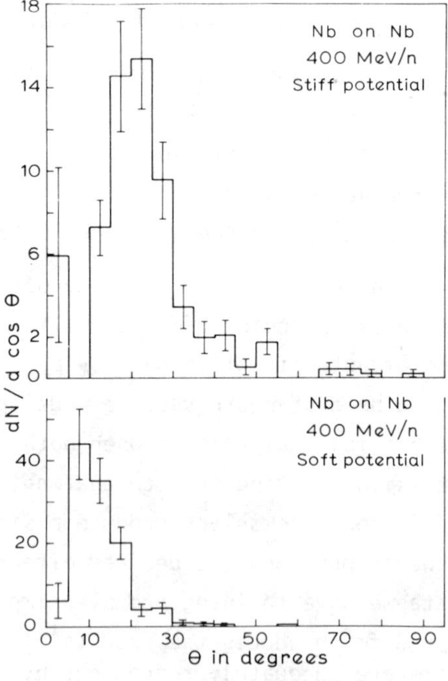

Fig. 4. The distribution of flow angles for Nb on Nb at 400 MeV/n using a stiff equation of state and a soft equation of state. The impact parameter b is integrated up to 3 fm.

It is interesting to check the beam energy dependence of this θ_{max}. We find that compared to Nb on Nb at 400 MeV/n, the value of θ_{max} drops at both higher and lower beam energy (Fig. 5). This prediction should be tested.

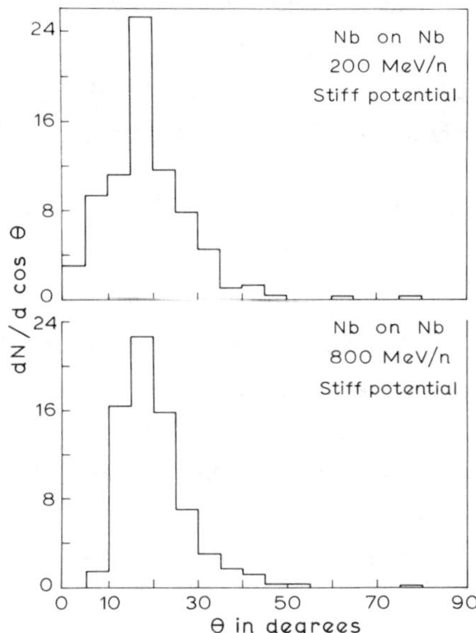

Fig. 5. For near central collisions (b integrated up to 3fm.) the distribution of flow angles for Nb on Nb collision at 200 MeV/n and 800 MeV/n.

In conclusion, the Boltzmann equation approach holds good promise for future calculations. At lower incident energy Pauli blocking prevents most of the collisions and the model resembles the TDHF model for heavy ion collision. At higher energies field effects become unimportant and hard collisions dominate the dynamics. We expect the Boltzmann equation to be a very useful tool in the beam energy regime 30 MeV/n to 1 GeV/n.

The work reported here was done in collaboration with G.F. Bertsch, C. Gale and H. Kruse.

REFERENCES

1) Nagamiya, S. and Gyulassy, M., Advances in Nuclear Physics, Vol. 13, 201 (1984).
2) Gustafsson, H.A., et. al., Phys. Rev. Lett. 18, 1590 (1984).
3) Siemens, P. J. and Kapusta, J.I., Phys. Rev. Lett. 43, 1486 (1979).
4) Stock, R., et. al., Phys. Rev. Lett. 49, 1236 (1982).
5) Cugnon, J., Mizubani, T. and Vandermeulen, J., Nucl. Phys. A352, 505 (1981).
6) Bertsch, G., Kruse, H. and Das Gupta, S., Phys. Rev. 29, 673 (1984).
7) Stocker, H., private communication.
8) Kapusta, J. I., and Strottman, D., to be published.
9) Bertsch, G. F. and Amsden, A. A., Phys. Rev. C18, 1293 (1978).
10) Kapusta, J. I. and Strottman, D., Phys. Lett. 106B, 33 (1981).
11) Gyulassy, M., Frankel, K. A. and Stocker, H., Phys. Lett. 110B, 185 (1982).
12) Gale, C., Das Gupta, S., and Bertsch, G. F., to be published.

A MANY BODY THEORY BASED NEW MASS RELATION FOR FINITE NUCLEI AND APPLICATION

L. Satpathy
Institute of Physics, Bhubaneswar - 751 005,
India

ABSTRACT

A new mass relation for finite nuclei have been obtained starting from the Hugenholtz-Van Hove theorem in many body theory. The mass relation combines nonperturbatively the two main features of nuclear dynamics namely the single -particle fẹature and the liquid drop feature. Its success in predicting masses far from stability is demonstrated. New saturation properties for nuclear matter is obtained through this study. This mass relation is applied to fission phenomena to describe fission barriers.

1. INTRODUCTION

The two main features of nuclear dynamics are the liquid drop feature and shell feature. However, the customary view of the nucleus has always been primarily a liquid drop with shell feature added only as a correction. The droplet model[1)] which is an improvement over the classic liquid drop model of Bethe and Weiszacker (BW) has such a picture as its basis. By now it is well known that the droplet model is not so successful in predicting masses far from the valley of stability. In fact these two features should be taken into account nonperturbatively. These two features have a common origin at the microscopic many-body level. Thus starting with a many-body theoretic approach, these two features can hopefully be combined non-perturbatively in describing the ground-state energies. In sec-2, we present the derivation of a new mass relation based on Hugenholtz-Van Hove [HVH] theorem[2] of many-body theory. In sec-3, we extract the new saturation properties of nuclear matter from such a mass relation. In sec-4, we apply this mass relation to strong deformation phenomena like fission.

2. NEW MASS RELATION

The material presented in this section was worked out in collaboration with Nayak[3]. The [HVH] theorem in general deals with the single-particle properties of a Fermi gas with interaction at the absolute zero of the temperature. In states that for a system with number of particles A and energy E

$$\frac{E}{A} + \rho \frac{\partial (E/A)}{\partial \rho}\Big/_{V} = \varepsilon \qquad .. (1)$$

where, ρ,V and ε are the number density, volume and Fermi energy respectively. Recently we have extended[3] this theorem to asymmetric nuclear matter and used it as a basis for new mass relation relevant feature of which are summarised here. The extended HVH theorem we had obtained assumes the following form for the ground state

$$\frac{E}{A} = \frac{1}{2}\left[(1+\beta)\varepsilon_n + (1-\beta)\varepsilon_p\right] \qquad .. (2)$$

where, $\beta = \frac{N - Z}{N + Z}$ is the asymmetry parameter and ε_n and ε_p are the neutron and proton Fermi energies. An asymmetric finite nucleus can be considered as perfect sphere composed of asymmetric nuclear matter cut out from infinite nuclear matter sea plus the effects due to deformation, shell, diffuseness, exchange Coulomb etc. Thus we can write the energy $E^F(A,Z)$ of a finite nucleus as

$$E^F(A,Z) = E(A,Z) + a_s A^{2/3} + a_c Z^2 A^{-1/3} - \delta(A,Z) + \eta(A,Z) \qquad .. (3)$$

The first 3 terms of the above equation represent the energy of sphere of asymmetric nuclear matter with radius $R(=r_o A^{1/3})$ related to the

Coulomb coefficient $a_c = 3e^2/5r_o$. a_s is the surface coefficient and the $\delta(A,Z)$ is the pairing term given by

$$\delta(A,Z) = \begin{cases} \Delta A^{-\alpha} & \text{for even-even} \\ 0 & \text{for odd A} \\ -\Delta A^{-\alpha} & \text{for odd-odd} \end{cases} \quad .. \ (4)$$

where Δ is a parameter and α the standard value 0.75or 0.5. $\eta(A,Z)$ represents the energy due to shell, deformation, exchange Coulomb etc. effects. With the help of Eq.(3), Eq.(2) reduces to

$$\frac{E^F(A,Z)}{A} - \frac{1}{2}\left[(1+\beta)\varepsilon_n^F + (1-\beta)\varepsilon_p^F\right] = S(A,Z) \quad .. \ (5)$$

where

$$S(A,Z) = (A^{-1} - 1)\, f(A,Z) + (N/A) f(A-1,Z) + (Z/A) f(A-1,Z-1) \ (6)$$

and

$$f(A,Z) = a_s A^{2/3} + a_c Z^2/A^{1/3} - \delta(A,Z) \quad .. \ (7)$$

The superscript 'F' denotes the quantities corresponding to finite nuclei. Here we have neglected terms due to $\eta(A,Z)$because of the following. The three terms that appear are $\{\eta(A,Z) - \eta(A-1,Z)\}$, $\{\eta(A,Z) - \eta(A-1,Z-1)\}$ and $\eta(A,Z)/A$. Since $\eta(A,Z)$ is smooth and much smaller than other smooth terms like surface and Coulomb term, the first two terms would be close to zero due to near cancellation. The last term can be safely neglected. In order to see how well Eq.(5) describes reality we took the experimental binding energies, the neutron and proton separation energies of 277nuclei ranging between Z = 10, A = 20 and Z = 101 and A = 253 in the valley of stability and per-performed a least square fit with Eq.(5) and obtained $a_s = 26.679$, $a_c = 0.827$ and $\Delta = 34.549$ MeV respectively. We had taken the standard value 0.75 for α. The r.m.s. derivation was found to be 390 KeV. Fig.1 shows the histogram of the deviation between the left hand side and the right hand side of Eq.(5) obtained in our fit. Eq.(5) can be used as a mass relation among three neighbouring nuclei with neutron and proton number (N,Z), (N-1,Z) and (N,Z-1) by expressing the ε_n^F and ε_p^F in terms of the binding energies. Then Eq.(5) would read

$$\frac{E^F(N,Z)}{A} - \frac{1}{2}\Big[(1+\beta)\ \{E^F(N,Z) - E^F(N-1,Z)\} + (1-\beta)\{E^F(N,Z) - E^F(N,Z-1)\}\Big] = S(A,Z) \qquad ..\ (8)$$

In this mass relation, the shell feature and the liquid drop feature have been combined nonperturbatively. The liquid drop feature is represented through the coefficients a_s and a_c and the single particle feature through the energy differences which are nothing but the neutron and proton Fermi energies. It must be recalled here, as has been shown by Hugenholtz and Van Hove[2)] the true single particle state in an interacting Fermi system is only the Fermi level which has infinite life time while other single particle states are metastable. Thus the present model contains the true single particle effect of the many-body system.

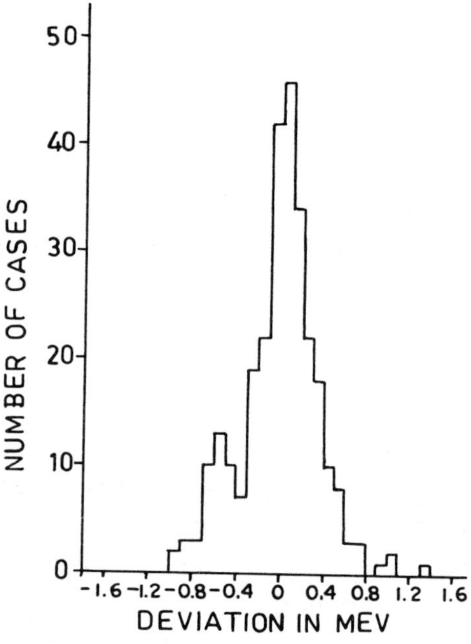

Fig.1. Histogram of the deviation between the left hand side and right hand side of Eq.(5).

As mentioned above we have used the data of 277 nuclei laying in the valley of stability in our least square fit and obtained the values of the 3 parameters a_s, a_c and Δ . To test the goodness of the model we use these values of the parameter to predict the binding energies of nuclei far away from the valley of stability whose data have not been used in the least square fit. The results of 21 such nuclei are presented in Table I as neutron-rich and ultra-neutron rich. The data for low mass nuclei have been taken from Jelley et al[4)] and for high mass nuclei from the mass table compiled by Viola et al[5)]. In this table of Viola et al the energies have been determined indirectly from the analysis of α-decay and β-decay and point energies. Since we are also using the same data for prediction, our agreement should be taken seriously. In Table-I, we have compared our predictions with those of Garvey Kelson[6)] and the droplet model calculated by Myers. We have also predictions (See Table I) for a third set of nuclei with N = Z and Z >N which were also not included in our fit. For these nuclei Garvey Kelson mass relations does not work. For all the three sets, our agreement with experiment is indeed better in most cases compared to those of Garvey-Kelson and droplet model. The main cause of its success with only 3 parameters is that the "energy differences" which appear in Eq.(8) cancels the other less important effects like shell corrections, deformation etc. to a great extent.

3. EMPIRICAL SATURATION PROPERTIES OF NUCLEAR MATTER

The two empirical properties of infinite nuclear matter are the binding energy per nucleon ω and the saturation density ρ_o which is related to the Fermi momentum k_f through $\rho_o = 2k_f^3/3\pi^2$. Traditionally ω is identified with the volume coefficient a_v of BW mass formula and its

Table-I: Prediction of binding energy in MeV
My ≡ droplet model[1)], GK ≡ Garvey et al[6)]

Category	Nucleus	Calc.-Expt.	GK-Expt.	My-Expt.	Expt.
Neutronrich	^{29}Mg	0.69	1.96	-0.45	235.28
	^{31}Al	-0.98	1.72	-0.19	256.06
	^{64}Co	-0.87	-0.15	0.07	555.57
	^{104}Mo	-0.13	0.16	1.52	886.78
	^{143}Cs	1.00	1.14	0.98	1178.43
	^{181}Yb	0.31	0.23	-0.38	1446.63
	^{204}Pt	-0.22	0.63	0.61	1603.17
	^{236}Ac	0.30	-0.44	2.18	1783.66
Ultra Neutronrich	^{143}Xe	0.84	2.39	2.16	1171.30
	^{180}Dy	-0.21	-1.65	4.37	1418.46
	^{223}Au	1.02	-8.66	6.83	1675.28
	^{228}Pb	-0.67	-7.47	2.94	1719.22
	^{238}At	1.22	-4.10	8.00	1769.21
	^{258}U	-0.54	-	14.56	1879.83
N = Z Odd-odd and Z > N	^{30}P	-1.14	-	0.11	250.60
	^{34}Cl	-0.97	-	-2.07	285.58
	^{46}V	0.50	-	0.43	390.35
	^{22}Mg	0.69	-	4.11	168.32
	^{28}P	0.15	-	3.53	221.92
	^{39}Ca	-0.19	-	-1.27	326.44
	^{42}Ti	-0.18	-	1.67	345.16

recent values obtained from droplet model is 15.97 MeV. The empirical value of ρ_o is obtained from electron scattering data. It is normally expected that the infinite nuclear matter density should be higher than the central density of a large nucleus, as in the latter there are surface and Coulomb effects. However, after Brandow[7] showed that these two effects oppose each other and nearly cancel, the infinite nuclear matter density is usually assumed to be the same as that of the central density of a finite large nucleus. The value of ρ_o thus obtained for nuclear matter is 0.170 fm^{-3} which corresponds to Fermi momentum k_f = 1.36 fm^{-1} and nuclear radius constant r_o = 1.12 fm.

Some disqueting features have continued to remain in this area the important one being the value of ρ_o which can be obtained from the Coulomb coefficient a_c of the mass formula through the relation $a_c = 3e^2/5r_o$. The liquid drop value for r_o is 1.22 fm and that of droplet model is 1.18 fm. Thus mass formula fit has always given a higher value of r_o. Further all nuclear matter calculations using the realistic interactions like PARIS, BONN etc. do not produce the saturation points near the empirical saturation point (see Fig.2). Recently Day[8] has summarised beautifully the present status of nuclear matter calculation with various forces[9-19]. Fig.2 is from his article in which the solid triangle is the result obtained by us which would be discussed later. The question arises, are the accepted empirical values ω = 15.96 and k_f = 1.36 fm^{-1} accurate enough? In view of the success of the present model, we would like to extract these quantities from the present study[20] as described below. Restricting to N = Z nuclei for which the energy E in Eq.(3) is $-a_1/A$ and using our above values of a_s, a_c

and Δ in the same equation, the only unknown can be uniquely determined if we neglect the term η. The average value of a_v so obtained by using the data of the known 22 symmetric nuclei above mass number 30 is 18.61 MeV. This differs from the accepted value in a major way. This encourages us to increase our data base by including N≠Z asymmetric nuclei. Now there will be another unknown namely the asymmetry coefficient a_a. We start with Eq.(2) in which the energy E corresponds now to the asymmetry nuclear matter energy and can be written as $-a_v A + a_a \beta^2 A$ which is the solution of Eq.(2). Then with help of Eq.(3), Eq.(2) reduces to

$$-a_v A + a_a \beta^2 A = \frac{1}{2}\left[(1+\beta)\varepsilon_n^F + (1-\beta)\varepsilon_p^F\right] \quad \cdot\cdot \quad (9)$$

$$- f(A,Z) + (N/A)\, f(A-1,Z) + (N/A)\, f(A-1,Z-1)$$

Using the above values of a_s, a_c and Δ determined in sec.1, we least square fit Ex.(9) with the data of same 277 nuclei and obtained a_v = 18.489 MeV and a_a = 37.55 MeV. The value of k_f obtained from our above value of a_c is 1.45 fm^{-1} surprisingly the saturation point corresponding to these values falls inside oval (see Fig.2) calculated by Day. This agreement should not be viewed as accidental because of following:

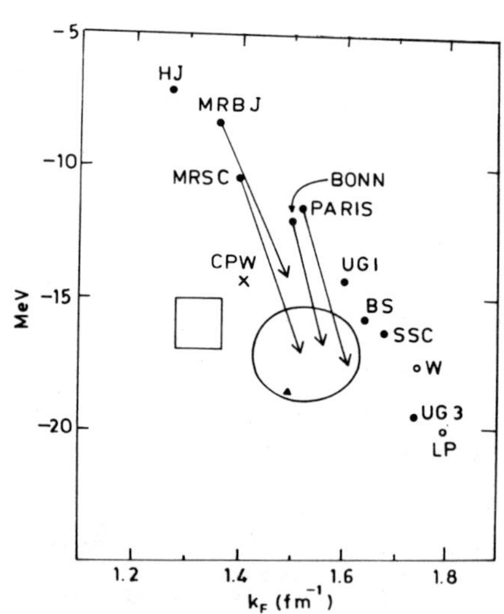

Fig.2. Nuclear matter saturation point for various potentials - see test.

1. Since the present model is based on many-body theory it is truely much better equiped than BW mass formula to determine the properties of nuclear matter. In fact it determines ω and k_f simultaneously.

2. The only physical property of nuclear matter contained in BW mass formula is the average energy per nucleus. The present model contains in addition to this the single particle property of an interacting Fermi system, as emphasised in sec-1.

3. Because of the "energy difference" in Eq.(8) the effects due to shell, deformation etc. nearly cancel. Thus our determination of a_s, a_c and Δ is not vitiated by these effects.

4. The smaller Coulomb terms due to exchange, difuse surface etc would nearly cancel due to the reason as above.

Thus the most important Coulomb term $a_c Z^2/A^{1/3}$ which contains information of the size of the sphere of nuclear matter gets accurately determined. To show this claim to be true we have compared in Table-II, the difference of the ground-state energies of eight pairs of mirror nuclei calculated in our model using our mass relation (8) with experiment. Hence our value of r_o and k_f obtained from a_c refer to nuclear matter.

In view of the above we study the stability of the numerical values of these parameters by increasing our data from 277 to 304 and 350 nuclei and also changing the pairing term through variation of α (Eq.4) from 0.75 to 0.5. As can be seen in Table-III, the values of the parameters have stabilised for both the pairing strength. The average saturation point corresponding to these six sets of values is represented as a solid triangle in Fig.2. To increase our confidence in the numerical values of those parameters we have calculated the uncertainties inherent in the least square fit procedure. As a representative we quote here result for the case one, i.e. corresponding to the first row of Table-III. These are δ_s=0.78 MeV, δ_c=0.017 MeV, $\delta\Delta$=0.99 MeV, δa_v=0.054 and δa_a=1.68 MeV. These uncertainties are very small (<3%) and show the reliability of the values obtained here. These six sets of result presented in Table-III could be combined by calculating the average and the deviation from the mean. Thus we obtained a_v=18.65±0.21 MeV, r_o=1.02±0.02 and k_f=1.49±0.03 fm^{-1}. Our value of k_f obtained here is of nuclear matter. This higher value k_f shows that the sufrace and Coulomb effect really do not cancel and the nuclear matter density is higher than the finite nuclear density which was believed to be the case by Brueckner[21]. Recent exhaustive analysis of electron scattering data by Friedrisch and Voegler[22] seems to support this view. Thus this study shows evidence in favour of a new empirical saturation point which is close to the theoretical prediction of recent improved many-body calculations[8] with several modern realistic two-body interactions. This will have relavance to the question[8] of 3-body force in nuclear matter and important bearing on our understanding of nuclear dynamics.

Table-II: Binding energy difference of mirror nuclei in MeV.

Nuclei			E_{cal}	$E_{expt.}$
^{22}Ne	-	^{22}Mg	9.20	9.20
^{32}P	-	^{32}Cl	12.56	12.54
^{41}Ca	-	^{41}Sc	7.48	7.28
^{47}V	-	^{47}Cr	8.29	8.23
^{50}Cr	-	^{50}Fe	17.46	17.35
^{53}Fe	-	^{53}Co	9.17	9.19
^{55}Co	-	^{55}Ni	9.45	9.47
^{58}Ni	-	^{58}Zn	19.44	19.58

4. APPLICATION TO FISSION

The material presented in this section was worked out in collaboration with S.K. Gupta[23]. In this section we show that the above mass relation can be applied to study the strong deformation phenomenon like fission process. We extend Eq.(8) to take into account deformation. For this purpose the appropriate relation similar to Eq.(3) would be

$$E^F(N,Z,\varepsilon) = E(N,Z) + a_s(\varepsilon)A^{2/3} + a_c(\varepsilon)Z^2A^{-1/3} - \delta(A,Z,\varepsilon) + \eta(A,Z) \quad ..(10)$$

where ε stands for the deformation, and the surface, Coulomb and the pairing coefficients are function of ε. Then using Eq.(10) in Eq.(2)

Table-III: Values of various coefficients of liquid drop - see text.

No. of Nuclei	α	a_s	a_c	Δ	χ	a_v	a_a	r_0 fm	k_f Fm^{-1}
277	0.75	26.679	0.827	34.549	0.398	18.489	37.955	1.04	1.46
304		26.964	0.849	34.679	0.485	18.636	37.358	1.02	1.50
350		26.527	0.842	34.939	0.481	18.476	36.256	1.03	1.48
277	0.50	27.413	0.846	11.959	0.388	18.747	38.666	1.02	1.49
304		27.647	0.865	11.915	0.477	18.867	38.048	1.00	1.52
350		27.235	0.858	11.817	0.475	18.713	37.015	1.01	1.51

we obtain

$$E^F(N,Z,\varepsilon) = \frac{N}{A-1} E^F(N-1,Z,\varepsilon) + \frac{Z}{A-1} E^F(N,Z-1,\varepsilon) - \frac{A}{A-1} S(A,Z,\varepsilon) \quad .. (11)$$

This equation relates the energies of three nuclei which are excited with deformation ε. If we assume the three nuclei with neutron and proton number (N,Z), (N-1,Z) and (N,Z-1) have same ground-state deformation ε_o and barrier deformation say ε then we obtain

$$F(N,Z,\varepsilon) = \frac{N}{A-1} F(N-1,Z,\varepsilon) + \frac{Z}{A-1} F(N,Z-1,\varepsilon) - \frac{A}{A-1} \quad S(A,Z,\varepsilon) - S(A,Z,\varepsilon_o) \quad .. (12)$$

where

$$F(N,Z,\varepsilon) = E^F(N,Z,\varepsilon) - E^F(N,Z,\varepsilon_o) \quad .. (12a)$$

is the fission barrier. The relevant deformation for the fission process is prolate. Thus one of the choices for the deformation can be taken as the ratio α of the semi-minor to semi-major axis. The appropriate values of α generally taken[24,26] for the ground-state, first barrier, second minimum and the second barrier are 0.8, 0.667, 0.5 and 0.4 for all the nuclei in this region. We have also adopted the same values in our calculation. We obtain the following fission barrier equations from the fission barrier relations (Eq.12)

$$F(N,Z,\varepsilon) = a_s \bar{B}_s(\varepsilon) A^{2/3} + a_c \bar{B}_c(\varepsilon) Z^2/A^{1/3} + \nu N + \pi Z + \mu \quad .. (13)$$

where, $\bar{B}_s(\varepsilon) = B_s(\varepsilon) - B_s(\varepsilon_o)$, $\bar{B}_c(\varepsilon) = B_c(\varepsilon) - B_c(\varepsilon_o)$

$\nu = \nu(\varepsilon) - \nu(\varepsilon_o)$, $\pi = \pi(\varepsilon) - \pi(\varepsilon_o)$, $\mu = \mu(\varepsilon) - \mu(\varepsilon_o)$

We have omitted the pairing term as it will cancel due to weak deformation dependence. Eq.(13) is a general solution of Eq.(12) with S function of Eq.(6) modified to S + μ/A. For the spheroidal deformation α defined by the ratio of semi-minor to semi-major axis, the functional form of $B_s(\alpha)$ and $B_c(\alpha)$ have been taken from Nix [25]. Thus in Eq.(13) the only three unknown are ν, π and μ for each extremum. We treat ν, π and μ as parameters and fit Eq.(13) with all the measured 47 data on the first barrier, 18 data on the second minimum and 35 data on the second barrier listed by Bjornholm and Lynn [26].

We got excellent fit with rms deviation of 0.32,0.32 and 0.29 MeV for the first barriers, the second minima and the second barriers respectively which are very much within the experimental uncertainties. We have performed calculation with 3 sets of parameters. The first set is the usual standard value [27] a_c=0.72 MeV and $a_s = a(1-k\beta^2)$ with a=19 MeV and k=2.84. The second set is with the value of a_s and a_c determined in this study in sec.-1. The third set is the parametrisation using the fission co-ordinate 'C' and the necking co-ordinate h. We used the values of {c,h} given by Bracket al [24]. We find equivalent description are obtained for all the three sets. This can be considered as a merit of Eq.(13) in the description of fission barrier. The results for set-I are presented in Figs.3&4 for all the nuclei ranging from Th to C_f. The continuous curves are the calculated ones which compare well with the data shown along with the errors. In all cases we have plotted our prediction for several nuclides for which data are not available. Though the thorium isotopes were excluded while fitting we explain the usually 'anomalous' thorium isotopes very well for both the inner and outer barriers while the previous calculations were discripant by about 1.5 MeV for the inner barrier height. For the second minima our values are around 2.5 MeV and as such are lower than the required values of about 4.5 MeV. In Eq.(13) the last three terms represent the shell energies while the first two terms are identified with those of liquid drop. In Table-IV we have presented for the set-I the shell energies with respect to the ground-state for the three representative nuclei ^{235}U, ^{238}Pu and ^{248}Cm. The sign and magnitude of these shell energies are in conformity with our expectation. Thus the present study show that the shell correction upto such large deformation like α =0.4 in the actinide region can be adequately represented by a sum of linear term in neutron number and a constant. This explains why our mass relation (8) works so well even though we had used the Eq.(3) without shell correction (neglecting $\eta(A,Z)$ as the linear terms exactly cancel. To our knowledge as yet no semi-phenomenological formula has been written down which could fit all the barrier data in the actinide region.

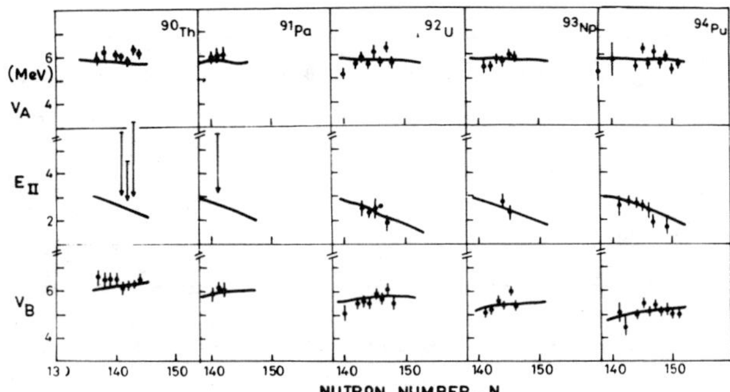

Fig.3. The inner barrier V_A, the second minimum E_{II} and outer barrier V_B as a function of neutron number for Z=90-94. Data are from Ref.26. Arrows indicate the cases where only an upper limit is listed in their tables. Data of thorium were excluded while fitting.

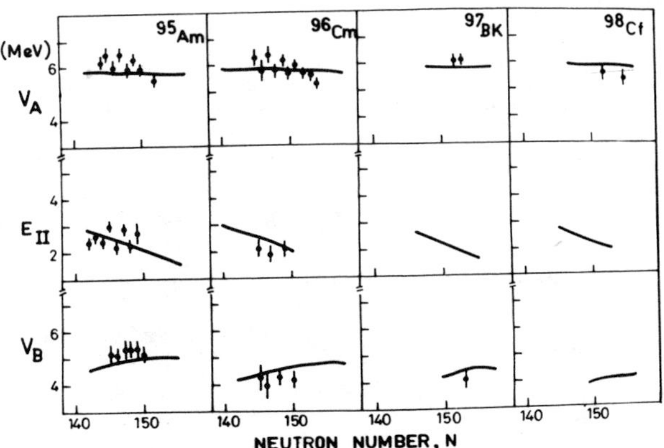

Fig. 4. Same as Fig.3 for Z=95-98.

Table-IV: Shell energies w.r.t. ground-state using set-I.

Nucleus	V_A	E_{II}	V_B
^{235}U	4.02	-3.06	-3.64
^{238}Pu	4.25	-2.32	-3.00
^{248}Cm	4.48	-2.03	-1.79

In conclusion we would like to say that the present mass relation being based on many-body approach and combining in its fold the two main features of nuclear dynamics seems to have potentiality for predicting masses and determining the saturation properties of nuclear matter in a novel way. It is also applicable to strong deformation phenomenon like fission. We feel for the first time a model based on many-body theory has been used to extract the properties of nuclear matter. This new saturation point deserves attention as it will have serious implication on our understanding of nuclear dynamics.

REFERENCES

1. Myers W.D. and Swiatecki W.J., Nucl. Phys. 81,1 (1968)
   Myers W.D., Droplet Model of Atomic Nuclei (Plenum, New York,1977).
2. Hugenholtz N.H. and Van Hove W., Physica (Urecht) 24, 363 (1958).
3. Satpathy L. and Nayak R., Phys. Rev. Lett. 51, 1243 (1983).
4. Jelley N.A. et.al., Phys. Rev. C11, 2049 (1975).
5. Voila V.E. et.al., At. Data Nucl. Data Tables 13, 35 (1974).
6. Garvey G.T. et. al., Rev. Mod. Phys. 41, S1 (1969)
7. Brandow B.H., Ph.D. thesis, Cornel Univ. 1964.
8. Dey B.D., Comments Nucl. Part. Phys. 11, 115 (1983).
9. Hamada T. and Johnston I.D., Nucl. Phys. 34, 382 (1962).
10. Dey B.D., Phys. Rev. C21, 1203 (1981).

11. Bethe H.A. and Johnson M.B., Nucl. Phys. A230, 1 (1978).

12. Carlson J. et.al., Nucl. Phys. A401, 591 (1983).

13. Holinde K. and Machleidt R., Nucl. Phys. A247, 497 (1975).

14. Lacombe M. et.al., Phys. Rev. C21, 861 (1980).

15. Ueda T. and Green A.E.S. Phy. Rev. 174, 1304 (1968).

16. Baryan R. and Scott B.L., Phys. Rev. 177, 1435 (1969).

17. Tourreil R. and Sprung D.W., Phys. Rev. A201, 193 (1973).

18. Lagaris I.E. and Pandharipande V.R. Nucl., Phys. A359, 349 (1981).

19. Wirringes R.B., Nucl. Phys. to be published.

20. Satpathy L., submitted for publication.

21. Brueckner K. et. al., Phys. Rev. 109, 1023 (1958).

22. Friedrich J., and Voegler N., Nucl. Phys. A373, 192 (1982).

23. Gupta S.K. and Satpathy L., submitted for publication.

24. Brack M. et. al., Rev. Mod. Phy. 52, 725 (1980).

25. Nix J.R., UCRL-11338, 179 (1964).

26. Bjornholm S., and Lynn J.E., Rev. Mod. Phys. 52, 725 (1980).

27. Pauli H.C. and Lederberger, Nucl. Phys. A175, 545 (1971).

FRAGMENTATION OF NUCLEI - SHATTERING OF GLASS OR CONDENSATION OF VAPOUR?

Jörg Hüfner

Institut für Theoretische Physik, Universität Heidelberg
Philosophenweg 19, D-6900 Heidelberg
GERMANY

ABSTRACT

We present some data observed for the fragmentation of nuclei in high-energy reactions and discuss two possible interpretations.

1. INTRODUCTION

A bullet is shot through a ball made of glass; the ball shatters into many pieces. A high-energy projectile is shot through a target nucleus; many little fragment nuclei are observed. Does a nucleus shatter like glass when hit by a high-energy projectile? Or are the fragments formed rather like droplets which condense from a vaporized nucleus? We shall try to summarize the present knowledge about this question. More complete information is contained in the review article[1)].

In a high-energy nucleus-nucleus collision the straight line motion of the projectile defines an overlap volume of projectile and target matter. Nucleons in the overlap zone are called participants, nucleons outside are called the spectators. Spectator nucleons belong either to the projectile if they are slow in the projectile rest frame or to the target if they are slow in the laboratory frame. This paper deals with medium mass fragments which are believed to originate when the spectator matter of the target breaks up.

2. EXPERIMENTAL SIGNATURES

We consider inclusive reactions of the form

$$p + A_T \rightarrow {}^{A}Z\ [\Omega, E] + X \tag{1}$$

where a high-energy projectile p (with energy above 500 MeV/u) strikes a target nucleus A_T and one fragment with mass A is observed with kinetic energy E and at an angle Ω. The basic experimental results can be summarized in the following way

(a) All particle stable nuclei are produced. The mass yield curve $d\sigma/dA$ (or $\sigma(A)$) decreases with increasing A like

$$\frac{d\sigma}{dA} = \sigma_0\ A^{-\tau} \tag{2}$$

for $A \lesssim 50$ (see Fig. 1). Then it increases again. The exponent τ lies between 2 and 3 and is nearly independent of the projectile, its energy and of the target.

(b) The angular distribution $d\sigma/d\Omega$ is essentially isotropic in a system which moves forward with a velocity which is of the order of 1% of the speed of light (Fig. 2).

(c) The energy distribution $d\sigma/dE$ in the moving system decays exponentially

$$\frac{d\sigma}{dE} = \sigma_0\ e^{-E/E_0} \tag{3}$$

for energies well above the Coulomb barrier. The slope parameter has a value $E_0 \approx 12$ MeV and is practically independent of the reaction and the observed fragment. For small energies, $d\sigma/dE$ shows the effect of Coulomb repulsion. However, $d\sigma/dE$ cannot be fitted by assuming a single Coulomb barrier. A distribution of barriers is required, whose average value is about one half of the one expected from two touching spheres.

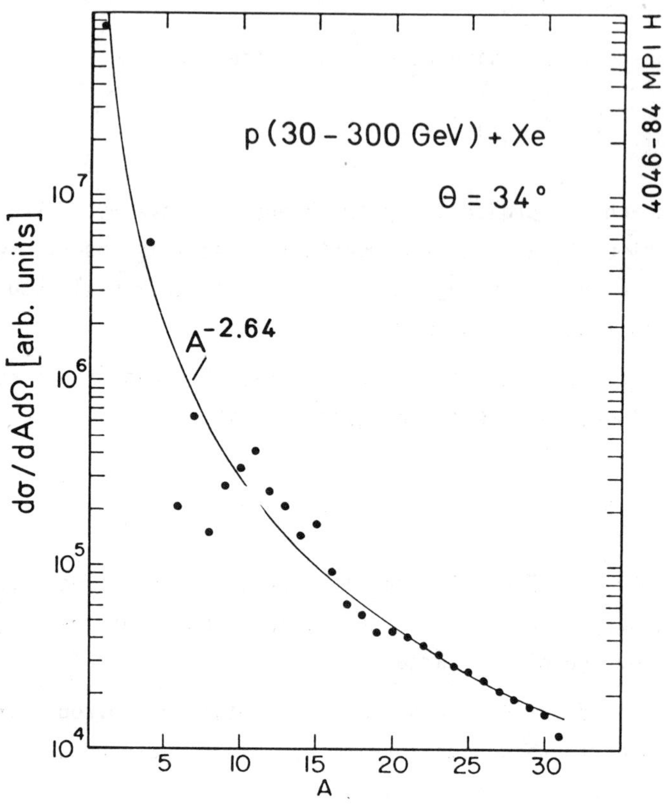

Fig. 1: Production of fragments in p + Xe collisions for energies between 30 and 300 GeV at an angle of 34^0. No energy dependence is observed, therefore the data represent the sum of the different runs. The solid line is the fit by a power law. The figure is taken from Hirsch et al.[7].

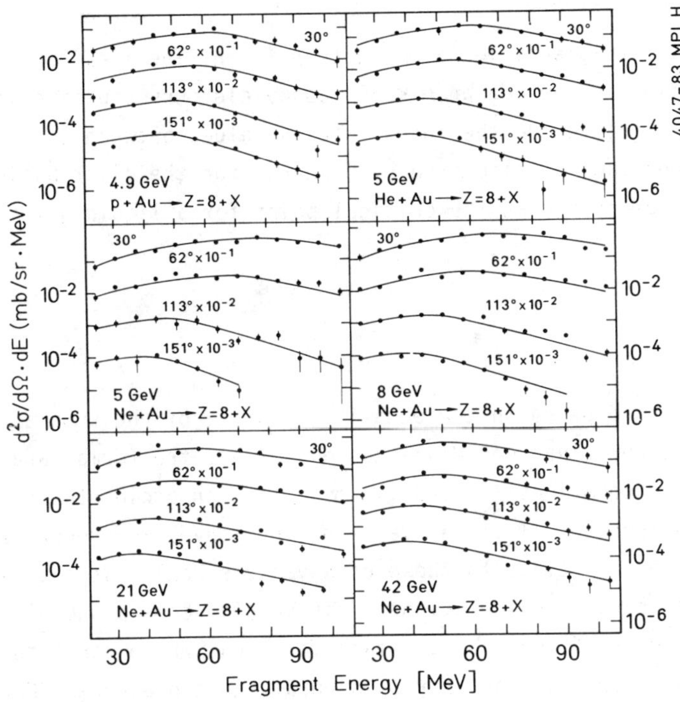

Fig. 2: Experimental double differential cross sections for various reactions leading to oxygen. The data are from Warwick et al.[8], the solid curves are fits by Aichelin et al.[5].

3. INTERPRETATION

Is multifragmentation a phase transition? The idea has the following origin. In the region $A \lesssim 50$ mass yield curves behave like power laws, eq. (2), where the index τ takes values between 2 and 3. A similar power law is calculated by Fisher[2)] for the distribution dP/dM of droplets of mass M at the critical point for a liquid-gas transition

$$\frac{dP}{dM} \propto M^{-\alpha}. \tag{4}$$

The exponent α depends on the equation of state of the substance which condensates. For a van der Waals gas $\alpha = 2\frac{1}{3}$. Is the formal analogy between the two equations (2) and (4) more than an accidental coincidence? Panagiotou et al.[3)] say yes. The thermodynamic properties of nuclear matter are close to those of a van der Waals gas, at least according to several calculations. This result seems plausible. Van der Waals derived "his" equation of state for a gas whose constituents feel a long-range attraction and a repulsion at short distances. Nuclear forces have exactly these properties, though on a very different scale of energies and densities. The numerical values for the critical quantities for nuclear matter are calculated to be

$$T_C = 18 \text{ MeV}, \ \rho_C = 0.4\ \rho_O, \ P_C = 0.4 \text{ MeV/fm}^3. \tag{5}$$

The critical temperature is close to the slope parameter E_0 in the distribution, eq. (3). Is the exponential to be interpreted as a Boltzmann factor and is E_0 a measure of the temperature?

It seems clear that nuclear matter behaves like a van der Waals gas. Therefore it must show condensation phenomena at the liquid-gas phase transition. At the critical point Fisher's equation for the distribution of droplet sizes applies. That is not where I see the problem. I rather doubt whether the experimental results on multifragmentation have anything to do with this phase transition. I have troubles to see why the experimental results eqs. (2) and (3) are so <u>insensitive</u>

to any variations in the initial conditions, e.g. changes in projectile, target and beam energy. In general, a phase transition is a quantity which is difficult to prepare.

A statistical but not thermodynamical explanation of multifragmentation has been proposed by Bondorf[4], by Aichelin et al.[5] and by Campi et al.[6]. These authors assume that the spectator matter remains essentially cold and breaks up like the shattering of glass under impact or like a drop of mercury which when hitting the ground splashes into several droplets. Neither glass nor mercury go through a phase transition before fragmentation. It is rather that internal stresses or hydrodynamical instabilities develop around the boundary between the participant and spectator zones in a high-energy collision and lead to the desintegration. Since little quantitative is known about the dynamics, one starts with a hypothesis: all possible partitions of the spectator matter are equally probable. This hypothesis leads to a mass yield in the multifragmentation regime of the form

$$\sigma(A) = \sigma_F \left(e^{0.9 \frac{A}{\sqrt{A_0}}} - 1\right)^{-1}. \qquad (6)$$

The shape is completely determined by the mass A_0 of the spectator fragment and no free parameter enters. Campi et al.[6] propose a related formula. Equation (6) is not of the form of a power law, but fits the data nearly equally well for $A \gtrsim 6$.

If multifragmentation is a chaotic break-up of cold spectator matter, the slope parameter E_0 is related to Fermi motion via Goldhaber's formula

$$E_0 = \frac{2}{3} \varepsilon_F \frac{A_s - A}{A_s - 1} \qquad (7)$$

where A_s is the mass of the spectator remnant. For a Fermi gas with a Fermi energy of ε_F = 40 MeV, which is typical for large nuclei, one finds

$$\frac{2}{5}\varepsilon_F = 16\ \text{MeV}, \tag{8}$$

which is in the range of observed slope parameters E_0.

I have limited this paper to a few essentials. The talk did not contain results which are not yet published and are easily accessible. In such a situation saving paper and therefore trees seemed important to me. The Conference at Bombay to which I was invited deeply impressed me through its lively exchange of ideas. I am very grateful to the organizers to have given me the chance of participating.

4. REFERENCES

1) Hüfner, J., "Heavy Fragments Produced in Proton-Nucleus and Nucleus-Nucleus Collisions at Relativistic Energies", to appear in Physics Reports.
2) Fisher, M.E., Physics 3, 255 (1967).
3) Panagiotou, A.D., Curtin, M.W., Toki, H., Scott, D.K., and Siemens, P.J., Phys. Rev. Lett. 52, 496 (1984).
4) Bondorf, J.P., Nucl. Phys. A387, 25c (1982).
5) Aichelin, J., Hüfner, J., and Ibarra, R., Phys. Rev. C30, 107 (1984).
6) Campi, X., Desbois, J., and Lipparini, E., Phys. Lett. 142B, 8 (1984).
7) Hirsch, A.S., Bujak, A., Finn, J.E., Gutay, L.J., Minich, R.W., Porile, N.T., Scharenberg, R.P., Stringfellow, B.C., and Turkot, F., Phys. Rev. C29, 508 (1984).
8) Warwick, A.I., Wieman, H.H., Gutbrod, H.H., Maier, M.R., Péter, J., Ritter, H.G., Stelzer, H., Weik, F., Freedman, M., Henderson, D.J., Kaufman, S.B., Steinberg, E.P., and Wilkins, B.D., Phys. Rev. C27, 1083 (1983).

HIGH ENERGY NUCLEUS-NUCLEUS INTERACTIONS

S.Lokanathan

Department of Physics
University of Rajasthan
Jaipur-302004(India)

1. INTRODUCTION

Although the title of the talk is imposing, I shall present some results mainly from experiments using nuclear emulsions. Even here, I shall only have time to discuss a few of the experiments.As in many other fields of high energy Physics, emulsions play the role of the front runner in this field i.e., the purpose of these experiments is largely to understand the gross features of nucleus-nucleus collissions which may help design other more exclusive experiments.

My talk will be mainly on the work of the collaboration of the Universities at Banaras, Chandigarh, Jaiipur, Jammu(India) and Lund(Sweden).(BCJJL). The work is based on interactions of Fe, Ar, and Kr in nuclear emulsions at beam energies of $\sim$1.8 GeV/nucleon. I shall present the results on (a) our search for anomalons in Ar interactions (b) relativistic α emission in Fe and Kr interactions.

2. ANOMALONS

2.1 Earlier Work

Let me remind you of the history of the anomalon problem. In the fifties, certain cosmic ray events in emulsions showed chains of interactions produced successively in projectile fragments[1,2,3]. Some examples of these are schematically shown in Fig. 1. The path lengths of these fragments prior to interacting were so much smaller than their normal mean free path (mean free paths in emulsion range from about 20 cm for He to about 8 cm for Fe) that there was a suspicion that these fragments were of anomalous character having abnormally large cross sections.

Systematic search for anomalons became possible with the advent of relativistic heavy ion beams at Berkeley at the Bevalac. An NRC-LBL collaboration using O and Fe beams studied 1760 interactions of secondaries produced by primary beam interactions in emulsions [4].

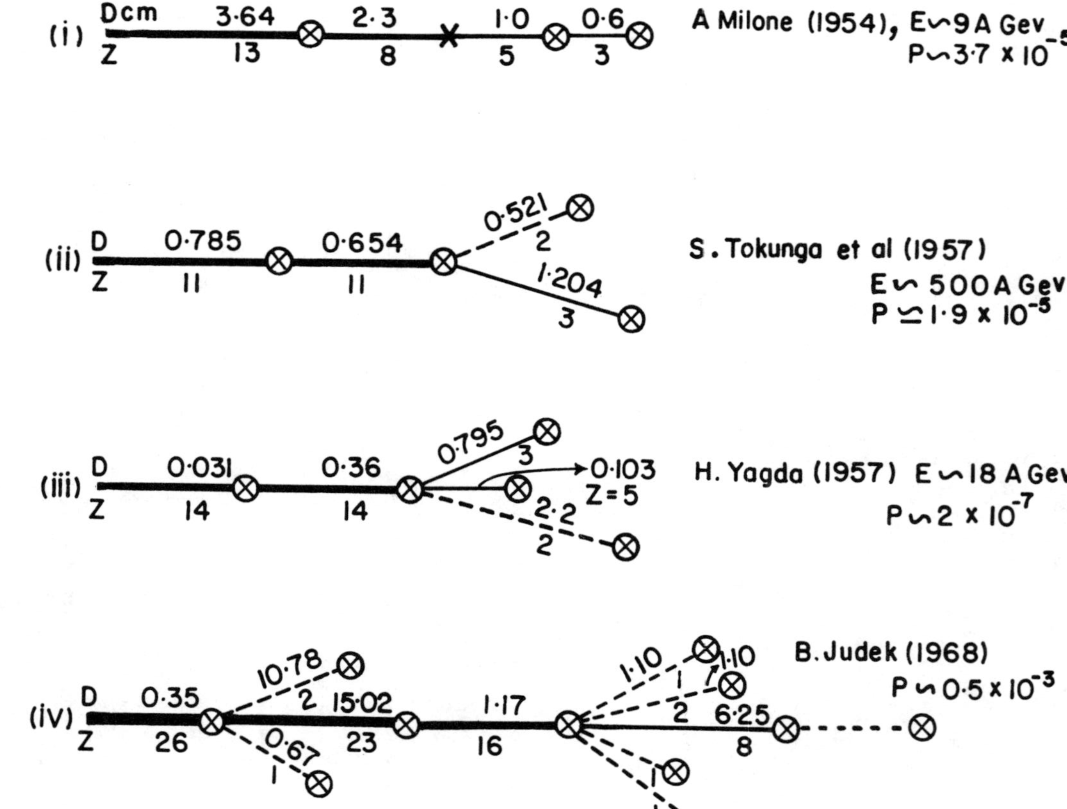

Fig 1. Multi chain events from Cosmic ray studies.

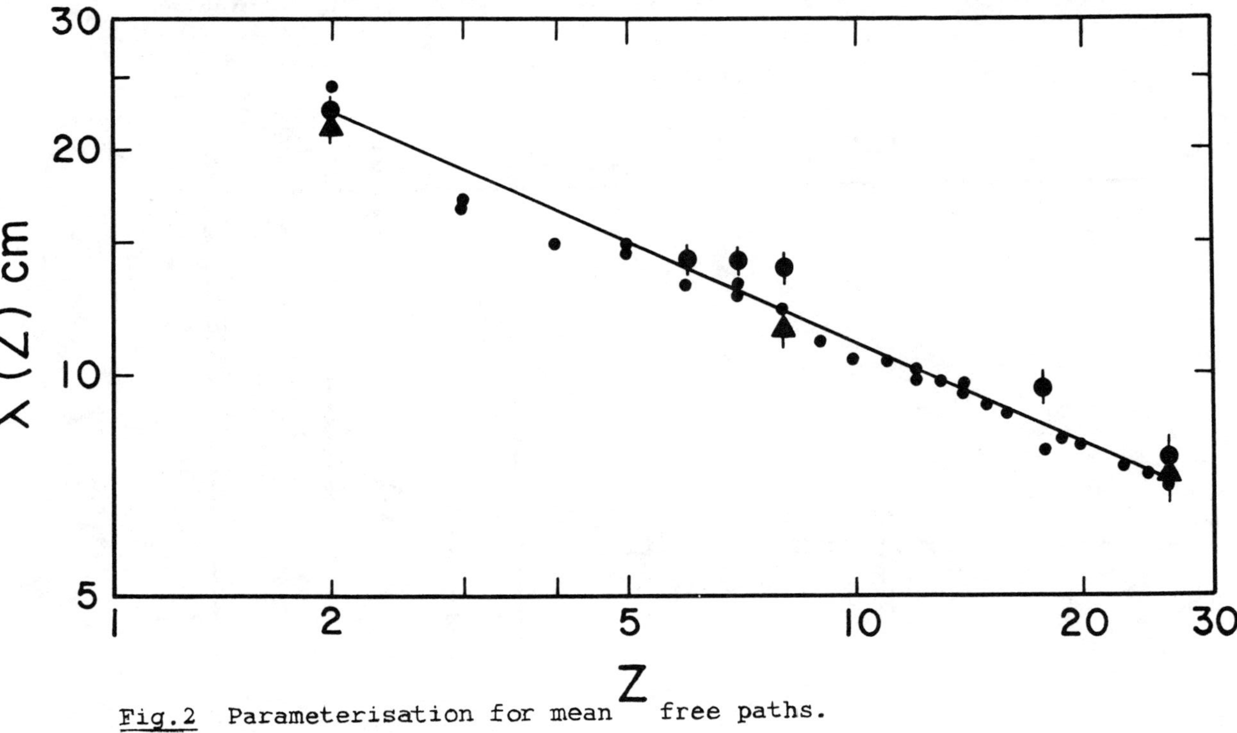

<u>Fig.2</u> Parameterisation for mean free paths.

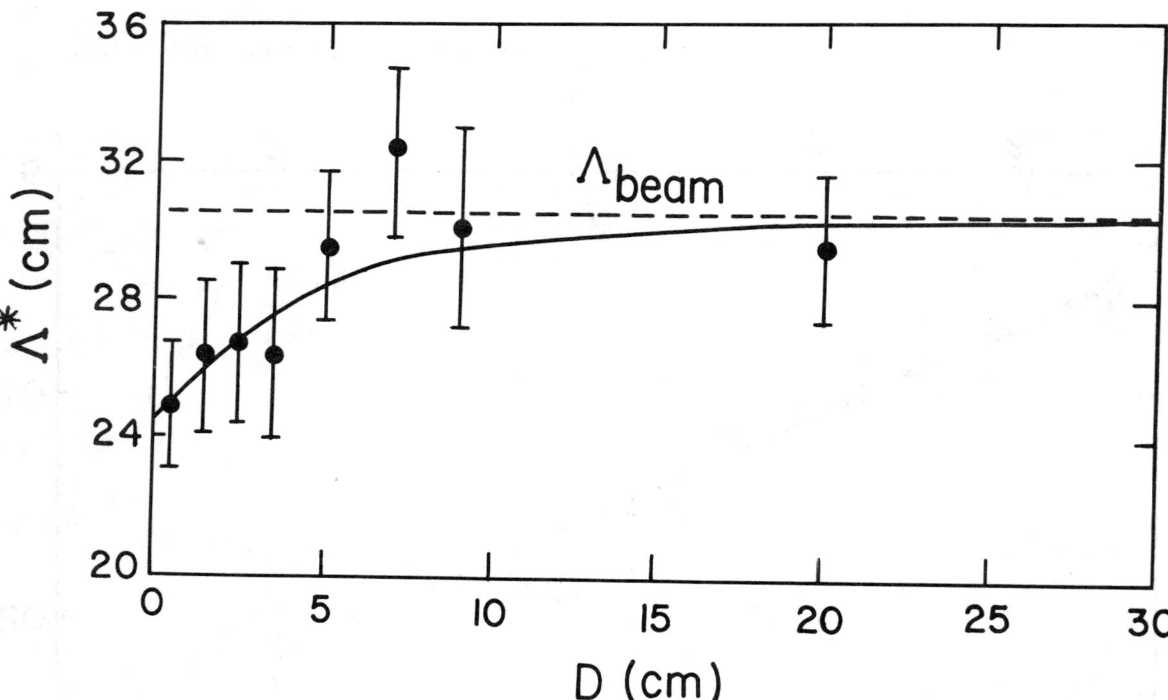

Fig.3 Variation of Λ^* with distance D from the origin of projectile fragments.

To search for anomalons in this sample, it is necessary first to look at normal mean free paths for fragments of various charges. (Fig. 2). This dependence is well represented by the parametrization

$$\lambda_z = \Lambda z^{-b}$$

where λ_z is the mean free path for a fragment of charge z and Λ and b are constants. The NRC-LBL work quoted best fits of

$$\Lambda = 30.4 \pm 1.6 \text{ and } b = 0.44 \pm 0.2$$

Isotope and other effects produce small deviations from this fit but hardly more than a few percent.

The importance of this parametrization(any other which fits as well would do) is that it is possible to look for anomalies in mfp of all the secondaries from the 1760 interactions lumped together. This was done by studying the variation of the parameter

$$\Lambda^* = \lambda_z \ z^{+b}$$

as a function of the distance that a fragment travels in emulsion before interacting. Note that this requires that the charges of the various secondaries must had to be determined in the experiment. Fig. 3 shows the results, the value of Λ^* being smaller at distances below 3 cm relaxing to normal values at larger distances.

Two other studies confirmed this effect[5,6], one of them being a compilation of earlier cosmic ray data and the other again a work based on emulsions exposed to Fe + Ar ions from Bevalac.

The NRC-LBL group, after a detailed statistical analysis of their data, concluded that about 6 percent of the O^{16} and Fe^{56} interactions produce an anomalous fragment (christened anomalons) of mfp $\simeq$ 2.5 cm, several times smaller than the normal mfp for even the heaviest ions in this sample.

2.2 Our experiment

The BCJJL experiment followed essentially the same method in its search. The beam, here, was Ar^{40} (1.8 GeV/ nucleon) and was based on 2192 interactions of secondaries (projectile fragments) from 6350 primary interactions found by line scanning. The mean free path for Ar (the beam) observed was 8.97 $\pm$ 0.16 cm. Charge determination was carried out by gap density and gap length measurements and by photometric track width measurements. Fig.4 shows the measured charge distribution of projectile fragments.

Since 902 of the secondary interactions are due to Z = 2 ions, it is possible to examine the data separately for Helium. Fig. 5 shows the variation of mfp with distance from production point. This result is particularly significant for two reasons. First, charge resolution of helium, based on gap density and gap length is relatively easy and reliable and secondly, the normal mfp is about 20 cm for helium against which any anomalous component of a few cm mfp would be easily discerned. The variation shown in Fig. 5 is consistent with no anomalons giving $\chi^2 = 15.2$ for 9 degrees of freedom.

For the remaining 1290 events representing the interactions of secondaries of charge $3 \leqslant Z \leqslant 18$, the sample is too small for a search for anomalons for each charge and we have to look at the data as a whole. This is done, as in the NRC-LBL experiement, by studying the distribution of $\Lambda^* = \Sigma\, S_z\, Z^b/(\Sigma\, n_z + 1)$ as a function of x. Our data, using the primaries and the fragments, give a best fit of $\Lambda = 24.1$ cm and $b = 0.34$. Fig. 6 shows the variation of Λ^* with x and the hypothesis of a single value of Λ (no anomalon) gives a $\chi^2 = 7.2$ for 9 degrees of freedom.

To get a feel for the level of significance of this experiment, we have plotted two curves in each of these figures for a 6 percent and 1 percent anomalon production with a 2.5 cm mean free path normalised to the actual number of interactions observed in this experiment.

As mentioned in my introductory remarks, one of the observations leading to the conjecture on anomalons, was the observation of chains of interactions in cosmic ray events. However, such a conjecture would be valid only if it can be shown that the frequency of such chains was beyond what one expects from fragments with normal mean free path.

We have tried to compare the observations of chains in our own sample with and without an anomalon component using a Monte Carlo procedure to simulate events within the geometrical and other constraints of this experiment.

Unfortunately, this Monte Carlo procedure requires first, an assumption of the model for anomalon production and behaviour and secondly, a set of assumptions about the normal charge distribution of fragments produced in

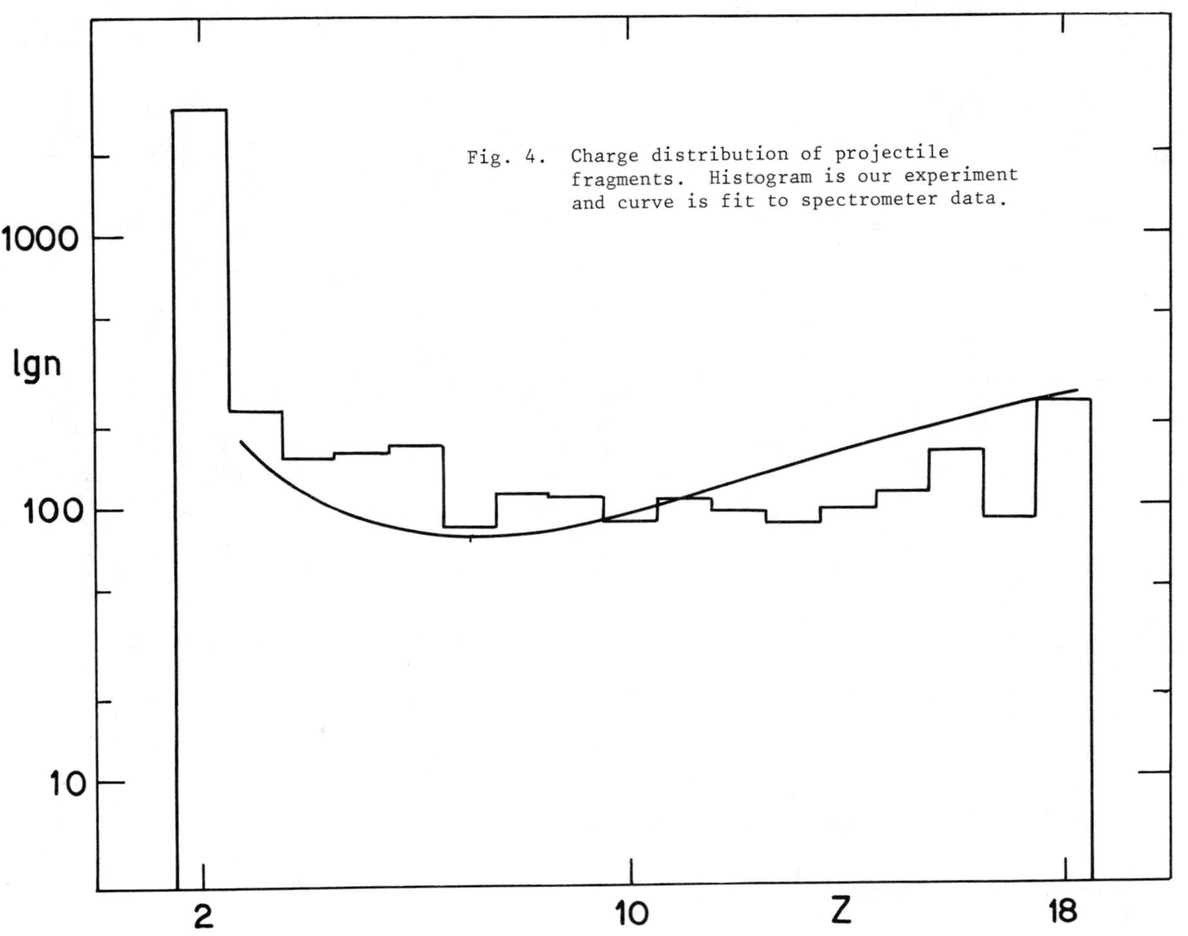

Fig. 4. Charge distribution of projectile fragments. Histogram is our experiment and curve is fit to spectrometer data.

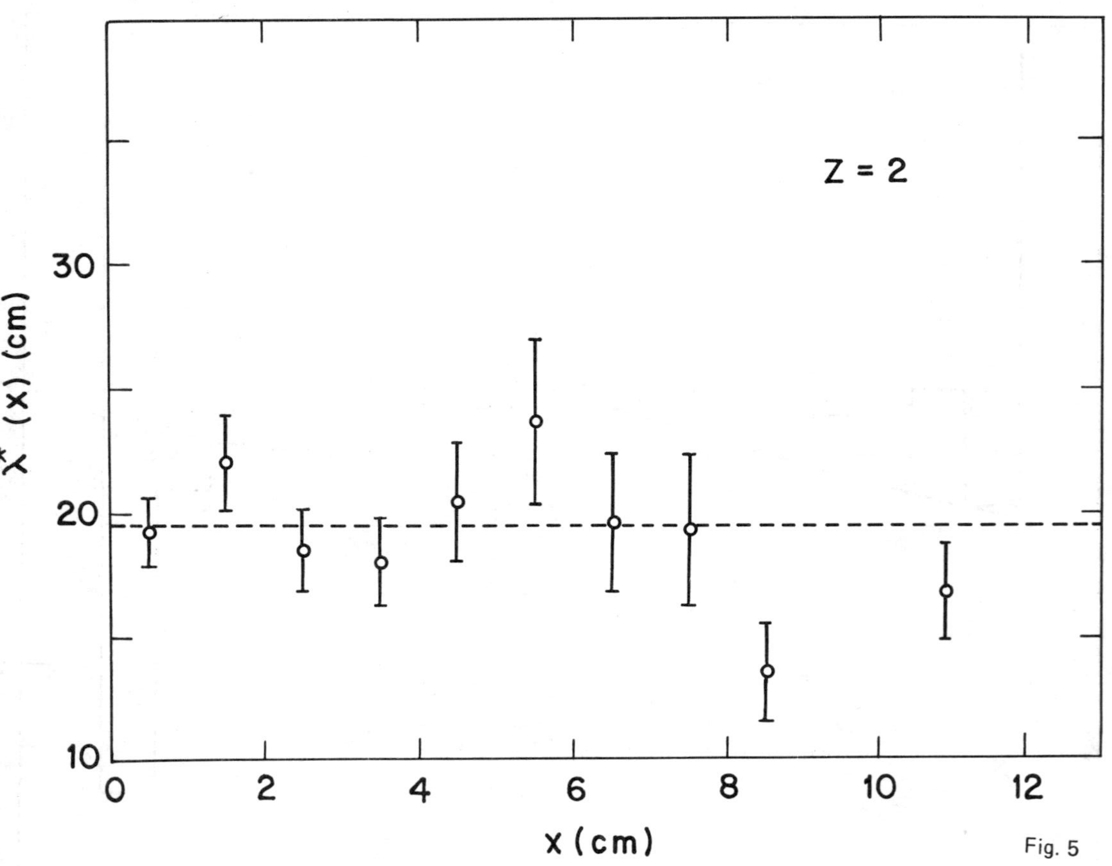

Fig. 5

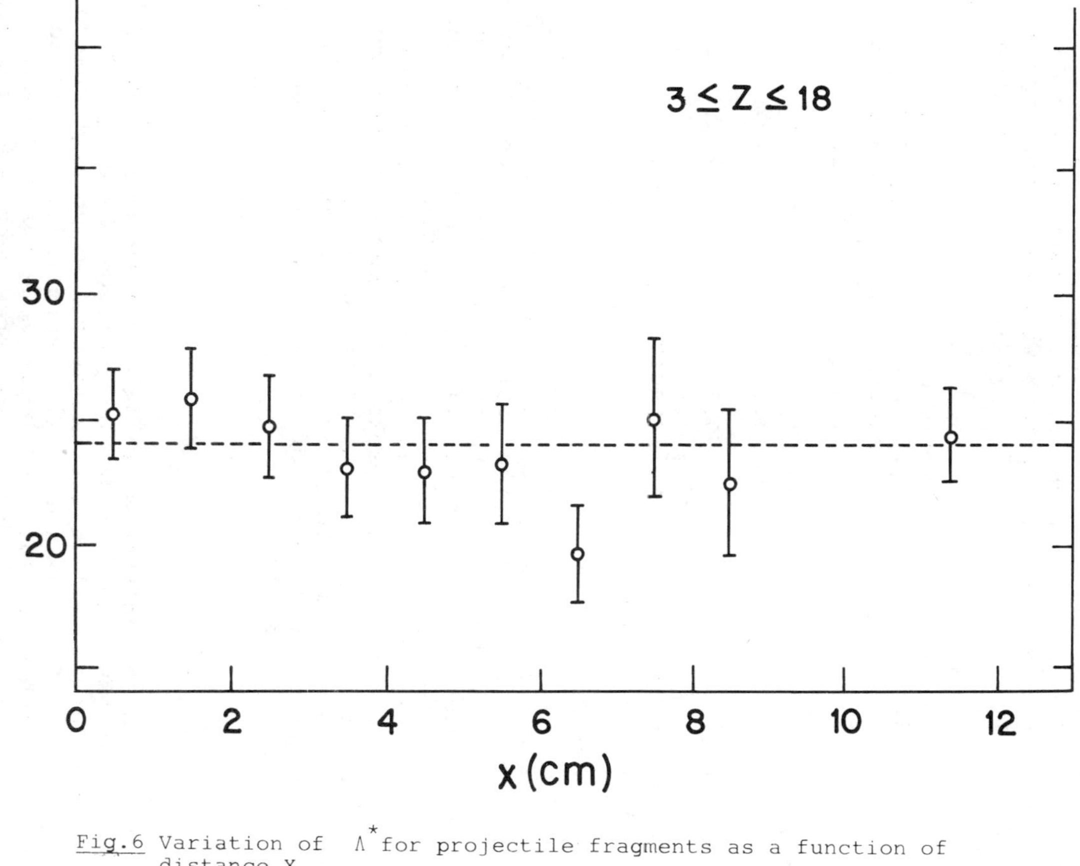

Fig.6 Variation of Λ^* for projectile fragments as a function of distance X.

interactions of nuclei of all charges (Z $\leqslant$18) as well as their angular distributions. Since our experimental data is too sparse certain assumptions were made about these distributions.The results of M.C.calculations are given below:-

Fraction of Anomalons	Ratio of $\frac{\text{III generation}}{\text{I generation}}$	Ratio of $\frac{\text{IV generation}}{\text{I generation}}$
0	7.3%	1.9%
5%	8.3%	2.4%
10%	9.1%	3.0%
Expt.	6.7%	2.0%

Comparison with experimental results again agrees more for no anomalon's picture.

2.3 Other Detectors:

I shall now review, briefly, a set of other experiments specifically directed to the search for anomalons carried out recently. These fall into two types. The first uses a plastic detector, CR39 ($C_6H_{12}O_7$)[7]. The basic technique was to expose a set of foils (230) each 0.6 mm thick, normally to a beam of Fe^{56} (1.7A GeV). When the plates are etched, the particle tracks are measured on the front and rear side of each foil using an automatic scanning system. The cross sectional area of the tracks is then used to measure the charge of the tracks. The charge resolution ranges from

$\Delta Z = 0.45$ for $Z = 20$ to

$\Delta Z = 0.56$ for $Z = 26$.

A change in charge then signifies an interaction. The experiment thus looks at (a) only charge changing interactions and (b) interactions of fragments of charge $20 \leqslant Z \leqslant 25$. For this range, the overall data is lumped together with no distinction being made for different charge fragments. A study of mean free path, as a function of x distance from production point, shows that the data is consistent with no anomalous behaviour at low x, the $\chi^2 = 2.05$ for 5 degrees of freedom for a single mean free path.

Two experiments using segmented Cherenkov detectors[8,9] also look at charge changing interactions and find no evidence for anomalons. Since the details of these will be presented by Symons at this Conference, I shall not go into it any further.

3. EMISSION OF RELATIVISTIC α PARTICLES :

In an earlier experiment, we (Jaipur-Jammu-Lund collaboration[10]) had studied relativistic α particles produced in Fe-emulsion interactions at 1.7A GeV. The main observations in this experiment were the following:

(a) Relativistic particles in Fe-emulsions are observed at angles larger than in ^{16}O-emulsion reactions at comparable energies. These angles are larger than can be understood for a pure fragmentation of projectile and projectile spectators.

(b) The $N_h = 0$ reactions of Fe-emulsion are quite similar to the general sample of ^{16}O-emulsion reactions. The general sample (all N_h events) of Fe reactions differs considerably from the general sample of ^{16}O reactions. In an attempt to understand the mechanism of production of large angle alphas, the observed angular distributions were compared with the predictions of Moving Boltzman Distributions[11]) with velocity, β , and Temperature T, of the thermal source as free parameters. From the comparison it was concluded that ^{16}O-emulsion reactions and Fe-emulsion reactions ($N_h = 0$ events) can be fitted to one M.B.D. with T ≃ 7 MeV, which is expected on the basis of pure fragmentation mechanism. On the other hand, for the general sample of Fe-emulsion reactions we need two M.B.D's one for the peak part of angular distribution, T ≃ 12 MeV, and another for the tail part, T≃ 50 MeV. This second sample of α's (T ≃ 50 MeV) is difficult to understand on the basis of a pure fragmentation mechanism.

These observations were confirmed by Baumgardt et al[12]) in an independent experiment following a different approach of analysis. They observed the existence of two type of Fe-emulsion reactions i.e. cold, at T ≃ 10 MeV and hot, at T ≃ 40 MeV.

Recently, we have recorded about 800 Kr reactions in an emulsion stack exposed to a Kr beam (≃ 1.5A GeV) from Bevalac. The results of preliminary analysis, based upon about 270 events, are as follows :

(a) Fig. 7 shows the observed N_α distribution for Kr reactions. In one event, we have 8 relativistic particles and 2 relativistic fragments (Z ≥ 3).

α - multiplicity distribution
282 events
522 α-particles
⟨Nα⟩ = 1·85 ± 0·11
dN/dNα
Nα
100
80
60
40
20
0
2
4
6
8
10

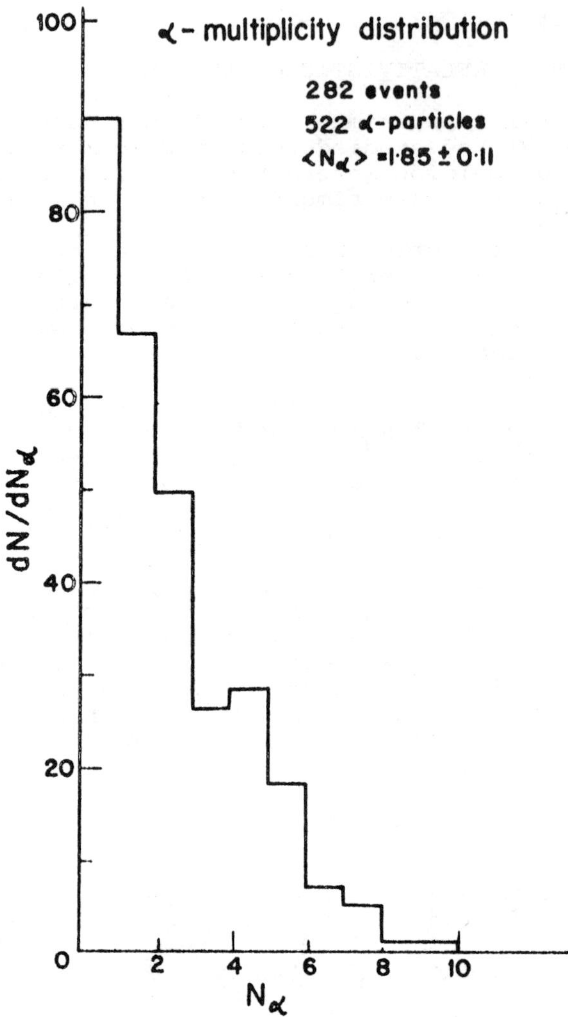

Fig. 7

Angular Distribution -(α-particles)
All events
16O
56Fe
84Kr
dNα/dθα
θα
0
20
40
60
80
100
2
4
6
8
10

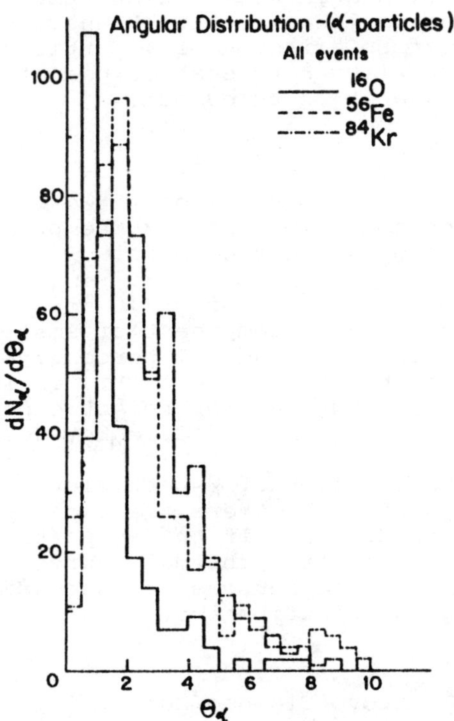

Fig. 8

(b) Fig. 8 shows the comparison of O, Fe and Kr reactions. It is observed that in Kr-emulsion reactions too α particles are observed at large angles as in Fe-reactions.

(c) It is observed that the fraction of larger angle α's increases with the degree of target and projectile fragmentation.

(d) A Baumgardt type of analysis shows the existence of cold and hot events in Kr-emulsion reactions also. One possible explanation is collective transverse flow of nuclear matter before emission of α particles, although other explanations like hard scatterings of α's from target nucleons cannot be ruled out.

4. CONCLUSIONS:

I have had time to discuss only two experiments, namely the search for anomalons and the production of alpha particles among projectile fragments.

The emulsion experiment of BCJJL does not show evidence of any anomalous component of small mean free path ($\lesssim$ 3 cm) at least beyond $\simeq$ 1 percent; the limit of the sensitivity of the experiment. This is in agreement with the findings of two sets of recent experiments using plastic detectors and segmented Cerenkov detectors.

Study of the emission of α-particles from Fe and Kr interactions suggests that there are two types of reaction mechanism for α-production. If the α-particles are to be thought of as emerging from thermal source, then at least two distinct temperature sources are necessary to fit the experimental angular distribution.

5. REFERENCES:

1. Milone,A. Nuovo Cimento Suppl. 12, 353 (1954).
2. Yagoda,H.Bull. Am. Phys. Soc. Series II, 64(1956); Nuovo Cimento 6, 559 (1957).
3. Tokunaga,S. et al., Nuovo Cimento 5, 517 (1957).
4. Friedlander,E.M. et al., Phys. Rev. Lett. 45 1084 (1980).Phys. Rev. C 27,1489 (1983).
5. Barbar,H.B. et al., Phys. Rev.Lett. 48,856 (1982).
6. Jain,P.L.,Das,G. Phys. Rev.Lett. 48,305 (1982).
7. Heinrich,W. et al., Nuclear Physics A 315C ,400(1983).

8. Symons,T.J.M. et al.,Phys.Rev. Lett. 52,982(1984).

9. Stevenson,J.D. et al., Phys.Rev.Lett. 52,515(1984).

10. Bhalla,K.B. et al., Nucl. Phys. A 367, 446 (1981).

11. Garpman,S.J.A. et al., Phys. Lett. 106B ,367 (1981)

12. Baumgardt,H.G. et al., J.Phys. G. Nucl. Phys. 7, L175 (1981).

REALISTIC SIGNATURES OF QUARK-GLUON PLASMA

Bikash Sinha

Variable Energy Cyclotron Centre
Bhabha Atomic Research Centre
1/AF, Bidhan Nagar
Calcutta 700 064
INDIA

ABSTRACT

It is shown that the ratio of the production rate of hard photons to lepton pairs saturates to a constant value beyond a certain critical temperature and turns independent of the baryonic chemical potential, signalling the phase transition from hadron matter to quark-gluon plasma; the most desirable kinematic window being the large transverse momentum.

1. INTRODUCTION

It is conventional wisdom now that with the deposition of sufficient energy density, normal hadronic matter of typical baryon density $\rho = 0 \cdot 145\ fm^{-3}$ will undergo a phase transition to the supposedly fundamental constituents of hadrons, quarks and gluons - the hot and/or dense hadronic matter becomes a plasma of quarks and gluons. Such a transition to the new state of matter is possible for

(Invited talk delivered at the International Conference on Nuclear Physics, Bombay December 27 - 31st, 1984)

temperatures beyond a certain critical temperature, Tc and/or hadronic densities beyond a certain critical value [1,2].

However, the exact nature of the phase transition seems to be shrouded by controversies - for a pure gluon field it is understood that one can have a phase transition [3,4] but with a field constituting of quarks and gluons the order or indeed the nature of the phase transition is not understood clearly.

Carruther [5], on the other hand has cautioned that Steefan-Bolt-Zmann ideal behavior of the quark-gluon plasma may not be definitive signature of the plasma but can be due to phonons and plasmons.

In this lecture however, I shall concentrate not so much on the theoretical aspects of the phase transition, but on the possible realistic signals of the quark-gluon plasma (QGP), if that be formed.

The problem of finding appropriate diagnostic signatures of QGP is amplified not only by the uncertain nature of the plasma and its transition back to hadronic matter but also by the complex dynamical enviornment in which the plasma is produced.

Indeed it is recognised by now [6,7] that there are two dynamical regimes of the scenarios, primarily defined by the incident energy. Concentrating at present for collisions of two nuclei moving at ultraelativistic energies the regimes, mentioned just now, are as follows: In the scaling regime, typically $E_{lab} \gtrsim 1$ TeV/A, the fragmentation region, containing the baryons, seperates leaving a rapidity space occupied only by mesons. This region is characterized by an approximate constancy of dN/dy, implying [8] that the evolution of the plasma is invariant under Lorentz boosts along the beam axis; it is also established [8,9] that the scaling regime axperiences a longitudunal expansion, independent of the nuclear size. Secondly, there is the stopping regime typically $E_{lab} \approx 10$ GeV/A dominated by high baryon density where the two nuclei can stop each other in the centre of mass system, the stopped matter can reach energy densities upto few GeV/fm^3, just like in the scaling regime, but the baryon densities can be as high as (8-10) times normal nuclear density. In this regime the signals from the baryonic enviornment get mixed up with the signals from the quark-gluon plasma.

In Fig. 1, it is shown that for $E_{lab} > 1$ TeV/A $> 2R/\tau_0$ here τ_0 is the initial time, the scenaros is characterised by the proper

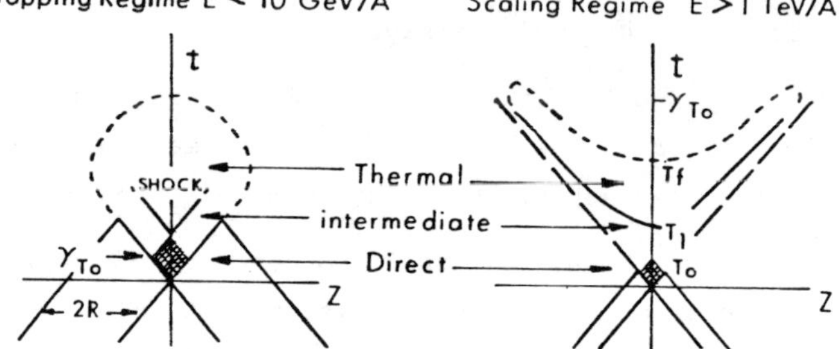

Fig. 1. The space-time evolution of the plasma in the stopped regime and the Scaling regime.

time τ ; the typical macroscopic variables like energy density, pressure and entropy are entirely determined by the proper time. Whereas for the stopped regime, in the c.m., the direct reactions occur in the relatively small space-time volume, Fig.1, unlike in the scaling regime where in the first fm/c all the partons of both nuclei pass through one another and have a chance at interacting directly.

2. SIGNALS OF THE QUARK-GLUON PLASMA

Possible signals of the QGP can be broadly classified as follows:

Thermometers: Typically dileptons, interacting weakly with the hadrons, hard photons product from the process $q\bar{q} \rightarrow \gamma g$ or $gq \rightarrow \gamma q$ and charm quarks tend to retain the information about the early epoch of space-time evolution of the plasma, the temperature in particular.

Barometers : Multiplicity as a function of average transverse momentum $\langle p_\perp \rangle$, high multiplicity at large tend to measure the pressure build up in the plasma [10,11].

Seismometer : Van-Hove [12] in particular has emphasized the importance of any violent fluctuation in the expansion phase of the plasma, as a possible signal of QGP.

Chronometer : By observing the ratio of various spacies of hadrons, strange mesons to pions or strange mesons to dilepton pairs for example one can quantitatively scan the space time evolution of the plasma.

Apart from the various meters just mentioned, arguably the most reliable signals of a phase transition of this kind can be traced back to couple of microseconds after the creation of the Universe (assuming the standard Big Bang model of the Universe). The phase transition from QGP to the hadronic world with its associated fluctuation could have left remnants still stored in the Cosmos - a subject of much interest.

The other cosmological object which can be in a state of QGP is the core of neutron star - cold but very ($10\,\rho_0$) dense in banryonic density, it is more than likely that the core of a neutron star is in QGP, luminosity measurements may indicate clearly such a conjecture [13,14]. In Fig. 2 the approximate nature of the phase transition is shown.

However, the central motivation of the present work is to device signals of QGP which are not only independent of the baryonic chemical potential beyond the critical temperature but also independent of the space-time evolution of the fireball. Independence of the signals with respect to the chemical potential implies that one can afford to ignore the immediate baryonic environment, for example, in the case of the stopped regime; independence of the space-time evolution also implies decoupling of the signals to the equation of state which determines the space-time evolution of either the fireball of pure quark-gluon plasma (scaling regime) or a mixed environment of QGP and baryons (stopped regime). The signal is a snap shot of the phase transition

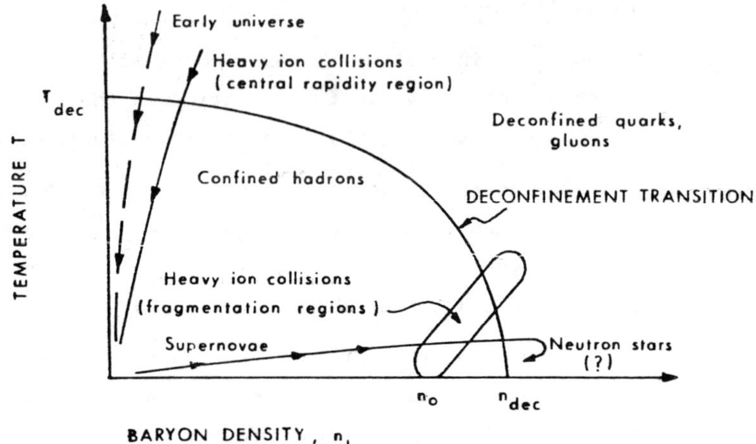

Fig. 2. Phase diagramme for transition from hadronic world to Quark-Gluon Plasma

independent of the baryons, independent of the space-time evolution of the fireball. What are those signals?

3. FORMALISM AND RESULTS

Leptons which interact weakly with hadrons retain the memory of a phase transition more efficiently, similarly hard photons interacting electromagnetically with the surrounding medium will also retain signatures of the plasma. Considering the most dominant channel of the lepton pair production from the QGP $q\bar{q} \rightarrow l\bar{l} + X$, clearly, in the naive non-relativistic limit of the Boltzman distribution, the production rate of lepton pairs which depends on the product of the Fermi distribution of the [15] $q\bar{q}$ pairs, $n(q)n(\bar{q}) \approx \ell^{-2\varepsilon/T}$ turns out to be independent of the baryonic chemical potential, the determining quantities being the temperature and the energy. The hard photon production rate [15,16] is by definition independent of the baryonic chemical potential. I wish to demonstrate that the ratio $(\ell\bar{\ell}/\gamma)$ is independent both of the baryonic chemical potential as well as the space-time evolution of the fireball, thus making the signal the ideal snap-shot of the phase transition to quark-gluon plasma in either the scaling regime or the stopped regime. For the "up" and "down" quarks, the number of lepton pairs in a space-timevolume element d^4X and of invariant mass M is given by [17]: (It is tacitly assumed that thermodynamic equilibrium is maintained throughout the space-time scenario.)

$$\frac{dN}{d^4x\, dM^2} = \frac{5}{3} \frac{M^2 T^2}{(2\pi)^4} \int_0^\infty d\, x\, dy \frac{1}{(\ell^{(X-Z)} + 1)} \frac{1}{(\ell^{(Y+Z)} + 1)} \Theta(xy - u^2/4) \quad .(1)$$

where $u = M/T$: $Z = \mu q/T$; $\mu_q = \frac{1}{3}\mu$, μ being the baryonic chemical potential and σ is the elementray annihilation cross-section into lepton pairs of mass m_ℓ from quark anti-quark pairs at temperature T

$$\sigma = (4\pi\alpha^2/M^2)(1 + 2m_\ell^2/M^2)(1 - 4m_\ell^2/M^2)^{1/2} \quad ..(2)$$

$m_\ell \equiv \mu^+, \mu^-$

After some algebraic manipulation, one can write without any approximation

$$\frac{dN}{d^4x\, dM^2} = \frac{5}{3} \frac{\sigma M^2 T^2}{(2\pi)^4} \sum_{n=1}^{\infty} \frac{(-1)^{n+1}}{n} \ell^{-nZ} \int_0^\infty \frac{\ell^{-nu^2/4x}}{(1 + \ell^{(x-z)})} \quad ..(3)$$

Clearly, in the lowest order $n = 1$ and $Z \to 0$ (no chemical potential) the integral is reduced to $uK_1(u)$, K_n's being the Mcdonald functions. I have computed dN/d^4X from eq.(3) by numerically integrating over dM^2 to find $\frac{1}{4\pi\alpha^2} \frac{1}{T^4} \frac{dN}{d^4X}$ saturate to a constant value for temperatures $T > 150$ MeV and indeed independent of the chemical potential over a wide range of values beyond a certain critical temperature as shown in Fig. 3.

The Maxwell-Boltzmann limit is also shown. Therefore, for the typical temperature regime $150 < T_C < 200$ MeV, evidently, the lepton pair production rate is independent of the baryonic chemical potential and is a function of T^4.

For lepton production rate from the hadronic matter, the most important channel is $\pi^+\pi^- \rightarrow \ell\bar{\ell} + X$ [17], given by the following expression:

$$\frac{dN}{d^4x\, dM^2} = \frac{1}{12}\left(\frac{\alpha^2}{\pi^3}\right)\left|F_\pi(M^2)\right|^2\left(1-\frac{4m_\pi^2}{M^2}\right)\left(1+\frac{2m_\ell^2}{M^2}\right)\left(1-\frac{4m_\ell^2}{M^2}\right)^{\frac{1}{2}} MT^2 G(W,\lambda) \quad ..(4)$$

where $W = \frac{M^2}{2m_\pi^2} - 1$, $\lambda = m_\pi/T$ and

$$G(W,\lambda) = \int_\lambda^\infty dx\,(\ell^X - 1)^{-1}\ell n\left[\frac{1-\exp(-WX - P(X^2-\lambda^2)^{\frac{1}{2}})}{1-\exp(-WX + P(X^2-\lambda^2)^{\frac{1}{2}})}\right] \quad ..(5)$$

Where $P = (W^2 - 1)^{\frac{1}{2}}$

the details of the pion form factor is given in ref. [18]. In Fig.2, the quantity $(1/4\pi\alpha^2)\,\frac{1}{T^4}\,\frac{dN}{d^4X}$ for lepton production from the hadronic is shown as a function of temperature. Clearly, the lepton production rate in the hadronic world rises more rapidly with temperatures than in the quark-gluon plasma. In Fig.4, I show the relation of $(dN/d^4X)_{hadron}/(dN/d^4X)_{QGP}$ as a function of temperature which shows clearly that for temperatures below the critical temperature the lepton production rate from the hadronic world as a function of the temperature varies much more rapidly than the corresponding quark-gluon plasma case

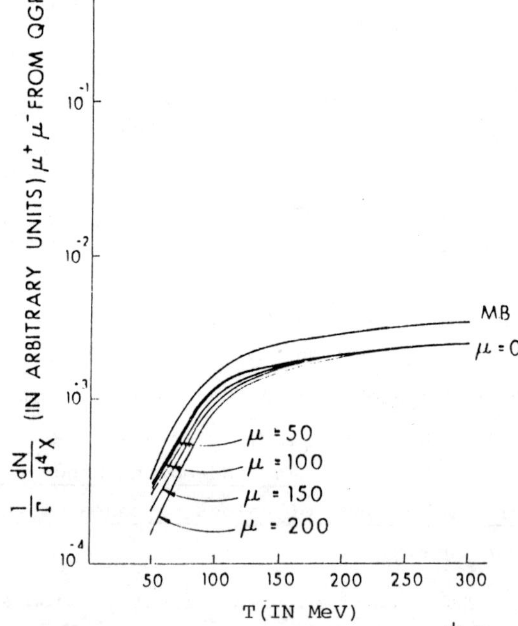

Fig. 3. The production rate of $\mu^+\mu^-$ pairs normalised by $\frac{1}{4\pi\alpha^2}\cdot\frac{1}{T^4} \equiv \Gamma$ as a function of the temperature; beyond a critical temperature the rate turns independent of the chemical potential.

and eventually saturating to a constant value beyond a certain temperature. However, for temperature below the critical temperature the production rate is certainly dependent on the baryonic chemical potential.

For completeness I have computed the $\mu^+\mu^-$ production rate from strange quarks. First, I wish to demonstrate [19] that there are as many $\bar{s}$ quarks as $\bar{u}$ or $\bar{d}$. It can easily be shown that [19]

$$n_q \simeq \exp(-\mu_q/T)\, T^3 (6/\pi^2) \qquad \text{.. (6)}$$

$$\bar{q} \equiv \bar{u} \text{ or } \bar{d}$$

similary, $n_s = n_{\bar{s}}$ (to conserve strangeness)

$\cong 3T(\frac{Ms^2}{\pi^2}) K_2(Ms/T)$, the chemical potential of strange quark being zero. Using the above expressions, we can get

$$(n_{\bar{s}}/n_{\bar{q}}) \approx \tfrac{1}{2}(\frac{Ms}{T})^2 K_2(Ms/T)\, \ell^{\mu/3T}$$

≈ 1.3 taking a value of Ms $\cong$ 300 MeV, T = 180 MeV and $\mu \approx 200$ MeV. The point one is asserting is that, approximately, there are as many strange quarks as antiquarks.

For $s\bar{s}$ pairs, the lepton pair production rate is given by (using essentially the same arguments as for $u\bar{u}$ and $d\bar{d}$ quarks) [20].

$$\frac{dN}{d^4x\, dM^2} = \frac{1}{3} \frac{M^2\sigma}{(2\pi)^4} \int_{Ms/T}^{\omega} \frac{dx}{1 + \ell^x} \left[\ell n\{1 + \ell^{-u^2/4x}\} \right] \qquad \text{.. (7)}$$

where once again, I have taken Ms ~ 300 MeV. In Fig.6, I show the quantity $(1/4\pi\alpha^2) . \frac{1}{T^4} . \frac{dN}{d^4x}$ as a function of the temperature. Beyond the typical critical temperature, it appears that the contribution from strange quark pairs $s\bar{s}$ for lepton production is significant upto 15% as conjectured before, [21,19].

It should be pointed out in this regard that if the chiral symmtry restoration temperature be the same as the deconfinment temperature, beyond this critical temperature the quarks become massless. I have therefore computed the integral in eq.(7) with $M_s = 0$; the contribution to lepton production with massless strange quarks is also shown in Fig.6. Although the absolute magnitude tends to increase, for deconfinment (chiral symmetry restoration temperature) temperature 150 - 200 MeV, even with massless strange quarks, the absolute contribution doès not increase more than 20% at the most. However, if the deconfinment temperature turns out to be rather low, less than 100 MeV, clearly the $s\bar{s}$ contribution to lepton pair production becomes rather large.

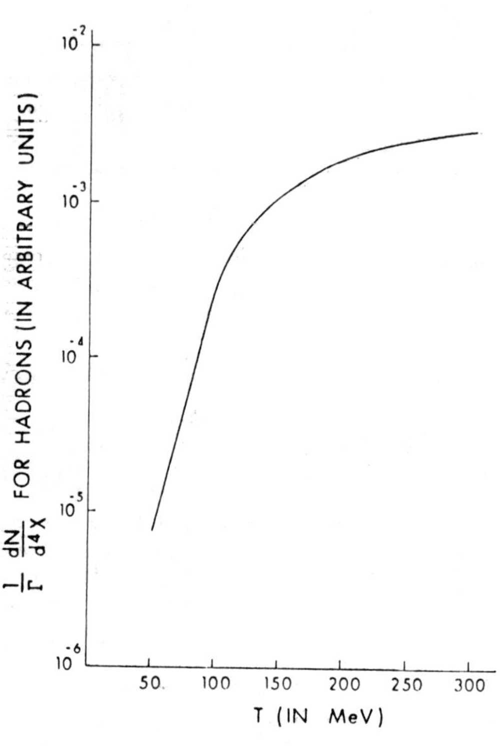

Fig.4. The production rate of $\mu^+\mu^-$ pairs normalised by $\frac{1}{4\pi\alpha^2} \cdot \frac{1}{T^4} \equiv \Gamma$ as a function of the temperature in the hadronic world.

The production of hard photons from a QGP can occur essentially in two process, the so called unusual Compton process $gq \to \gamma q$ where g refers to a gluon and/or $q\bar{q} \to \gamma g$. Taking into account the contribution from u, d and s quarks, the elementary cross section of the above two process are given by [22,23]

$$\sigma(gq \to \gamma q) = (16\pi\alpha\alpha_s/81s)\ \ln\ (s/4mq^2)$$
$$\sigma(q\bar{q} \to \gamma g) = (4\pi\alpha\alpha_s/3s)\ \ln\ (s/4mq^2) \quad \text{.. (8)}$$

where α_s is the running coupling constant and mq, the mass of the quark. Going through essentially the same steps as in the case of

$\mu^+\mu^-$ production, it is straightforward to show that [19]

$$\left(\frac{d\sigma}{d^4x}\right) \cong \left[\frac{\pi}{36}\alpha\alpha_s \ln(T/m_{u,d}) + \frac{\pi}{144}\alpha\alpha_s \ln(T/ms)\right] T^4 \qquad \text{.. (9)}$$

Depending on the duration of the expansion phase, $s\bar{s}$ pairs either go to $\gamma\gamma$ channel or gg channel, the first channel being dominant if the expansion phase is long enough. In this analysis I have ignored the saturation of $s\bar{s}$ pairs. Clearly in eq.(9) we still have the problem of quark mass singularity - following Kajantie and McLarren [24,25], due to finite temperature and finite size effects, $\ell_n(T/m) \tilde{\rightarrow} \ln(1/ds)$; thus,

$$(d\sigma/d^4x)_\gamma \approx \frac{5}{144}\pi\alpha\alpha_s \ln(1/\alpha_s)\, T^4 \qquad \text{.. (10)}$$

Thus taking into account eq.(10) for hard photon production and $\mu^+\mu^-$ production from "u", "d" and "s" quarks, I have computed the ratio $R(\mu^+\mu^-/\gamma)$ for various values of the baryonic chemical potential and temperature. The results are shown in Fig.7. For temperatures below the critical temperature $\mu^+\mu^-$ production rate as a function of the temperature rises rather rapidly and then beyond a critical temperature the ratio turns independent of the baryonic chemical potential and saturates to a constant value, approximately given by $|(3.5/\alpha)\alpha_s \ \ln(1/\alpha_s)|^{-1} \approx 3 \times 10^{-3}$ (using a value of

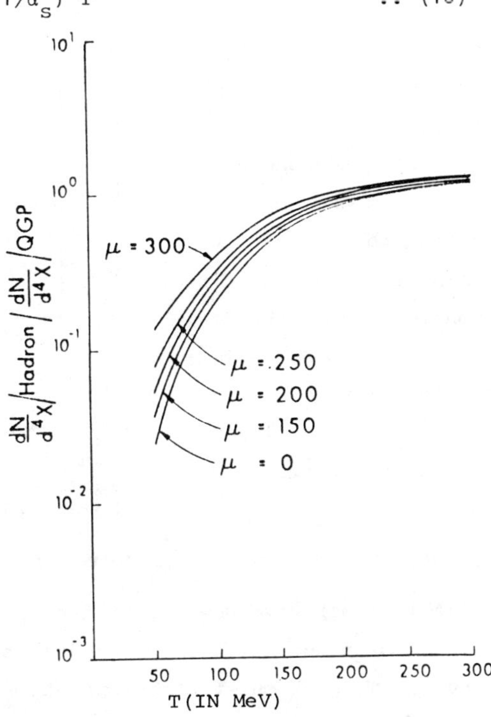

Fig. 5. The ratio of $\mu^+\mu^-$ production rates in the hadronic world to the Quark-Gluon plasma as a function of the temperature.

$\alpha_s \approx 0.6$). The above observation is for quark-gluon plasma. The dashed curve in Fig.6 shows the ratio as observed in the hadronic world where the photons are produced primarily from pi mesons and therefore goes as T^4 and $\mu^+\mu^-$ are produced as shown in eq.(4); this ratio goes on increasing monotonically as a function of the temperature unlike the quark-gluon phase.

With this scenario, the ratio $R(\mu^+\mu^-/\gamma)$ in the hadronic world rises rather rapidly as a function of temperature and once the critical temperature is reached and the hadronic world goes over to the quark-gluon plasma the ratio turns independent of the baryonic chemical potential thus in effect becoming independent of the kinematic regime, the scaling or the stopped regime; the saturation to a constant value is the signal of the phase transition.

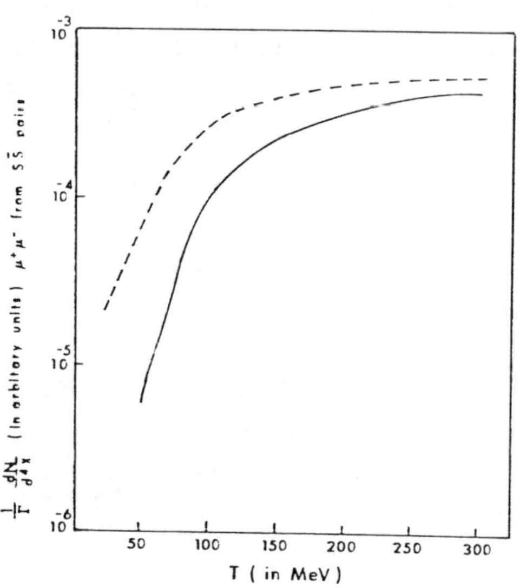

Fig.6. The production rate of $\mu^+\mu^-$ pairs from the $s\bar{s}$ pairs in the Quark-Gluon plasma as a function of the temperature; the dashed curve for $m_s = 0$.

4. PION DECAY AND BACKGROUND GAMMA

It is imperative however to make sure that signals of QGP which involve photons do not get polluted by the photons, arising out of π^o decay. To seek out the appropriate kinematic window, I evaluate first the ratio of the production rate $\pi^o/\mu^+\mu^-$ both from the hadronic world as well as from QGP. For pions, the important observation is that, in contrast to leptons or photons which freely leave the plasma, hadron bakage is a surface and not a volume effect - the flux of hadrons with energy E per unit time per unit surface being given by [22,24]

$$E \frac{dJ_\pi}{dp^3} = \left| g_\pi p_\perp /(2\pi)^3 \right| \exp\left(\frac{-m_\perp ch y}{T}\right) \quad .. (11)$$

For a QGP, the above "leakage" corresponds to quarks and gluons while for a non ideal gas it refers to hadrons, pi mesons for example; $m_\perp$ is the transverse mass given by $m_\perp = (M^2 + p_\perp^2)^{1/2}$ for transverse momentum $p_\perp$, y being the repidity of the dadron concerned. Assuming cylindrical shape it can be shown [19,22] , for pions

$$E \frac{d\sigma}{dp^3} \cong \frac{g_\pi p_\perp}{2 r_o (2\pi)^3} \cdot f(m_\perp , y)$$

where $$f(m_\perp , y) = \int \exp\left(\frac{-m \ ch(y - \hat{y}(x,t))}{\hat{T}(x,t)}\right) ds_\perp \, dt \quad .. (12)$$

is a universal function, independent of the kind of hadron as long as local thermodynamical equilibrium is maintained during the process of evolution; $\hat{y}(x,t)$ and $\hat{T}(x,t)$ are respectively the local rapidity and the temperature of the system and $V = 2r_o S$, r_o being the transverse radius of the system. Similarly for $\mu^+\mu^-$ production one can show [19, 22]

$$E \frac{d\sigma}{dM^2 dp^3} = \frac{\alpha^2}{8\pi^4} G(M,m\mu) f(m_\perp , y) \quad .. (13)$$

where $G(M,m\mu) = (1 + 2m\mu^2/M^2)(1 - 4m\mu^2/M^2)^{1/2}$

Comparing eq.(12) and eq.(13), one gets the ratio

$$R(\pi^o/\mu^+\mu^-) \approx \frac{1.5 \, g_\pi p_\perp}{r_o \alpha^2} \quad .. (14)$$

For hadron "leakage" $g_\pi = 1.0$ and for quark "leakage" $g_\pi = 0.2$. Therefore, if one observes $R(\pi^0/\mu^+\mu^-)$ as a function of $p_\perp$, the signal of the phase transtion will show up in the gradient of $R(\pi^0/\mu^+\mu^-)$ as a function of $p_\perp$. Indeed the experimental results of Branson et.al.[25] clearly shows this branching; beyond $p_\perp = 4$ GeV/C hadron "leakage" becomes negligible and the data predominantly shows quark "leakage" gradient of $g_\pi = 0.2$.

Evidently, comparing eq.(14) and $R(\mu^+\mu^-/\gamma)_{h.p.}$ one gets $R(\pi^0/\gamma)_{h.p.} \approx 3\text{-}5$ for $p > 4$ GeV/C; the pi mesons in this comparison arises from the "leakage" from the QGP.

However, we are really concerned with decay products of $\pi^0 \to 2\gamma$ compared to $\gamma_{h.p}$ from the QGP. Two important points emerge: for high $p_\perp > 4$ GeV/C even the π^0s are products of QGP, so the γ's, the decay products of π^0, arise from the π^0 formed from the QGP, the γ's which are products from pions, belonging originally to the hadronic world will not show up in the high $p_\perp$ window. The fact that at high $p_\perp$, the γ's are predominantly hard photons from QGP has yet another compelling reason. From simple momentum conservation the γ's coming out π^0, to have the same $p_\perp$ as the γ's from QGP, the parent π^0 has to have a momentum of $2p_\perp$; since $n(\pi_0) \approx \ell^{-10p_\perp}$, the number of π° with $p > 2(4\text{GeV/C})$ will be extremely small rendering all the γ's for $p > 4$ GeV/C primarily the hard photons from QGP and not the decay products of π°'s, even for the π°'s originating from QGP.

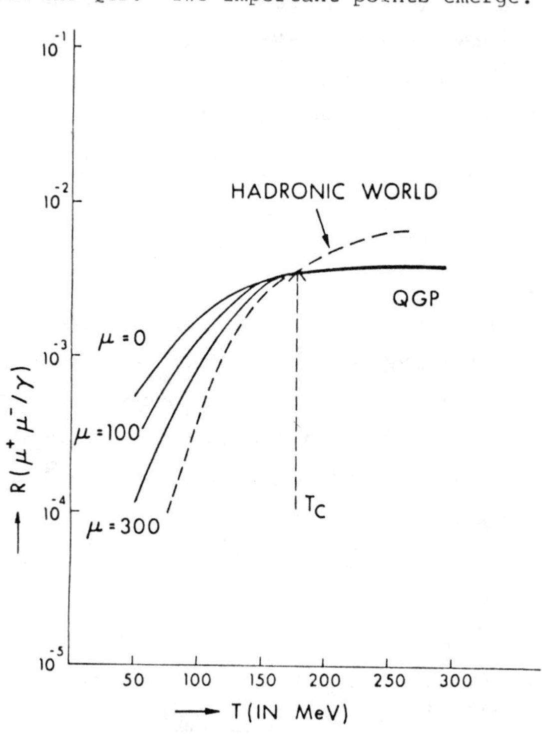

Fig.7. The ratio of the production rate of $\mu^+\mu^-$ pairs to hard photons as a function of the temperature; beyond T_C the ratio saturates to a constant value and turns independent of the baryonic chemical potential; for the hadronic world the ratio increases monotonically.

Finally therefore the ratio ($\mu^+\mu^-/\gamma$) as a function of the temperature saturates to a constant value signalling the phase transition to Quark-Gluon Plasma from the hadronic world, the most desirable kinematic window for observing this ratio, essentially arising from the QGP is for high transverse momentum beyond 4 GeV/C.

The author would like to thank A.K. Chaudhury and D.N. Basu for invaluable computational assistance and D. Shyam for most interesting discussion on the transverse momentum window for hard photons.

REFERENCES

1 M. Jacob and H. Satz, Proc. Bielefeld Workshop (1982) (World Scientific, Singapore, 1982)

2 G. Baym, Prog. Part. Nucl. Phys. 8 (1982) 73

3 J.B. Kogut, Nucl. Phys. A418 (1984) 38/C

4 H. Satz, Nucl. Phys. A418 (1984) 447C

5 P. Carruthers, Nucl. Phys. A418 (1984) 501C

6 M. Gyulassy, Nucl. Phys. A418 (1984) 59C

7 S. Nagamiya and M. Gyulassy, Adv. in Nucl. Phys. 13(1983) 199

8 J.D. Bjorkln, Phys. Rev. D27 (1983) 140

9 M. Gyulassy, Proc. Bielefeld Workshop (1982) 81 (World Scientific, Singapore 1982)

10 G. Baym et.al. Urbana Preprint 1983 and Nucl. Phys. A418 (1983) 525C

11 L.D. Landau and E.F. Lifshitz, Fluid Mechanics (Pergamon Press, New York, 1982) P 503

12 L. Van Hove, Cern Th. 3592 (1983) Preprint

13 G. Glen and P. Sutherland, Ap. J. 239 (1980) 671

14 K. Van Riper and D.Q. Lamb, Sp. J. (Lett.), L13 (1981) 244

15 K. Kajantie and H.I. Miettinen, Z. Phys. C14 (1982) 357

16 B. Sinha, Phys. Letts 128B (1983) 91

17 G. Domokos and G.I. Goldman, Phys. Rev. D23 (1981) 203

18 K. Kajantie and H.I. Miettinen Z. Phys. C9 (1981) 341

19 J. Rafelski, Nucl. Phys. A418 (1984) 215C

20 B. Sinha, Phys. Letts. B to be published

21 B. Sinha, Phys. Letts. 135B (1984) 169

22 E.V. Shuryak, Phys. Rep. 61 (1980) 71

23 E.V. Shuryak, Yad. Fiz. 28 (1978) 796

24 O.V. Zhirov, Yad. Fiz. 30 (1979) 1098

25 J.G. Branson et.al. Phys. Rev. Letts. 38 (1977) 1331

EXCITATION OF GIANT RESONANCES BY PROTON INELASTIC SCATTERING

N. Marty

Institut de Physique Nucléaire
BP N 1, 91406 Orsay
France

ABSTRACT

The interest of medium energy protons for studying giant resonances is discussed. The excitations in ^{208}Pb of parity favoured resonances E2, E0, E1(ΔT=0) and E1(ΔT=1) are taken as examples. For the spin-flip M1 resonance, special attention is given to the N = 28 nuclei. For proton transitions, the comparison between strengths measured in (e,e') and (p,p') inelastic scattering is a severe test of the models ; those used until now cannot explain the experimental results. At least part of the quenching of the ΔL=0 ΔS=1 transitions may be due to a deficiency of the models. Low energy 1+ states are briefly discussed.

1. INTRODUCTION

The giant resonances have been the subject of much theoretical and experimental work during the last ten years. Many results have been obtained, but all the predicted giant resonances are not yet localized and the spreading of their strength is not perfectly known. This is not surprising due to their complexity. In Fig. 1 is given the spectrum predicted theoretically for 90 MeV electron inelastic scattering at an angle θ = 75°, and its decomposition on resonances of different angular momentum, parity and isospin taken from Ref. 1. Therefore different probes, which excite selectively some of the resonances, are needed to disentangle the problem. The advantages and disadvantages of 200 MeV proton inelastic scattering will be discussed.

The main features of the proton-nuclei interaction come directly from those of the nucleon-nucleon interaction which can be parametrized as :

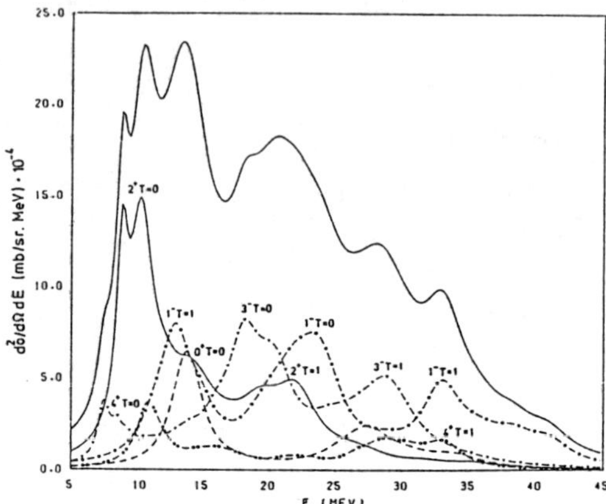

Fig. 1 : ee' scattering($E = 90$ MeV $\theta = 75°$)as predicted theoretically (Ref. 1).

$$V = V_C + V_\circ + V_\sigma(\vec{\sigma}_1\vec{\sigma}_2) + V_\tau(\vec{\tau}_1\vec{\tau}_2) + V_{\sigma\tau}(\vec{\sigma}_1\vec{\sigma}_2)(\vec{\tau}_1\vec{\tau}_2) +$$

$$\{V_{LS} + V_{LS\tau}(\vec{\tau}_1\vec{\tau}_2)\}\,\vec{L}\vec{S} + \{V_T + V_{T\tau}(\vec{\tau}_1\vec{\tau}_2)\}S_{12}$$

V_C is the coulomb interaction. At 200 MeV proton energy and small momentum transfers, the orbital V_{LS} and tensor V_T terms can be neglected. In Fig. 2 are represented as a function of energy the scattering amplitude t(2a) the ratio :

$$(\frac{t_{\sigma\tau}}{t_\tau})^2 \text{ (2b) and } (\frac{t_{\sigma\tau}}{t_\circ})^2 \text{ (2c)}$$

in the parametrization of Ref. 2 taken at $q = 0$.

$V_\circ$ has a strong minimum for energies between 200 and 400 MeV ; in this energy range multiple scattering processes can be neglected and experimental results can be analyzed in the distorted wave impulse approximation (DWIA) ; the proton penetrability goes through a maximum at 200 MeV, form factors until four fermis can be tested. As $V_{\sigma\tau}$ is nearly energy independent, there is a maximum of :

$$(\frac{V_{\sigma\tau}}{V_\circ})^2 \text{ and of } (\frac{V_{\sigma\tau}}{V_\tau})^2$$

(Fig. 2b, 2c). Therefore a strong excitation of $\Delta S = 1$ $\Delta T = 1$ transitions, induced by the $V_{\sigma\tau}$ part of the interaction, is expected.

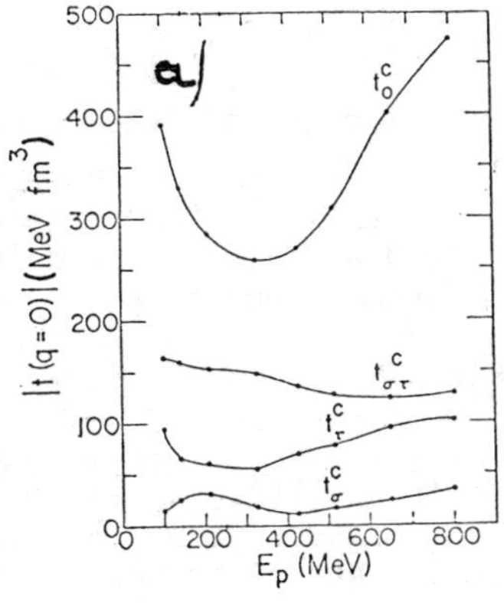

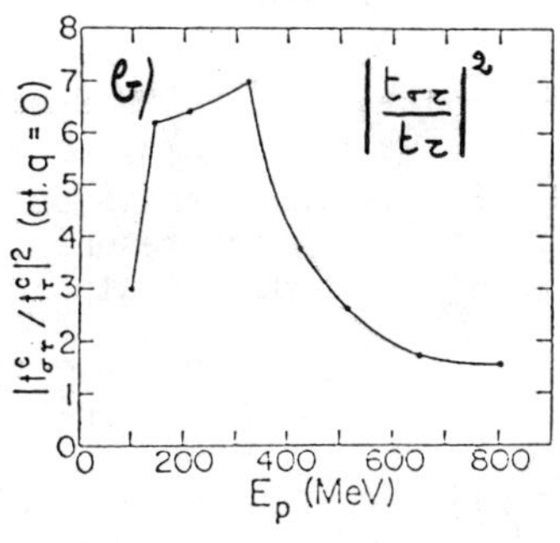

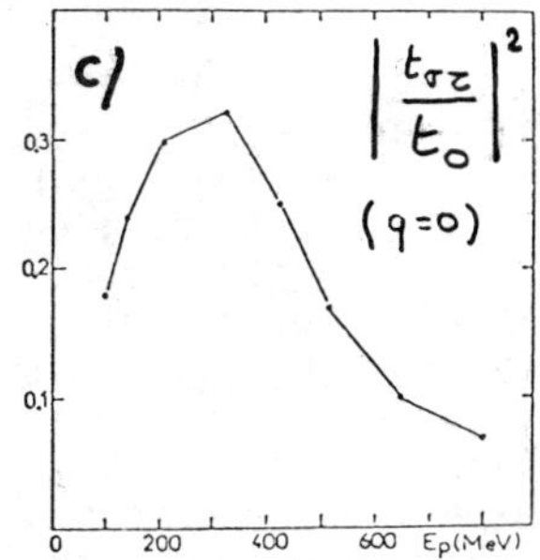

Fig. 2 : See text.

Protons are especially suited for exciting states of low angular momentum through $\Delta L = 0$ transitions at forward angles, through $\Delta L = 2$ transitions at the first maximum for the corresponding angular distribution.

As 200 MeV protons are relativistic particles ($\beta = v/c = 0.6$) ; they strongly excite isovector dipole resonances and also isovector quadrupole resonances through coulomb interaction and this can be considered as a disadvantage. Three examples will be discussed, parity favoured resonances in ^{208}Pb, spin-flip transitions in N = 28 nuclei and low energy 1+ states.

2. EXCITATION OF PARITY FAVOURED RESONANCES IN ^{208}Pb

2.1 Sum Rules

The total strength for exciting states of multipolarity L is given by a sum rule. The most frequently used is the energy weighted sum rule :

$$\Sigma(Q_L) = \sum_n (E_n - E_\circ) |\langle nL \mid Q_L \mid 00 \rangle|^2 = \sum_n (E_n - E_\circ) B_n(EL)$$

nL, 00, E_n, $E_\circ$ denote the excited states nL, the ground state 00 and their respective energy.

$B_n(EL)$ is the transition probability for exciting the state n. In this analysis, the resonances are described as macroscopic deformations of the ground state, Q is a multipole mass operator

$$Q_{LM} = \sum_{L=1}^{A} f(r_i)\, Y_{LM}(\theta_i\ \phi_i)$$

with $f(ri) = r_i^L$ for $L \geq 2$. For hadron scattering B(EL) is directly related to the mass deformation parameter β_L measured experimentally if one assumes that the procedure used for low energy probes or strongly absorbed particles is always valid.

2.2 The Giant Isoscalar Quadrupole Resonance (GQR)

It is generally considered that for heavy nuclei as ^{208}Pb strenghts measured with different probes agree and are compatible with 100 % of the (EWSR)[2] . When looking in more detail to the different experimental results, first (pp') data[3], (α,α')[4] (e,e')[5] and more recent (^{3}He, ^{3}He')[6] data are indeed compatible with a strength exhausting the (EWSR) ; but recent $(\alpha\alpha')$ results[7] can only be analyzed by assuming that the (GQR) exhausts 70 % of the (EWSR) and that some $\Delta L = 4$ strength is present in the same excitation energy range. High resolution (e,e') experiments find only 30 % of the (EWSR)[8].In our (pp') experiments performed at 155 and 201 MeV, due to the selectivity of the protons, the excitation of L = 4 states is small. If the results were analyzed as described in 2.1, only 30 % of the sum rule would be found[9]. But it is no more justified to use mass operators for high energy protons. When DWIA calculations are performed using RPA wave functions which predict that the GQR exhausts 60 to 70 % of the sum rule, a very good agreement is obtained with experimental cross sections. The same explanation can be given for the results of Ref. 10.

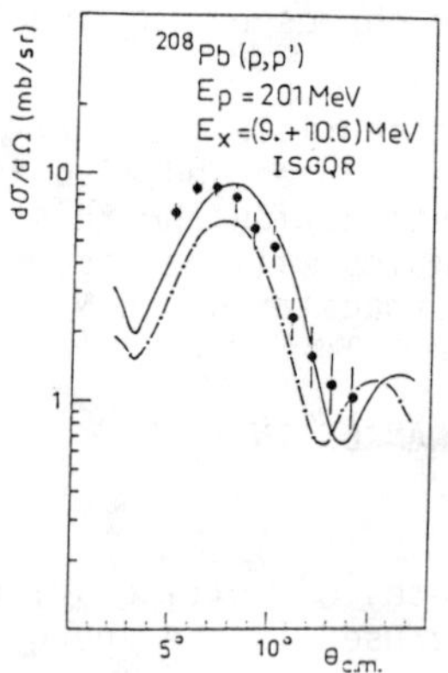

Fig. 3 : Experimental cross sections for the GQR compared to RPA calculations.

2.3 The Giant Isoscalar Monopole Resonance (G.M.R.)

This example shows the disadvantages of 200 MeV protons. The localization of the (GMR) or breathing mode has been a challenge ten years ago : from the energy of this mode the compression modulus of the nuclei and the nuclear matter can be extracted. In 1975, by comparing , in the region of the quadrupole resonance,the spectra obtained on ^{208}Pb and ^{90}Zr by deuteron inelastic scattering to the spectra obtained by (α,α') scattering[4] we suggested that the difference could be explained by the monopole resonance. This resonance was predicted to be favoured at the scattering angle chosen for the (dd') measurement[11].

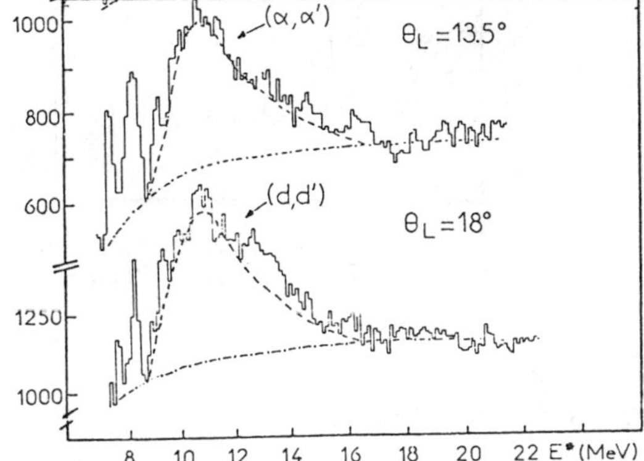

Fig. 4 : Region of the GQR excitated by ($\alpha\alpha$') and (dd') scattering.

The existence of the monopole resonance has been clearly settled afterwards by (α,α')[12] and (^{3}He, ^{3}He')[6] measurements.

We tried without success to study the (GMR) by 200 MeV proton inelastic scattering. The monopole resonance is hindered by the isovector giant dipole resonance which energy (13.5 MeV in ^{208}Pb) differs only slightly from that of the GMR (13.9 MeV).

2.4 Giant Dipole Isoscalar and Quadrupole Isovector Resonances

In Fig. 5, one can see a large resonance centered at about 21.5 MeV, with a width of 5.7 ± 0.2 MeV. Two different resonances have previously been reported at about this energy, the isovector giant quadrupole resonance excited in (e,e') scattering[13] and a compression isoscalar mode of angular momentum L = 1 (ISGDR)[14] which can hardly be distinguished from other states in (α,α') experiments. We could analyze the 21.5 MeV resonance as the sum of the two resonances. The IVGQR is excited through coulomb interaction and exhausts the whole sum rule. The $\Delta L = 1$ compression mode exhausts 70 % of the EWSR in agreement with predictions of RPA models[15] 200 MeV protons are particularly suited for exciting this dipole mode whose form factor is peaked at 5 fm in ^{208}Pb. Strongly absorbed particles will only be sensitive to the surface part of the form factor.

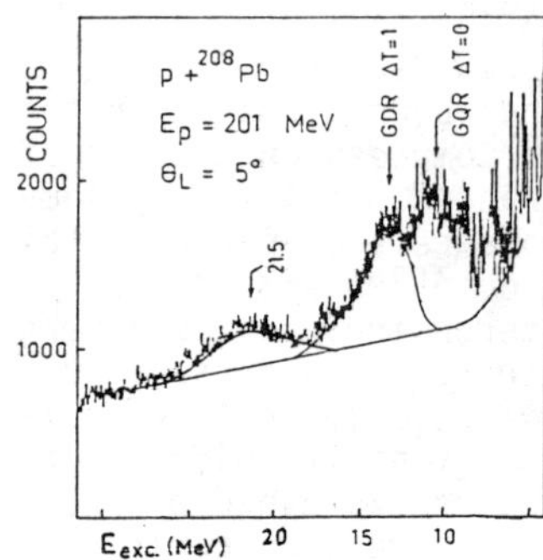

Fig. 5 : Isoscalar GDR and isovector GQR in ^{208}Pb located at 21.5 MeV.

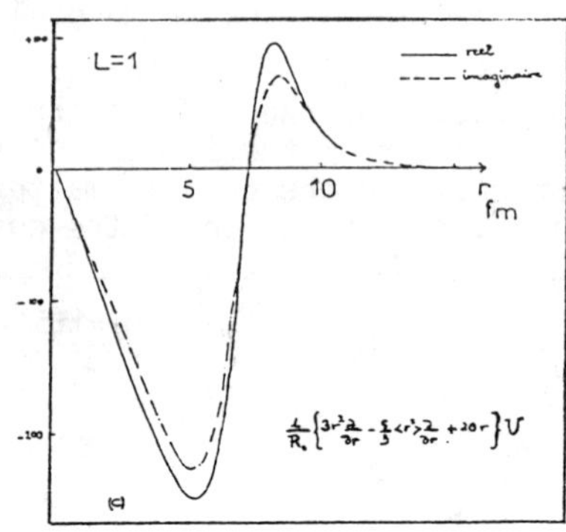

Fig. 6 : Form factor for the 1^- T = 0 state in ^{208}Pb.

3. ΔL = 0 SPIN-FLIP TRANSITIONS

3.1 Interaction and Sum Rules for Spin-Flip Transitions

Spin-flip transitions can be excited by charge exchange reaction: they are called Gamow-Teller transitions by analogy with β decay or by inelastic (e,e') or (p,p') scattering : they are called M1 transitions.

This field has been very active during the last years. An answer has been given to the question : why 1+ states were not seen in (e,e') or low energy (p,p') scattering when their analogs or antianalogs were excited in (p,n) reactions[16] ? However the quenching of the spin-flip transitions is still not explained.

In section 1,200 MeV(p,p') scattering was shown to be suited for exciting such transitions. An example of their selectivity at forward direction is given in Fig. 7 a)where angular distributions are given for ΔL = 0, 1 and 2 ; they are normalized arbitrarily to the same maximum cross section.

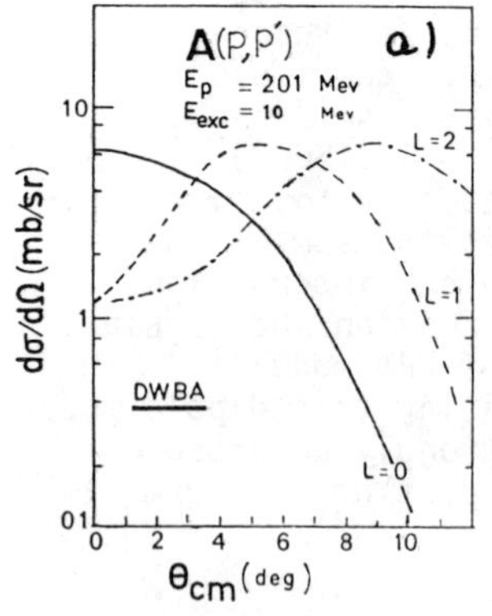

Fig. 7 a) : See text

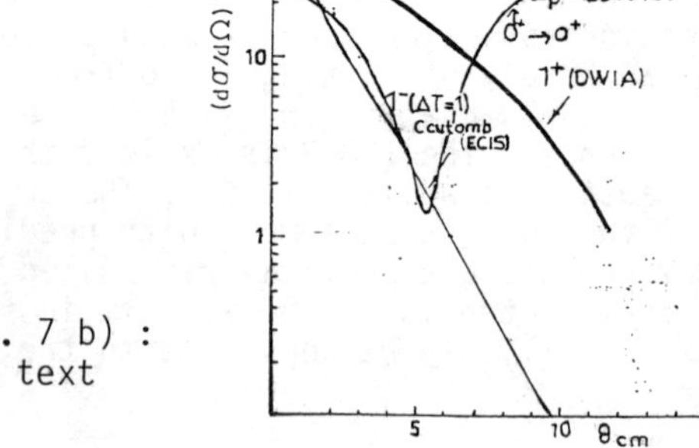

Fig. 7 b) : See text

Fig. 7 b)illustrates, for ^{48}Ca, how $\Delta L = 0$ spin-flip angular distributions differ from $0^+ \rightarrow 0^+$ and from coulomb excited 1^- angular distributions.

In a simple model, 1+ transitions are particle-hole (p,h) transitions between $j = \ell + 1/2$ and $j = \ell - 1/2$ configurations. In (p,n) reactions, these transitions are induced by the $V_{\sigma\tau}$ interaction. In (p,p') scattering the central V_σ and $V_{\sigma\tau}$ interactions can both excite such states, giving only T_o states for the V_σ interaction, T_o and eventually $T_o + 1$ states for $V_{\sigma\tau}$.

For electromagnetic interactions, the M1 operator can be written:

$$\sum_{k=1}^{A} \vec{\mu}(k) = \sum_{k=1}^{A} \left\{ \left(\frac{g_p^\ell}{2} \vec{\ell} + \frac{g_n^s + g_p^s}{2} \vec{s}\right) + \left(\frac{g_n^s - g_p^s}{2} \vec{s} - \frac{g_p^\ell}{2} \vec{\ell}\right) \tau_Z \right\}$$

Because of the value of $g_p^s = 5.59\ \mu_N$ and $g_n^s = -3.83\ \mu_N$, the isovector interaction is dominant. As a difference with the proton nucleus interaction, a proton transition is induced by the spin as well as by the orbital term $g_p^\ell \vec{\ell}$.

In Gamow-Teller transitions a sum rule is defined as :

$$S^- - S^+ = \Sigma|< n|\sigma\tau_- \ |0>|^2 - \Sigma|< n|\sigma\tau_+ \ |0>|^2 = 3(N - Z)$$

the quenching Q is measured by reference to this sum rule. In inelastic scattering no model independent sum rule can be defined, Q is the ratio of measured to predicted B(M1) transition probabilities or differential cross sections.

3.2 Excitation of M1 resonances in N = 28 nuclei

Because of its simple shell structure, a strong $\nu(f\,5/2, f\,7/2^{-1})$ transition is expected in ^{48}Ca. It has been effectively seen, at an excitation energy of 10.2 MeV in electron scattering at backward angles[17]. Its analogous is strongly excited in a charge exchange (p,n) reaction at 160 MeV[18]. The 10.2 MeV state is also strongly excited in our (p,p') experiments[19].

If two protons are added to ^{48}Ca to give ^{50}Ti, the 10.2 MeV state is split in more than seven states which centroid energy is 9.9 MeV ; a $(T_o + 1)$ 1+ state appears at 15.4 MeV and there is a concentration of strength near 8.5 MeV which can be interpreted as due to a proton $\pi(f\,5/2, f\,7/2^{-1})$ transition[20].

If two more protons are added, one get ^{52}Cr. The number of states near 10 MeV increases, they can hardly be separated from states near 8.5 MeV, levels located between 12 and 13 MeV may be considered as

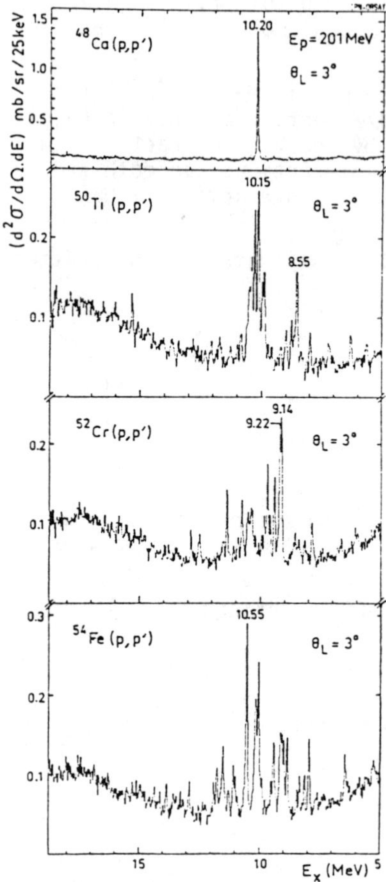

Fig. 8 : (p,p') spectra on even N = 28 isotopes

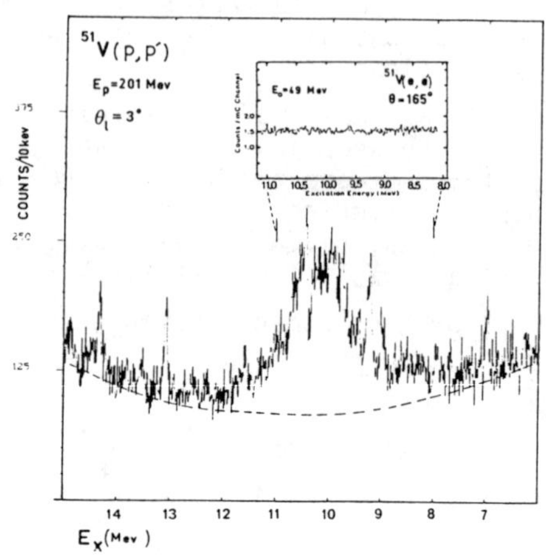

Fig. 9 : (p,p') and (e,e') spectra on the odd N = 28 nucleus ^{51}V

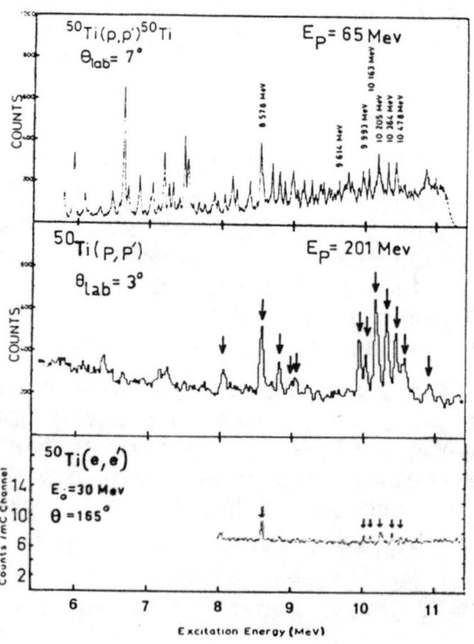

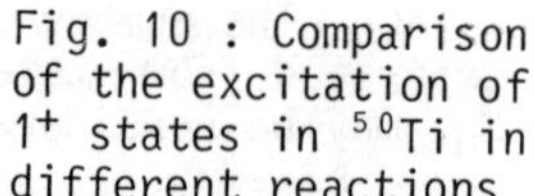

Fig. 10 : Comparison of the excitation of 1^+ states in ^{50}Ti in different reactions

T_o + 1 levels. In ^{54}Fe it is no more possible to distinguish T_o from T_o + 1 states. For a N = 28 odd Z nucleus as ^{51}V, the splitting of the 1+ states is so large that they appear as a resonance centered at 10.15 MeV (Fig. 9) ; no M1 strength could by extracted from a (e,e') experiment on ^{51}V[21].

As an example of the selectivity of 200 MeV protons for exciting the 1+ states, in Fig. 10, our results are compared with 65 MeV (p,p') data, and 30 MeV (e,e') results at backward angle.

The strengths and quenchings obtained in (p,p') and (e,e') experiments will now be compared.

For ^{48}Ca, if the 10.2 MeV transition is assumed to be ν(f 5/2, f $7/2^{-1}$), the measured (p,p') cross section $(d\sigma/d\Omega)pp'$ can directly be compared to an "*equivalent*" cross section extracted from the (p,n) experiment at the same momentum transfer[22]. The results are $(d\sigma/d\Omega)pp'$ = 4.7 ± 0.5 mb sr^{-1} at 2.5 ^{0} and $(d\sigma/d\Omega)p,n$ equiv. = 4.9 ± 0.6 mb sr^{-1}.

From the (p,p') measurement a $\mathcal{B}$(M1) value can be extracted and compared with the B(M1) measured by (e,e') scattering : the results B(M1)e,e' = 3.9 ± 0.3 μ_N^2 and $\mathcal{B}$(M1) = 3.8 ± 0.4 μ_N^2 are in perfect agreement.

For ^{50}Ti the particle hole configurations are more complex and such calculations cannot be performed but one can compare the experimental relative cross sections and transition probabilities as well as the relative cross sections and B(M1) predicted by different models taking a transition as reference[20]. Such a comparison is given in Fig. 11. The ratio of the cross sections and B(M1) are very similar for transitions which can be considered as the splitting of the 10.2 MeV transition in ^{48}Ca. In contrast there is a strong discrepancy for the state at 8.56 MeV. The large B(M1) measured value can be explained if the transition is mainly a proton transition and if orbital and spin excitations add coherently. An enhancement of the B(M1) is predicted by both models for states lying in the region of 5 to 8.5 MeV ; the experimental results are in better agreement with predictions from model (Ref. 24) but the experimental enhancement is larger than predicted.

In the table the experimental cross sections and B(M1) summed over all the excited states of ^{50}Ti are compared with those predicted by the different models and the corresponding quenching factors are given.

Table

Measured	predicted	B(M1)	Q[B(M1)]	$(d\sigma/d\Omega)$	Q[$d\sigma/d\Omega$]
$B(M1)_{e,e'}=3.4\mu_N^2$	model Ref.23	15.65	0.22	13.6	0.4
$d\sigma/d\Omega$=5.3 mb sr^{-1}	Ref. 24	11.46	0.3	10.2	0.5
	Ref. 25	~ 2.5	~1.3		

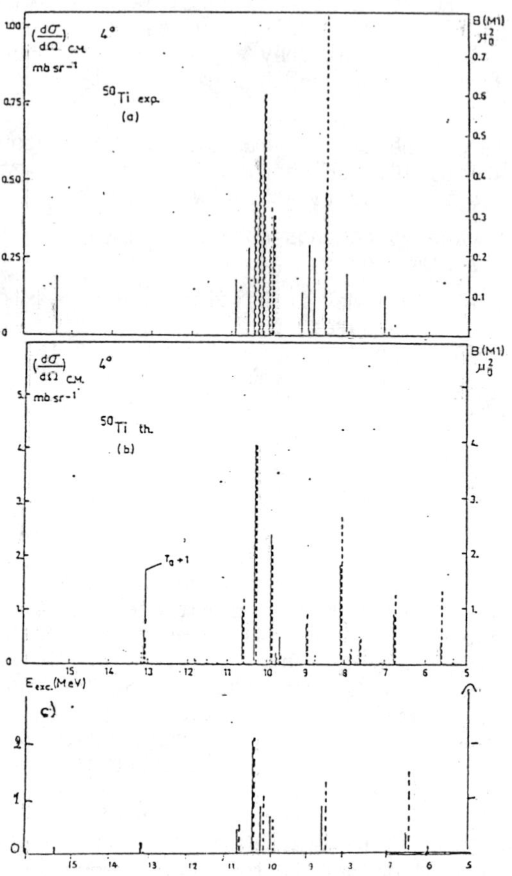

Fig. 11 : M1 strength distribution and cross sections in ^{50}Ti
a) experimental results b) model of Ref. 23
c) model of Ref. 24. Full lines : cross sections ; dashed lines B(M1). Only the strongest predicted states are represented.

The (p,p') cross section have not yet been calculated with model Ref. 25 in order to see if the enhancement of B(M1) for the "*proton*" transition agrees with experiment and if the quenching for the cross section is nearly one as it is for the B(M1).

3.3 Low energy 1+ states in Deformed Nuclei and in Nuclei of the f 7/2 Shell

Different models predict the existence of low energy collective 1+ states in heavy deformed nuclei corresponding to the oscillation, round an axis perpendicular to the symmetry axes, of deformed bodies of protons and neutrons. Such states must be excited only through the orbital part of the electromagnetic interaction. They have been detected by (e,e') scattering[26] and afterwards by (γ,γ') fluorescence[27].

Very recently Zamick predicted low energy 1+ states in light nuclei especially in nuclei of the f 7/2 shell as ^{46}Ti. Effectively a 1+ state has been localized by (e,e') scattering at 4.32 MeV in ^{46}Ti[28]. This transition can be induced not only by the orbital part but also by the spin part of the electromagnetic interaction.

It has been shown in section 1 that the spin-orbit part of the nucleon-nucleon interaction is negligible. Then, from the theoretical predictions, one expects not to excite any low energy collective 1+ state in deformed nuclei, and perhaps to excite the 1+ state in ^{46}Ti. Preliminary results entirely confirm these predictions. No low energy 1+ state is seen in ^{154}Sm, ^{156}Gd, ^{164}Dy ; the 4.32 level is weakly excited in ^{46}Ti.

4. SUMMARY AND CONCLUSIONS

Inelastic scattering of 200 MeV protons is shown to be a very good probe to study giant resonances of low angular momentum L. The large mean free path of the protons makes them suitable for testing transition densities not peaked at the surface of the nuclei as the compression mode $\Delta T = 0$ $\Delta L = 1$. But at this energy coulomb excitation of the isovector $\Delta L = 1$ and $\Delta L = 2$ resonances is strong.

200 MeV (p,p') scattering at forward angles is especially selective for exciting spin-flip transitions. The comparison of the quenchings measured in (e,e') and (p,p'), when proton transitions are involved, is a severe test of the orbital effects and consequently of the wave functions of the ground and excited states. The models used until now are not able to fit the experimental results. The inadequacy of the models may at least partially explain the quenching of the $\Delta L = 0$ spin-flip transitions.

The physicists involved in these experiments are in Orsay C. Djalali, M. Morlet, A. Willis and J.C. Jourdain. The 1+ resonances were studied in a collaboration Orsay, MSU (N. Anantaraman, G.M. Crawley, A. Galonsky) and the low energy 1+ excitations in a collaboration Orsay, Darmstadt (A. Richter, D. Bohle, H. Hartmann) MSU.

5. REFERENCES

1) Speth J. and Van de Woude A. Rep. on Prog. in Phys. 44, 719 (1981)

2) Love W.G. and Franey M.A. Phys. Rev. C24, 1077 (1981)

3) Bertrand F. Nucl. Phys. A354, 129c (1981)

4) Youngblood D.H. et al. Phys. Rev. C13, 994 (1976)

5) Pitthan R. et al. Phys. Rev. Let. 33, 849 (1974)

6) Buenerd M. Journ. de Phys. Col. C4 45, 115 (1984)

7) Morsch H.P. Phys. Rev. C29, 2075 (1984)

8) Kühner G. et al. Phys. Let. 104B, 189 (1981)

9) Djalali C. et al. Nucl. Phys. A380, 42 (1982)

10) Kailas S. et al. Phys. Rev. C 29, 2075 (1984)

11) Marty N. et al. Proc. Int. Symp. on highly excited states Julich 1975 and Orsay report IPNO 76.03

12) Youngblood D.H. Giant multipole Res. Oak Ridge 113(1979)

13) Pithan R. Giant multipole Res. Oak Ridge 161 (1979)

14) Morsch H.P. et al. Phys. Rev. Let. 45, 337 (1980)

15) Djalali C. et al. Nucl. Phys. A380, 42 (1982)

16) Gaarde C. Nucl. Phys. A396, 127c (1983)

17) Steffen W. et al. Phys. Let. 95B, 23 (1980)

18) Anderson B.D. et al. Phys. Let. 114B, 15 (1982)

19) Crawley G.M. et al. Phys. Let. 127B, 322 (1983)

20) Djalali C. et al. Nucl. Phys. A410, 399 (1983) and A417, 564 (1984)

21) Benda D. et al. Nucl. Phys. A398, 408 (1983)

22) Djalali C. Thesis Orsay 1984 unpublished

23) Metsch B.C. and W. Knüpfer, private communication

24) Muto K. and Horie H. Phys. Let. 138B, 9 (1984)

25) Zybert L. et al. Z für Physik A Atoms and nuclei 318, 363 (1984)

26) Bohle D et al. Phys. Let. 137B, 27 (1984), and 148B, 260 (1984)

27 Berg U.E.P. et al. preprint

28) Zamick L. preprint

29) Richter A. and Bohle D. private communication.

INELASTIC PROTON SCATTERING AND THE QUENCHING OF ΔL=0 SPIN-FLIP EXCITATIONS IN NUCLEI

N. Anantaraman

National Superconducting Cyclotron Laboratory
Michigan State University
East Lansing, Michigan 48824-1321

Abstract

The results of 201-MeV (p,p') studies of ΔL=0, ΔS=1 transitions at very forward angles are discussed. Typically, the observed strength is 25% to 30% of that predicted. This is illustrated by data for the Ca isotopes; comparisons with (p,n) and (e,e') measurements are made. The observation, in ^{28}Si, of comparable reduction factors for isoscalar transitions, where Δ-isobar effects must be insignificant, indicates the importance of configuration mixing as a quenching mechanism. This conclusion is consistent with the results of spin-flip cross section measurements using $(\vec{p},\vec{p}')$.

1. INTRODUCTION

During the 1970's, new types of giant electric resonances were discovered and studied[1]), in addition to the giant dipole which had been known for a long time. These resonances added to our knowledge of nuclear collective motion in which only the spatial degrees of freedom of the nucleons are involved. In the 1980's, the emphasis has shifted to the study of magnetic resonances, viz. those that involve the spin degrees of freedom of the nucleons. They have been excited by a variety of probes, including (e,e'), (γ,γ'), (p,p') and (p,n), every one of which has added to our knowledge of spin

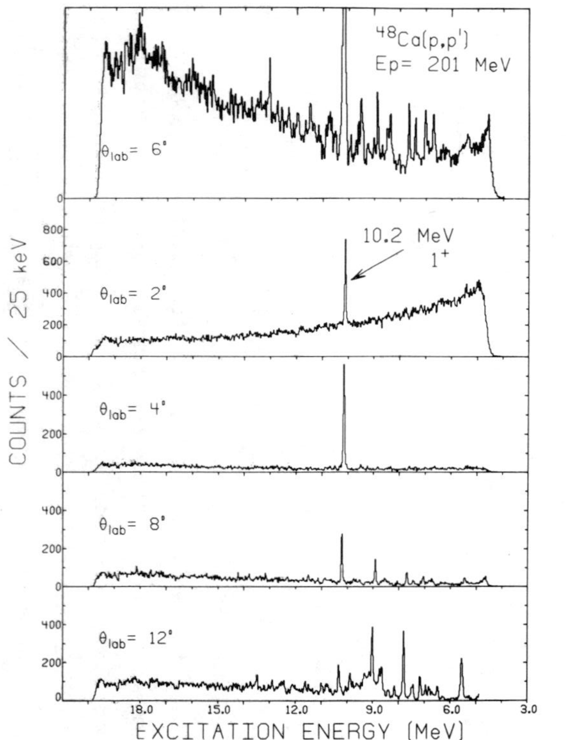

Fig. 1 - *Spectra of protons inelastically scattered from ^{48}Ca. The uppermost spectrum has been scaled to show the weakly excited states.*

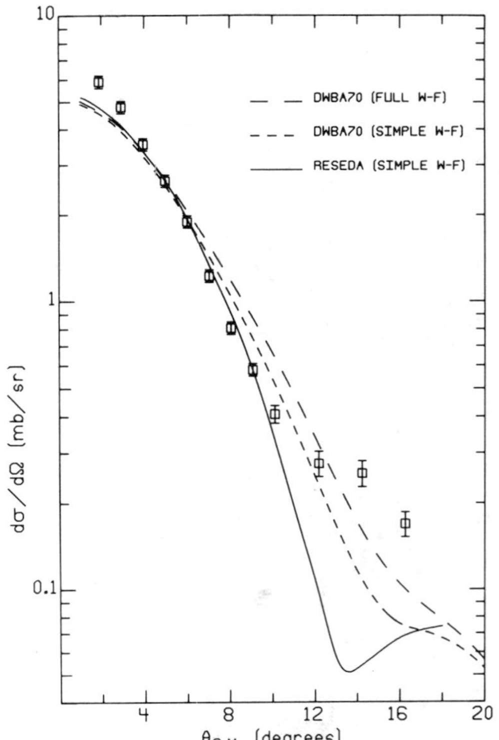

Fig. 2 - *Angular distribution for ^{48}Ca (p,p') to the 10.2 MeV state. The points are the measured values and the curves are from calculations described in the text.*

excitations [2,3]. This paper will focus on the results obtained with inelastic proton scattering at 201 and 319 MeV. The transitions of interest (for an even-even target) are of the type $0^+\to 1^+$, which involves the transfer to the nucleus of one unit of spin angular momentum but of no orbital angular momentum [4]).

Some of the reasons for the interest in studying 1^+ states may be understood by considering a spin-unsaturated nucleus (one with an occupied $j_> = \ell + 1/2$ orbit and an empty $j_< = \ell - 1/2$ orbit). In such a nucleus, the 1^+ state is a simple mode of excitation in which a nucleon from the $j_>$ orbit is excited to the $j_<$ orbit, the particle and hole then coupling to 1^+. An example is ^{48}Ca, the ground state of which consists to a good approximation of 8 neutrons in the $1f_{7/2}$ orbit. The 1^+ state is formed by exciting one of the neutrons to the spin-orbit partner level $1f_{5/2}$. Spin pairings (ground-state correlations) in ^{48}Ca leads to partial occupancy of the $f_{5/2}$ level in the ground state, which turns out to have the effect of reducing the $0^+\to 1^+$ strength. The energy of the 1^+ state depends on the spin-orbit splitting of the $f_{7/2}$ and $f_{5/2}$ orbits plus some residual interaction. Thus the <u>location</u> of the state measures the effective spin-spin interaction [5]), while the <u>strength</u> with which the state is excited measures the degree of ground-state correlations [6]). It has also been proposed [7]) that the strength reflects the admixture of Δ-isobar components in the low-lying 1^+ wave function; this will be discussed in Sec. 3. It is this possibility, involving the coupling of nucleonic with non-nucleonic degrees of freedom in the nucleus, which accounts for a large part of the interest in spin excitations.

When we began the (p,p') work in 1981 at Orsay [8]), the initial motivation was to find the kinematic conditions under which $0^+\to 1^+$ transitions are strongly excited. These conditions were found to be a high bombarding energy (E_p = 200 MeV) and very forward angles (θ_{lab} = 2°-5°). Fig. 1, which shows ^{48}Ca (p,p') spectra [9]) at a number of angles, is a beautiful illustration of how the (p,p') reaction at small angles selectively excites 1^+ states. Fig. 2 shows the angular distribution of the 1^+ state at 10.2 MeV. The way in which such (p,p') data are analyzed and the strengths of the $0^+\to 1^+$ transitions obtained is presented in Sec. 2 by reference to our data on the Ca isotopes. Typically, only about one-third to one-fourth of the predicted strength is found, not only for the Ca isotopes but for several other nuclei in the mass range $40 \leq A \leq 140$ for which we could reasonably analyze the data using simple wave functions[10]). For the closely related Gamow-Teller transitions excited by the (p,n) reaction in a wide variety of nuclei spread over the periodic table, only about 60% of the expected strength is observed[2,3]).

Thus the quenching of the spin transfer strength to 1^+ states appears to be a general feature in nuclei. The elucidation of the mechanism(s) responsible for the quenching is perhaps the most important open problem in the field of spin excitations. The mechanisms proposed so far will be surveyed in Sec. 3. In Sec. 4, we shall see how a comparison of quenching in isoscalar and isovector channels in (p,p') throws some light on the problem. Additional light is shed by measurements of spin-flip probability as a function of excitation energy; such data [12]) have been obtained using the polarized proton beam at LAMPF and are discussed in Sec. 5.

2. THE Ca ISOTOPES

The isotopes of Ca are expected to show a wide range of behavior. In ^{48}Ca, the 1^+ state is predominantly a one-particle--one-hole (1p-1h) excitation between spin-orbit partner levels. Moreover, it is almost entirely a neutron excitation, so that good correspondence between (p,p') and (e,e') strengths should be expected. [In general, when both proton and neutron excitations contribute, there will not be such a corespondence, because only the spin term contributes to (p,p') at small momentum transfer, while both orbital and spin terms contribute to (e,e'). But in cases where only neutron excitations are involved, there is no orbital part to (e,e').]

In ^{40}Ca, on the other hand, there would be no 1^+ strength in the simplest model of a closed core; and any observed strength must be due to a correlated many-particle--many-hole ground state. Both neutrons and protons would contribute to such a correlated ground state. Thus no correspondence between (p,p') and (e,e') strengths should be expected.

For the Ca isotopes with masses between 40 and 48, we expect[13]) a combination of the above two types of excitation: core excitation as in the case of ^{40}Ca and valence neutron excitation as in the case of ^{48}Ca. Again, no strict correspondence between (p,p') and (e,e') strengths is to be expected.

With these considerations in mind, we have made a comparison study of ^{40}Ca, ^{42}Ca, ^{44}Ca and ^{48}Ca with the (p,p') probe at E_p = 201 MeV. The data were obtained with the highly developed detection system of the Orsay spectrometer magnet, which allowed us to take

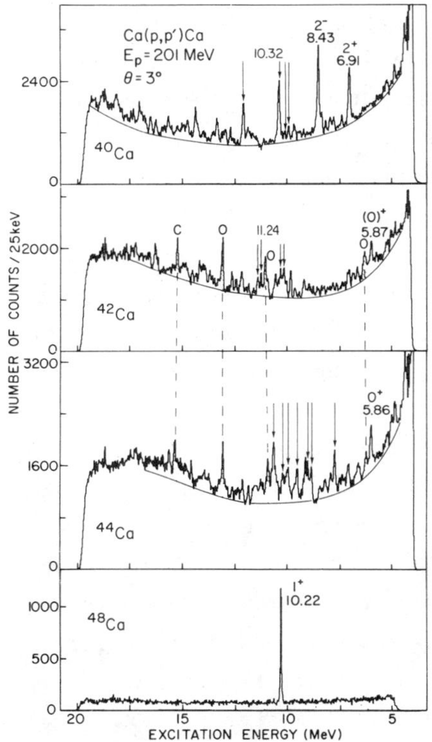

Fig. 3 - Spectra of protons inelastically scattered from ^{40}Ca, ^{42}Ca, ^{44}Ca and ^{48}Ca.

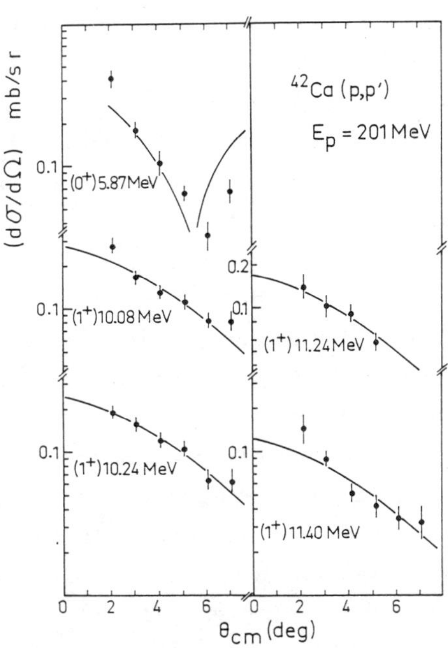

Fig. 4 - Angular distributions for ^{42}Ca(p,p') to a known 0^+ state at 5.87 MeV and to four possible 1^+ states.

spectra at angles as small as 2° with only moderate background from the elastic tail. Spectra measured at 3° are shown in Fig. 3. We have extracted the $0^+ \rightarrow 1^+$ cross sections in each isotope and compared them to the large-angle (e,e') data taken at Darmstadt[14]), where the 1^+ states were first observed, and to the predictions of microscopic distorted wave impulse approximation (DWIA) calculations using detailed shell-model wave functions[6]).

We start with ^{48}Ca. Referring back to Fig. 1, we note that the 1^+ state at 10.2 MeV is sharp, so that the cross section for exciting it can be obtained quite accurately, with little ambiguity in the background subtraction. The resulting angular distribution is shown in Fig. 2. It is very sharply forward peaked, as is characterisic of an orbital angular momentum transfer ΔL of zero at a bombarding energy of 201 MeV. ΔL=0 implies that $\Delta J^\pi = 0^+$ or 1^+. But the 0^+ possibility can be ruled out because empirically we have found for nuclei up to mass 60 that transitions to known 0^+ states have a more sharply falling angular distribution than transitions to 1^+ states. This is illustrated in Fig. 4, where the angular distributions for one known 0^+ level at 5.87 MeV and four possible 1^+ levels in ^{42}Ca are shown. These four levels have nearly the same angular distribution shape, which is the same as for the strong 10.2 MeV state in ^{48}Ca, and it is on this basis that they are identified as 1^+ candidates.

Microscopic DWIA calculations have been carried out for the 10.2 MeV state in ^{48}Ca using the codes DWBA70 and RESEDA. The results, normalized to the data, are shown as the curves in Fig. 2. Two of the curves were computed assuming a closed $\nu f_{7/2}$ shell for the ground state and a simple $\nu(f_{5/2} f_{7/2}^{-1})$ configuration for the 1^+ state (simple W-F); and one with the eight excess neutrons distributed in the full f-p shell[6]) (full W-F). Details of the calculation are given in Ref. 9. The main result is that the simple W-F gives a quenching factor Q, defined as the ratio of experimental to calculated cross section, of 0.21 with DWBA70 and 0.24 with RESEDA. The full W-F gives Q=0.30; this is not an unexpected result because typical calculations indicate that about 20% of the 1^+ strength predicted by the independent particle model disappears due to the influence of ground-state correlations. This still leaves a large gap between theory and experiment.

Since we are dealing with an almost pure neutron excitation, quenching should appear in (p,p'), (p,n)[15]) and (e,e')[14,16]) reactions in comparable proportions. Only the sharp peak at 10.2 MeV is observed consistently in the different reactions. The quenching factors for excitation of this state are given in Table 1. The (p,p') values are the lowest, but they are indeed comparable to the other two.

Table 1: Quenching factors for excitation of the 10.2 MeV state of ^{48}Ca.

	(p,p')	(p,n)	(e,e')
Closed-shell wave function	0.21	0.26	0.33
Full f-p shell wave function[a)]	0.30	0.35	0.47

a) Ref. 6

When we turn to the other Ca isotopes (Fig. 3), significant differences are observed between (p,p') and (e,e') results. In ^{40}Ca(p,p'), the well known 1^+ state at 10.32 MeV, first seen in (e,e'), is excited. However, an additional state seen in (p,p') at ≈12.06 MeV (indicated by an arrow in Fig. 3), which we believe is a 1^+ state, is not seen in (e,e'). Of course, if this state corresponds predominantly to a proton excitation, there might be a cancellation between the orbital and spin parts of the B(M1) operator which might suppress the transition in (e,e').

In ^{42}Ca, the 1^+ state at 11.24 MeV that is so prominent in (e,e') is seen only weakly in (p,p'). Whereas in (e,e') this state has a strength about 16% of that of the 10.2 MeV state in ^{48}Ca, the corresponding ratio in (p,p') is only about 3.3%. Moreover, there are three other possible 1^+ states excited in (p,p') to a comparable extent. One must again invoke a proton excitation component for the transitions and appeal to an orbital contribution in (e,e') to explain the lack of detailed agreement with (p,p').

The spectrum of ^{44}Ca is quite a contrast to the ^{48}Ca spectrum (Fig. 3) in that there is no dominating peak. The 1^+ strength distribution is very fragmented. But the (p,p') probe is still selective of 1^+ states. The selectivity is good enough to produce several peaks (indicated by arrows) which are relatively prominent, more so than in the (e,e') experiment.

Fig. 5 compares the observed and predicted 1^+ strength distributions for the Ca isotopes. The predictions are the results of microscopic DWIA calculations using the full f-p shell basis neutron wave functions of Ref. 6 for ^{42}Ca, ^{44}Ca and ^{48}Ca and the wave functions of B.A. Brown[17]) for ^{40}Ca. In Brown's model, the core is

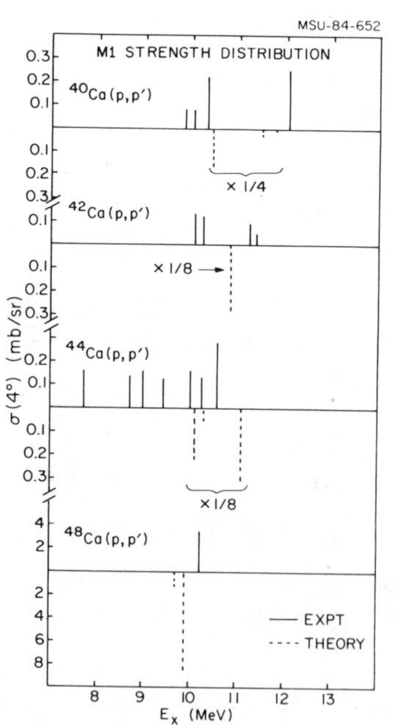

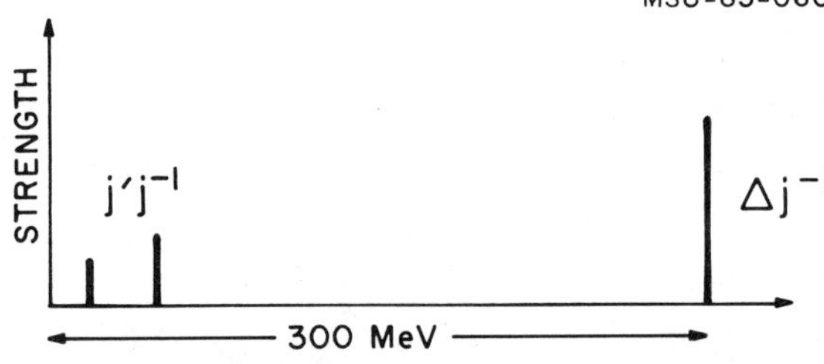

Fig. 6 - The independent particle spin-isospin strength function, including the delta state of the nucleon.

Fig. 5 - Experimental 1^+ strength distributions in the Ca(p,p') reactions compared to calculations described in the text.

^{32}S and the valence nucleons are in the $d_{3/2}$ and $f_{7/2}$ orbits. Except for ^{48}Ca, the correspondence between the observed and predicted excitation energies is poor. The ratio of the total observed strength to the total predicted strength for each nucleus, i.e. the overall quenching factor Q, is 0.77 for ^{40}Ca, 0.18 for ^{42}Ca, 0.24 for ^{44}Ca and 0.30 for ^{48}Ca. We disregard the value for ^{40}Ca on the grounds that Brown's model does not include $f_{7/2} \rightarrow f_{5/2}$ transitions, which are important. The quenching factors for the other Ca isotopes are in line with what we have come to expect from our (p,p') data over a wide range of nuclei[10]).

3. QUENCHING OF ISOVECTOR SPIN EXCITATIONS

The quenching factors for $0^+ \rightarrow 1^+$ transitions obtained from (p,p') data depend upon model wave functions used in distorted wave calculations to make predictions of the expected strength. There are uncertainties both in these model wave functions and to a lesser extent in the reaction calculations because of uncertainties in the interaction used. There is also the possibility that some of the 1^+ strength resides in the typical "background" region and is therefore not identifiable.

All the above arguments, however, lose their force in the case of ^{48}Ca. There is little uncertainty about the wave function of the 1^+ state in ^{48}Ca, and the excitation is very concentrated in energy. This is therefore an ideal case to show that there really is less cross section than predicted. Such quenching is predominantly isovector quenching because (p,p') transitions at 200 MeV in general are predominantly isovector transitions ($V_{\sigma\tau} >> V_{\sigma}$).

Even more compelling evidence for quenching of isovector 1^+ strength comes from (p,n) studies of Gamow-Teller (GT) transitions. Unlike (p,p'), (p,n) reactions have a rigorous lower limit on the total GT strength which derives from a sum rule based on fairly general principles. The total observed GT strength is only about 60% of that expected[2,3]). Nuclear spins appear to be quenched in other phenomena as well. The GT β-decay rate of the heavier mirror nuclei is reduced by about a factor of two[18]).

A number of explanations have been proposed to account for this quenching. One suggestion[19]) is that ordinary nuclear structure effects, such as configuration mixing between 1p-1h and 2p-2h states, might spread out the 1^+ strength in the low excitation energy region over several tens of MeV. This strength would be difficult to detect in cross section measurements since no peak structure would be apparent. A more exotic suggestion[7]) is that Δ-isobar admixtures enter into the nuclear wave functions in first order. This would have the effect of shifting part of the 1^+ strength to much higher excitation energy (≈300 MeV, see Fig. 6). Since all the nucleons can participate in the Δ excitation, the effect might be substantial. At present, there is much debate as to the relative importance of the two effects and whether both can explain the observed quenching.

4. QUENCHING IN ISOSCALAR $0^+ \rightarrow 1^+$ TRANSITIONS: ^{28}Si(p,p')

Since the isobar-to-nucleon coupling is isovector, the Δ-isobar mechanism is blocked from playing a significant role in isoscalar processes. Any quenching of isoscalar strength must thus be taken as strongly indicative of an important role for higher-order configuration mixing, not only in isoscalar processes but perhaps in isovector as well.

The probe of choice for exciting isoscalar transitions is (p,p'), since the alternative, (e,e'), is very much dominated by the isovector component of the M1 operator. One needs a T=0 target, so that transitions to T=0 excited states will be purely isoscalar and transitions to T=1 excited states will be purely isovector. Other considerations governing the choice of target are: (a) the isoscalar 1^+ states must be well separated from isovector 1^+ states, so that isospin mixing is minimized; and (b) good wave functions must be available for analyzing the data. The nucleus ^{28}Si satisfies all these criteria.

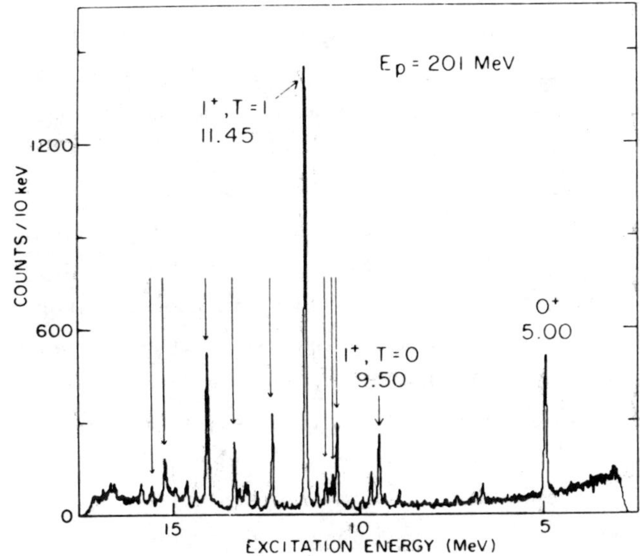

Fig. 7- Spectrum of protons inelastically scattered from ^{28}Si at 3.2°.

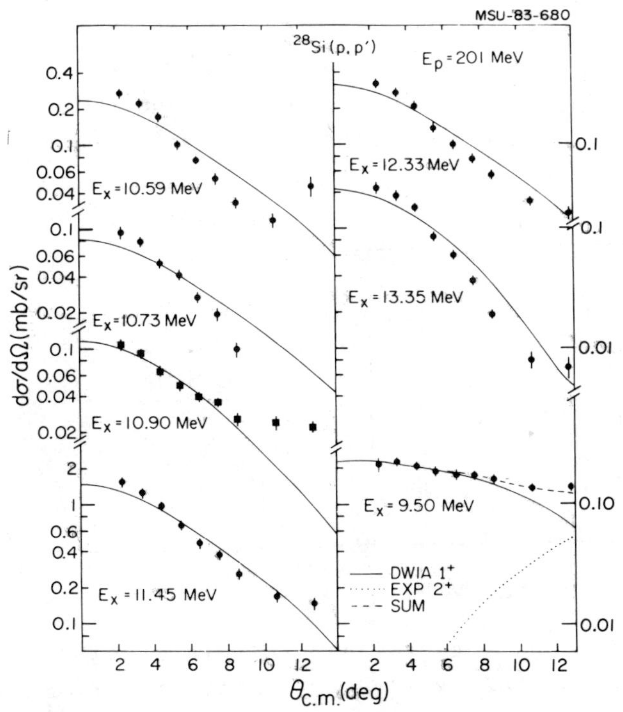

Fig. 8- Angular distributions for seven 1⁺ states, one (at E_x = 9.50 MeV) with T = 0 and the others with T = 1.

A spectrum[11]) of ^{28}Si(p,p') is shown in Fig. 7. The peaks with arrows above them are T=1, 1^+ states; the peak at 9.50 MeV is a T=0, 1^+ state. Angular distributions for six of the nine observed T=1 states and for the T=0 state are shown in Fig. 8. The curves were computed with the code DWBA70 in the usual manner, with wave functions in which the twelve nucleons outside of a ^{16}O core were unrestricted in the s-d shell. Hence, a good deal of configuration mixing is already included in the theoretical cross sections. The angular distribution for the T=0 state (E_x=9.50 MeV) is much flatter than the angular distributions observed for the T=1 1^+ transitions. The DWIA calculations reproduce both the T=0 and T=1 data. The difference in the shapes can be understood qualitatively as being due to the strong, attractive $V_{\sigma\tau}$ interaction present only in the T=1 channel. The good fit obtained for the angular distribution of the 9.50 MeV state is an indication of the pure isoscalar nature of this state.

All of the calculated curves have been normalized downward in order to fit the data. The experimental and theoretical strength distributions, specifically the differential cross sections at 4°, are given in Fig. 9. It is clear that for both isoscalar and

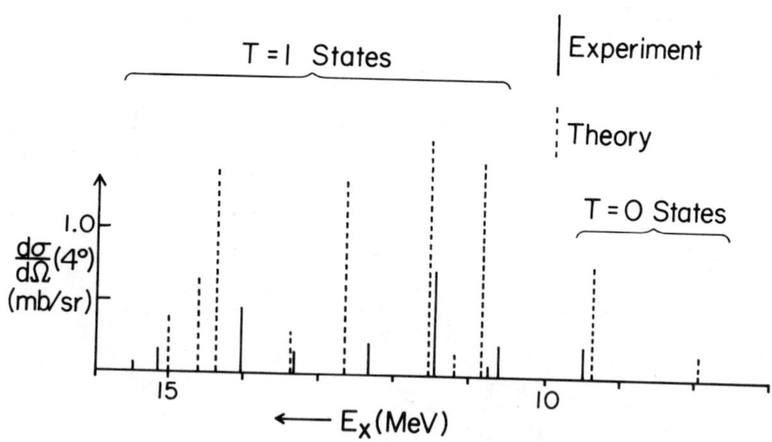

Fig. 9 - *Experimental and theoretical strength distributions for 1^+ states in ^{28}Si excited by (p,p') at 201 MeV.*

isovector states the experimental strength is much less than the theoretical. The ratios, that is, the quenching factors, are 0.24 for the isoscalar channel and 0.33 for the isovector channel. While the actual magnitude of this quenching is dependent on the details of the calculations we have used to extract it, the relative amounts of quenching in the two channels should be less sensitive to these details. The comparable quenching in the two channels indicates that the Δ-isobar admixture mechanism alone is not sufficient to explain the quenching. Our result, while not ruling out this mechanism in isovector transitions, points to the importance of higher-order configuration mixing as a quenching mechanism.

5. POLARIZATION TRANSFER MEASUREMENTS AT E_p=319 MeV

The polarization of the outgoing proton has been measured at LAMPF for small-angle inelastic scattering from ^{90}Zr and ^{51}V with a 319-MeV incident polarized proton beam [12,20]. The spin-flip cross section σS_{nn} (where S_{nn} is the probability that an incoming proton with spin up relative to the scattering plane goes out with spin down after the scattering) is a direct measure of spin excitations in the nucleus. In the plane wave limit, $S_{nn}=0$ for $\Delta S=0$ transitions and $=2/3$ for $\Delta S=1$ transitions. Thus S_{nn} provides a unique signature for spin-transfer reactions.

The results for ^{90}Zr (Fig. 10) show a large σS_{nn} which is approximately evenly distributed between 8 and 25 MeV excitation.

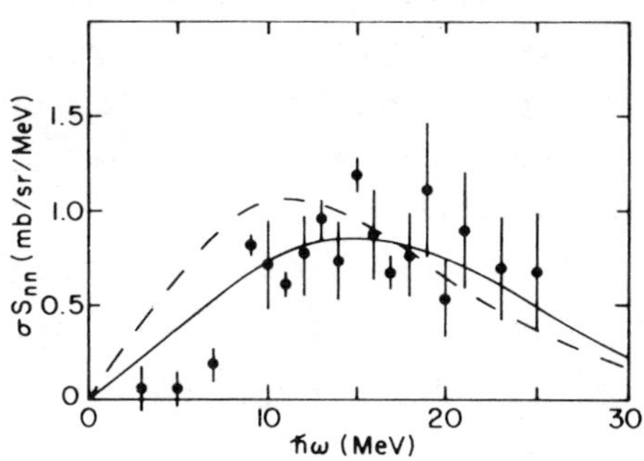

Fig. 10 - *Spin-flip cross sections for $^{90}Zr(\vec{p},\vec{p}')$ at E_p = 319 MeV and θ = 3.5°. The solid (dashed) line is the prediction of Ref. 21 with (without) collective effects.*

While the σ spectrum shows a definite bump at E_x=8-9 MeV (the 1^+ resonance), this region is not prominent in the σS_{nn} spectrum. The observation of large spin-flip strength in the continuum above the 1^+ resonance suggests that the missing strength might well be found here, rather than in the region near 300 MeV excitation. This would be in accord with the predictions of the configuration-mixing model. Of course, while the small-angle σS_{nn} spectrum arises from low-multipole spin excitations, it is probable that not all of it corresponds to 1^+ excitation. A calculation which avoids the problem of a multipole decomposition has been carried out recently by Esbensen and Bertsch[21]). This is a parameter-free calculation of the response of a semi-infinite Fermi liquid to a spin-isospin dependent probe and gives an extremely good qualitative fit to the σS_{nn} data (Fig. 10). The model appears rough but, taken at face value, the comparison of theory and experiment indicates that there is little missing overall ΔS=1 strength and thus no room for strong effects of the Δ-isobar.

6. CONCLUDING THOUGHTS

The (p,p') reaction at 200 MeV is a good probe for exciting ΔL=0, ΔS=1 transitions. The (p,p') work has brought forth much interesting data and some open problems as well. The data do not always agree with (e,e') data, both as regards the transition strength and the number of 1^+ states excited. The difference may be due to orbital magnetic contribution in (e,e'). The quenching of strength is a general feature; typically, the observed (p,p') strength is only 25% to 30% of that predicted. Our finding of similar quenching in isoscalar and isovector transitions in ^{28}Si indicates the importance of configuration mixing. It is important not to oversimplify the problem by evaluating one or two effects independently of each other or of others, but to consider quantitatively all the related effects. Ground-state correlations, np-nh contributions, Δ-isobar admixture effects, effective nucleon-nucleon interaction, and non-one-step processes must all be included in a comprehensive treatment.

ACKNOWLEDGEMENTS

This work was partly supported by the U.S. National Science Foundation under grants PHY-80-17605, INT-81-16064 and INT-82-13242. The 201-MeV (p,p') work was carried out at Orsay in collaboration with C. Djalali, N. Marty, M. Morlet, A. Willis and J.C. Jourdain of the IPN, Orsay and G.M. Crawley and A. Galonsky of the NSCL at Michigan State University. However, any errors or lack of clarity in the present text are my own doing.

REFERENCES

1. Marty, N., Invited paper in these Proceedings.
2. "Spin Excitations in Nuclei", edited by Petrovich, F. et al. (Plenum Press, New York, 1984).
3. Proc. Intl. Symp. Highly Excited States and Nuclear Structure (Journal de Physique, Colloque C4, 1984).
4. Strictly, both ΔL=0 and ΔL=2 contribute, but at the very forward angles where the data are taken, the ΔL=2 part is negligible.
5. Bertsch, G.F. Nucl. Phys A354, 157c (1981).
6. McGrory, J.B. and Wildenthal, B.H., Phys. Lett. 103B, 173 (1981).
7. Ericson, M. et al., Phys. Lett. 45B, 19 (1973); Oset, E. and Rho, M., Phys. Rev. Lett. 42, 47 (1979); Bohr, A. and Mottelson, B.R., Phys. Lett. 100B, 10 (1981); Weise, W., Nucl. Phys. A396, 373c (1983).
8. Anantaraman, N. et al., Phys. Rev. Lett. 46, 1318 (1981).
9. Crawley, G.M. et al., Phys. Lett. 127B, 322 (1983).
10. Djalali, C. et al., Nucl. Phys. A388, 1 (1982).
11. Anantaraman, N. et al., Phys. Rev. Lett. 52, 1409 (1984).
12. Nanda, S.K. et al., Phys. Rev. Lett. 51, 1526 (1983).
13. Brown, B.A. et al., Phys. Lett. 127B, 151 (1983).

14. Steffen, W. et al., Phys. Lett. 95B, 23 (1980).
15. Osterfeld, F. et al., Phys. Rev. Lett. 49, 11 (1982).
16. Eulenberg, G. et al., Phys. Lett. 116B, 113 (1982) and private communications from W. Steffen and A. Richter.
17. Reported in Rapaport, J. et al., Nucl. Phys. A427, 332 (1984).
18. Raman, S. et al, Atomic Data and Nuclear Data Tables 21, 567 (1978).
19. Shimizu, K. et al., Nucl. Phys. A226, 282 (1974);
Bertsch, G.F. and Hamamoto, I., Phys Rev. C26, 1323 (1982).
20. Glashausser, C., preprint prepared for Comments on Nuclear and Particle Physics (1984).
21. Esbensen, H. and Bertsch, G.F., Ann. Physics 157, 255 (1984).

SPIN EXCITATIONS IN NUCLEI AND NUCLEAR STRUCTURE STUDIES WITH COOLER STORAGE RINGS

G.P.A. Berg
Institut für Kernphysik, Kernforschungsanlage Jülich
D-5170 Jülich, F.R. Germany

ABSTRACT

High resolution experiments of nuclear spin excitations will be presented with emphasis on the energy dependence and selectivity of different hadronic probes. These allow to define requirements for future experimental studies of spin excitations which are met by appropriately designed cooler storage rings, such as the Cooler-Synchrotron COSY, a project at the Institute of Nuclear Physics of the KFA Jülich.

1. INTRODUCTION

The title of this contribution might be misleading in the sense that indeed there are no cooler storage rings in operation where nuclear structure studies can be performed with the exception of the LEAR ring at CERN which is exclusively dedicated to antiproton studies[1]. But there are cooler storage rings under construction at IUCF[2], in Uppsala[3] and at INS Tokyo[4] which will make use of phase space cooling and other advantages of storage rings for nuclear physics experiments. Several other cooler storage rings are planned[5-7] and in the last part of this contribution I will report on the cooler synchrotron COSY[8,9] planned at the Institute of Nuclear Physics at the KFA Jülich.

At first the experimental situation of high resolution studies of spin excitations in nuclei will be discussed with emphasis on the energy dependence, the selectivity of different hadronic probes and some aspects of charge transfer reactions. Then it will be discussed how cooler storage rings can be helpful in future experimental investigations of nuclear spin excitations and nuclear structure.

2. PROBES TO STUDY NUCLEAR SPIN EXCITATIONS

The motivation of the experimental and theoretical investigations of nuclear spin excitations is given by the possibility to obtain information on the spin and the spin-isospin dependent parts of the residual particle hole interaction which can be given in the zero range Landau-Migdal parametrization[10]:

$$F_{ph}(\vec{r}_1-\vec{r}_2) = C_o(g_o\vec{\sigma}_1\cdot\vec{\sigma}_2+g_o'\ \vec{\sigma}_1\cdot\vec{\sigma}_2\ \vec{\tau}_1\cdot\vec{\tau}_2)\delta(\vec{r}_1-\vec{r}_2) \qquad (1)$$

with C_o = 302 m/m* (MeV fm^3)

The strength parameter g_o' is related to the observed quenching of the total M1- and Gamow Teller (GT) strength which is discussed in the contribution to this conference by Anantaraman[11]. Spin excitations in nuclei have been very actively investigated by many groups using highly selective reactions: charge exchange[12] by (p,n) and recently[13] (^{3}He,t), inelastic (e,e') scattering[14], nuclear resonance fluorescence[15] ($\vec{\gamma},\vec{\gamma}$) and inelastic hadron scattering[11,16-18]. Fig. 1 shows schematically the isospin selection rules for different

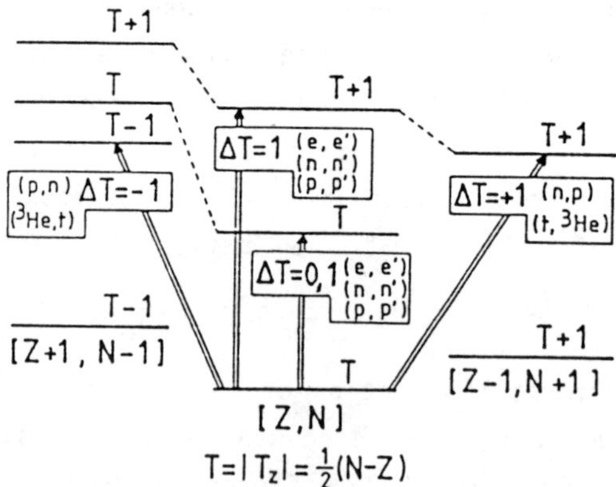

Fig. 1: Isospin selection rules in isobaric nuclei for some reactions

reactions in three isobaric nuclei. Because of experimental difficulties the (n,p) and (t,^{3}He) reaction has not yet been exploited extensively for spin excitation studies. The resolution in (n,p) reactions is typically 1 MeV and high energy tritons (⩾ 30 MeV) are not available. In table 1 the spin-isospin selection rules for inelastic hadron scattering are summarized showing that by measurements with different inelastic probes the isoscalar and isovector, spin-flip and non-spin flip nuclear modes can be excited selectively. Using polarized projectiles (e.g. polarized protons[19]) the spin change ΔS can be determined unambiguously. Calculations[20] of the energy dependence of the volume integral of nucleon-nucleon interactions show that the isoscalar spin independent central part V_0 is small compared to the spin-isospin interaction $V_{\sigma\tau}$ in the energy range from about 150 - 400 MeV.

Table 1: Allowed inelastic scattering transitions

projectile	isospin: ΔT	spin: ΔS
p, d, ^{3}He, α	0	0 (isoscalar, non-spin flip)
p, d, ^{3}He	0	1 (isoscalar, spin flip)
p, ^{3}He	1	0 (isovector, non-spin flip)
p, ^{3}He	1	1 (isovector, spin flip)

3) ENERGY DEPENDENCE OF SPIN EXCITATIONS

The energy dependence of nuclear spin excitations has been studied most extensively for the 1^+ state at 10.2 MeV in ^{48}Ca(p,p'). In fig. 2 measured angular distributions[19,21] at several energies are depicted as function of the transfered momentum. This figure shows very clearly the enhancement of the high energy data at forward angles, a result of the enhancement of the $\Delta\ell=0$ as compared to the $\Delta\ell=2$ contribution at higher energies which can be described by microscopic DWBA calculations[21]. Both contributions can populate 1^+ states if a spin flip $\Delta S=1$ takes place. For large momentum transfer the data points for all energies follow the same characteristics.

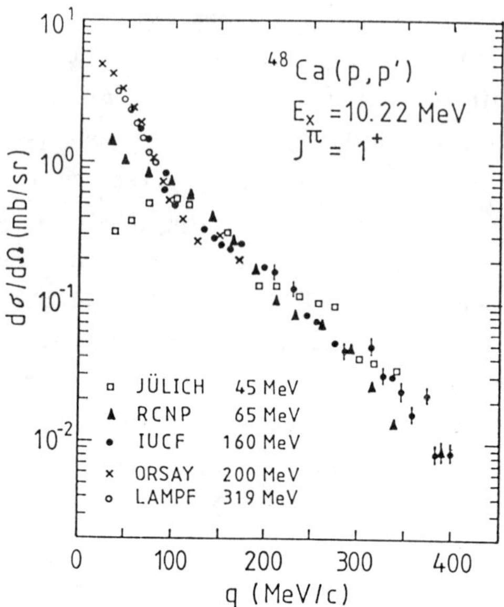

Fig. 2: Angular distributions of the strongly excited 1^+ state in ^{48}Ca for various incident proton energies plotted as function of the transfered momentum q.

The same behaviour is seen[22] for the 1^+ state in ^{58}Ni at 10.66 MeV by comparing 45, 65 and 201 MeV measurements. The 1^+ state at 3.594 MeV in ^{58}Ni(p,p') shows a peculiar energy dependence. At E_p = 25 MeV it is clearly excited as shown in fig. 3 with a maximum cross section in the angular distribution of about 20 µb/sr. The lower part of fig. 3 shows a (d,d') spectrum which will be discussed later. At E_p = 45 MeV the cross section drops to about 10 µb/sr but the peaks[22] are still visible (Fig. 4) while the 65 MeV data show no indication[23] of the population of this state, with an upper limit of a few µb/sr. Inelastic proton scattering on ^{58}Ni at 200 MeV measured at Orsay[24] shows a broad 1^+ structure in the excitation energy range E_x = 7 - 11 MeV interpreted as T = 0 and T = 1 components.

Figs. 3 and 4 show ^{58}Ni(p,p') sample spectra[22] measured with the high resolution spectrometer[25] BIG KARL. It can be seen that individual 1^+ states can be isolated with the resolution being about 10 keV at E_x=2.901, 3.594, 6.715 and 10.662 MeV. Fig. 5 shows the angular dis-

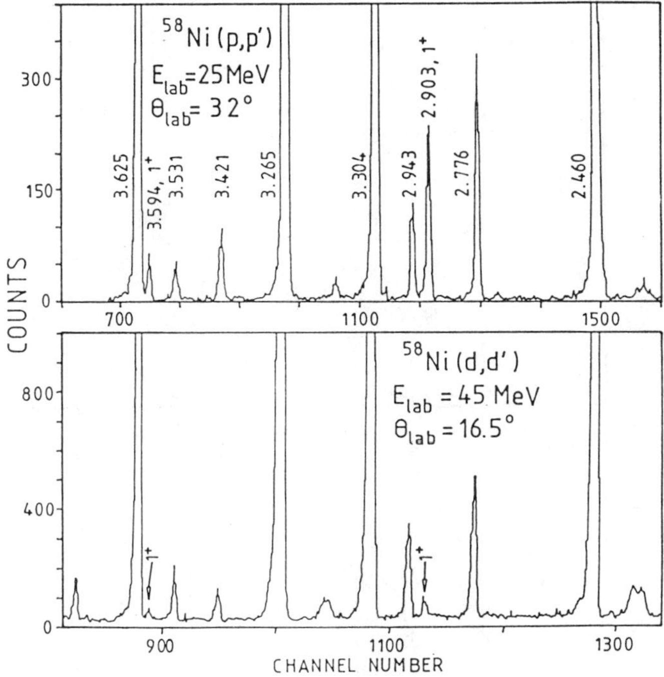

Fig. 3: Sample spectra of (p,p') and (d,d') leading to two low lying 1^+ states in ^{58}Ni.

tributions with microscopic DWBA calculations except for the 1^+ state at 3.594 MeV which shows an anomalous energy dependence. Calculations of the ^{48}Ca(p,p') scattering[17)] leading to the 1^+ state at 10.2 MeV show differences in the differential cross sections of the order of up to 30 % when the Δ_{33} resonance excitation is included. The existing data[17)] do not allow a decision on the importance of this effect.

These examples show that investigations of strongly excited individual 1^+ states at excitation energies up to about 10 MeV with resolutions of $\leqslant$ 10 keV are experimentally possible up to about 65 MeV incident energies with existing facilities. But the measurement of weakly excited 1^+ states necessitates higher incident energies in order to take advantage of the mentioned enhancement at foreward angles. At energies $\geqslant$ 160 MeV the presently available resolution of 50-100 keV is often not sufficient to resolve individual 1^+ states.

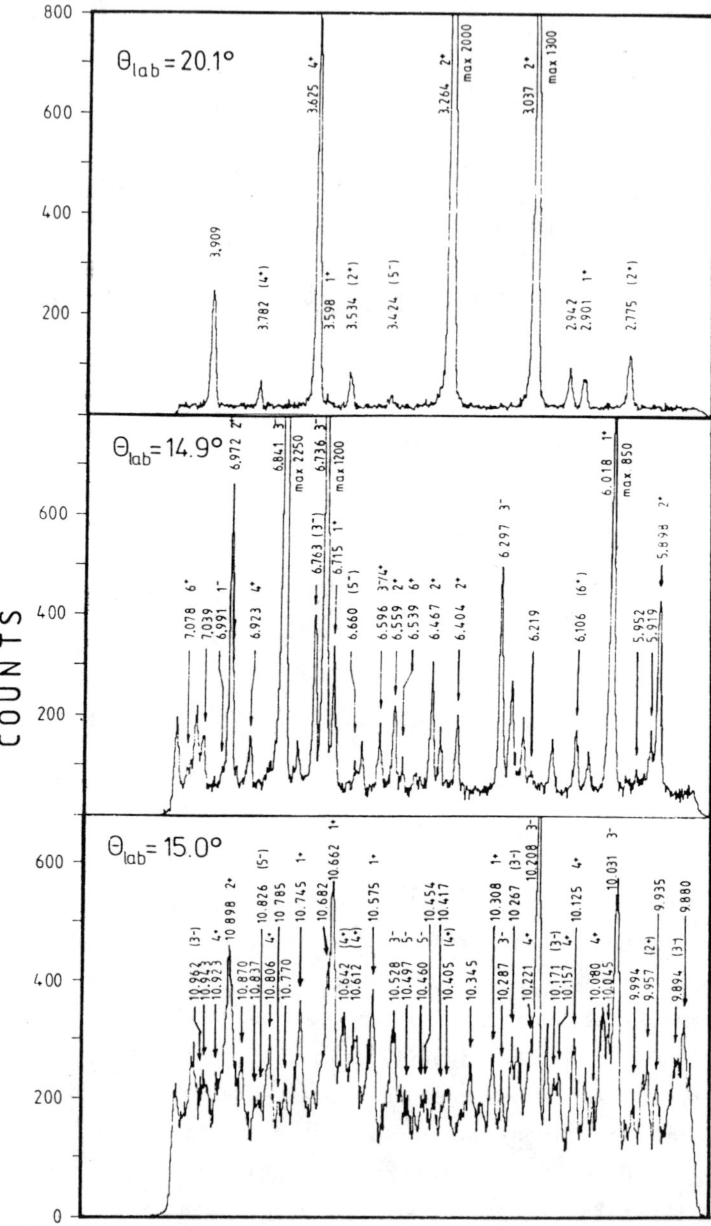

Fig. 4: Sample spectra of $^{58}Ni(p,p')$ at 45 MeV incident energy showing the excitation of several 1^+ states

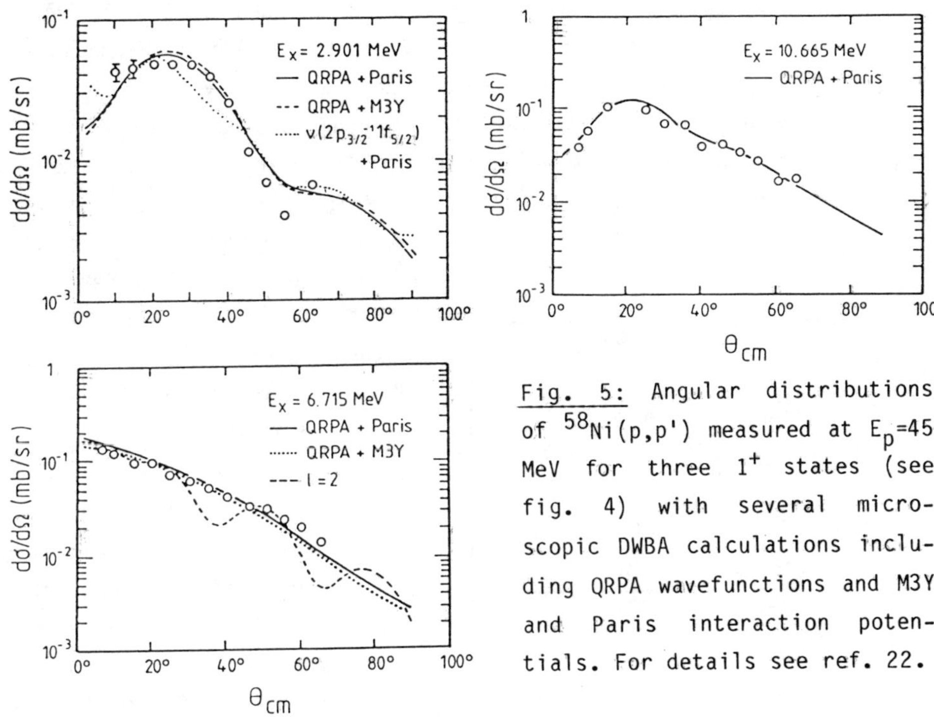

Fig. 5: Angular distributions of ^{58}Ni(p,p') measured at E_p=45 MeV for three 1^+ states (see fig. 4) with several microscopic DWBA calculations including QRPA wavefunctions and M3Y and Paris interaction potentials. For details see ref. 22.

4. SELECTIVITY OF DIFFERENT PROBES

The selectivity of different nuclear probes can be used to identify different types of spin excitation modes and nuclear structure properties. A comparison[26] of the (p,p') and (α,α') scattering on ^{48}Ca leading to the 1^+ state at 10.2 MeV measured with the Raiden spectrometer shows clearly the spin flip character of this excitation. As expected there is no population of this state in (α,α') except for near lying weakly excited states. The (p,p') angular distributions[27] of several nearby states are different from the 1^+ state and are therefore most likely non-spin flip states.

A comparison of (p,p') and (d,d') measurements[18] of the 1^+ state at 5.844 MeV in ^{208}Pb demonstrates its isoscalar character. The angular distributions of both reactions plotted as function of the transfered momentum show remarkable agreement.

In ^{28}Si at excitation energies above E_x = 8 MeV several T = 0, T = 1 and T = 0+1 natural and unnatural parity states are known. We measured at 25 MeV (p,p') and at 45 MeV (d,d') angular distributions in this excitation energy range. High resolution proton spectra show clearly the excitation of the following states: 1^+, T = 0 at 8.328 MeV; 3^+, T = 1 at 9.316 MeV; 2^+, T = 1 at 9.381 MeV; $(1,2)^+$, T = 0 at 9.497 MeV; 1^+, T = 0+1 at 10.597 MeV; 1^+, T = 0+1 at 10.725 MeV; and 1^+, T = 1 at 11.446 MeV. The measured deuteron spectra show no excitation of the T=1 state and reduced excitation of the other states. Fig. 6 shows a comparison of spectra for two of the mentioned states at $E_x \sim 9.4$ MeV. Fig. 3 shows the comparison of (p,p') and (d,d') measurements of the low lying 1^+ states at 2.903 and 3.594 MeV. Microscopic (p,p') and (d,d') DWBA calculations[28] are in progress in order to obtain information on the isocalar or isovector character of these states and in order to test the usefulness of this method.

These examples show that inelastic hadron scattering can provide important nuclear structure information in spin excitation studies.

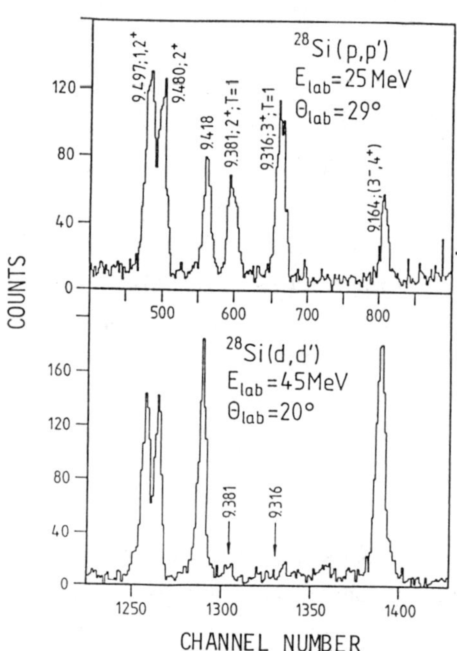

Fig. 6: Comparison of (p,p') and (d,d') spectra of several T=0 and and T=1 states in ^{28}Si clearly showing the suppression of T=1 states in (d,d').

5. CHARGE TRANSFER REACTIONS

The (p,n) charge transfer reaction[12)] has been used quite extensively employing beam swinger systems because of the remarkable selectivity for spin-isospin excitations. It has been shown[29)] that also the (^{3}He,t) reaction shows the same selectivity.

We utilized the ^{3}He beam of 135 MeV of the JULIC cyclotron and the spectrometer BIG KARL to measure the (^{3}He,t) reaction[30,31)] on ^{208}Pb, ^{120}Sn and ^{90}Zr at very forward angles including the 0° measurement which are feasible because of the high magnetic rigidity of the tritons. The ^{90}Zr(^{3}He,t) spectrum at 7° in fig. 7 shows the strong excitation of the IAS at 5.5 MeV, the broad GT resonance at around 8.3 MeV, and the two low lying unnatural parity states: the 1^+ state at 2.1 MeV and the 3^+ level at = 0.65 MeV. A review of the experimental situation of the (n,p) reaction with tagged neutrons was given at the IUCF cooler workshop by Bowman[32)]. The resolution is about 1 MeV which is in general insufficient to separate individual levels. Because of radiation safety problems there are only a few laboratories where triton beams are available. The available energies below 30 MeV are not high enough to populate effectively spin excitations in the (t,^{3}He) reaction. In a cooler storage ring the circulating 10^9 - 10^{11} particles pass typically 10^6 times/s through the thin target until the beam is consumed by nuclear reactions and backward elastic scattering. This allows a very efficient use of the tritons.

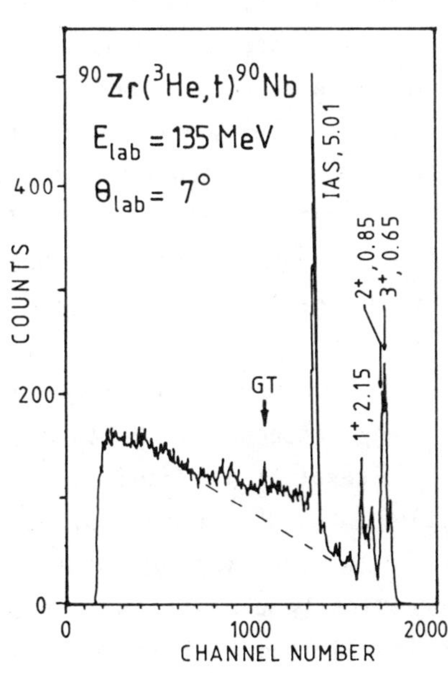

Fig.7: Spectrum showing the strong selectivity of (^{3}He,t) exciting unnatural parity states

6. THE COOLER SYNCHROTRON COSY

The discussion above suggests ideal requirements for the investigation of spin excitations in nuclei:

- beams of $\vec{p}$, $\vec{d}$, ^{3}He, t, α-particles and monoenergetic neutrons
- energy variability up to about 300 MeV
- high resolution ($E/\Delta E \geq 10^4$) to resolve individual states

Cooler storage rings as e.g. COSY[8,9] have the potentiality to meet these requirements.

6.1 General Aspects of Cooler Storage Rings

In conventional nuclear reaction measurements the beam from an accelerator impinges onto a target, undergoes with a certain probability a nuclear reaction but most of the particles interact only with electrons causing energy and angle straggling. For high resolution experiments[25] the targets have to be very thin, typically ≤50 μg/cm. In addition for limited beam quality the beam resolution can be improved by reducing the phase space ellipse e.g. in a monochromator system[25]. These requirements reduce the luminosity L (= target thickness x beam current) so that high resolution usually means small counting rates.

In a storage ring a number N of particles is injected into a magnetic system where the beam may circulate for many seconds. In the case of COSY a 200 MeV proton beam has a revolution frequency of $\nu = 1 \cdot 10^6$ Hz. Therefore the circulating beam $N \cdot \nu$ seen by the target can be of the order of several mA, e. g. for $N = 10^{10}$ about 1.7 mA. This increases the luminosity drastically even if the target thickness has to be reduced because of the limited cooling speed of the electron or stochastic cooling system[33].

Another advantage of stored beams is the possibility to manipulate the beam by slow processes which require more than about a μsec. Examples are:

- phase space cooling[33] by electron and stochastic cooling

- beam bunching and stretching
- single cavity RF acceleration and deceleration
- accumulation and efficient use of rare particles (e.g. tritons produced in a nuclear reaction)

6.2 Properties of the COSY-Ring

The planned COSY-project developed from discussions[9,34-36] of physicists of the nuclear physics institutes at the KFA Jülich and the universities of Bochum, Bonn, Eindhoven, Köln and Münster. Fig. 8

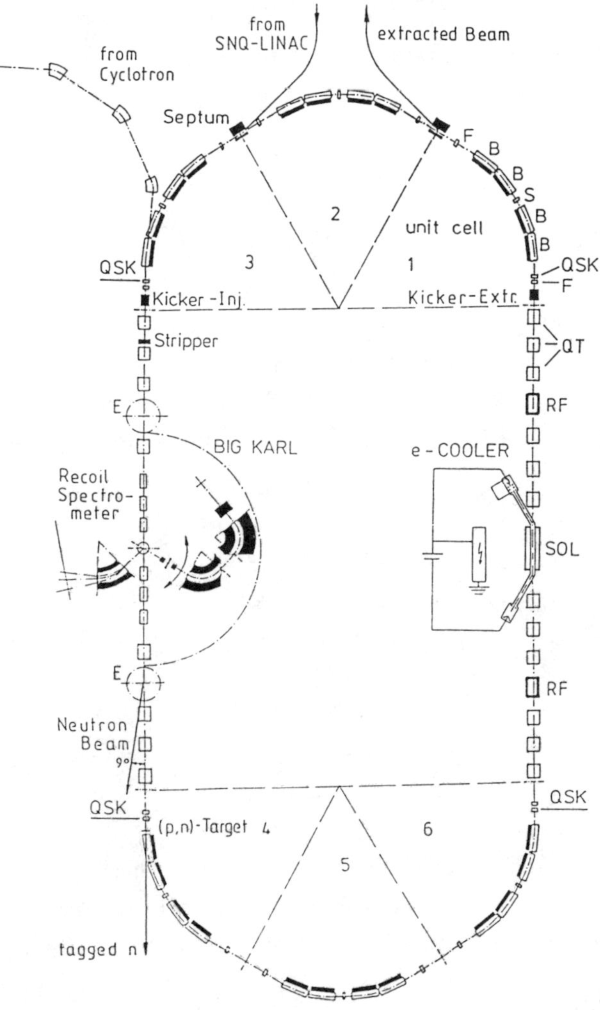

Fig. 8: Lay-out of the cooler-synchrotron COSY with the magnetic elements (B, F, S, QSK), the RF-cavities and two experimental areas E.

shows the lay-out of the storage ring, consisting of a magnet system with 6 unit cells and two long straight telescopic sections. The design parameters are listed in table 2.

Table 2: Parameters of COSY

6 unit cells of the structure Quadrupole Dipole Dipole Quadrupole	
Two double telescopes of 35 m length each and with magnification M=+1	
Circumference:	163.6 m
Bending radius	7.0 m
Dipole field for 500 MeV protons	0.5 T (Stage I)
Dipole field for 1.5 GeV protons (p=2.25 GeV/c)	1.1 T (Stage II)
Free length for cooling section:	4.5 m
Free length for BIG KARL:	2.6 m
Acceptance ($\Delta p=0$): $\varepsilon_x = 200\ \pi$ mm mrad	$\varepsilon_z = 30\ \pi$ mm mrad
Emittance cooled:	$\varepsilon \leqslant 0.15\ \pi$ mm mrad
Dispersion at the target:	variable: 1 - 20 m
Vacuum:	$\leqslant 10^{-9}$ mbar

The following values correspond to the dispersion $|D| = 10$ m at the target

β-function:	β_{target}:	$0.27\ m \leqslant \beta \leqslant 10.25$ m
Maxima in the unit-cell:	$\beta_{horizontal}$:	$\beta_x = 20.8$ m
	$\beta_{vertical}$:	$\beta_z = 13.3$ m
Q-values:		$Q_x = 2.75$
		$Q_z = 3.30$
γ_{tr}		1.002

In the first stage particles will be injected from the existing cyclotron JULIC which delivers p, d, ^{3}He and α-particles of E = 22.5 - 45 MeV/nucleon. An external injection with an ECR-source[37] system (ISIS project) is under construction and will provide also heavier

particles up to about ^{20}Ne. The external injection system allows also the later installation of a polarized ion source. For the storage of protons up to the space charge limit of the ring of about 10^{11} particles injection is planned from the LINAC of the spallation neutron source SNQ[38)] planned at the KFA Jülich.
High resolution of $E/\Delta E \sim 10^{-4}$ for 10^9 particles in the ring for proton energies up to about 500 MeV will be accomplished by an electron-cooling system[9)], where the circulating beam interacts via Coulomb scattering with an intense electron beam with the same velocity. After about one second cooling time the phase space ellipse of the beam is drastically reduced. In order to make profit of the high beam quality the high resolution spectrometer BIG KARL will be installed in one telescopic section of the ring which is designed to allow the matching[8,25)] between the ring and the spectrometer. The spectrometer allows the momentum analysis of protons up to 500 MeV with a tested resolution of $E/\Delta E \sim 2 \cdot 10^4$.

Besides the high resolution of $\Delta E \sim 10$ keV at proton energies up to 200 MeV there are the following possibilities which are interesting for future measurements of spin-isospin excitations:

- Vertically polarized beam in the ring are possible. The spin precession in the longitudinal magnetic field (SOL,~1 kG) of the cooler can be compensated.
- Production of tagged neutrons[39)] by internal ^{12}C or ^{7}Li targets via the (p,n) reaction leading to the strong GT peak at 0°.
- Storage and efficient use of tritons in the ring for (t,^{3}He) reactions using internal targets. The tritons can be produced[36)] via e.g. the nuclear reaction ^{7}Li(d,t).

The COSY magnet system is designed to allow also operation at higher energies up to about 1.5 GeV. Experimental areas (E) are provided in the ring to study subnuclear degrees of freedom by e.g. coherent meson production. Here we want to point out that if appropriately equipped

COSY and other storage rings can contribute to the investigation of spin excitations and the nuclear structure of the atomic nucleus.

The author gratefully acknowledges the collaboration of A. Magiera, J. Meissburger, B. Brinkmöller, D. Paul, D. Prasuhn, J.G.M. Römer, P. von Rossen and J.L. Tain in the experiments at BIG KARL.

References

1) Physics at LEAR with Cooled Low Energy Antiprotons, ed. Gastaldi, U., Klapisch, R.; Plenum Press (1984), Proceedings of the Workshop on Physics with LEAR, Erice, May 1982
2) Proceedings of the 4th annual IUCF Fall Workshop on "Nuclear Physics with Stored Cooled Beams", Bloomington, Indiana, Oct. 15-17, 1984
3) Kaivola, M., Nielsen, U., Poulsen, O., Workshop on the Physics Program at CELSIUS, Uppsala, November 1983, CELSIUS-Note 83-17, 1983, ed. Karlsson, B.R., Tibell, G.
4) Noda, A., Lattice design of TARN II Cooler Ring, ECOOL 84, Karlsruhe, Sept. 1984 and Tanabe, T., TARN II, Electron Cooling Project at INS, ECOOL 84, Karlsruhe, Sept. 1984
5) Franzke, B., Eickhoff, H., Franczak, B., Langenbeck, B., Zwischenbericht zur Planung des Experimentierspeicherrings (ESR) der GSI, GSI-SIS-INT/84-5, August 1984, GSI Darmstadt
6) Jaeschke,E.,"Study on a Heidelberg Heavy Ion Cooler Ring",Workshop on the Physics with Heavy Ion Cooler Rings,1984,MPI Heidelberg
7) Herlander, C.J., "The Stockholm Storage Ring Project", Workshop on the Physics with Heavy Ion Cooler Rings, May 1984, MPI Heidelberg, and Stensgaard,R.,"The Aarhus Storage Ring", private communication
8) Martin, S., "First ideas about a cooler-synchrotron (COSY) at Jülich", Workshop on Electron Cooling, Bad Honnef, May 1982, Jül-Spez-159, pp 2.1 - 2.12 (1982) and Berg, G.P.A., "COSY: A Cooler-Synchrotron for the KFA Jülich", Proceedings of the 4th annual IUCF Fall Workshop on "Nuclear Physics with Stored Cooled Beams", Bloomington, Indiana, Oct. 15 - 17, 1984
9) Berg, G., Gaul, G., Günther, Ch., Hacker, U., Hagedoorn, H.L., Hardt, A., van der Heide, J.A., Hinterberger, F., Huber, M., Inoue, M., Jahn, R., Kilian, K., Köhler, M., Martin, S., Mayer-Kuckuk, T., Meissburger, J., Osterfeld, F., Paetz gen. Schieck, H., Poth, H., Prasuhn, D., Riepe, G., Rogge, M., von Rossen, P., Schult, O.W.B., Speth, J., Turek, P. and Wagner, G.J., "COSY: Proposal for a Cooler-Synchrotron as a Nuclear Physics Facility at the KFA Jülich", IKP-KFA Jülich, January 1985
10) Migdal, A.B., "Theory of Finite Fermi Systems and Applications to

Atomic Nuclei", Wiley, New York, (1967) and Speth, J., Werner, E., Wild, W., Phys. Rep. 33, 127 (1977)

11) Anantaraman, N., "Inelastic Proton Scattering and the Quenching of ΔL=0 Spin Flip Excitations in Nuclei", contribution to this conference

12) Doering, R.R., Galonsky, A., Patterson, D., Bertsch, G.F., Phys. Rev. Lett. 35, 1961 (1975) and Horen, D.J., et al., Phys. Lett. 99B, 383 (1981)

13) Galonsky, A., Didelez, J.P., Djaloeis, A., Oelert, W., Phys. Lett. 74B, 176 (1978) and Gaarde, C., Physica Scripta, Vol. T5, 55-62 (1983)

14) Steffen, W., Gräf, H.D., Gross, W., Meuer, D., Richter, A., Spamer, E., Titze, O., Knüpfer, W., Phys. Lett. 95B, 23 (1980) and Lindgren, R.A., et al., Phys. Rev. C14, 1789 (1976)

15) Berg, U.E.P., Rück, D., Ackermann, K., Bangert, K., Bläsing, C., Kobras, K., Naatz, W., Schneider, R.K.M., Stock, R., Wienhard, K., Phys. Lett. 103B, 301 (1981)

16) Marty, N., "Giant Resonances - Some Old and New Results", contribution to this conference

17) Berg, G.P.A., Hürlimann, W., Katayama, I., Martin, S.A., Meissburger, J., Osterfeld, F., Römer, J.G.M., Styczen, B., Tain, J.L., Gaul, G., Santo, R., Sondermann, G., "Spin Flip 1^+ Excitations by Inelastic Hadron Scattering", Spin Excitations in Nuclei, ed. Petrovich, Brown, Garvey, Goodmann, Lindgren, and Love, Telluride 1982, Plenum Publ. Corp., 311-321 (1984)

18) Berg, G.P.A., Hürlimann, W., Katayama, I., Martin, S.A., Meissburger, J., Römer, J.G.M., Styczen, B., Tain, J.L., "Isoscalar Character of the 1^+ State at 5.844 MeV in ^{208}Pb", Proc. 1983 RCNP Int. Symp. Light Ion Reaction Mechanism, RCNP Osaka, May 1983, ed. Ogata, H., Kammuri, T., Katayama, I.

19) Nanda, S.K., et al., Phys. Rev. C29, 660 (1984)

20) Petrovich, F. and Love, W.G., Nucl. Phys. A354, 499c (1981)

21) Fujiwara, M., Imanishi, S., Fujita, Y., Morinobu, S., Katayama, I., Yamazaki, T., Itahoshi, T., Ikegami, H., Katori, K., Hayakawa, S.I., Proceed. 1983 RCNP Int. Symp. Light Ion Reaction Mechanism, RCNP Osaka, May 1983, ed. Ogata, H., Kammuri, T., Katayama, I.

22) Sondermann, G., "Spin-Flip Anregungen in ^{48}Ca and ^{58}Ni durch inelastische Protonenstreuung", Diss. Univ. Münster, Mai 1984

23) Fujiwara, M., Fujita, Y., Katayama, I., Morinobu, S., Yamazaki, T., Ikegami, H., Hayakawa, S.I., Katori, K., Nucl. Phys. A410, 137 (1983)

24) Djalali, C., Marty, N., Morlet, M., Willis, A., Jourdain, J.C., Anantaraman, N., Crawley, G.M., Galonsky, A., Kitching, P., Nucl. Phys. A388, 1 (1982)

25) Martin, S.A., Hardt, A., Meissburger, J., Berg, G.P.A., Hacker, U., Hürlimann, W., Römer, J.G.M., Sagefka, T., Retz, A., Schult,

O.W.B., Brown K.L., Halbach, K., Nucl.Instr.Meth. 214, 281 (1983)
26) Fujita, Y., Fujiwara, M., Morinobu, S., Yamazaki, T., Itahashi, S., Ikegami, H., Nayakawa, S.I., Phys. Rev. C25, 678 (1982)
27) Berg, G.P.A., Hürlimann, W., Katayama, I., Martin, S.A., Meissburger, J., Römer, J., Styczen, B., Osterfeld, F., Gaul, G., Santo, R., Sondermann, G., Phys. Rev. C25, 2100 (1982)
28) Lenske, H., Baur, G., Nucl. Phys. A344, 151 (1980)
29) Ellegaard, C., Gaarde, C., Larsen, J.S., Goodman, C., Bergquist, I., Carlen, L., Ekström, P., Jakobsson, B., Lyttkens, J., Bedjidian, M., Chamecham, M., Grossiord, J.Y., Guichard, A., Gusakow, M., Haroutinian, R., Pizzi, J.R., Bachelier, D.,Boyard, J.L., Hennino, T., Jourdain, J.C., Roy-Stephan, M., Boivin, M., Radvanyi, P., Phys. Rev. Lett. 50, 1745 (1983)
30) Morsch, H.P., Berg, G.P.A., Decowski, P., Tain, J.L., Rogge, M., Zemlo, L., Turek, P., Meissburger, J., Römer, J.G.M., "Study of the (^{3}He,t) Reaction on Heavy Nuclei", Proceed. 1983 RCNP Int. Symp. Light Ion Reaction Mechanism, RCNP Osaka, May 1983, ed. Ogata, H., Kammuri, T., Katayama, I.
31) Tain, J.L., Brinkmöller, B., Berg, G.P.A., Hlawatsch, G., Magiera, A., Meissburger, J., Oelert, W., Palla, G., Römer, J.G.M., Sondermann, G., Int. Symp. on Highly Excited States and Nuclear Structure, Orsay, Sept. 1983, ed. Marty, N., Van Giai, N.
32) Bowman, D., "Tagged neutrons", Proceedings of the 4th annual IUCF Fall Workshop on "Nuclear Physics with Stored Cooled Beams", Bloomington, Indiana, Oct. 15 - 17, 1984
33) Cole, F.T., Mills, F.E., "Increasing the phase-space density of high energy particle beams", Ann. Rev. Nucl. Part. Sci. 31, 295 - 335 (1981)
34) Berg, G.P.A., Hürlimann, W., Römer, J.G.M., "Workshop on Electron Cooling", Bad Honnef, May 1982, Jül-Spez-159, July 1982
35) Berg, G.P.A., Prasuhn, D., Römer, J.G.M., "COSY-Arbeitstreffen", Institut für Kernphysik, KFA Jülich, 26.-27. Oktober 1983, Jül-Spez-253, April 1984
36) Berg, G., Gaul, G., Hagedoorn, H., Hardt, A., van der Heide, J.A., Hinterberger, F., Huber, M., Jahn, R., Martin, S., Mayer-Kuckuk, T., Osterfeld, F., Paetz gen. Schieck, H., Prasuhn, D., Riepe, G., Rogge, M., von Rossen, P., Schult, O.W.B., Speth, J., Turek, P., "Studie zum Bau eines kombinierten Kühler-Synchrotron-Ringes an der KFA Jülich", Jül-Spez-242, February 1984
37) Mathews, H.-G., Beuscher, H., Krauss-Vogt, W., "ECR-Ion-Source Development at the Jülich Cyclotron", Proc. Int. Ion-Engineering Congress-ISIAT '83 & IPAT '83, Kyoto (1983)
38) SNQ Projektbericht zum Abschluß von Phase B, Spallations-Neutronenquelle, KFA Jülich, Februar 1984
39) Gaarde, C., Workshop on the Physics Program at CELSIUS, Uppsala, November 1983, CELSIUS-Note 83-17, ed. Karlsson, B.R., Tibell, G.

ON THE INTERPRETATION OF THE EMC EFFECT

A.W. Thomas
Department of Physics
University of Adelaide
ADELAIDE SOUTH AUSTRALIA 5001.

ABSTRACT

Since the discovery of a significant change in the structure function of a "nucleon" inside a nucleus there has been a great deal of interest in deep inelastic scattering within the nuclear community. We shall outline a number of proposals which have been made to account for this data. Special emphasis is given to the proposition that the virtual pion field of the nucleus may be enhanced, and to the idea of colour conductivity.

1. INTRODUCTION

Our modern belief in the quark model of hadron structure dates from the discovery of scaling at SLAC in the late 60's. Even though we have been unable so far to liberate quarks from infrared slavery, we have been able to observe them inside hadrons by a judicious choice of leptonic probes.[1)] We know by direct observation that a nucleon consists of three valence quarks (carrying about 36% of its momentum[*]), a sea of virtual quark-antiquark pairs (10% of its momentum), plus glue (carrying the remaining 54% of its momentum).[2,3)] In view of this success it is surprising how little effort has been made to use deep-inelastic scattering (DIS) to constrain models of hadronic - let alone nuclear - structure. The current popularity of Skyrmion-like models of nuclear structure is testimony enough to this.[4,5)]

However, since the revelations of the European Muon Collaboration just two years ago,[6)] this has begun to change. What

[*] All these numbers refer to measurements at $Q^2 = 5$ GeV^2.

they found, and this has since been confirmed by three independent experiments,[7] is that the structure function of a nucleon in Fe deviates by as much as 15% from that of a free nucleon. (This becomes 30% after the relatively reliable correction for fermi motion[8].) At present the number of theoretical papers almost outstrips the number of data points. However our task is made easier by the fact that the number of genuinely different ideas is quite small. In Sect. 2 we briefly outline these key ideas. Sections 3 and 4 are devoted to a more comprehensive discussion of two of the ideas on which our group has had something original to say. In Sect. 5 we make some concluding remarks.

2. OVERVIEW

Perhaps the best capsule summary of the EMC data is that given by Close and collaborators[9] (see also Nachtman and Pirner[10]). From the observation that the structure function in Fe at (x, Q^2) is approximately equal to that in d at $(x, 2Q^2)$ $(0.2 \leqslant x \leqslant 0.8)$ they concluded that there is a change in scale in the nuclear system. In itself this observation was not very satisfying - there was no suggestion by Close et al of how this change of scale might vary with atomic number, nor whether it necessarily involved colour.

Amongst the early models for this change of scale was the possibility that any given quark might find itself in an exotic (6q) state in a nucleus.[11] In either the MIT or cloudy bag models[12], where the nucleon bag has a radius of about (0.8 - 1.0)fm, it is perfectly natural to think of the short range N-N force arising from bag overlap. As the radius of a 6q-bag is about 30% greater than that of a 3q-bag it is clear that this provides one mechanism for a change in the confinement scale for the quarks.

A more extreme version of this idea was actually proposed by Krzywicki before the discovery of the EMC effect.[13] He suggested that for the purposes of describing DIS the nucleus could be treated as one large bag containing 3A valence quarks. Within this framework he actually predicted the enhancement of the nuclear sea seen in the small-x region by EMC. (We shall discuss at greater length in Sect. 3 the evidence for and against such an enhancement in view of the apparent contradiction between the EMC and SLAC data). While we feel that this picture of the nucleus as one large bag is hard to believe (given our present beliefs about nucleon structure), some progress has been made by the Bonn group in deriving conventional shell model behaviour from it.[14] A somewhat more attractive picture,[15] in which the quarks only begin to spread throughout the nucleus at higher Q^2, will be critically discussed in Sect. 4.

An alternative to the multi-quark bag picture, which nevertheless leads to an increase in scale in the nucleus, is the idea that the nucleon itself may swell inside the nucleus.[16,17] Phenomenologically it seems that an increase of some 15% in Fe would be needed to explain the EMC data. As bizarre as this idea seems to nuclear physicists it is amazing how difficult it is to find hard evidence to refute it. Sick has shown that the success of y-scaling in low-energy, inclusive electron scattering from ^{3}He is incompatible with more than a 6% increase in the proton radius there.[18] However, since the SLAC data shows an A-dependence proportional to the <u>average</u> nuclear density (which is quite low for ^{4}He) Sick's limit does not contradict the possibility of a 15% increase in Fe.

One possible scenario wherein such swelling could take place involves two nucleons, described as Friedberg-Lee solitons,[19] which approach each other in a nucleus. Clearly at some point the self-consistent scalar field will no longer reach its asymptotic (free space) value in between the nucleons and the quarks will be less effectively confined. Several groups have estimated this effect using a mean field approximation,[17,20] and claim that a scale change of the required magnitude is quite likely.

From our point of view the main dissatisfaction with such a model is that the quarks were not really confined in the first place. It may be reasonable to argue that giving quarks outside the nucleon a mass of 800 MeV is not unreasonable if one aims to describe the properties of an isolated nucleon.[21] However, in considering the leakage of quarks in a many-body system it is not clear that using such a model does not beg the question.

Finally, in anticipation of the following section, let us state the obvious. Even in a decent quark model of nucleon structure the longest range structure is its pion cloud.[12] In a many-body environment it is that cloud which one would expect to be altered first. Such an alteration would necessarily change the structure function inside a nucleus. We shall see in Sect. 3 that with a little phenomenology, and parameters within the range acceptable to low energy physics, this idea is consistent with the SLAC data. At first sight, however, it does not seem to be consistent with the EMC data which originally suggested the calculation. We defer further discussion of these data sets to the end of Sect. 3.

3. MORE VIRTUAL PIONS IN THE NUCLEUS

The enhancement of the nuclear structure function at small x, where the sea dominates, naturally suggests that there has been an increase in the number of virtual $q\text{-}\bar{q}$ pairs. Llewellyn-Smith first suggested that if the number of virtual pions per nucleon in a nucleus were about 0.15 higher than in free space, the EMC data would be explained.[22] That such an increase is not unreasonable was soon established by Ericson and Thomas.[23] Indeed this data may constitute the first evidence for the enhancement of the pion propagator in the region of (300-400) MeV/c momentum transfer once associated with pion condensation.[24,25]

While we know now that the short range repulsion in the spin-isospin channel (often described by the Landau-Migdal parameters g'_{NN}, $g'_{N\Delta}$, $g'_{\Delta\Delta}$) is such that actual pion condensation does not occur, there is still room for a substantial bump in the pion propagator.[24,25] The size of this enhancement is controlled by the process of Δ-h formation which is proportional to the effective nuclear density. Figure 1 shows the variation of fermi momentum,

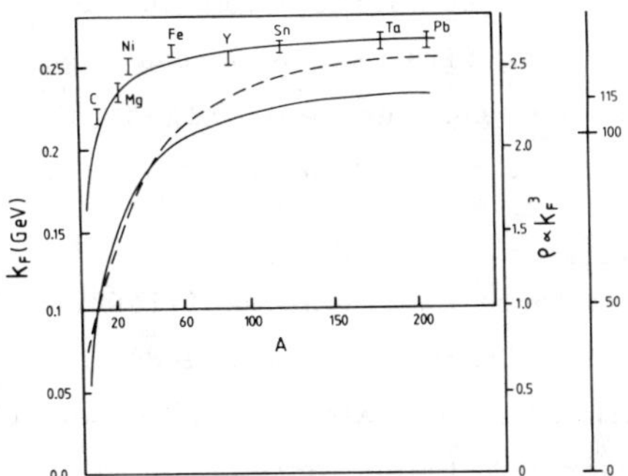

Figure 1: Variation of effective fermi momentum with atomic number [26] (on left), as well as k_F^3 and EMC effect as a percentage of the effect in Fe (right axes)[27].

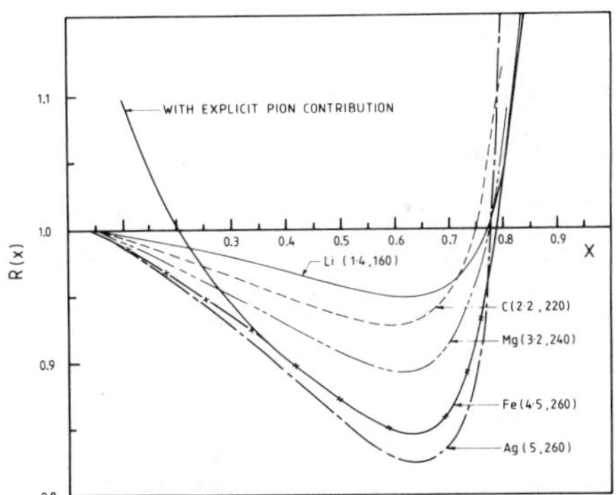

Figure 2: Predicted A-dependence of the contribution to the EMC effect because of loss of momentum to the virtual pion field.[27] The numbers in brackets indicate the extra percentage of momentum carried by pions and the effective fermi momentum used.

k_F, with atomic number as determined from low energy quasi-elastic electron scattering.[26] The first scale on the right shows k_F^3, while the scale on the extreme right shows the expected variation of the EMC effect in such a model - normalised to 100% in Fe.[27]

Whereas the A-dependence should be as illustrated in Fig. 2, the actual magnitude of the effect is controlled by $g'_{N\Delta}$. Unfortunately there is little information on this parameter at (300-400) MeV/c, however a value of order 0.6 is not unreasonable. Recent studies of the (p, Δ) reaction by Jain[28] would actually suggest a number nearer 0.5, and hence a much bigger effect than we calculate.

In order to compare with data one must not only compute the sea enhancement, but also balance momentum overall. Following the simple procedure suggested by Llewellyn-Smith[22] we obtain the predictions for the A-dependence of the EMC effect shown in Fig 2.[27] (The $g'_{n\Delta}$ parameter was fine tuned to give a fit to Fe, after which there is no further freedom in the model).

As an example, we show the predictions for Fe in Fig. 3. It is somewhat bizarre that even though the calculation was prompted by the EMC data, the fit is better for the SLAC data! Investigations are presently underway to see whether this is pure chance, or whether there is some physics in this. For example, it is possible[4] that the picture of a pion cloud about a quark core makes sense at $Q^2 \sim 1$ GeV^2, but not at (30-40) GeV^2.

In the remainder of this section we shall briefly mention some other tests of this idea. However to finish the theoretical discussion we recall the argument in favour of the pionic model which was given by Pandharipande et al.[29] They argued that the pion exchange force between nucleons and deltas is relatively well known, because it is the longest range piece. Thus the pion exchange contribution to nuclear binding energies is relatively model independent. However, the pion number operator is just equal to the potential energy associated with one pion exchange, divided by ω_π. In

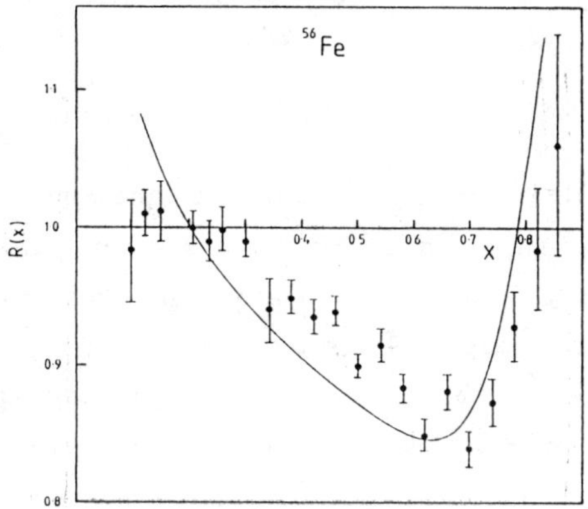

Figure 3: Comparison of the pionic model for the EMC effect in Fe 27), with the data of Arnold et al.[7)]

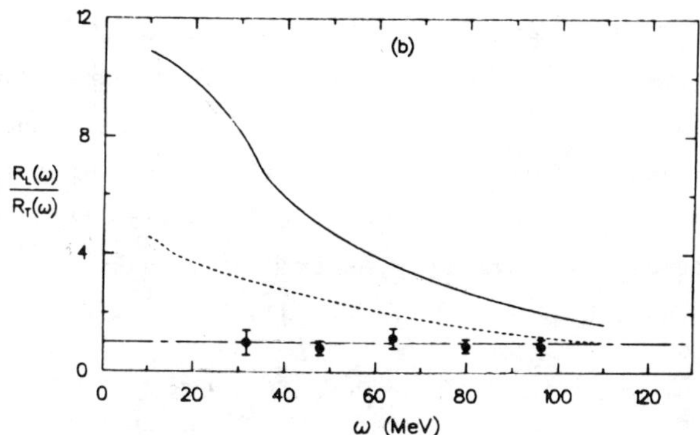

Figure 4: Ratio of the longitudinal to transverse response functions in Pb - see text for details.[35)]

this way they estimate an excess of 0.12 pions per nucleon in Fe (for the Argonne V_{28} interaction).

One obvious test of this idea that the sea is enhanced in Fe is to exploit the sensitivity of $\bar{\nu}$ beams to anti-quarks. The CDHS group recently measured the ratio of $\bar{d}$ and $\bar{s}$ quarks in Fe and H to be of the order 1.1 ± 0.2 at low x, and this has been claimed to contradict the pionic model of EMC.[30)] In fact, because Fe is an isoscalar target, even a relatively small asymmetry in the $\bar{u}$ and $\bar{d}$ distributions in H[31] enables one to reproduce the CDHS data.[32)]

A second experiment at BEBC measured the y-distributions for $\bar{\nu}$ scattering from Ne and d.[33)] Despite claims that the relatively flat y-distributions contradict the pionic model for EMC, Bickerstaff, Birse and Miller[34)] have shown that it gives quite a good fit (as do several other models).

A rather different test of the pion model alone has been carried out at LAMPF.[35)] By a judicious measurement of polarisation observables in proton-nucleus inelastic scattering it is possible to isolate the longitudinal response in the region of (300-400) MeV/c momentum transfer. This would isolate the response of the nucleus to a virtual pion if the exchange were pure isovector. Such a reaction would be possible through the $(\vec{p}, \vec{n})$ reaction, which should have a high priority on the attention of our experimental colleagues in future. Unfortunately, the only experiment performed so far involves the $(\vec{p}, \vec{p}')$ reaction, which mixes the isoscalar and isovector contributions.

Figure 4 shows the ratio of the longitudinal to transverse response for $(\vec{p}, \vec{p}')$ on ^{208}Pb at 340 MeV/c, in comparison with the calculations of Ericson et al for isovector alone at central density (solid line),[36)] and weighted according to the effective density probed by the protons (which are strongly surface absorbed) - dashed line.[35)] Allowing for some uncertainty because of the isoscalar contribution, the agreement above w = 60 MeV is not bad, whereas the low energy region looks terrible. However, the calculations referred to ignore nucleon binding energies which kill the response at low-w. Since the EMC calculation strongly weights the region of w ~ 80-100 MeV, this experiment certainly does not rule out the pionic explanation for EMC.

In concluding this section we mention that the 2p-2h contribution to the pionic response has not yet been included in the EMC calculations[25] - nor has the possible shadowing of the Δ contribution at very small x (≤ 0.05). It should also be said that the apparent dichotomy between the pionic and other explanations of the EMC effect may be illusory. In a quark model the Landau-Migdal parameters, g', would be associated with the regions where two bags overlap. Thus when we learn enough to treat the whole problem consistently, it may be that momentum balance will arise naturally from a softening of the valence quark distribution in a 6q bag, which in turn controls the enhancement of the virtual pion field.

4. DO NUCLEONS SWELL ?

We have already remarked on the unsatisfactory nature of the Friedberg-Lee type of solitons[19] for treating quark deconfinement in nuclei. One model which does not suffer this problem is the model of Nielsen and Patkos where the confining scalar field, χ , is no longer the chiral partner of the pion.[37] In their model true quark confinement is achieved through a term $-(m_q/\chi)\,\bar{q}\,q$ in the Langrangian density. It is assumed that χ feels an effective potential $U(\chi)$ which has an absolute minimum at $\chi=0$, so that m_q/χ becomes infinite outside the "bag".) One's first thought would be to identify m_q with the running quark mass).

Now as pointed out by Pirner and collaborators,[15] if one ignores soliton overlap by approximating a nucleus as a bunch of colourless 3q-clusters ("nucleons") each inside a Wigner-Seitz cell, the lowest energy configuration will not have $\chi=0$ between the "nucleons". Each quark can reduce its kinetic energy by a percentage $\chi_N/(2m_q R)$ by having a non- zero χ-field, χ_N, between the nucleons.

The penalty for not being at the minimum of $U(\chi)$ is $\frac{4\pi}{3}(r_0{}^3 - R^3)\,\frac{1}{2}\,m^2_{GB}\,\sigma_v{}^2\,\chi_N{}^2$ (where r_0 = Wigner-Seitz radius, R = bag radius, m_{GB} = glueball mass and σ_v a scale parameter). Minimising the total energy with respect to χ_N solves the problem.

Pirner et al. estimate $\chi_N = 0.024$ which, with a quark mass of 20 MeV, would give the quarks an effective mass of only 800 MeV between the nucleon-like clusters. Thus the quarks leak out of their bags and the nucleon effectively swells (by about 10% in Fe). Even more beautiful, if m_q is the running mass which goes to zero logarithmically as $Q^2 \to \infty$ the quarks become less confined as Q^2 goes up - hence the name "colour conductivity".

Unfortunately, the very feature which makes the model attractive, namely the freedom to identify m_q as a running mass, is its downfall.[38] In particular the appearance of $1/m_q$ in the change of the quark kinetic energy gives an enormous splitting between neutron and proton energy levels (after correcting for the coulomb force). For example, with $m_d - m_u = 5$ MeV, and $\chi = 0.024$ we find an 8.9 MeV shift at nuclear matter density!

The approach of Williams and Thomas[38] has been to use the observed equality of n and p energy levels (within say 1 MeV[39]) to put an upper limit on the allowed value of χ_N. Using more realistic values[40] for $\Delta m_q/m_q = 33\%$ and $(m_u + m_d)/2 = 15$ MeV (at say 1 GeV^2), they found $\chi_N(1\ GeV^2) \leqslant 0.0019$. Thus the effective quark mass between nucleons in a heavy nucleus should be at least 8 GeV! This leads to less than a one percent change in the nucleon size, which is too small to explain the EMC effect by itself.

This discussion does not eliminate the possibility that nucleon swelling as proposed by Pirner et al contributes to the EMC effect. (At very high Q^2 it may eventually dominate.) However, other mechanisms must be invoked to explain most of the presently observed effect. On the positive side, this shift in neutron and proton levels may contribute substantially to the explanation of the Nolen-Schiffer anomaly in low energy nuclear physics.

5. CONCLUSION

Whatever message the EMC effect has for us, there is no doubt that we are being led to profound new insights in strong interaction physics. We can look forward to an exciting interplay between theoretical ideas and new experiments for at least three or four years before we will be sure of exactly what that message may be.

ACKNOWLEDGMENTS

It is a pleasure to acknowledge the stimulating discussions with M. Ericson, C.H. Llewellyn-Smith and A.G. Williams with whom I have collaborated on various aspects of the work described here. I am also grateful to B.K. Jain, V.R. Pandharipande and H. Pirner for discussions on some of these issues. I would like to thank Mrs P. Coe for her efforts in preparing this manuscript very rapidly.

This work was supported by grants from The University of Adelaide and the ARGS.

REFERENCES

1. F.E. Close, An Introduction to Quarks and Partons (Academic, N.Y. 1979).
2. F. Eisele, J. de Physique (Colloque) C3, 337 (1982).
3. A.W. Thomas, Prog. Part. Nucl. Phys. 11, 325 (1984).
4. A.W. Thomas, Invited paper at 10 PANIC, ADP-312 (July 1984) - to appear.
5. H. Sonoda and M.B. Wise, preprint CALT-68-1157 (July 1984).
6. J.J. Aubert et al., Phys. Lett. 123B, 275 (1983).
7. R.G. Arnold et al., Phys. Rev. Lett. 52, 727 (1984); S. Wimpenny, Invited paper at 10 PANIC - to appear.
8. A. Bodek and J.L. Ritchie, Phys. Rev. D23, 1070 (1981).
9. F.E. Close et al, Phys. Lett. 129B, 346 (1983).
10. O. Nachtmann and H.J. Pirner, Z. Phys. C21, 277 (1984).
11. R.L. Jaffe, Phys. Rev. Lett. 50, 128 (1983).
12. A.W. Thomas, Adv. Nucl. Phys. 13, 1 (1984); G.A. Miller,

Int. Rev. Nucl. Phys. 2, (1984).

13. A. Kryzwicki, Phys. Rev. D14, 152 (1976).

14. Lecture Notes in Physics 197, 236 (1984).

15. G.Chanfray, O. Nachtmann and H.J. Pirner, Phys. Lett. 147B, 249 (1984).

16. J.V. Noble, Phys. Rev. Lett. 46, 421 (1981).

17. E.M. Levin and M.G. Ryskin, Leningrad preprint 888 (1983); L.L. Frankfurt and M.I. Strikman, Leningrad preprint 886 (1983); M. Jandel and G. Peters, Stockholm preprint TRITA-TF4-83-28.

18. I. Sick, Invited paper at 10 PANIC, Basel preprint (1984) to appear.

19. R. Friedberg and T.D. Lee, Phys. Rev. D18, 2623 (1978); T.D. Lee, Phys. Rev. D19, 1802 (1979).

20. L.S. Celenza, A. Rosenthal and C.M. Shakin, Brooklyn College B.C.I.N.T. 84/051/124.

21. R. Goldflam and L. Wilets, Phys. Rev. D25, 1951 (1982).

22. C.H. Llewellyn Smith, CERN 83-02, 180 (1983); Phys. Lett. 128B, 107 (1983)

23. M. Ericson and A.W. Thomas, Phys. Lett. 128B, 112 (1983).

24. E. Oset et al, Phys. Rep. 83, 281 (1981).

25. M. Ericson, Prog. Nucl. Part. Phys. 11, 277 (1984).

26. R.A. Smith and E.J. Moniz, Nucl. Phys. B43, 605 (1972).

27. M. Ericson, C.H. Llewellyn Smith and A.W. Thomas, to be published.

28. B.K. Jain, Invited talk at this conference.

29. B. Friman, V.R. Pandharipande and R.B. Wiringa, Phys. Rev. Lett. 51, 763 (1983); E.L. Berger, F. Coester and R.B. Wiringa, Phys. Lett. D29, 398 (1984).

30. H. Abramowitz et al, CERN-EP/84-57.

31. A.W. Thomas, Phys. Lett. 126B, 97 (1983).

32. M. Ericson and A.W. Thomas, Phys. Lett. 148B, 191 (1984).

33. A.M. Cooper et al CERN-EP/84-37.

34. Private communication.

35. T.A. Carey et al. Phys. Rev. Lett. 53, 144 (1984).

36. W.M. Alberico, M. Ericson and A. Molinari, Nucl. Phys. A379, 429 (1982).

37. H.B. Nielsen and A. Patkos, Nucl. Phys. B195, 137 (1982).

38. A.G. Williams and A.W. Thomas, Adelaide preprint ADP-324/T8 (Dec 1984).

39. J.A. Nolen Jr. and J.P. Schiffer, Ann. Rev. Nucl. Sci. 19, 471 (1969).

40. J. Gasser and H. Leutwyler, Phys. Rep. 87, 77 (1982).

ROTATIONAL BANDS AND SPIN-ORBIT SPLITTING IN BARYON SPECTRA*

R.K. Bhaduri

Physics Department, McMaster University
Hamilton, Ontario, Canada L8S 4M1

ABSTRACT

The absence of appreciable spin-orbit splitting in the low-lying even and odd parity states of the nucleon and delta is puzzling in conventional quark models. A constituent quark model, in which the quarks interact through gluon as well as pion exchange, and the baryon is allowed to deform in the excited states, is shown to resolve this puzzle.

1. INTRODUCTION

There is a wealth of experimental data [1] in baryon spectroscopy that should constitute a good testing ground for nonperturbative QCD. At the present time, however, it is not possible to confront the data from a fundamental theoretical point of view. Rather, a number of quark-models have been developed [2-6] with some ingredients of the theory put in phenomenologically. These are either nonrelativistic constituent quark models [2,3] with a suitably chosen interaction potential, or relativistic bag like models [4,5] where the interaction is treated perturbatively. There are also variants of the constituent quark model with relativistic kinematics included in a variational calculation for the excited states [6]. A comprehensive analysis of the data has been made with constituent quark models [2,3,6,7] and, to a lesser degree, with M.I.T. [4] and chiral [5] bag models. Recently, the nonrelativistic three-body Faddeev equations have been also solved [8] for the excited states of baryons with a phenomenological two-body interaction. All models should attempt to answer some glaring puzzles in the data (we shall concentrate on the N and Δ states here):

---

*Work done in collaboration with M.V.N. Murthy (McMaster), M. Brack (Regensburg), and B.K. Jennings (TRIUMF).

1) the occurrence of the low-lying even-parity excitation $N(1440)1/2^+$ at an energy lower than the lowest lying odd-parity excitation. In the Δ, the situation is similar, but the state $\Delta(1600)3/2^+$ is weak and it has a large mass uncertainty.
2) The appearance of the odd-parity states $\Delta(1900)1/2^-$, $\Delta(1930)5/2^-$ at relatively low excitations. Such states, both in relativistic and nonrelativistic models, should come at much higher energies. There is some evidence [3,9] that such states are also present in the nucleon.
3) The near-absence of spin-orbit splitting in the odd and even parity excitations of both the N and Δ.
4) The qurk-models predict many more low-lying states than are experimentally detected [2,3,4].

These problems are present both in nonrelativistic and relativistic quark models. The relatavistic models face the additional problem of centre-of-mass spurious states. Let us briefly review why the above four points constitute a challenge to the models in order to set the stage for the present paper. In the bag model, the energy spacing between the even and odd-parity state are governed by the bag radius R, and in the constituent quark model by the oscillator spacing $\hbar\omega$ or by the slope of the confining potential. The relevant parameter (R or $\hbar\omega$) may be adjusted and the spurious CM states eliminated to yield the observed number of low lying odd-parity excitations at the right position, but then the next set of even parity excitations appear at nearly twice the excitation energy. It is important to explain both the even- and odd-parity excitations <u>without</u> introducing additional parameters. The simple bag model approach has failed in this respect. The Isgur-Karl model has apparently been more successful. In this model, the constituent quarks move nonrelativistically in a harmonic confinement, and anharmonic residual interactions are treated by first order perturbation theory in a truncated model space. The matrix-element of the anharmonic interaction in the ground state is roughly of the same magnitude as the oscillator spacing. In first order perturbation, the diagonal matrix-element of this interaction brings down the energy of the $L=0$, $N=2$ harmonic excitation close to the odd-parity $N = 1$ state. This explains the observation of $N(1440)1/2^+$ around the same energy as the low-lying odd-parity states. If, however, this interaction is exactly diagonalised in the truncated model space of the spherical oscillator, the large off-diagonal element between the ground state and its nodal excitation neutralises much of this effect. The exact Faddeev calculation [8] with a Coulomb plus linear potential confirms that the first excited $1/2^+$ state is considerably above the odd-parity excitation. The Isgur-Karl prescription of neglecting the off-diagonal matrix-element should then be viewed as taking account of some missing ingredient in the model, which would naturally suppress this off-diagonal matrix-element. We suggest that this may be the case if the ground state is spherical but the excited $N = 2$ state considerably deformed.

Forsyth and Cutkosky [3] have examined question 2) regarding the odd-parity $N = 3$ states of the Δ and N in an extended harmonic oscillator model. The difficulties get compounded in this case and one needs an additional parameter in the anharmonic matrix-element to lower these states to the observed range. On the other hand, if one assumes that the baryon gets more and more deformed with excitation, then the position of these states can be explained without introducing any additional parameter [10-12].

The experimental spectrum shows practically no spin-orbit splitting, except between the $\Delta(1620)1/2^-$ and $\Delta(1700)3/2^-$ states. This is a serious problem both in the bag model and the constituent quark model. We shall again concentrate on the nonrelativistic models. In spectroscopic calculations, the q-q interaction that is diagonalised is assumed to arise from the one-gluon exchange mechanism. This force has spin-spin, tensor and spin-orbit components whose strengths are determined by the quark-gluon coupling constant, α_s. With an oscillator spacing of $\hbar\omega \approx 500$ MeV and the constituent u-d quark mass $m = 330$ MeV, one has to take $\alpha_s \approx 1$ to reproduce the observed N-Δ ground state splitting. This value of α_s is about three times larger than the estimate of the QCD running coupling constant at $q^2 = 1$ Gev^2. It has also the very undesirable effect of introducing large spin-orbit splittings.

For example, although the odd-parity (N = 1) Δ states ($1/2^-$ and $3/2^-$) get split by about the right amount, the nucleon L = 1, S = 3/2 states with differing J's become separated by more than 500 MeV. Isgur and Karl [2] pointed out that there may be delicate cancellation between the one-gluon-exchange spin-orbit force and the spin-orbit interaction arising from the Thomas term of the scalar confining potential. This is true, but it does not solve the problem. An appropriate strength of the Thomas term may indeed cancel the spin-orbit splitting from the one-gluon exchange in the odd parity nucleon states, but it also enhances the splitting between the odd parity Δ states to an unacceptable degree. Moreover, it is unlikely that such delicate cancellations should persist in the N = 2 even-parity states. Isgur and Karl [2] recognised these difficulties and decided to drop the spin-orbit interaction altogether. The spin-orbit puxzle in the odd-parity states has been recently carefully studied by Gromes [13]. He suggested that spin-orbit forces may also arise from the nonlocality of the confining scalar potential. An additional parameter characterising this nonlocality may consistently explain the puzzle for the N = 1 states. However, more parameters are necessary to specify the nonlocality for the even-parity N = 2 states, which were not examined by Gromes.

We examine both the even and the odd-parity excitations of the nucleon and the delta, and offer a solution to the spin-orbit puzzle. We assume that the quarks couple not only to the gluons, but also to pions. This point of view has been advocated both in the bag model context, and in relativistic potential models by various authors [14-16]. It has also been examined in the nonrelativistic constituent quark model approach [17-19] for the ground state properties of baryons, and in the derivation of the N-N potential from q-q interaction [20]. So far as spectroscopy is concerned, the effect of pion coupling has only been examined for the low lying odd-parity states of the N and Δ in the chiral bag model [5], but no one has examined both the odd- and even-parity states in a single framework. In this paper we do so, with particular emphasis on the spin-orbit problem. The constituent quarks interact not only via the one-gluon-exchange potential (OGEP), but also through one-pion exchange (OPEP). An appropriate pion form-factor is chosen to take account of the pion size, although this is not crucial. The contributions of the self-energy effects, due to the emission and absorption of a pion (or a gluon) by the same quark, are neglected: these are all lumped together in a constant constituent mass of the quark. This is not justifiable, as self-energy effects are state-dependent, and should therefore have a bearing on the spectroscopy of the states. Their calculations, however, have inherent difficulties [21], and we choose to retain only the lowest order interquark potential in our model. With this assumption, somewhat more than half the N-Δ ground state splitting arises from OPEP, which has no spin-orbit component. We then require the quark-gluon coupling $\alpha_s \gtrsim 0.35$, consistent with the QCD estimate, to reproduce the observed N-Δ mass difference. This cuts down the one-gluon-exchange spin-orbit strength by about a factor of three. The scalar one-body confinement also gives rise to a spin-orbit splitting, whose magnitude is estimated from bag model calculations for the odd- and even-parity states. When deformed orbitals are used, we find a further suppression in the spin-orbit matrix-elements, specially for the even parity excited states which are more deformed. These two ingredients - deformation and pion coupling to quarks, may consistently reproduce the observed N-Δ spectrum.

We assume that the quarks are moving in a mean field which is spherical in the ground state of the baryon, but is deformed when the baryon is excited. Appreciable deformation of the ground state is anyway ruled out from the experimental data, which should otherwise give rise to low-lying rotational states built on the ground state. The mean field is arising due to the confinement potential, which is universal, and also due to many-body effects like the polarization of the $q\bar{q}$-sea with excitation, and multigluon effects. The deformation of the mean field is not arising [8] from the OGEP and OPEP potentials, which are diagonalised in a set of basis states generated by the deformed mean field. Our view is that the experimental data may be naturally explained if deformation of the excited states is assumed.

We do not know how to generate this deformation from a microscopic theory. Rather, in section II, we describe a model [11] in which the mean field is assumed to be deformable, and the deformation parameter for each state is determined by minimising the energy of the state, subject to some constraints.

Before we go on to the next point it is worth mentioning that although the baryonic spectra seem to show collective many-body effects like deformation, there is no such evidence in the meson spectra. With the exception of the pion [16], the mesonic spectra seems to be well explained by assuming it to be a simple $q\bar{q}$ bound state [24-26]. Indeed, one expects this simple structure of mesons from 1/N-expansion of QCD, where N is the number of colours; but the baryon structure is more complicated [27].

We now make a few remarks regarding the question 4), which pertains to the point that quark models in general generate many more states than observed by πN scattering. The answer to this problem has been successfully given in the context of the Isgur-Karl model [7]. It has been pointed out that spin-dependent interactions give rise to mixing of states of different symmetry in a way that decouples many states from the πN channel. These are not excited in the πN scattering, but may be observable in photonuclear excitations. This constitutes a triumph for the Isgur-Karl model. It should be pointed out, however, that the photo-decay helicity amplitudes of $N(1440)1/2^+$ to the ground state is not obtained too well by taking spherical orbitals for this state. Somewhat better agreement is obtained with deformed orbitals in our model [12], although a dynamical treatment of the pion current may be needed to fully resolve this problem [16].

In the next section, we briefly describe the mean-field model and the interaction Hamiltonian. We shall omit all the details of calculation, and present the final results and compare these with the experimental data in the first section. The interested reader should look up refs. [11], [12] and [28] for more details.

2. THE DEFORMED MEAN-FIELD MODEL

In this model the valence quarks are assumed to be moving in a deformable mean-field. The deformed eigenfunctions of this mean-field provide the basis for diagonalising the residual two quark interaction due to the one gluon exchange (OGE) and the one pion exchange (OPE) including the non-central components.

The mean field is state dependent and therefore some of the states could be non-orthogonal. The diagonalisation of the residual interaction in this nonorthogonal basis restores orthogonality through configuration mixings. The resulting eigenvalues and eigenstates in our model represent the masses and the structure of resonances in the baryon spectrum.

The deformed mean field in which the valence quark is moving is assumed to be of the form

$$V(\vec{r}) = \frac{m}{2}(\omega_x^2 x^2 + \omega_y^2 y^2 + \omega_z^2 z^2) \ , \tag{2.1}$$

where $\vec{r} = \vec{r}(x,y,z)$ denotes the position of the constituent quark in the body fixed frame of reference. The mass of the quark is m, where

$$m = m_u = m_d \tag{2.2}$$

for u- and d-type quarks, as we are only interested here in the non-strange sector. The Hamiltonian of the three quark system, after eliminating the centre of mass terms, can then be written as

$$H_0 = \frac{1}{2m}(p_\rho^2 + p_\lambda^2) + \frac{m}{2} \sum_{i=x,y,z} \omega_i^2(\rho_i^2 + \lambda_i^2) \tag{2.3}$$

in terms of the relative coordinates,

$$\vec{\rho} = \frac{1}{\sqrt{2}}(\vec{r}_1 - \vec{r}_2) \qquad ; \qquad \vec{\lambda} = \frac{1}{\sqrt{6}}(\vec{r}_1 + \vec{r}_2 - 2\vec{r}_3) \ , \tag{2.4}$$

and their conjugate momenta $\vec{p}_\rho$ and $\vec{p}_\lambda$ respectively. The corresponding energy of the three quark system is then given by

$$E_0 = (N_x+1)\hbar\omega_x + (N_y+1)\hbar\omega_y + (N_z+1)\hbar\omega_z \ , \tag{2.5}$$

where

$$N_i = N_{\rho_i} + N_{\lambda_i} \qquad ; \qquad i = x,y,z \ . \tag{2.6}$$

In order to minimize the energy given by Rq. (2.5) we impose the volume constraint, namely,

$$\omega_x \omega_y \omega_z = \omega_0^3 \quad , \tag{2.7}$$

The minimization condition may also be written as [28]

$$\omega_x (N_x+1) = \omega_y (N_y+1) = \omega_z (N_z+1) \quad . \tag{2.8}$$

The constraint (2.7) simply means that the volume of the baryon in any state is the same as in the spherical model. The volume of a deformed baryon in this model grows with the excitation quanta N, the same as in conventional oscillator models.

The intrinsic energy E_0 at equilibrium configuration is reproduced in Table 1 for $N=0,1,2$ and the lowest $N=3$ excitations of the quark orbitals. Note that both for $N=1$ and $N=2$, only the axially symmetric solutions yield equilibrium configuration, while for $N=0$ the shape is spherical. For $N=3$, the lowest configuration remains axially symmetric while the triaxial and spherical solutions occur at higher energies. We shall throughout take the z-axis to be the body-fixed axis of symmetry.

We may now construct the intrinsic wave functions in the ρ-coordinate as

$$\psi_{N_{\rho_x} N_{\rho_y} N_{\rho_z}}(\vec{\rho}) = \phi_{N_{\rho_x}}(\omega_x, \rho_x)\phi_{N_{\rho_y}}(\omega_y, \rho_y)\phi_{N_{\rho_z}}(\omega_z, \rho_z) \tag{2.9}$$

where $\phi_{N_{\rho_i}}(\omega_i, \rho_i)$ is the one-dimensional oscillator eigenfunction with enery $(N_{\rho_i} + \frac{1}{2})\hbar\omega_i$, and similarly for the $\vec{\lambda}$-coordinate. The total number of excitation quanta in deformed orbitals is $N = \sum_{i=x,y,z} (N_{\rho_i} + N_{\lambda_i})$. By constructing spatial wave functions of appropriate permutation symmetry, and combining these with the spin-isospin wavefunctions of the three quarks, we obtain totally symmetric states for a given N in space, spin and isospin space. This is because the colour part of the 3-quark system is a singlet and is totally antisymmetric under the exchange of any two quarks. These intrinsic wavefunctions are given in Ref. [11]. These form the basis in

Table 1

The intrinsic states in the deformed oscillator basis. The total number of excitation quanta of the $\vec{\rho}$ and $\vec{\lambda}$ oscillators is $N = N_x + N_y + N_z$. A given set of N_x, N_y, N_z defines the intrinsic state, the first few of these are listed in the first column. The SU(6) multiplet structure to which the spatial part couples is shown in the second column. The fourth column shows the corresponding eigenvalues of H_0. Note that these intrinsic states are not all orthogonal.

Number of Excitation, N	Multiplet Structure (parity)	Equilibrium Configuration	Intrinsic Energy
N=0 $N_x = N_y + N_z = 0$	$[56^+]$	$\omega_x = \omega_y = \omega_z$ (spherical)	$3\hbar\omega_0$
N=1 $N_x = N_y = 0, N_z = 1$	$[70^-]$	$\omega_x = \omega_y = 2\omega_z$ (prolate)	$3.780\hbar\omega_0$
N=2 $N_x = N_y = 0, N_z = 2$	$[56^+], [70^+]$	$\omega_x = \omega_y = 3\omega_z$ (prolate)	$4.327\hbar\eta_0$
$N_x = N_y = 1, N_z = 0$	$[20^+]$	$\omega_x = \omega_y = \frac{\omega_z}{2}$ (oblate)	$4.762\hbar\omega_0$
N=3 $N_x = N_y = 0, N_z = 3$	$[56^-], [70^-]$	$\omega_x = \omega_y = 4\omega_z$ (prolate)	$4.762\hbar\omega_0$

which the residual interactions (except the tensor and spin-orbit parts) are diagonalised to obtain the intrinsic energies and the configuration mixed wavefunctions. States of good angular momenta J are then projected out from these states, and the tensor and spin-orbit matrix-elements are obtained. A further diagonalisation of the appropriate matrices then yields the energies that may be compared with the experimental data. For details, you should see Refs.[11] and [28].

We now briefly outline the various terms in the complete Hamiltonian H. This may be written as

$$H = H_0 + V_{OGE} + V_{OPE} + V_{SO} \ . \tag{2.10}$$

The first term is the mean field Hamiltonian already discussed earlier. The second term, V_{OGE}, is obtained by a Breif-Fermi reduction of the

OGE amplitude [29]. If the momentum-dependent terms are dropped, it takes the form

$$V_{OGE} = -\sqrt{2}\,\frac{\alpha_s}{\rho} + \frac{\pi\alpha_s}{\sqrt{2}\,m^2}\,\delta^3(\vec{\rho}) + \frac{\sqrt{2}\pi\alpha_s}{3m^2}\,\vec{\sigma}_1\cdot\vec{\sigma}_2\,\delta^3(\vec{\rho})$$

$$+\frac{\alpha_s}{4\sqrt{2}m^2}\,\frac{1}{\rho^3}\,S_{12} + \frac{3\alpha_s}{4\sqrt{2}m^2}\,\frac{1}{\rho^3}\,\{(\vec{\sigma}_1+\vec{\sigma}_2)\vec{\rho}\times\vec{p}_\rho - \frac{1}{3\sqrt{3}}\,(\vec{\sigma}_1-\vec{\sigma}_2)\cdot(\rho\times\vec{p}_\lambda)\} \quad (2.11)$$

Here α_s is the quark-gluon coupling constant and $S_{12} = 3\vec{\sigma}_1\cdot\hat{\rho}\sigma_2\cdot\hat{\rho}-\sigma_1\cdot\sigma_2$ is the tensor operator in the interaction. The last term contains two-body and three-body spin-orbit interactions, the latter arising due to the elimination of the CM coordinate [13]. For the interaction V_{OPE}, we take

$$V_{OPE}(\vec{r}_{ij}) = -\frac{1}{(2\pi)^3}\,\frac{\pi\alpha_\pi}{m^2}\,\vec{\tau}_i\cdot\vec{\tau}_j \int d^3q\; e^{iq\,r_{ij}}\,\frac{\vec{\sigma}_i\vec{q}\vec{\sigma}_j\cdot\vec{q}}{q^2+m_\pi^2}\,|F_\pi(q)|^2. \quad (2.12)$$

Here $\alpha_\pi = g^2_{qq\pi}/4\pi$, the pseudoscalar coupling constant, and $F_\pi(q^2)$ is the form factor at the $qq\pi$-vertex to account for the size of the pion. We take the simple form

$$F_\pi(q^2) = \frac{m_\rho^2}{m_\rho^2+q^2}\ , \quad (2.13)$$

and with m_ρ = 770 MeV, this yields the rms radius of the pion to be 0.63 fm.

The last term, V_{SO}, is the one-body spin-orbit potential put in phenomenologically to simulate the spin-orbit splitting that would arise from a relativistic treatment of the scalar confinement potential. We estimate this term by directly calculating the single-particle spectrum of a zero-mass current quark in the Dirac equation, and then choosing the strength of V_{SO} to reproduce this splitting. This approach is in the spirit of ref. [30] which attempted to justify the constituent quark model. We take

$$V_{SO} = -\frac{C}{2}\sum_i \vec{\ell}_i\cdot\vec{\sigma}_i\ , \quad (2.14)$$

which goes over to the form $V_{SO} = -C\vec{\sigma}_3\cdot(\vec{\lambda}\times\vec{p}_\lambda)$. The strength C is a constant in a given shell.

3. RESULTS AND DISCUSSION

The energy spectra of the nucleon and the delta obtained in our model are compared with the available experimental data in Figs. (1-3). The parameters of the model are given by

$$\alpha_s = 0.35,\ \alpha_\pi = 0.95,\ \hbar\omega_0 = 550\ \text{MeV},\ m = 330\ \text{MeV},$$

$$C(N=1) = 100\ \text{MeV},\ C(N=2) = 50\ \text{MeV}\ . \qquad (2.15)$$

In addition, an overall constant shift of -1262 MeV is added to the total Hamiltonian to fix the nucleon ground state at 940 MeV. The quark-gluon coupling constant $\alpha_s = 0.35$ is consistent with the running

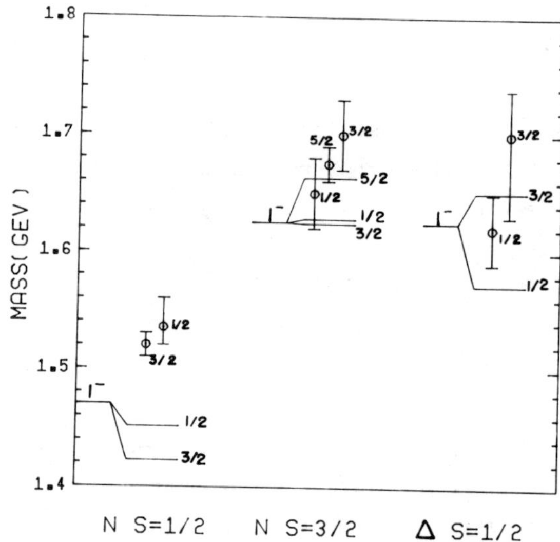

Fig. 1 The low-lying odd-parity states of the nucleon and delta. The experimental data are taken from Ref. [1] and are shown by open circles. The error bars indicate the uncertainty in the nominal masses. The position of the projected $\ell = 1$ states is indicated in the figure for reference. The calculated positions of the good-J states are shown by horizontal lines. The corresponding wavefunctions for these states in the deformed oscillator basis are given in Table 3, (a) and (b)

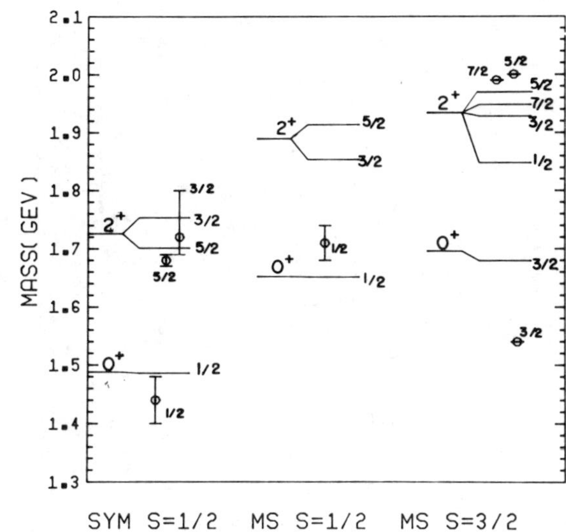

Fig. 2 The low-lying even-parity states of the nucleon. The legend is in the same style as Fig. 1. The open circles without error bars denote experimental states which are not well established. The first two members of the rotational band, 0^+ and 2^+, which are projected from the deformed intrinsic states, are also indicated for reference. The structure of the wavefunctions of the states is given in Table III, (c).

coupling constant of QCD if $q^2 = 1\ \text{GeV}^2$. The qqπ-coupling α_π is chosen to obtain the required N-Δ ground state splitting when α_s is fixed at 0.35. A value of α_π close to unity is also obtained by fitting the pion-decay widths of the resonances in a simple quark model [31]. The constituent mass m = 330 MeV is the standard value used to reproduce the nucleon magnetic moment. Our calculated spectrum is not, however, sensitive to the choice of m. Except for the Coulomb term, the matrix-elements of the residual interactions scale as $\alpha\sqrt{\omega_0^3/m}$, where α is α_s or α_π. An increase in m by a factor of 2, for example, may be compensated for by increasing α by a corresponding factor of $\sqrt{2}$, or even less if ω_0 is changed slightly. The oscillator parameter ω_0, which determines the effective moment of inertia in the projected rotational

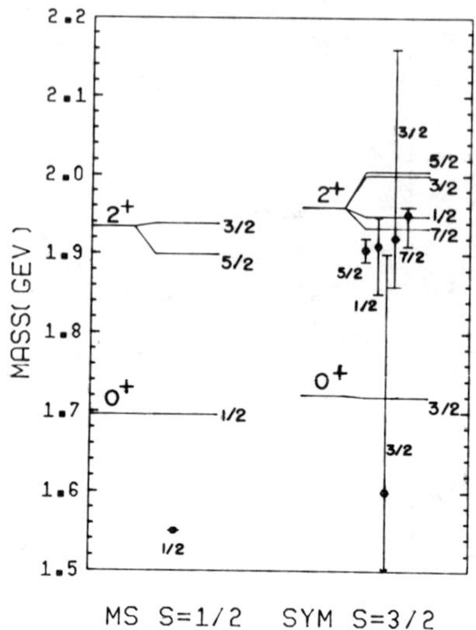

Fig. 3 The low-lying even-parity states of the delta. The legend is in the same style as in Fig. 2. The structure of the wavefunctions is given in Table III, (d). Note, as in Fig. 2, that even-parity states whose dominant component is mixed symmetric (70) are mostly weakly seen, or not seen experimentally.

band, is chosen to reproduce the spacings between the members of the band. The calculated results indicate that $\hbar\omega_0$ should be in the range 500-600 MeV. The deformations of the excited states are determined through Eq. (2.11), and are not free parameters. For the justification of the choice for C in the N = 1 and N = 2 shells, see [28].

From the figs. (1-3), it will be seen that we get fair agreement with experimental data for both the odd- and even parity excited states. On the whole, we find that the odd-parity states are about 50-70 MeV lower than the experimental values, and a similar situation exists for the even parity states that are dominantly of mixed-symmetry (MS) character. Actually, unlike the conventional models, our deformed model gives a strong mixing between the S = 1/2 symmetric and mixed symmetric states of the nucleon. As emphasized in [10], this helps to explain the radiative decay amplitude of $N(1440)1/2^+ \rightarrow N(940)1/2^+ + \gamma$, and also the near vanishing amplitude for $N(1710)1/2^+ \rightarrow N(940)1/2^+ + \gamma$.

In conclusion, in this talk I have proposed a model to resolve the spin-orbit puzzle in the odd- and even-parity states of the nucleon and the delta. Two important physical inputs help to explain the data - the deformation of the baryon in the excited states, and the pion-coupling to the quarks. Our model, of course, is an oversimplified one. In particular, one should be able to derive the mean field H_0 microscopically through a rotationally invariant Hamiltonian. Nevertheless, we think that in a more sophisticated approach, deformation of the excited states and pion-quark coupling should still play important roles in explaining the data on baryon spectroscopy.

REFERENCES

1. Review of Particle Properties, Rev. Mod. Phys. 56, no. 2 (1984).
2. N. Isgur and G. Karl, Phys. Rev. D18, 4187 (1978); D19, 2653 (1979);
   N. Isgur, in "The New Aspects of Subnuclear Physics, Erice, (1978), edited by A. Zichichi (Plenum, New York, 1980), p. 107.
3. C.P. Forsyth and R.E. Cutkosky, Z. Phys. C18, 219 (1983).
4. T.A. DeGrand and R.L. Jaffe, Ann. Phys. (N.Y.) 100, 425 (1976).
   T.A. DeGrand, in Proc. of the IVth International Conference on Baryon Resonances, Toronto, (1980), edited by N. Isgur (University of Toronto Press).
5. F. Myhrer and J. Wroldsen, Z. Phys. C25, 281 (1984).
6. J. Carlson, J.B. Kogut and V.R. Pandharipande, Phys. Rev. D27, 233 (1983); D28, 2807 (1983).
7. R. Koniuk and N. Isgur, Phys. Rev. D21, 1868 (1980).
8. B. Silvestre-Brac and C. Gignoux, I.S.N. Grenoble preprint (1984), to appear in Phys. Rev. D.
9. R.E. Cutkosky, C.P. Forsyth, R.E. Hendrick and R.K. Kelly, Phys. Rev. D20, 2839 (1980).

10. R.K. Bhaduri, B. Jennings and J.C. Waddington, Phys. Rev. D29, 205 (1984).

11. M.V.N. Murthy, M. Dey, J. Dey and R.K. Bhaduri, Phys. Rev. D30, 152 (1984).

12. M.V.N. Murthy and R.K. Bhaduri, McMaster University preprint (1984), to appear in Phys. Rev. Lett.

13. D. Gromes, Z. Phys. C18, 249 (1983).

14. G.E. Brown, M. Rho and V. Vento, Phys. Lett. 84B, 383 (1979).

15. G.A. Miller, A.W. Thomas and S. Theberg, Phys. Lett. 91B, 192 (1980); Phys. Rev. D22, 2823 (1980);
A.W. Thomas, Advances in Nucl. Phys. 13, 1 (1983).

16. W. Weise, "Quarks, Chiral Symmetry and Dynamics of Nuclear Constituents", TPR-84-8 (University of Regensburg preprint), to be published in International Review of Nucl. Phys., Vol. I (1984), World Scientific, Singapore.

17. Y. Nogami and N. Ohtsuka, Phys. Rev. D26, 261 (1982).

18. J. Navarro and V. Vento, Phys. Lett. 140B, 6 (1984).

19. T. Hastuda, Contribution B11 in the Tenth International Conference on Particles and Nuclei, Heidelberg, 1984.

20. K. Bräuer, A. Faessler, F. Fernandez and K. Shimizu, Tuebingen University preprint (1984).

21. S.A. Chin, Phys. Lett. 109B, 161 (1982), Nucl. Phys. A382, 355 (1982).
Y. Nogami and A. Suzuki, Prog. Theor. Phys. 69, 1184 (1983).

22. C. Hajduk and B. Schwesinger, SUNY Stony Brook preprint (1984).

23. Z.Y. Ma and J. Wambach, Phys. Lett. 132B, 1 (1983).

24. S. Godfrey and N. Isgur, Toronto Univ. preprint (1984).

25. H.W. Crater and P. Van Alstine, Phys. Rev. Lett. 53, 1527 (1984).

26. S. Godfrey, TRIUMF preprint (1984).

27. E. Whitten, Nucl. Phys. B160, 57 (1979).

28. M.V.N. Murthy, M. Brack, R.K. Bhaduri and B.K. Jennings, Z. Physik C (to be published).

29. A.J.G. Hey and R.L. Kelly, Phys. Rep. 96, 71 (1983).

30. R.K. Bhaduri and M. Brack, Phys. Rev. D25, 1443 (1982).

31. A.W. Hendry, Ann. Phys. (N.Y.) 140, 65 (1982).
M.A. Preston and A.K. Dutta, Phys. Rev. D27, 2780 (1983).

EXCITED NUCLEONS IN NUCLEAR COLLISIONS

B.K. Jain
Nuclear Physics Division
Bhabha Atomic Research Centre
Bombay 400 085

ABSTRACT

The ^{6}Li(p, Δ^{++}) ^{6}He and A(^{3}He, t)B_Δ reactions have been investigated in the general framework of the distorted wave Born approximation. It has been shown that the experimental data on these reactions are reproduced well, both in magnitude and shape, using one-pion-and-one-rho-exchange for the transition interaction, $V_{\sigma\tau}$ (NN → NΔ). Using only one-pion-exchange potential with the Landau-Migdal repulsive term it is found that the parameter g'_Δ lies below 0.4. The ^{6}Li(p, Δ^{++})^{6}He data are also used to determine the spin-isospin quark-quark interaction.

1. INTRODUCTION

Considering general interest in the role of baryon excitations in nuclear physics, recently some pretty experiments have been done at Saturne, Saclay, where real Δ-isobars are produced in the nuclear reactions. These experiments correspond to the excitation of the projectile in ^{6}Li (p, Δ^{++})^{6}He reaction[1] at 1.04 GeV incident energy and of a nucleon in several target nuclei in A(^{3}He, t)B_Δ reaction[2] at 2GeV incident energy. Theoretically, these reactions carry rich information. In the (p, Δ^{++}) reaction, because the Δ^{++} corresponds to the S = 3/2, T = 3/2 and T_z=+3/2 state of the baryon, like the (p,n) reaction it can transfer ΔS = 1 and ΔT = 1 to the nucleus. In addition, it also transfers large linear momentum ($\gtrsim$ 250 MeV/c) and can excite nuclear states differing from the ground state by even two units of spin and/or isospin. In the (^{3}He, t) reaction, since the Δ is produced in the nucleus, the data are of immense importance to learn the Δ-nucleus dynamics, specifically the question relating to the nature of excitation in the nucleus, i.e. whether they are the collective

ΔN^{-1} excitations or the quasi-free. These reactions also provide an opportunity to learn, in a direct way, about the spin-isospin $NN \rightarrow N\Delta$ coupling interaction. The knowledge of this interaction, which is quite unsettled at present, is of prime importance in determining the effect of the Δ (1232) isobar in the nuclear dynamics. In this paper I first identify an appropriate theoretical framework to describe these reactions and then use it to determine some important parameters pertaining to the $NN \rightarrow N\Delta$ interaction.

For the description of the reaction mechanism we start with the DWBA as a candidate theoretical framework which assumes that these reactions proceed via one step quasi-free mechanism. The interaction before and after the $NN \rightarrow N\Delta$ event, is described in this theory by a mean field. In Section 2 we apply this framework to the (p, Δ^{++}) reaction and in Section 3 to the $(^3He, t)$.

2. $^6Li(p, \Delta^{++})$ He REACTION

2.1 Formalism

Considering the Δ^{++} as an elementary particle, the differential cross section for the $A(p, \Delta^{++})B$ reaction is written as

$$\frac{d\sigma}{dt} = E_p E_A E_\Delta E_B / [4\pi(\hbar^3 c^3 E_c k_p)^2] \langle |T_{BA}|^2 \rangle, \qquad (1)$$

where the angular brackets around $/T_{BA}/^2$ denote the sum and average over the spins in the final and initial states, respectively. The transition amplitude, T_{BA}, is given by

$$T_{BA} = \left[\chi^-_{\underset{\sim}{k}_\Delta}, \langle B, \Delta^{++} | \sum_i V_{\sigma\tau}(i) | A, p \rangle, \chi^+_{\underset{\sim}{k}_p} \right], \qquad (2)$$

where $E_x(\underset{\sim}{k}_x)$ is the total energy (wave vector) of the particle x, t is four-momentum transfer squared, and E_c is the total energy in the center of mass. χ represents the distorted wave in the initial and final states. $V_{\sigma\tau}$ is the effective spin-isospin transition potential for $pp \rightarrow n\Delta^{++}$, summed over the "active" nucleons in the target nucleus. This potential, in principle, should depend on both the incident energy and the transferred energy-momentum. However, for the present, since we are going to analyze the experimental data at a single energy (1.04 GeV) only, we will not worry about the incident energy dependence. For the rest, the general form of the interaction is

$$V_{\sigma\tau}(\omega, \underset{\sim}{Q}) = \left[V_L(\omega, Q)\, \vec{S}\cdot\hat{Q}\, \vec{\sigma}\cdot\hat{Q} + V_T(\omega, Q)(\vec{S}\times\hat{Q})\cdot(\vec{\sigma}\times\hat{Q}) \right] \vec{T}\cdot\vec{\tau}$$

$$= \left[V_C(\omega, Q)\, \vec{S}\cdot\vec{\sigma} + V_{NC}(\omega, Q)\, S_{12}(\hat{Q}) \right] \vec{T}\cdot\vec{\tau}\,, \quad (3)$$

where the central (V_C) and the non central (V_{NC}) terms are

$$V_C = (V_L + 2V_T)/3 \quad , \quad V_{NC} = (V_L - V_T)/3$$

w(Q) is the energy (momentum) transfer from the incident proton to the ith nucleon in the nucleus, $\vec{S}$ ($\vec{T}$) is the spin (isospin) transition operator for $p \to \Delta^{++}$, S_{12} is the tensor operator, and V_L and V_T represent the (w, Q) dependence of the longitudinal and transverse part of the interaction respectively. Using this form of the interaction the final expression for the $/T_{BA}/^2$, as reported elsewhere,[3] is given by

$$\langle |T_{BA}|^2 \rangle = \frac{2}{3}(2J_A+1)^{-1} \sum_{M_A M_B} \sum_{m=-1}^{+1} \Big| V_C(\omega, Q)\, F^{BA}_{-m,-1}(Q)$$

$$+ \left(\frac{24\pi}{5}\right)^{1/2} V_{NC}(\omega, Q) \sum_{\mu} (-1)^{\mu} (1 1 \mu m | 2 M)$$

$$\times\; Y_{2,-M}(\hat{Q})\, F^{BA}_{\mu,-1}(Q) \Big|^2 , \quad (4)$$

where F^{BA}, the "distorted" nuclear structure factor, is given by

$$F^{BA}_{\nu,-1}(Q) = \int d\underset{\sim}{r}\; \chi^{-*}_{\underset{\sim}{k}_\Delta}(\underset{\sim}{r})\, \chi^{+}_{\underset{\sim}{k}_P}(\underset{\sim}{r})\, \rho^{BA}_{\nu,-1}(\underset{\sim}{r}) , \quad (5)$$

In our paper, Phys. Rev. C 29 1396 (1984), the non-central term should be multiplied by an additional factor of 3. The reported results are correct, however.

with $\rho^{BA}_{\nu,-1}$, the spin-isospin transition density, given by

$$\rho^{BA}_{\nu,-1}(\underset{\sim}{r}) = \langle B | \sum_i \delta(\underset{\sim}{r} - \underset{\sim}{r}_i)\, \sigma_\nu(i)\, \tau_{-1}(i) | A \rangle . \quad (6)$$

Using $(1p)^2$ configuration to calculate the spin-isospin transition density, the "distorated" nuclear structure factor F^{BA}(Eq. 5)) for $^6Li \rightarrow {}^6He$ is given by

$$F^{BA}_{\nu,-1}(\underset{\sim}{Q}) = C_0\, \delta_{M_A,-\nu}\, G_{00}(Q) + C_2\, (1\,2\, M_A - M_L | 1 - \nu)\, G_{2M_L}(Q) , \quad (7)$$

where

$$G_{LM}(Q) = (4\pi)^{-1/2} \int d\underset{\sim}{r}\; \chi^{-*}_{\underset{\sim}{k}_\Delta}(\underset{\sim}{r})\, \chi^{+}_{\underset{\sim}{k}_p}(\underset{\sim}{r})\, \rho_\ell(r)\, Y_{LM}(\hat{r}). \quad (8)$$

Here, ρ_ℓ (r) is the radial transition density and is normalized to unity. L is the transferred orbital angular momentum. For a $(1p)^2$ configuration its values are limited to 0 and 2. C_0 and C_2, the weights for the transfer of these two values of L, are determined by the configuration mixing coefficients for the description of the initial and final states of the nuclei. For the following configurations,

$$\Psi_{^6Li}(1^+, T_A = 0) = (\alpha_A\, {}^3S_1 + \beta_A\, {}^1P_1 + \gamma_A\, {}^3D_1)\, | T_A = 0, M_{T_A} = 0 \rangle$$

$$\Psi_{^6He}(0^+, T_B = 1) = (\alpha_B\, {}^1S_0 + \beta_B\, {}^3P_0)\, | T_B = 1, M_{T_B} = -1 \rangle , \quad (9)$$

the C coefficients are given by

$$C_0 = -2\alpha_A \alpha_B + 2(3)^{-1/2} \beta_A \beta_B$$

$$C_2 = 4(10)^{-1/2} \gamma_A \alpha_B + 2(6)^{-1/2} \beta_A \beta_B + 3(5)^{-1/2} \gamma_A \beta_B . \quad (10)$$

The best choice for the values of the mixing parameters and $\rho_\ell(r)$ comes from the inelastic scattering of medium energy electrons in the three-momentum transfer ranger 1.0-3.0 fm^{-1} from the ground state of ^6Li to its second excited state (0^+, T = 1) at 3.56 MeV. This excited state is the isobaric analog of the ground state (0^+, T = 1) of ^6He. From their measured experimental data Bergstrom et al[4] have extracted the radial transition density $\rho_\ell(r)$, which is parametrized as

$$\rho_\ell(r) = \exp(-r^2/b^2)(0.1063r^2 - 0.05091r^3 - 0.008433r^4 - 0.0001126r^6 + 0.0.000001407r^3), \tag{11}$$

with b = 2.324 fm. and

$$\alpha_A = 0.924, \beta_A = 0.2, \gamma_A = 0.102$$
$$\alpha_B = 1.00, \beta_B = 0.08 \tag{12}$$

These parameters, apart from reproducting the (e,e') form factors for the 3.56 MeV (0^+, T = 1) and 5.37 MeV (2^+, T = 1) states in ^6Li, also fit the ground moments.

In addition to ρ^{BA} the numerical evaluation of $/T_{BA}/^2$, as seen from Eq. (4,5), requires the knowledge of the optical potentials to generate disorted waves for protons and isobars and the transition potential functions V_L and V_T.

The required optical potential for a proton on ^6Li at 1.04 GeV is generated by use of the high-energy ansatz of folding in the elementary nucleon-nucleon scattering amplitude with the nuclear density. We write

$$V_p(r) = (i + \alpha) W_0 \rho(r)/\rho(0), \tag{13}$$

with W_0, the imaginary part of the potential, given by

$$W_0 = -(\hbar^2 c^2 k/2E)\rho_0 \sigma_T . \tag{14}$$

Here, σ_T is the total proton-nucleon cross section and ρ the nuclear density. α is the ratio of the real to imaginary part of the nucleon-nucleon scattering amplitude.

The values of σ_T and α are taken to be 44 mb and -0.073, respectively, from the scattering of 1.05 GeV protons on nucleons[5]. The function

$\rho(r)/\rho(0)$ describes the radial dependence of $V_p(r)$ in terms of the density. $\rho(r)$ is taken from the elastic electron scattering analysis on ^{6}Li of Li et al[6].

For the Δ -nucleus potential (V_Δ) in the intermediate-energy range, unfortunately, no information exists. At low energies, some estimates have been made[7], but they are not of much relevance for the present purpose. In view of such a lack of knowledge about V_Δ, we take guidance from the fact that the Δ is an excited state of the nucleon and thus carries, intrinsically, about 300 MeV more energy than the proton/neutron at the same kinetic energy. Therefore it is plausible that W_o for the Δ at a kinetic energy T may be close to that for the proton around T + 300 MeV. This correspondance with the W_o at an enhanced energy for protons to a certain extent takes care of the additional channels (like π N, etc.) which are open to the Δ. Hence, as a resonable guess, to start with, we have taken W_o for Δ to correspond to that for the proton at the incident energy. The energy of the incident proton approximately equals that of the Δ plus 300 MeV. For the real part of the Δ -nucleus potential we are again guided by that of protons. The proton optical potential (the real part) goes to zero around 600 MeV laboratory energy and becomes repulsive beyond it. Since the laboratory energy of Δ for a 1.04 GeV incident proton energy is around 700 MeV, we guess that the real part of the Δ -nucleus potential should be small. It is taken equal to zero (i.e. $\alpha_\Delta = 0.0$). We would, however, investigate the dependence of the cross section on V_Δ later on. For the choice of the $V_{\sigma\tau}$ (pp $\rightarrow$ n Δ^{++}) to begin with, it is necessary that we start with some ansatz for it. For this purpose, we take guidance from the (p,n) reaction for the corresponding interaction in the NN $\rightarrow$ NN channel. For this interaction, it has been shown by Brown et al[8] that the $V_{\sigma\tau}$ (pn $\rightarrow$ np) of Love and Petrovich[9], which fit the intermediate energy (p,n) data, can be reproduced very well by the one-pion-and-rho-exchange potentials. It is also found that, in the spin-isospin channel, the dominant contribution to this interaction comes only from the first order Born term. Following this we have identified V_L and V_T of $V_{\sigma\tau}$(pp$\rightarrow$n Δ^{++}) with the one-pion and one-rho exchange potential, respectively i.e.

$$V(\omega, Q) = -\frac{4\pi\hbar c\, f f^*}{m^2} F(Q) F^*(Q) \frac{Q^2}{Q^2 + m^2 - \omega^2}, \qquad (15)$$

where f(F) and f*(F*) are the coupling constants (form factors) at the π NN and πNΔ vertices,respectively. m is the mass of the exchanged boson. The structure of the form factors was chosen to be of the monopole type with the assumption that F = F*. The values of the cutoff momenta, Λ, for them and the coupling constants in Eq.(15) are taken from those required for the description of π N scattering, pp and np charge-exchange scattering, one boson exchange nucleon-nucleon potentials, etc. These values are

$$f_\pi^2 = 0.081 \,, \quad f_\pi^{*2} = 0.37 \,, \quad \Lambda = 1.2 \text{ GeV}$$
$$f_\rho^2 = 4.86 \,, \quad f_\rho^* = 1.85 f_\rho \,, \quad \Lambda = 2 \text{ GeV} \,. \tag{16}$$

2.2 RESULTS

The calculated cross-sections (continuous curve) as a function of t, along with the measured values, are shown in Fig. 1. It is remarkable that the calculated results agree very well, in magnitude as well as in shape, with the experimental data. The only source of uncertainty in these results arises from the uncertainty in the Δ-nucleus optical potential. In order to see the changes in the results due to the variation of this potential, we show in Fig. 1 two more curves. These curves correspond to W_0 for Δ increased and decreased by 30% from that used for the continuous curve. We observe that the results do not change much. Variation in the real part of the potential, to the extent of ± 10 MeV also does not change the results much (see Fig. 2).

Therefore, the agreement shown in Fig. 1 is practically parameter-free. It determines, in a direct way, that the effective spin-isospin coupling potential $V_{\sigma\tau}$(NN $\rightarrow$ NΔ) is correctly described by he one-pion-and-one-rho-exchange interaction.

In order to convince ourselves further that the ($1\pi + 1\rho$)-exchange interaction is the correct representation of the spin-isospin coupling potential $V_{\sigma\tau}$ (NN $\rightarrow$ ΔN), we have also tried some other prescriptions for it. We have calculated the cross section using one-pion-and-one-rho-exchange interactions with their central part modified to $\tilde{V}_C$ as

$$\tilde{V}_c(Q) = V_c(Q) - (4\pi Q_c^2)^{-1} \int d\underset{\sim}{k}\, \delta\left[|\underset{\sim}{Q} - \underset{\sim}{k}| - Q_c\right] V_c(k) \,, \tag{17}$$

where Q_C is the correlation parameter. It is taken equal to 3.93 fm^{-1}.

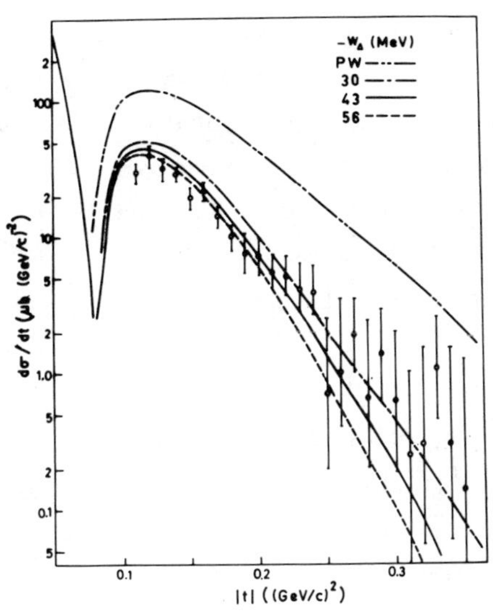

Fig. 1 Differential cross-section $d\sigma/dt$ vs /t/ for $^6Li\ (p, \Delta^{++})^6He$ for various values of W_Δ. The experimental points are from Ref.1. The PW curve corresponds to the plane wave approximation for protons and deltas.

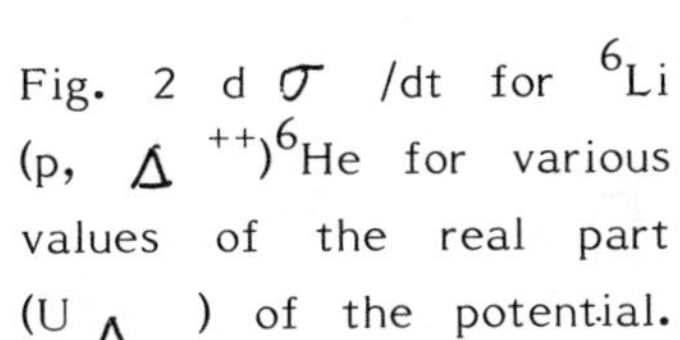

Fig. 2 $d\sigma/dt$ for $^6Li\ (p, \Delta^{++})^6He$ for various values of the real part (U_Δ) of the potential.

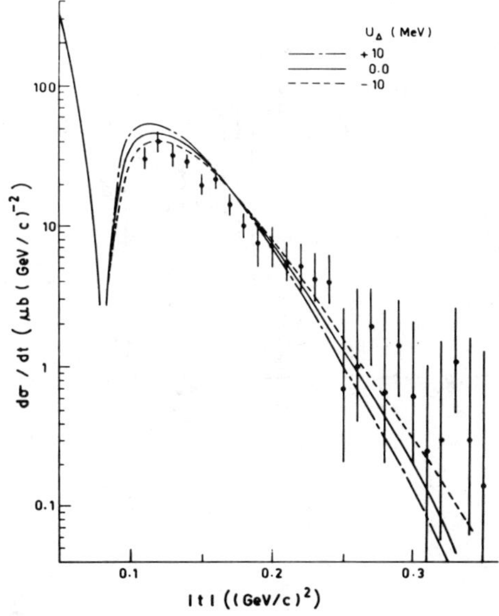

We have not modified the noncentral part of the interaction as it is already cut, at high momentum transfers, due to opposite signs of the one-pion-and-one rho exchange interactions (see Eq. (3)). For the other choice of interaction we have taken the potential consisting of one-pion-exchange only. The effect of the other pieces of the interaction is included via the form factor with the appropriate choice of the cutoff momentum Λ . Values of Λ are taken equal to 2 and 4 fm^{-1}. The former value is consistent with the πN shifts', while the latter, using single-nucleon and two nucleon absorption/production vertices, provides a reasonable description of the total cross section for $\pi^+ d \rightarrow pp$ and the inverse reaction $\vec{p}p \rightarrow \pi^+ d$[10].

The results corresponding to these potentials are shown in Fig.3. It is evident that, except for the $(1\pi + 1\rho)$-exchange interaction, the results due to all other interactions do not agree with the experimental data.

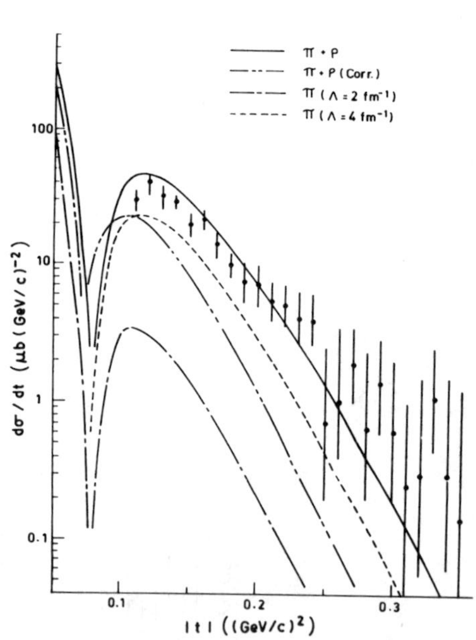

Fig.3 $d\sigma/dt$ for $^6Li(p, \Delta^{++})\,^6He$ for various choices of the coupling potential $V_{\sigma\tau}$ ($pp \rightarrow n\Delta^{++}$).

3. A(^{3}He, t)B$_\Delta$ REACTION

3.1 Formalism

Using similar procedure as given in the earlier Section the cross-section for A(^{3}He, t)B$_\Delta$ is written as

$$\frac{d\sigma}{d\Omega} = \left[E_{^3He} E_A E_t E_B / (2\pi\hbar^2 c^2 E_c)^2\right] k_t / k_{^3He} \langle |T_{BA}|^2 \rangle, \quad (18)$$

where T_{BA}, the transition amplitude, is given by

$$T_{BA} = \left[\chi^-_{\underset{\sim}{k}_t}, \langle \{B_\Delta, t\} | \sum_{i,j} V_{\sigma\tau}(i,j) | \{A, {}^3He\}, \chi^+_{\underset{\sim}{k}_{^3He}}\right]. \quad (19)$$

$V_{\sigma\tau}$ is the effective spin-isospin transition potential given by the one-pion and one-rho exchange. The delta produced in the nucleus would be Δ^+ or Δ^{++} depending upon whether the struck nucleon is a neutron or a proton. Sum i, j goes over the nucleons in the projectile and target nuclei, respectively. The curly brackets in Eq.(19) represent the anti-symmetrization between the nucleons in projectile (ejectile) and the target (residual) nuclei. Ignoring the difference, which should be small, between the phase shifts for ^{3}He and triton in the region of 2 GeV, for doubly closed shell nuclei (like ^{12}C, ^{16}O, ^{40}Ca, etc.), the direct part of the T_{BA} is given by (details will be reported elsewhere)

$$\langle |T_{BA}|^2 \rangle = \frac{2}{9} \frac{A}{A+3} (T_N T_z \nu_N \nu_z | T_A \nu_A)^2 |\rho(Q)|^2$$

$$\times \sum_{M_B} \sum_{\nu=-1}^{+1} |G_{\nu M_B}(\omega, Q)|^2, \quad (20)$$

where (T_x, ν_x) are the isospin and its projection. $\rho(Q)$ is the transition density for ^{3}He $\rightarrow$ t, and is normalized such that $\rho(0) = 3$. For the conversion of a nucleon in "$n \ell_n j_n$" shell to a delta in "$n_\Delta \ell_\Delta j_\Delta$" shell $G(\omega, Q)$ is given by

$$G_{\nu M_B}(\omega, \underset{\sim}{Q}) = (N + \sqrt{3}\, Z) \left[\frac{(2j_\Delta+1)(2\ell_n+1)(2\ell_\Delta+1)}{3\pi} \right]^{1/2}$$

$$\times \sum_{\ell} (-1)^{\ell} \begin{Bmatrix} \ell_n & \ell_\Delta & \ell \\ \frac{1}{2} & \frac{3}{2} & 1 \\ j_n & j_\Delta & j \end{Bmatrix} (\ell_\Delta \ell_n 0 0 | \ell 0)$$

$$\times \Big[(-1)^{m_\ell} V_C(\omega, Q) (\ell 1 m_\ell - \nu | J_B M_B) I_{\ell m_\ell}(Q)$$

$$+ \left(\frac{24\pi}{5}\right)^{1/2} V_{NC}(\omega, Q) \sum_{\mu} (-1)^{m'_\ell} Y^{*}_{2,-m}(\hat{Q})$$

$$\times (1 1 \mu \nu | 2 -m)(\ell 1 m'_\ell \mu | J_B M_B) I_{\ell m'_\ell}(Q) \Big]. \tag{21}$$

Integral $I_{\ell m}$ is defined as

$$I_{\ell m}(Q) = 2\pi i^m \int db\, dz\, b\, e^{i Q_{\parallel} z} J_m(Q_{\perp} b)\, e^{2i\delta(b)}$$

$$\times R_{n \ell_n j_n}(r) R_{n_\Delta \ell_\Delta j_\Delta}(r)\, \text{Ⓗ}_{\ell,-m}(\theta), \tag{22}$$

where $\delta(b)$ is the phase-shift for the mass three particles and is complex. $Q_{\parallel}$ and $Q_{\perp}$ are the longitudinal and transverse components of the momentum transfer $\underset{\sim}{Q}$. R's represent the radial wavefunctions of nucleons and delta in the nucleus. N and Z in Eq. (21), denote the neutron and proton number in the particular shell of the target nucleus, respectively. These two terms arise due to the conversion of neutrons to Δ^+ and proton to Δ^{++} in the nucleus. All other symbols have their usual meanings.

For ^{12}C we consider that the nucleons in any of the two shells, $1s_{1/2}$ and $1p_{3/2}$, of the nucleus is converted into delta. The cross section is then calculated for various bound states of delta in ^{12}C. For comparison with the experimental data the cross-section is summed over all the possible bound states. At this point it may, of course, be mentioned that for the description of individual Δ -nucleon hole states the assumption of ^{12}C as 100% $1s_{1/2}^{4}$ $1p_{3/2}^{8}$ configuration is not quite correct. However, for the summed cross-section, it is probably not very unsatisfactory.

The radial nucleon wave function is generated in the Woods-Saxon well whose parameters are taken from the work of Elton and Swift[11]. For the delta wavefunction, of course, only meager information exists. Taking guidance from Moniz[7] the delta radial wavefunctions are also generated in a Woods-Saxon potential. Its depth is taken corresponding to the average binding potential, 55 MeV, the radius parameter, r_0', equal to 1.15 fm and diffuseness, a', equal to 0.65 fm. With this potential the delta states are bound upto 2s in ^{12}C.

For the phase shift function, δ (b), between 1-2 GeV, information exists only for alpha particles at 1.37 GeV on Calcium isotopes[12]. In this case, $\exp(2i\delta(b))$, which gives a good description of the α -elastic scattering data, is found to be purely real and has a Woods-Saxon form. The radius parameter, r_0 $(R = r_0 A^{1/3})$, and diffuseness, a, are found to be 1.45 fm and 0.68 fm, respectively. In view of the lack of availability of any information on mass 3 nuclei, for over purpose also we have taken same form except that the value of r_0 has been reduced because the mass 3 particles are known to have lesser absorption than alpha.

The projectile-ejectile form factory ρ (Q), is approximated by the magnetic form factor obtained from electron scattering on ^{3}He at large momentum transfer[13], i.e.

$$\rho(Q) = \exp(-\alpha^2 Q^2) - \beta^2 Q^2 \exp(-\gamma^2 Q^2), \tag{23}$$

with $\alpha = 0.654$ fm, $\beta = 0.456$ and $\gamma = 0.821$ fm

3.2 RESULTS

The calculated angular distributions for values of r_0 equal to 1.2 1.1 and 1.0 fm alongwith the experimental data are shown in Fig. 4. The two results are seen to be in appreciable agreement with each other,

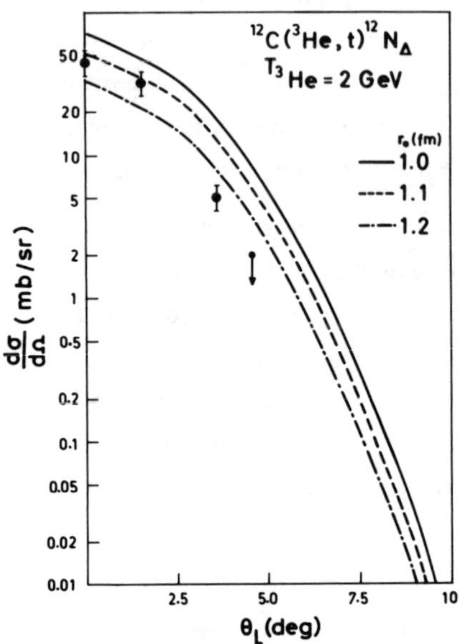

Fig. 4. Angular distribution for the $^{12}C(^3He, t)^{12}N_\Delta$ reaction. The experimental points are from Ref. 2.

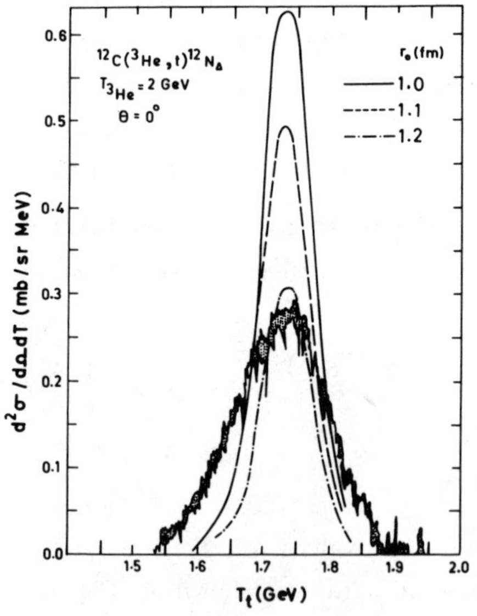

Fig. 5. Triton energy spectrum for the $^{12}C(^3He, t)^{12}N_\Delta$ at 0°. The experimental points are from Ref. 2.

both in magnitude and shape.

Experimentally, in the triton spectrum at 0°, it has also been found that the centroid energy for the Δ -bump is around 1720 MeV, and remains approximately same for all the nuclei studied. Considering that the lab. energy of the triton at 0° and 300 MeV excitation energy in the proton is 1603 MeV, this observation amounts to a large shift in the resonance energy of Δ in the nucleus. In order to see as how far the approach followed here reproduces this shift, we first modify the expression (18) for the cross-section to incorporate the width of Δ . Taking the Breit-Wigner form for it we write

$$\frac{d^2\sigma}{d\Omega dE_t} = \frac{\Gamma(\omega)}{2\pi\left[(\omega + m - E^*) + \Gamma^2(\omega)/4\right]} \frac{d\sigma}{d\Omega} , \qquad (24)$$

where E* is the resonance mass, Γ is the width and m the mass of a nucleon. In Fig. 5 we show the calculated triton energy spectra for the phase shift radius parameter r_0 equal to 1.2, 1.1 and 1.0 fm alongwith the measured energy spectrum. It is seen that for all the values of r_0 the centroid of the calculated spectrum appear at the same point, and is in agreement with the experiment.

Considering that the results shown in Figs. 4-5 do not contain any really arbitrary free parameter, the extent of their agreement with the measured data is quite remarkable. It again demonstrates that the underlying model used in getting these results is essentially correct.

4. SUMMARY

Thus, in summary, it can be concluded that the results of Section 2-3 establish that (i) the DWBA is the correct theoretical framework to analyze these reactions and (ii) using this framework it is possible to directly determine the spin-isospin $NN \rightarrow N\Delta$ interaction.

Exploiting this observation, now we use the ^{6}Li(p, Δ^{++})^{6}He data to determine the Landau-Migdal parameter, g'_Δ , and the spin-isospin quark-quark interaction.

5. LANDAU-MIGDAL PARAMETER, g'_Δ

Recently in nuclear physics the effect of isobar-nucleon hole excitations in the nucleus has attracted much attention. In these studies $V_{\sigma\tau}$ is normally written as a sum of the one-pion-exchange potential without any form factor and a purely phenomenological repulsive spin-spin term, i.e.

$$V_{\sigma\tau}(NN \rightarrow N\Delta) = 4\pi\hbar c \frac{f_\pi f_\pi^*}{m_\pi^2}\left[-\frac{\vec{\sigma}\cdot\hat{Q}\,\vec{S}\cdot\hat{Q}}{Q^2+m_\pi^2} + g'_\Delta\vec{\sigma}\cdot\vec{S}\right] \times \vec{T}\cdot\vec{\tau} \,, \qquad (25)$$

where g'_Δ is called the Landau-Migdal parameter. The motivation for writing $V_{\sigma\tau}$ in this form arises from the basic philosophy that, in the boson-exchange theory of nuclear forces, the essential bosons which are exchanged are pions; ρ exchange is really the pole term in the exchange of a continuous mass spectrum of two interacting pions. Secondly, the repulsion in the interaction at short distances arises from several sources. Therefore, instead of putting the screening effects due to these sources separately, it is more economical to summarize them in one parameter, g'. If the repulsive effects are sufficiently short ranged, g' can be taken as a constant, independent of momentum and energy transfer. For the nucleon-nucleon case, the value of this parameter (g'_N) seems to be around 0.6-0.7. For g'_Δ, however, there does not exist any information, although it is often taken to be equal to g'_N. It is, therefore, of considerable interest if this parameter could be determined from the data on the (p, Δ^{++}) reaction. With this motivation, in Fig.6, using the interaction given by Eq. (25), we present results for various values of g'_Δ, ranging from 0.2 to 0.7. Comparison of these results with the experimental data shows that the value of g'_Δ lies below 0.4. This value of g'_Δ is obviously lower than the normally used value (0.5 - 0.7) (except for the recent conclusions of Arima et al[14]).

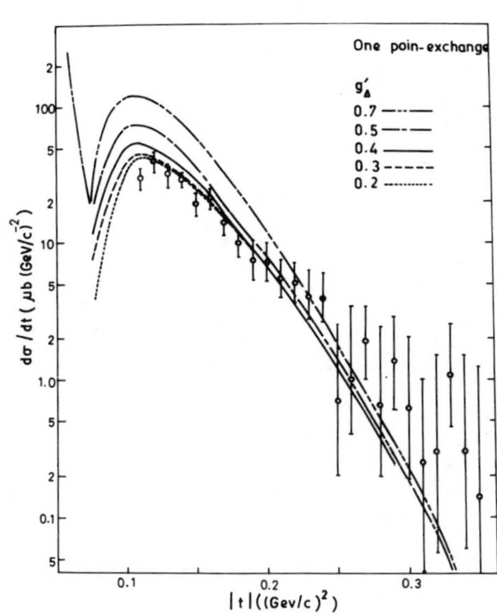

Fig. 6. Sensitivity of the ^{6}Li (p, Δ^{++})^{6}He differential cross-section to g'_Δ using the one-pion-exchange potential (without form factor only).

6. SPIN-ISOSPIN QUARK-QUARK INTERACTION*

The ^{6}Li(p, Δ^{++})^{6}He reaction can also be used to determine the spin-isospin quark-quark interaction. As shown earlier the essential process for the ^{6}Li(p, Δ^{++})^{6}He reaction is the elementary spin-isospin flip pp $\rightarrow$ n Δ^{++} reaction. In the quark model of nucleons this can be visualized as two bags, of three quarks each, coming together and flipping spins and isospins of two quarks, one in each nucleon bag, without changing radial or orbital motion. Phenomenologically, effective interaction for this process can be written as

$$V_{\sigma\tau} = C \sum_{\substack{i\in a \\ j\in b}} \vec{\sigma}_i \cdot \vec{\sigma}_j \; \vec{\tau}_i \cdot \vec{\tau}_j \,, \qquad (26)$$

where sum runs over quarks in proton a and b. Due to short range nature of the process, C, the strength of the interaction can be taken as independent of the momentum transfer. Though this interaction is the simplest

---

* This work was done in collaboration with S.K. Gupta, BARC.

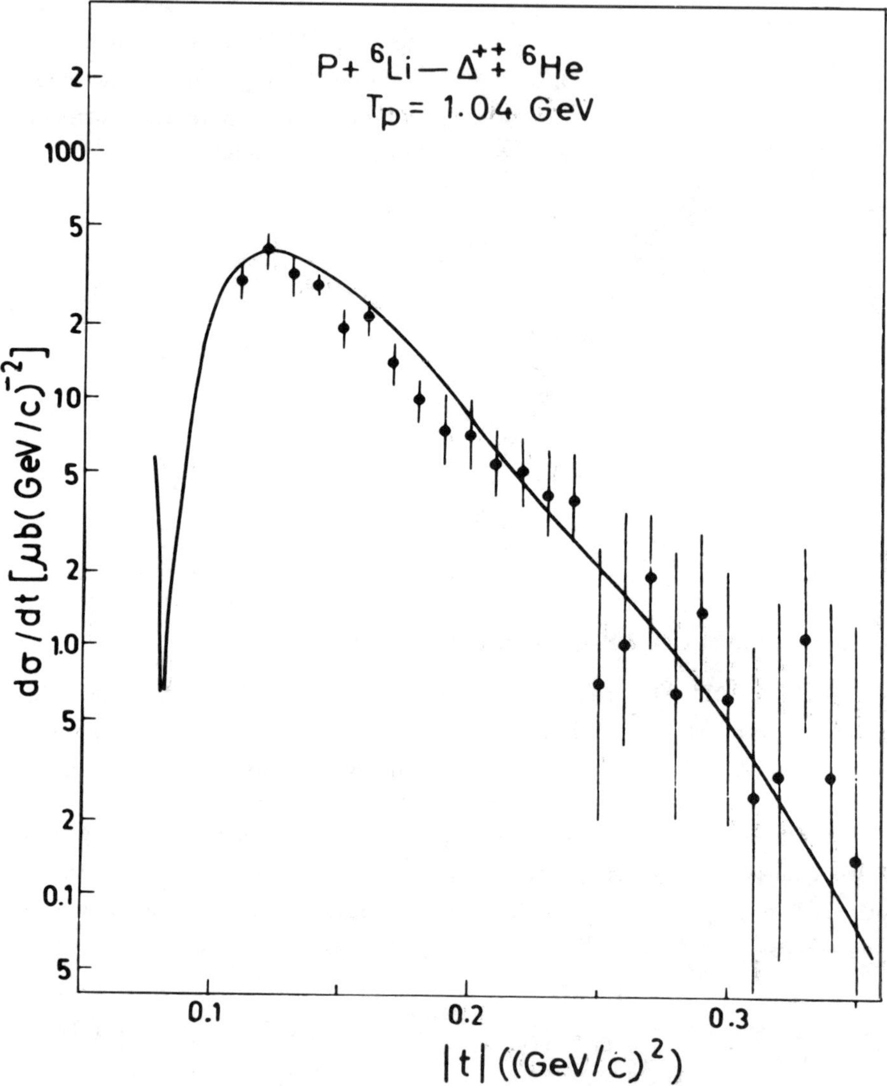

Fig. 7. $d\sigma/dt$ for the ^{6}Li (p, Δ^{++}) ^{6}He. The continuous curve is the fit due to quark model.

possible interaction between two quark bags it is not at all evident as how it would arise from theories motivated by the QCD. As shown in the recent work of Phatak[15], the simplest diagram which involves the exchange of a gluon alongwith the exchange of one quark contributes negligibly to this process. In view of this situation it is quite useful that C be determined phenomenologically from the ^{6}Li (p, Δ^{++})^{6}He reaction. For this purpose, using the dominent $^3S_1$ and ^1S_O configuration s for ^{6}Li and ^{6}He, respectively and the interaction given by Eq. 26 the cross-section in the DWBA is written as

$$\frac{d\sigma}{dt} = \frac{1}{6} F_{Kin} \left| G_{OO}(Q) \right|^2 \sum_{\substack{M M_{\Delta^{++}} \\ M_p}} \Big| \sum (-1)^{1/2 - m_n} \times \left(\tfrac{1}{2} \tfrac{1}{2} - m_n m_p \,|\, 1 M\right) \times \langle M_{\Delta^{++}}(a)\, m_n(b) \,|\, V_{\sigma\tau} \,|\, M_p(a)\, m_p(b) \rangle \Big|^2 , \quad (27)$$

where G(Q) is given by Eq. (8). Wave functions written in the angular brackets are the intrinsic wave functions of nucleons and isobars using three quarks. Using this expression the calculated cross-section is compared with the experimental data. In order to get an agreement as shown in Fig. 7 the value of C is found to be around 70 MeV fm^3.

REFERENCES

1. T.Hennino et al., Phys. Rev. Lett. 48, 997 (1982).

2. C. Ellegaard et al., Phys. Rev. Lett. 50, 1745 (1983). C. Gaarde et al., Proc. of the Symposium on Detla-Nucleus Dynamics, ed. by T.S.H. Lee et al., May, 1983, A.N.L., U.S.A., p. 395 .

3. B.K. Jain, Phys. Rev. C 29, 1396 (1984).

4. G. Bergstrom, U. Deutschmann, and R. Neuhausen, Nucl. Phys. A327, 439 (1979).

5. D.V. Bugg et al., Phys. Rev. 146, 980 (1966); S. Barshay, C.B. Dover and J.P. Vary, Phys. Rev. C11, 360 (1975).

6. G.C. Li, I. Sick, R.R. Whitney and M.R. Yearian, Nucl. Phys. A162, 583 (1971).

7. E.J. Moniz, Nucl. Phys. A374, 557c (1982).

8. G.E. Brown, J. Speth, and J. Wambach, Phys. Rev. Lett. 46, 1057 (1979).

9. F. Petrovich, W.G. Love, and R.J. Mc Carthy, Phys. Rev. C21, 1718 (1980)

10. J. Chai and D.O. Riska, Nucl. Phys. A338, 349 (1980)

11. L.R.B. Elton and Swift, Nucl. Phys. A94, 52(1967).

12. D.C. Choudhury, Phys. Rev. C 22, 1848 (1980)

13. J.S. Mc Carthy, I. Sick and R.R. Whitney, Phys. Rev. C 15, 1396 (1977).

14. A. Arima et al., Phys. Lett. 122B, 126 (1983).

15. S.C. Phatak, preprint and contribution to this Conference.

SCATTERING BEHAVIOR OF CLUSTERS OF CONFINED QUARKS *

J.T. Londergan

Dept. of Physics and Nuclear Theory Center,
Indiana University,
Bloomington, Indiana, 47405,
USA

ABSTRACT

One of the most important problems facing nuclear physics today is how to incorporate quark degrees of freedom into a consistent picture of many-hadron systems. We wish to study systems of completely confined quarks which can only separate in color-singlet clusters. The confining nature of the quark-quark interaction makes it extremely difficult to formulate a picture of multi-quark systems in terms of two-body interactions. Although the bag model has been quite successful in describing the properties of an isolated hadron, in multi-hadron systems this picture ignores the identity of quarks in different clusters until they completely overlap. We will outline a multi-quark non-relativistic potential model due to F. Lenz et al. which offers many insights into the relationship between many-quark systems and traditional nuclear physics. This model is a 'string-flip' potential model for systems of quarks: it is straightforwardly solvable, and it allows one to investigate the role of quark-exchange between clusters. This simple model suggests that, at low energies, quark-exchange processes are extremely important and they help to give rise to many of the qualitative features which we associate with nuclear physics phenomena. Specifically, we will show that: (1) some qualitative features of the two-nucleon interaction and the properties of the deuteron can be understood as a consequence of quark-exchange processes; (2) we can draw several analogies between many-quark systems and atomic physics processes in which electron exchange processes play a major role (e.g., the e-H system, and the binding and scattering of two H atoms); (3) we can straightforwardly introduce color into this model, and we can show the importance of 'hidden-color' states in hadron-hadron interactions at short distances. Finally, we will discuss a search for phenomena which provide a 'signature' for quark effects in nuclei. The most important effect appears to be a 'softening' of the quark momentum distribution in the nucleus. We discuss the experimental consequences of this result, and the relation between this effect and the 'EMC effect' which has been observed in deep inelastic scattering from nuclei with electroweak probes.

*Work supported in part by the National Science Foundation.

1. INTRODUCTION

It is generally accepted that the underlying theory of the strong interactions is QCD. Individual hadrons are made up of quarks, which interact via exchange of colored gluons. It is extremely difficult to describe low-energy phenomena with QCD. For an isolated hadron, bag models [1,2] or constituent-quark potential models [3,4] can successfully describe many properties of the hadron. For interactions between two or more hadrons, nuclear physics has traditionally described these in terms of 'elementary' hadrons exchanging mesons. An important problem in nuclear physics is to incorporate quark degrees of freedom into a many-hadron system. It has proven rather difficult to find a quark 'signature' in a system of nucleons; that is, some observable phenomena where a picture which explicitly relies on an underlying quark structure is manifestly superior to the 'traditional' description of nuclear forces.

In this talk, we will outline a simple non-relativistic potential model which allows us to treat many-hadron systems as clusters of confined quarks with confining string potential interactions between pairs of quarks (or antiquarks, for mesons). In a multi-hadron system, the string configuration is determined to be the shortest string combination, i.e. the lowest energy, for all possible quark interactions. This model satisfies the following conditions for a multi-quark system: (1) individual quarks are confined such that only color-singlet clusters can separate asymptotically; (2) exchange symmetry between quarks is guaranteed; (3) long-range color van der Waals forces between color-singlet hadrons are avoided.

The model we will describe was proposed by F. Lenz et al.[5-7] We will discuss a very simple version of this model, which can be straightforwardly solved and which gives a rather clear qualitative picture of the physics involved. This model has been extended to treat meson-meson and baryon-baryon interactions by Oka and collaborators[8-10], and has been generalized to the 'nuclear matter' problem by Horowitz et al. [11-12] This model offers a convenient way to introduce many-body confining forces which are necessary to satisfy condition (3) listed above. It is very similar to flux-tube models of quark confinement [13]; the adiabatic approximation in flux tube models gives a model equivalent to the 'string-flip' model.

In this talk we will concentrate on a very simple system: two 'mesons', each a quark-antiquark pair, interact via harmonic-oscillator confining potentials and the 'string-flip' condition. With this model, we will investigate the following questions: how important are quark-exchange forces in low-energy interactions of these two clusters of confined quarks? How does this quark-cluster picture give rise to qualitative features of 'traditional' nuclear physics? How can we relate the results of this confined-quark system to some features of atomic physics systems? And finally, can we identify some observable effect where quark effects in many-hadron systems can be clearly seen?

2. HAMILTONIAN FOR 2-MESON SYSTEM

We consider two 'mesons', each meson being a confined $q-\overline{q}$ pair. At first, we ignore the color degree of freedom. A $q-\overline{q}$ interaction v which increases as the separation will confine the quarks. For the examples in this talk, we will use harmonic-oscillator (h.o.) forces. For two interacting mesons, we must determine the minimum string potential energy. For our simple example, we consider only quark-antiquark interactions. A non-relativistic Hamiltonian satisfying the three criteria in part 1 is

$$H = \sum_i T_i + U \qquad (1)$$

where the potential energy is determined by the smaller of two partial energies: $U = \min(U_1, U_2)$, where $U_1 = V_{1\overline{1}} + V_{2\overline{2}}$ and $U_2 = V_{1\overline{2}} + V_{2\overline{1}}$. For oscillator potentials, U takes a particularly simple form when expressed in terms of relative coordinates between cluster centers,

$$U = \frac{m\omega^2}{2} [x^2 + y^2\theta(z-y) + z^2\theta(y-z)] \qquad (2)$$

In Eq. 2, we have introduced the relative coordinates shown in Fig. (1),

$$x \equiv \frac{1}{2}(\vec{r}_1 + \vec{r}_2 - \vec{r}_{\overline{1}} - \vec{r}_{\overline{2}}) \qquad Z \equiv \frac{1}{2}(\vec{r}_1 + \vec{r}_{\overline{2}} - \vec{r}_2 - \vec{r}_{\overline{1}})$$

$$(3)$$

$$y \equiv \frac{1}{2}(\vec{r}_1 + \vec{r}_{\overline{1}} - \vec{r}_2 - \vec{r}_{\overline{2}}) \qquad R \equiv \frac{1}{2}(\vec{r}_1 + \vec{r}_2 + \vec{r}_{\overline{1}} + \vec{r}_{\overline{2}})$$

The potential energy U has the following properties: once the clusters $(1\overline{1})$ and $(2\overline{2})$ begin to separate ($y > z$), there is no interaction between the clusters in the relative coordinate y. A more realistic interaction would be one which included the long-range part of the two-body interaction as the separation became large. For the

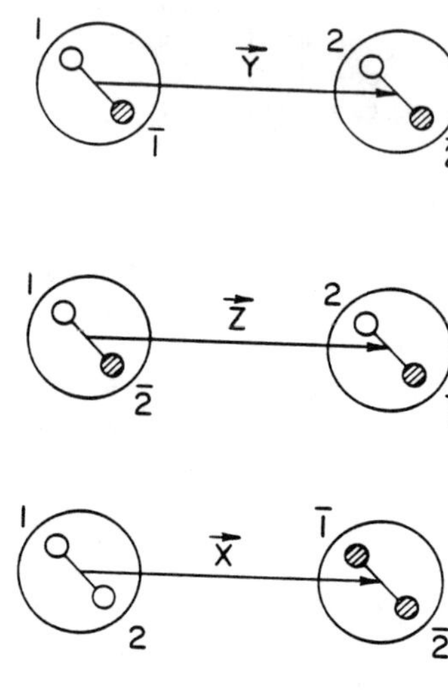

Figure 1. Relative coordinated for the two-meson problem. Here, $\vec{y}$ is the coordinate between the center of the q-$\bar{q}$ pairs ($1\bar{1}$) and ($2\bar{2}$).

baryon-baryon interaction, Oka and Yazaki [9] have included an effective meson-exchange potential for the longer-ranged part of the interaction.

At first sight, the appearance of θ-functions in the potential energy may seem to give a non-continuous potential, and also to lack any analog in more 'well-known' systems. In fact, as is easily verified, the potential energy of Eq. 2 is continuous for continuous variation of any of the coordinates. Furthermore, we can produce a very similar potential energy for a rather well known atomic system: the H^- ion. If we consider an H^- ion with electrons at positions r_1 and r_2, as shown in Fig. 2, and a (infinitely massive) proton at the origin, the potential has the form

$$U_c = -e^2\left[\frac{1}{r_1} + \frac{1}{r_2} - \frac{1}{|\vec{r}_1-\vec{r}_2|} \right] \qquad (4a)$$

$$U_c = -e^2 \left[\frac{1}{r_1}\,\theta\,(r_2-r_1) + \frac{1}{r_2}\,\theta\,(r_1-r_2) - \sum_{\ell=1}^{\infty} \frac{r_<^{\ell}}{r_>^{\ell+1}} P_\ell(\theta_{12}) \right] \qquad (4b)$$

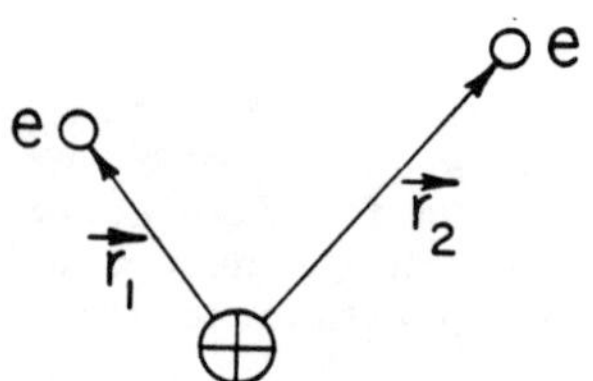

Figure 2. The H^- ion, with a massive proton at the origin and electrons at r_1 and r_2.

Eq. 4b is obtained by expanding the e-e term in Legendre polynomials. For the H^- ion, the monopole part of the potential has just the same form as for our string-flip model, except that in the quark case the potential varies as r^2, while for the atomic case the potential varies as r^{-1}. In the H^- ion, the monopole part of the potential is such that the farther electron does not 'see' the proton because it is shielded by the closer electron.

Because of the similar shielding behavior in both the atomic and quark cases, the H^- ion has many features very similar to the meson-meson case we will be studying. For example, the binding of the second electron in the H^- ion is only 6% of the binding energy of the first electron. Also, in e-H scattering, sharp resonances are observed when the energy crosses an inelastic threshold energy.[14] We will observe these features also in our string-flip model.

We can straightforwardly include color in this string-flip model with the following potential energy:

$$U = [2x^2 + y^2 + z^2 - \theta(y-z)P_y\{(2-\rho)x^2 + y^2 + (1-\rho)z^2\} - \theta(z-y)P_z\{(2-\rho)x^2 + (1-\rho)y^2 + z^2\}] \quad (5)$$

In Eq. 5 we have introduced color projection operators P_y and P_z, where $P_y = P_{1\bar{1}}P_{2\bar{2}}$, and $P_z = P_{1\bar{2}}P_{2\bar{1}}$; here $P_{i\bar{j}}$ is the projection operator onto the color-singlet state for quark i and antiquark j. In Eq. 5, we can see that the additional parameter ρ changes the strength of the interaction in the 'hidden-color' states without altering the interaction in the color-singlet sector. By varying ρ, we can then shift the energies of the hidden-color states relative to the color-singlet sector.

3. QUALITATIVE RESULTS FOR TWO-BODY SYSTEMS IN THE STRING-FLIP MODEL

Let us now consider the low-energy properties of two 'mesons', each being a $q\text{-}\bar{q}$ confined state, with the interaction of Eq. 2. The variable x is always confined, in fact with Eq. 2 the x-dependence of the wave function can just be factored out. The Hamiltonian is symmetric with respect to exchange of quarks or antiquarks; consequently, in this simple model the resulting wave functions can be either symmetric or antisymmetric under exchange of y and z (this corresponds to exchange of quarks or antiquarks). Our Hamiltonian has only one energy scale: the parameter ω, which determines the internal excitation energy of an isolated meson. For scattering of two mesons, we have two characteristic times: the first is the time scale τ for quark exchange between clusters (τ is proportional to $1/\omega$); the second is the cluster overlap time, given by $\tau_o = 2R/v$, where $R=\sqrt{3/(8m\omega)}$ is the radius of a single meson, and v is the relative velocity between the mesons.

For low energies, $\tau_0 << \tau$, and we can use an adiabatic approximation (a variation of the Born-Oppenheimer method) to decouple 'fast' and 'slow' motion and to derive an effective local potential between the centers of the clusters. The adiabatic potentials for the symmetric (S) and antisymmetric (A) cases are shown in Fig. 3. At short distances, both effective potentials are repulsive, whereas at intermediate distances the symmetric potential becomes attractive. We warn that the potentials shown here are equivalent local potentials for highly non-local interactions, a fact to which we will return later.

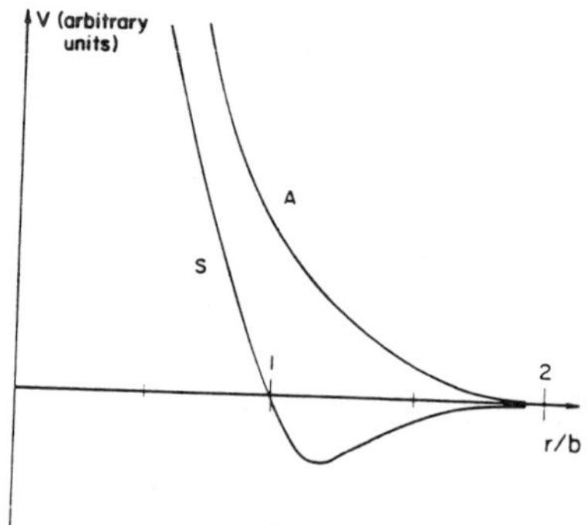

Figure 3. Adiabatic potentials for the symmetric (S) and antisymetric (A) two-meson interaction with no color. The distance is given in units r/b, where b is the h.o. parameter for a single meson. These potentials are very similar to the Heitler-London potentials for H-H scattering (Ref. 16, p. 225).

The adiabatic potentials we show in Fig. 3 are interesting for two reasons. First, the potentials shown here are very similar to those for the scattering of two hydrogen atoms. The atomic adiabatic potentials were first derived in 1927 by Heitler and London[15,16] (in the atomic case, S and A refer to the spatial symmetry of the two electrons). The large-r behavior of the atomic and meson potentials is quite different, but at short distances the two potentials are extremely similar. This reflects the fact that the short-distance behavior is dominated by the particle-exchange (quark or electron), and depends only slightly on the exact form of the potential. The second interesting feature of the symmetric potential is that it has a qualitative similarity to a 'typical' nucleon-nucleon potential, with a strong short-range repulsion and an intermediate-range attraction.

In the symmetric case, the intermediate-range attraction is sufficient to support a weakly-bound state of the two mesons. The binding energy is $E_b = 0.04\omega$, and is a result of the quark-exchange forces. We see that the quark-exchange forces can give us a 'nuclear' binding energy from a model with only a 'high-energy' scale ω (for example, if the energy scale ω were 500 MeV, then the meson-meson binding energy would be 20 MeV). The two bound particles form a state like the deuteron, i.e. where the average separation of the clusters is much larger than the size of either cluster. The mean radius of the bound state is 4.7R. Although the binding is a result only of the exchange forces of the quarks, because of the large size of the bound state, the two mesons are almost never overlapping.

We have mentioned that the symmetric potential here has some qualitative similarities with a typical N-N potential, having a short-range repulsion and an intermediate-range attraction. There is an important difference between these two potentials, however. In the one-boson-exchange (OBE) picture of nuclear forces, the repulsion is primarily due to ω exchange, while two-pion (or scalar) exchange give the medium-range attraction. Since the couplings of these mesons are presumably independent, the weak binding of the deuteron appears as a result of an 'accidental' cancellation between a large repulsion and a large attraction. In our model, since both the attraction and repulsion both arise from the same effect (quark exchange), the two pieces are correlated. We can see this by noting that since the binding energy scales as the scale ω, decreasing the potential strength does not remove the bound state. In this sense, the effective local adiabatic potential may be misleading, since the weak binding does not arise from an 'accidental' near-cancellation between two opposing pieces of the potential.

Consequently, we find that the quark-exchange forces in this model can qualitatively reproduce the short and medium-ranged parts of the nuclear force. Now, since we have ignored both spin and color in this simple model, we must ask whether these results will persist when we make our model more realistic, and attempt to calculate N-N forces. Oka and Horowitz[10] have calculated the $^1S_0$ N-N phase shift using a short-ranged quark-quark potential and and effective meson-exchange interaction for the longer-ranged part. They find a good fit to the energy dependence of this phase shift. Oka and Yazaki[11] have also investigated the contribution of the inter-quark potential to the short-ranged baryon-baryon interaction. They find that the short-range interaction depends rather sensitively on both the quark-exchange dynamics and the spin and color structure. In particular, they find that the short-range repulsion varies considerably for different baryons, and that in a few cases there may not be any short-range repulsion. This differs from OBE models in which the short-range repulsion from ω exchange is assumed to be universal for the baryon octet. Harvey and collaborators[17,18] have also investigated the relation between quark-exchange forces and the short-ranged nuclear force, and have reached similar conclusions.

Because we have used a power-law confining potential, we have a virial theorem which relates the expectation value of the potential and kinetic energies,

$$\langle T\rangle = \langle V\rangle \quad (6)$$

[more generally, if the two-body potential varies like r^{α}, we would have $\langle T\rangle = \alpha/2\langle V\rangle$].

The virial theorem persists despite the presence of the θ-functions in the potential energy of Eq. 2. In the symmetric two-meson state, the presence of a bound state means that the energy of an interacting two-particle state is smaller than the energy of two isolated particles. Because of the virial theorem, this means that the expectation value of the kinetic energy of the quarks must be smaller in the two-particle system than for a single meson: in other words, we find a 'softening' of the quark momentum distribution in a two (or many-) particle system.

The softening of the quark momentum distribution in a many-particle system is a bit surprising. If we consider n(p) to be the probability of finding a quark with momentum between p and p+dp in a system, then we might guess that the expectation value of the quark kinetic energy (which is proportional to $\langle p^2 n(p)\rangle$) would be <u>increased</u> in a many-particle system. This is due to Fermi motion effects. If the quark momentum distribution inside an individual particle were identical to that for an isolated particle, then for large p we will have an increased probability of an external probe finding a quark with this value of p, due to the internal motion of the particles in the nucleus (the momentum of the quarks is the sum of the momentum of the quark inside the particle, plus the momentum of the particle in the system). The softening of the quark momentum distribution in this many-particle system is due to quark-exchange effects. It will occur for power-law confining two-body forces in any model similar to the string-flip model we have outlined; we will later argue that this change in the quark momentum distribution can lead to observable effects, and may give us the most direct way to see quark-exchange effects in 'nuclear' systems.

Having discussed the bound state of the two-meson symmetric state in our model with no color, we discuss the scattering of two 'mesons' in this model. In Fig. 4, we show the S-wave elastic S-matrix for scattering of two mesons with no color. In the symmetric case, the phase shift (the dashed curve) starts at 180° because of the bound state. The phase shift falls with energy, but then rises by 180° due to a very narrow resonance which occurs just below the first inelastic threshold (i.e., the energy needed to internally excite one of the mesons into its first excited state). Such a resonance occurs because there is a strong attractive force between the ground state of one meson and the first excited state of the other. However, the coupling

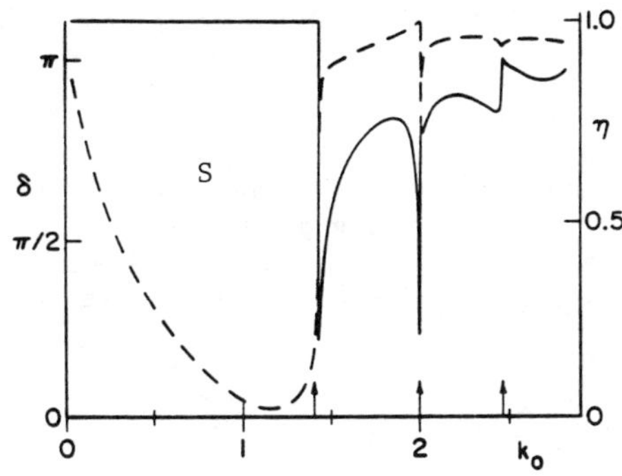

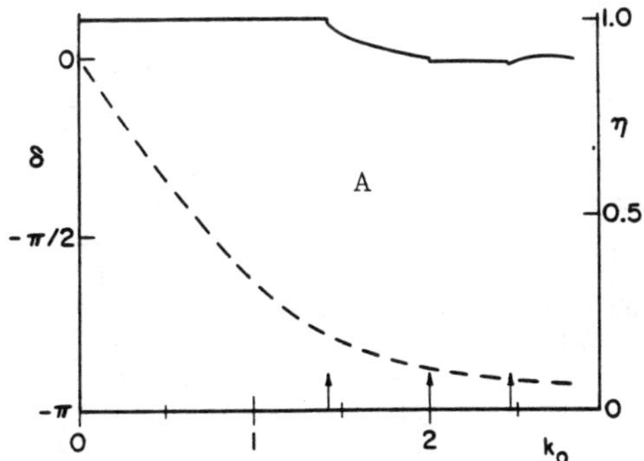

Figure 4. The elastic S-matrix, $S=\eta e^{2i\delta}$, for symmetric (S) or anti-symmetric (A) S-wave meson-meson scattering with no color. k_o is the elastic channel wave number in units b^{-1}; arrows indicate the location of inelastic thresholds.

of this state back to the ground state is weak. These two features produce the sharp resonance just below the threshold energy for this excited state. The arrows denote the positions of the inelastic thresholds, and we note that in the symmetric case a resonance occurs just below the threshold for each excited state of the system. We have previously noted that, in the e-H scattering system, sharp resonances also occur just below inelastic threshold energies [14].

In the antisymmetric case, however, the effective adiabatic potential is everywhere repulsive. Therefore, the phase shift is negative. Furthermore, at low energies the phase shift is reasonably linear with the momentum k, and behaves very much like 'hard-sphere' scattering. If we fit the scattering length and effective range at low energies, we find that a good representation of the antisymmetric scattering phase is that for 'hard-sphere' scattering with a hard-sphere radius of 1.8R. There are no resonances as we cross inelastic scattering thresholds, contrary to the symmetric case.

If we include the color degree of freedom, then we have considerable freedom since we can vary the 'string tension' in the colored states without making any change in the color-singlet potentials. We can thus move the 'hidden-color' (HC) states up or down in energy relative to the color-singlet states. One result for scattering of two mesons with color was given in Ref. 7. Oka[8] has discussed in detail a more sophisticated version of this model for

S-wave scattering of two mesons. A general feature of these string-flip models is the appearance of HC resonances, as the energy crosses color thresholds. The widths of these states vary from a few MeV to a few hundred MeV. These resonances are dominated by the confined HC component.

Jaffe[19] investigated $q^2\bar{q}^2$ states in the MIT bag model. For some values of the coupling strength ρ in Eq. 5, Oka [8] obtained rather similar results to those for the MIT bag. Meson-meson resonances have also been predicted from confining potential models[19,20]. However, their results are radically different from those of Oka, since in the potential picture the HC component is negligible.

4. OBSERVABLE CONSEQUENCES OF QUARK-EXCHANGE FORCES

In the simple model we have outlined, we can solve for the wave functions, form factors, and other observables, and we can compare the differences between the properties of isolated particles and of interacting particles. We would like to find some property which showed recognizable differences from a collection of 'free' elementary constituents bound together with effective local potentials. As we have mentioned, it may not be easy to see 'hidden-color' states unless they lie rather low in energy, since high-lying 'hidden-color' states may be extremely broad structures. The one property which we have found to give a characteristic 'signature' of quark properties in many-body systems is the 'softening' of the quark momentum distribution, which we alluded to earlier. This means that the expectation value of $\langle p^2/2m \rangle$ for quarks in a two (or more) particle bound state is smaller than that for a single isolated particle.

In the two-meson model we have used here, we find a 3-4% decrease in the quark kinetic energies for two particles in a bound state. If the quark momentum distribution in each particle was identical with that in an isolated meson, we would expect about a 9% increase in this value for the bound state, due to Fermi motion effects. We might ask whether these differences could be made to disappear by 'fine-tuning' a phenomenological potential between two hadrons. Within an 'elementary particle' description of this bound state, we must change the properties of the individual particles in order to achieve the softening of the momentum distribution.

The most efficient way to reproduce the observed change in the momentum distribution is to increase the size of the hadrons in the nucleus, relative to the size of an isolated hadron. For h.o. wave functions, an increase in the size parameter b will correspondingly decrease the kinetic energy. In our model, we find that if the mesons in the bound state increase their size parameter by 5% relative to a free meson, then we can reproduce the observed decrease in the kinetic energies. This means that the particles in a many-body system appear to be somewhat 'larger' than isolated free particles. This apparent increase in size for particles in a 'nucleus' directly reflects the quark-exchange processes which produce this result. In the language of the bag model, this occurs because in a many-hadron system the individual bags may overlap, and quarks in different bags can interact with one another. Then we have 'deconfinement transitions' which effectively make the bag larger in the many-body environment than an isolated bag.

This has been discussed recently in the context of bag models by Jaffe et al. [21] Also, Shakin et al. [22] have recently proposed that the nucleon electromagnetic form factors should be different for nucleons in a nucleus than for free nucleons; these changes come about through effects of interactions between quarks in different bags. Now, the quark momentum distribution function is not itself an observable. How can we find observables which will show the 'signature' of these changes in the quark momentum distribution? This can be seen in deep inelastic scattering from systems of hadrons at sufficiently high energy loss and momentum transfer. For large enough energy transfer, we can relate the deep inelastic structure function directly with the quark momentum distribution n(p). The details are given in Close [23]; if we neglect the quark mass and interactions in the final state, and the binding energy and recoil in the initial state, and we work in the limit where both the momentum transfer q and energy loss ν go to infinity while their ratio

$$x = (q^2-\nu^2)/2M\nu \tag{7}$$

remains constant (here M is the hadron mass, and m is the constituent mass), then we obtain

$$F(x) \cong 2\pi M \int_{M\left|\frac{m}{M}-x\right|}^{\infty} dp\; pn(p) \tag{8}$$

In order to see the changes which occur in the structure function, we need to know the qualitative changes which occur in the quark momentum distribution n(p). Let n(p) be the momentum distribution for quarks in a many-body system, and $n_o(p)$ be the quark momentum distribution in an isolated particle. If both of these are normalized to unity, and if we define $\delta n(p) = n(p)-n_o(p)$, then we have the conditions : (1) $\langle\delta n(p)\rangle = 0$ (since both distributions are normalized); (2) $\langle p^2\delta n(p)\rangle < 0$ (this is a consequence of the virial theorem for the many-body system); (3) as $p \to \infty$, $\delta n(p) > 0$ (due to Fermi-motion effects). The simplest form for $\delta n(p)$ which satisfies conditions (1)-(3) is shown in Fig. 5. Qualitatively, it is similar to the results we obtain with our model. It must have at least two nodes.

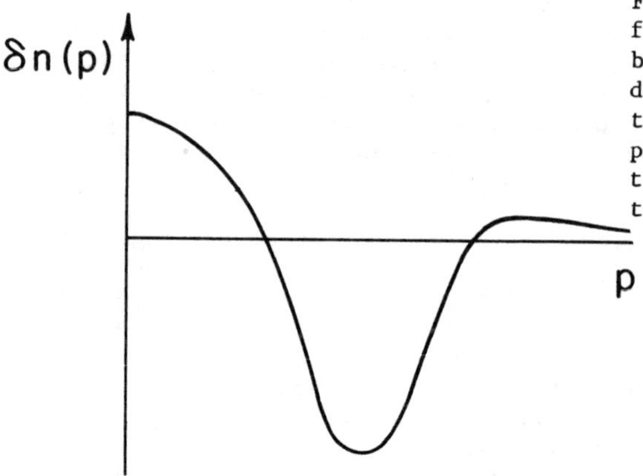

Figure 5. Qualitative form for δn, the difference between the quark momentum distributions for the two-meson and one-meson problem. δn satisfies the three conditions listed in the text.

For our model, we can insert our values for n(p) and no(p) into Eq. 8 and we can obtain the structure functions for a single meson ($F_o(x)$) and for a two-meson bound state (F(x)). We are particularly interested in the difference between these two structure functions, $\delta F(x) = F(x)-F_o(x)$. In Fig. 6, we have plotted the quantity $\delta F(x)/F_o(x)$ vs. x for our model. The upper values of x (for the number of quark constituents N=2) are given above the x-axis. However, we wish to relate these structure functions to analogous results which are seen for nuclear targets. In this case, we have three constituent quarks in a hadron. If we use the same momentum distribution n(p) which we obtain from our model, but assume the ratio $M/m = N = 3$ in Eq. 8, then the respective values of x are given below the x-axis in Fig. 6. The difference between the single-particle and the 'nuclear' structure functions shows up as an oscillatory behavior for δF(x) vs. x. For $x \geqslant 0.8$ (here we are using the x-values appropriate to a nucleon), δF(x)/F(x) becomes increasingly positive. This is due to Fermi motion effects. However, the characteristic feature of this curve is a negative value for δF(x) for x between about 0.5 and 0.7. This is clearly a result of the quark-exchange effects in our model.

Such an effect in the structure functions for deep inelastic scattering seems to be experimentally verified. The European Muon Collaboration (EMC) [24] has measured structure functions for muon scattering from deuterium and iron at energies from 100 to 280 GeV. Deep inelastic electron scattering has been carried out on several targets at SLAC [25]. Although there may not be complete consistency between the various results, significant differences were found between the structure functions for heavier nuclei and those for deuterium. This is called the 'EMC Effect'. The ratio of cross sections for Fe to those from D is shown in Fig. 7. The open squares in this plot are from the EMC work, and the solid circles are SLAC results. There is a prominent dip in this ratio between x= 0.4 and x= 0.8 .

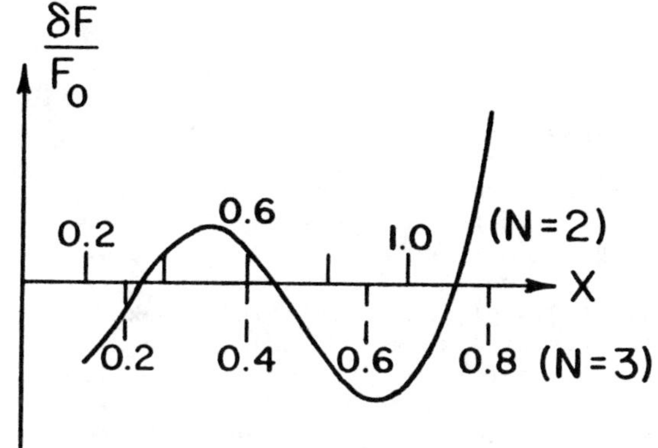

Figure 6. $\delta F/F_o$ vs x. $\delta F \equiv F-F_o$, where F and F_o are the quark structure functions for the two-particle and one-particle case, respectively. N is the ratio of the hadron mass to the constituent mass. The upper scale is for N=2 (meson) and the lower scale for N=3 (baryon).

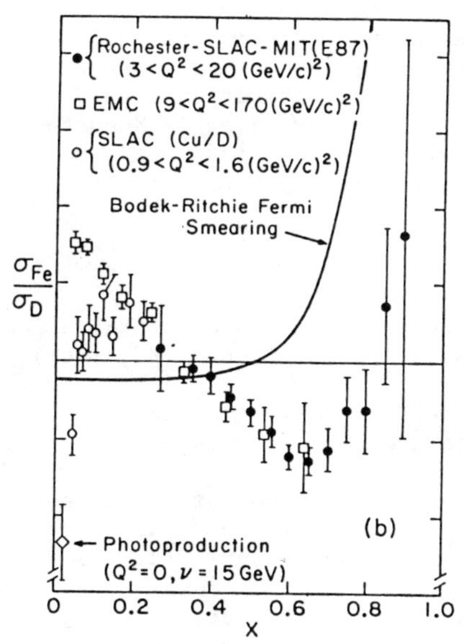

Figure 7. The 'EMC Effect'. Open squares: σ_{Fe}/σ_D for muon scattering, Ref. 24; open circles: σ_{Cu}/σ_D for electron scattering, Ref. 25; solid circles: σ_{Fe}/σ_D for electron scattering, Ref. 25.

From our simple model, we suggest that the A-dependence of the structure functions for deep-inelastic scattering can straightforwardly arise from quark-exchange effects. Although there are many alternative explanations of the EMC effect [26], our result is very similar to the bag-model description of Jaffe et al. [21].

5. CONCLUSIONS

We have presented here a simple model which illustrates the physics of the string-flip models. These offer promising ways to incorporate the interactions of clusters of confined quarks, while retaining the identity of quarks in different clusters, and without introducing some of the pathological difficulties which plague so many potential models of quark effects in nuclei. We showed the qualitative features of our simple model, and were able to make analogy to a well-known atomic physics system. We showed how many of the qualitative features of hadron-hadron interactions are obtained by explicitly including quark-exchange effects between clusters. We discussed briefly the bound state and low-energy scattering properties of our 'meson-meson' system in the string-flip model. Finally, we showed that the 'softening' of the quark momentum distribution, predicted by this model, may be the cause of the EMC Effect seen in deep inelastic scattering by electroweak probes from nuclei.

Although we have concentrated on this one simple model, a considerable amount of work has been done on extensions of this approach to more realistic meson-meson and baryon-baryon interactions. The work we have described here was done in collaboration with F. Lenz (SIN), E. Moniz (MIT), R. Rosenfelder (Mainz), M. Stingl (Muenster), and K. Yazaki (Tokyo). We also would like to acknowledge very useful discussions with M. Oka and C. Horowitz (MIT).

6. REFERENCES

1. R.L. Jaffe, Proceedings of XVII International School of Subnuclear Physics, Erice, Italy, 1979, ed. A. Zichichi (Plenum, NY, 1982), p. 69.
2. A.W. Thomas, Chiral Symmetry and the Bag Model, Advances in Nuclear Physics 13, ed. J.W. Negele and E. Vogt (Plenum, NY, 1983).
3. N. Isgur, Proceedings of XVI International School of Subnuclear Physics, Erice, Italy, 1980, ed. A. Zichichi (Plenum, NY, 1983).
4. A. duRujula, H. Georgi and S.L. Glashow, Phys. Rev. D12, 147 (1975).
5. F. Lenz, J.T. Londergan, E.J. Moniz, R. Rosenfelder, M. Stingl and K. Yazaki, to be published.
6. K. Yazaki, Nucl. Phys. A416, 87c (1984).
7. J.T. Londergan, in Hadron Substructure in Nuclear Physics, AIP Conference Proceedings No. 110, ed. W-Y.P. Hwang (AIP, NY, 1984), p. 285.

8. M. Oka, CTP preprint #1194, 1984, submitted to Phys. Rev. D.
9. M. Oka and K. Yazaki, in Quarks and Nuclei, ed. W. Weise (World Publishing Co., Singapore, 1984).
10. M. Oka and C.J. Horowitz, CTP preprint #1206, 1984, submitted to Phys. Rev. Dl.
11. C.J. Horowitz, E.J. Moniz and J.W. Negele, to be published.
12. C.J. Horowitz, in Hadron Substructure in Nuclear Physics, AIP Conference Proceedings No. 110, ed. W-Y.P. Hwang, (AIP,NY,1984), p. 312; C.J. Horowitz and R. Panoff, in Intersections Between Particle and Nuclear Physics, AIP Conference Proceedings No. 123, ed. R. Mischke (AIP, NY, 1984), p. 527.
13. N. Isgur and J.E. Paton, Phys. Lett. 124B, 247 (1983).
14. S. Geltman, Topics in Atomic Collision Theory, (Academic Press, NY, 1969), pp. 142-158.
15. W. Heitler and F. London, Z. Physik 44, 455 (1927).
16. G. Herzberg, Atomic Spectra and Atomic Structure, (Dover Publishing, NY, 1944), pp. 223-228.
17. M. Harvey, Nucl. Phys. A352, 301,326 (1981).
18. M. Harvey, J. LeTourneux and B. Lorazo, Nucl. Phys. A424, 412 (1984).
19. R.L. Jaffe, Phys. Rev. D15, 267,281 (1977).
20. J. Weinstein and N. Isgur, Phys. Rev. Lett. 48, 659 (1982); Phys. Rev. D27, 588 (1983).
21. R.L. Jaffe, F.E. Close, R.G. Roberts and G.G. Ross, Phys. Lett. 134B, 449 (1984).
22. L.S. Celenza, A. Rosenthal and C.M. Shakin, Phys. Rev. Lett. 53, 892 (1984).
23. F.E. Close, An Introduction to Quarks and Partons, (Academic Press, NY, 1979).
24. J.J. Aubert et al. (EMC Collaboration), Phys. Lett. 105B, 315, 322 (1981); Phys. Lett. 123B, 275 (1983).
25. A Bodek et al., Phys. Rev. Lett. 50, 1431 (1983); Phys. Rev. Lett. 51, 534 (1983); R.G. Arnold et al., Phys. Rev. Lett. 52, 727 (1984).
26. C.H. Llewellyn Smith, Phys. Lett. 128B, 107 (1983); M. Ericson and A.W. Thomas, Phys. Lett. 128B, 112 (1983); C. Carlson and T. Havens, Phys. Rev. Lett. 51, 261 (1983).

NUCLEAR REACTIONS WITH POLARIZED BEAMS

R.C. Johnson
Department of Physics, University of Surrey,
Guildford, Surrey, GU2 5XH
ENGLAND

1. INTRODUCTION

In this talk I want to discuss several topics which illustrate the power of polarized beam studies of nuclear reactions. As I shall show, even measurements as comparatively simple as elastic scattering take on a new richness when performed with polarized particles of spin greater than $\frac{1}{2}$. As an example of elastic scattering I shall mention recent developments in the interpretation of polarized ^{6}Li and ^{7}Li scattering from heavy targets. Some new results on transfer reactions with polarized deuterons will be discussed as another example of the novel information obtainable from polarization measurements.

2. SPIN-DEPENDENCE IN HEAVY-ION SCATTERING

2.1 Description of Polarized Beam Data

All the experimental data I shall refer to was obtained by the Heidelberg-Marburg collaboration using the polarized ^{6}Li and ^{7}Li source at Heidelberg [1)], which until recently was unique. New sources are being constructed at Daresbury, Florida, and Wisconsin and I have recently seen preliminary data from a new source at Heidelberg.

Most people are familiar with the description of polarization experiments involving spin- $\frac{1}{2}$ particles. A key feature of the Li results is the fact that both isotopes have spin greater than $\frac{1}{2}$: ^{6}Li has $I = 1$, ^{7}Li has I = $\frac{3}{2}$. An important consequence of this is that, e.g., in the elastic scattering of ^{6}Li from a spin zero target, there are 5 independent observables at each scattering angle and incident energy, even if no measurements are made of the polarization of the beam after scattering. The corresponding number of observables in the spin- $\frac{1}{2}$ case is 2: the unpolarized differential cross-section and the left-right asymetry (vector analyzing power). The larger number of observables in the spin-1 and spin- $\frac{3}{2}$ cases reflects the fact that the general state of polarization of a beam of such particles can not be specified by giving a single direction and a set of populations of magnetic sub-states with respect to that direction *. The 5 observables are, of course, merely combinations of scattered intensities corresponding to different choices of the spin state of the incident beam. A crucial feature of the Heidelberg polarized ion source is that this spin state can be selected at will and in a controlled fashion.

The 5 real observables appropriate to a spin-1 or spin $\frac{3}{2}$ particle are usually denoted:

$\sigma_0(\theta)$: differential cross-sections for an unpolarized beam.

$iT_{11}(\theta), T_{20}(\theta), T_{21}(\theta), T_{22}(\theta)$: analyzing powers for various choices of the incident polarization state.

I refer to reference [2] for the details of how these observables are related to intensities. A key point is the fact that they have been defined to form components of tensors: σ_0 is a tensor of rank zero, the $T_{1q}(\theta)$ (q = ± 1,0) are the components of a vector (rank-1) tensor, and the T_{2q} (q = ± 2, ± 1,0) are the components of a 2nd rank tensor. If parity is conserved, and if the tensor components are referred to the conventional Madison [3] co-ordinate system (z-axis along incident beam, y-axis along normal to scattering plane) then $T_{10} = 0$, T_{11} is pure imaginary, and the T_{2q} are purely real and satisfy $T_{2q} = (-)^q T_{2-q}$.

---

* An elegant way of describing the general situation is given by G. Ramachandran and V. Ravishankar in a contribution to this Conference.

The tensor analyzing powers are related to the (2I + 1) × (2I + 1) scattering matrix, M(θ), by

$$T_{kq}(\theta) = \frac{\mathrm{Trace}(M\tau_{kq}M^{+})}{\mathrm{Trace}(MM^{+})} , \tag{1}$$

where the τ_{kq} are a standard set [3] of irreducible tensor operators constructed out of the components of the spin operator $\underline{I}$. The denominator in eq. (1) is proportional to $\sigma_o(\theta)$. (For I = $\frac{3}{2}$ (e.g. ^{7}Li) observables with k = 3 are also possible, but I will have little to say about them).

The importance of this way of organising the observables is that it corresponds to a useful way of classifying the spin-dependence of the Hamiltonian of the system. In an optical potential description the usual $\underline{L}.\underline{I}$ interaction is a vector in spin-space; for I > $\frac{1}{2}$, spin-dependent forces involving tensors of higher rank are possible. The best known example of the latter is the analogue of the well-known tensor force in the nucleon-nucleon interaction. In terms of the separation, $\underline{R}$, between projectile and target, this tensor force can be written

$$U(R)\ T_R , \tag{2}$$

where

$$T_R = (\underline{I}.\hat{\underline{R}})^2 - \tfrac{1}{3}\underline{I}^2$$

$$\equiv T_2(\underline{R}).\tau_2(\underline{I}) . \tag{3}$$

and τ_{2q} is an irreducible rank-2 spin tensor. The tensor $T_{2q}(\underline{R})$ in eq. (3) is proportional to $Y_{2q}(\underline{R})$. Tensor forces involving the projectile-nucleus momentum (T_p interaction), or angular momentum (T_L interaction), are also possible as I shall mention later.

To a very good first approximation for heavy-ions (the situation is somewhat more complicated for light ions such as deuterons) the vector observable is dominated by the interference between central and $\underline{L}.\underline{I}$ potentials, and the 2nd rank tensor observables are dominated by the interference between central and 2nd rank tensor forces. The differential cross-section arises almost entirely from the central forces. The different components of the analyzing power are therefore sensitive to quite different parts of the interaction Hamiltonian. As I will explain below, the different types of spin dependence are in turn associated with different elements of the dynamics of the collision process.

2.2 Differential Cross-section and Vector Analyzing Powers

The experimental data for σ_o/σ_R in the case of Li + ^{58}Ni at 18.1 MeV are shown in the upper parts of Figs. 1 and 2. They display typical Fresnel-like diffraction patterns and are very similar for the two Li isotopes. Clearly the central parts of the interaction between ^{58}Ni and ^{6}Li and ^{7}Li are very similar and σ_o contains very limited detailed information about the structure of the nuclei involved.

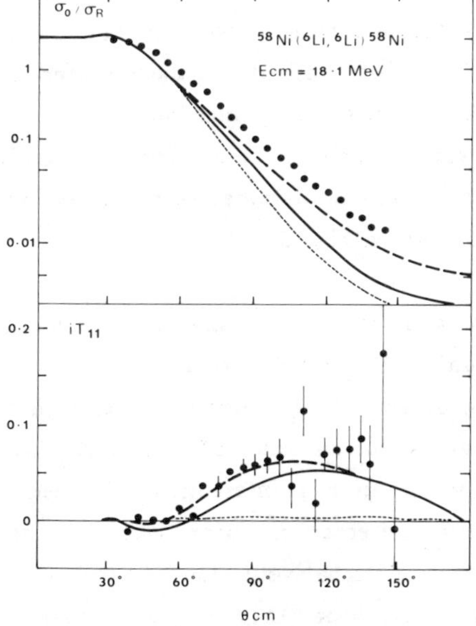

Fig. 1. Data [4)] and calculations [5)] for the differential cross-section (ratio to Rutherford) and vector analyzing power for ^{6}Li + ^{58}Ni elastic scattering at E_{cm} = 18.1 MeV. The short-dashed, solid and dashed curves are the results of one-channel, two-channel, and four-channel calculations, respectively, included the ^{6}Li ground and excited states shown in Fig. 3.

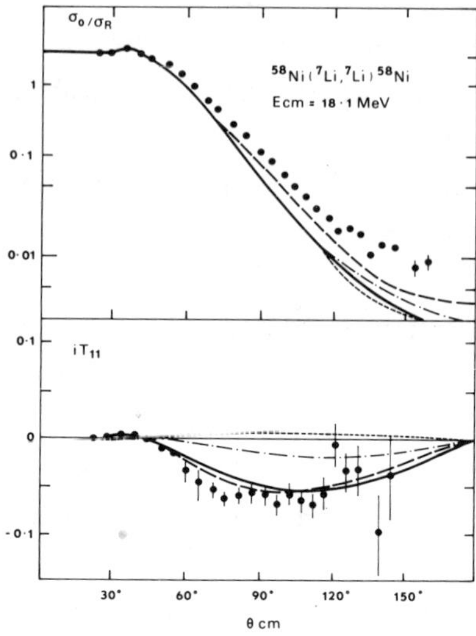

Fig. 2. Data [6)] and calculations [5)] for the differential cross-section (ratio for Rutherford) and vector analyzing power for ^{7}Li + ^{58}Ni elastic scattering at E_{cm} = 18.1 MeV. The short-dashed and dash-dotted curves are the results of one-channel calculations neglecting and including the T_R tensor contribution to the ^{7}Li ground state matrix element. The solid and dashed curves are the results of 2 and 4 channel calculations, respectively, including the ^{7}Li ground and excited states shown in Fig. 3.

The situation with the vector analyzing powers, shown in the lower parts of Figs. 1 and 2, is very different:

(a) The analyzing power $iT_{11}(\theta)$ has opposite sign for the two isotopes.

(b) The $iT_{11}(\theta)$ are about an order of magnitude larger than expected, i.e. as estimated using an $\underline{L}.\underline{I}$ interaction obtained by double-folding the nucleon-nucleon interaction over the Li and Ni ground state wavefunctions. Alternatively, the $\underline{L}.\underline{I}$ interaction can be estimated in a cluster model picture of ^{6}Li(d + α) and ^{7}Li(α + t), in which case the Li-Ni $\underline{L}.\underline{I}$ interactions are generated by the d-Ni and t-Ni $\underline{\ell}.\underline{s}$ interactions. The results of this single-folding procedure are very similar to the double-folding calculations and are shown by the short-dashed curves in Figs. 1 and 2.

An explanation of the source of the discrepancies and an excellent consistent account of the data without invoking any freely adjustable parameters has been given independently by two Groups [5,7]. These

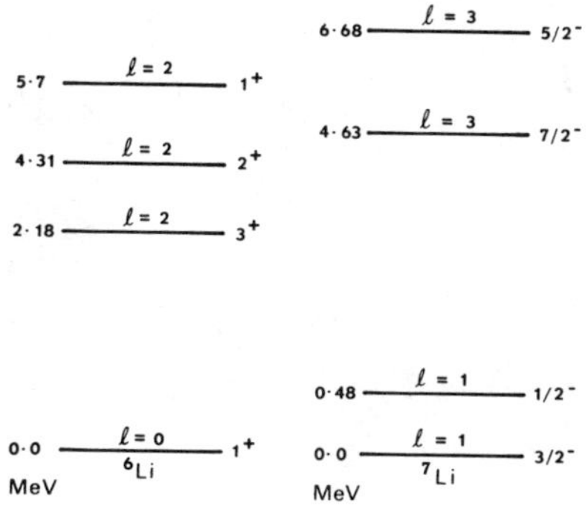

6·68 $\ell = 3$ 5/2⁻
5·7 $\ell = 2$ 1^+
4·63 $\ell = 3$ 7/2⁻
4·31 $\ell = 2$ 2^+
2·18 $\ell = 2$ 3^+
0·48 $\ell = 1$ 1/2⁻
0·0 $\ell = 0$ 1^+ ^{6}Li MeV
0·0 $\ell = 1$ 3/2⁻ ^{7}Li MeV

Fig. 3. Low-lying levels of ^{6}Li and ^{7}Li taken into account in the projectile-excitation calculations of refs. 5) and 7). At E_{cm} = 18.1 MeV the projectile excitation effects are dominated by the 3^+ state in ^{6}Li and the $\frac{1}{2}^-$ state in ^{7}Li.

calculations show that both features (a) and (b) arise because of the extreme sensitivity of the effective $\underline{L}.\underline{I}$ interaction, and hence $iT_{11}(\theta)$, to the detailed nature of the low-lying spectra of the two Li isotopes (Fig. 3). Two step processes involving the $\frac{1}{2}^-$ state (0.48 MeV) in ^{7}Li and the 3^+ (2.18 MeV) state in ^{6}Li turn out to be strongly spin-dependent. A simple explanation of this spin-dependence, including the sign difference as observed, has been given in ref. 5) within a simple (t + α) and (d + α) cluster description of these states. It is important to note that this mechanism does not involve any intrinsic spin-dependence of the cluster-target potentials whatsoever. Effects due, e.g., to the d-Ni $\underline{\ell}.\underline{s}$ interaction in the ^{6}Li case are very small. In the cluster model picture of refs. 5) and 7) the effective $\underline{L}.\underline{I}$ force is induced by quadrupole nuclear tidal forces generated by the target nucleus and acting on the relative motion of the clusters in the projectile.

It can be shown [2] that the effective $\underline{L}.\underline{I}$ interaction generated through a two step process involving a state of spin I' and induced by a quadrupole interaction has the general form $V_{so}(R,R',L)\underline{L}.\underline{I}$, where

$$V_{so}(R,R',L) = -\frac{3}{4}\frac{[6+I(I+1)-I'(I'+1)]}{I(I+1)(2I+1)}$$

(4)

$$\times \frac{L-1}{(2L+1)(2L-1)} \bar{V}^{(2)}_{I'I}(R)\ \bar{G}_{L,I'}(R,R')\ \bar{V}^{(2)}_{I'I}(R'),$$

and where

$$\bar{G}_{L,I'}(R,R') = G_{L-2,I'} + \frac{2L+1}{(L-1)(2L+3)} G_{L,I'}$$

(5)

$$- \frac{(L+2)(2L-1)}{(L-1)(2L+3)} G_{L+2,I'}\ ,$$

is a weighted sum of Green functions for propagation in the intermediate excited state with orbital angular momentum L, L-2, L+2, and $\bar{V}^{(2)}_{I'I}$ (R) is an excitation form factor as defined in ref. 2). This expression clearly shows the dependence of the sign of the effective $\underline{L}.\underline{I}$ interaction on the spin of the excited state (the square bracket has the value 9 for I = $\frac{3}{2}$, I' = $\frac{1}{2}$ (^{7}Li case) and - 4 for I = 1, I' = 3 (^{6}Li case)), and the crucial rôle played by the L-dependence of the Green function (note that $\bar{G}_{L,I'} \equiv 0$ if $G_{L',I'}$ is independent of L). The expression (4) gives the quantum mechanical counterpart of the semi-classical arguments of ref. 5).

The importance of the "projectile excitation effect" for the $\underline{L}.\underline{I}$ interaction for ^{6}Li and ^{7}Li scattering at other low energies (≤ 20 MeV) and for other targets has also been demonstrated [8].

2.3 Tensor Analyzing Forces

The tensor analyzing powers $T_{2q}(\theta)$ for ^{6}Li and ^{7}Li on the same target are also quite different: large for ^{7}Li and very small for ^{6}Li (see Fig. 4). This has a simple interpretation in terms of the very different shapes of the two isotopes. ^{7}Li has a large negative spectroscopic quadrupole moment (Q^{Charge} = - 3.4 ± 0.6 efm^{2}). For ^{6}Li, Q^{Charge} is also negative, but very much smaller in magnitude (Q^{Charge} = - 0.064 efm^{2}). A simple qualitative picture of the situation is shown in Fig. 5 in terms of the component $^{T}T_{20}(\theta)$ of the tensor analyzing power referred to a z-axis along the normal to the scattering plane. This 'shape-effect' idea works very well for ^{7}Li scattering [10]. The fact that $^{T}T_{20}(\theta)$ for ^{6}Li is small and positive, rather than small and negative as expected on the simple picture, can be understood in terms of a simple α + d cluster model of ^{6}Li. According to refs. 5) and 12), the main contribution to the observed ^{6}Li tensor analyzing power at energies near the Coulomb barrier arises from the positive quadrupole moment of the deuteron cluster in the ^{6}Li ground state.

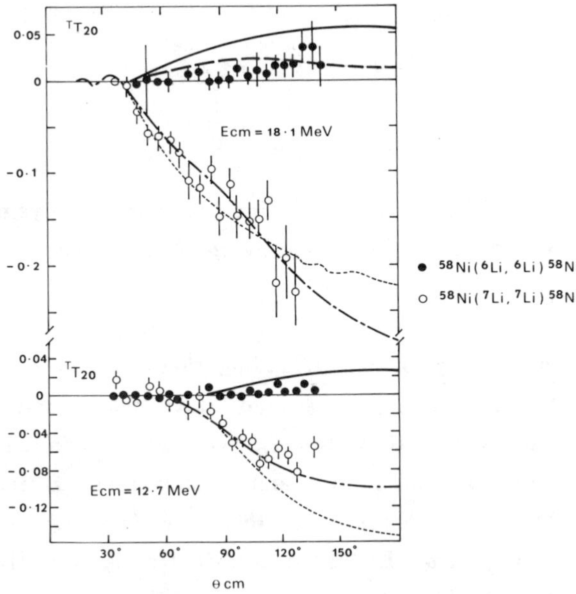

Fig. 4. The tensor analyzing power component $^{T}T_{20}$ for beams aligned perpendicular to the scattering plane. The data for ^{6}Li + ^{58}Ni (solid circles) are from refs. 4) and 9) and for ^{7}Li + ^{58}Ni (open circles) are from refs. 10) and 11). The calculations are from ref. 5). The solid and dashed curves in the ^{6}Li case, and the short-dashed and dash-dotted curves in the ^{7}Li case, correspond respectively to one channel (i.e. folding model with T_R terms) and multichannel calculations including the states shown in Fig. 3.

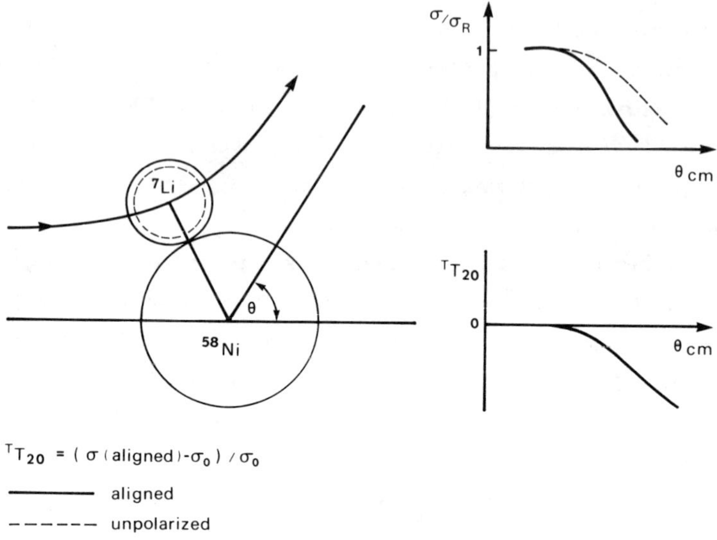

Fig. 5. "Shape effect" picture [10] for the tensor analyzing power in ^{7}Li scattering, in which ^{7}Li is represented by an oblate spheroid.

A convenient way of understanding these results quantitatively is to recognise that the oblate ^{7}Li ground state wavefunction automatically generates a tensor potential of the T_R type when a folding model is used for the ^{7}Li optical potential, and quantitative agreement with the Heidelberg data can be obtained in this way (see Fig. 4). This tensor force also is so large for ^{7}Li that it contributes significantly to iT_{11} in 2nd order (see caption to Fig. 2).

Further insight into the situation can be obtained by using a co-ordinate system with z-axis along $(\underline{K}_i + \underline{K}_f)$ and x-axis along the direction of momentum transfer. The latter direction coincides with the direction, $\underline{R}_o(\theta)$, to the turning point of the classical Coulomb orbit (see

Fig. 6).

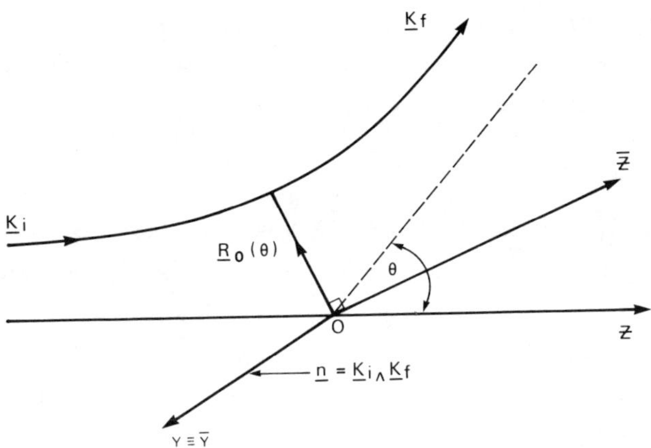

Fig. 6. Madison Convention (OXYZ) and 'barred' ($O\overline{XYZ}$) co-ordinate systems.

Quantities referred to this co-ordinate system are denoted by bars: $\overline{T}_{21}$, $\overline{T}_{22}$, $\overline{T}_{20}$, etc. The second order effects of the $\underline{L}.\underline{I}$ interaction have a very small effect on the T_{2q} and can be ignored in the first instance. To first order in the T_R force

$$3\sigma_o(\Theta)\overline{T}_{2q}(\Theta) = 2\mathrm{Re}(\overline{M}_{oo}\overline{M}^*_{2q}) \ , \tag{6}$$

where

$$\sigma_o(\Theta) = \frac{1}{3}\ |\overline{M}_{oo}|^2 \ . \tag{7}$$

In these expressions the $\overline{M}_{kq}(\Theta)$ $(0 \leqslant k \leqslant 2I)$ are defined [13] as the coefficients in the expansion of the scattering matrix in terms of irreducible spin-tensors $\tau_{kq}(\underline{I})$. The $\overline{M}_{kq}$ are irreducible tensors constructed from $\underline{K}_i$

and $\underline{K}_f$ and expression in terms of S-matrix elements are given in ref. 13). To first order in spin-dependent terms in the Hamiltonian, $\bar{M}_{oo}$ is determined by the central forces alone and $\bar{M}_{2q}$ is proportional to the strength of any tensor forces of rank-2 in spin space.

A special property of the barred co-ordinate system is that $\bar{M}_{21} = 0$ (if general symmetry requirements are satisfied), and hence that

$$\bar{T}_{21}(\theta) = 0 \ . \tag{8}$$

This prediction is satisfied very well by the data (see Fig. 7) and verifies that the spin-dependent forces can be treated perturbatively at these energies.

In the same approximation

$$\bar{M}_{2q} = \langle \underline{K}^{(-)}_{out} | \bar{V}_{2q} | \underline{K}^{(+)}_{in} \rangle \ , \tag{9}$$

where V_{2q} is the coefficient of $\tau^{+}_{2q}(\underline{I})$ in the expression for the tensor force, and the $|\underline{K}^{\pm}\rangle$ are distorted waves generated by the central potential only. For a T_R-force V_{2q} is just $u(R)Y_{2q}(\underline{R})$ where $u(R)$ is a radial form factor.

Following Nishioka and Johnson [14,15] we use the 'turning point model' which assumes that the matrix element $\bar{M}_{2q}$ is dominated by contributions from $\underline{R}$ close to the turning point $\underline{R}_o$ of the classical orbit. It is easy to see from Fig. 6 that this means that in the barred co-ordinate system the polar angles of $\underline{R}_o$ are $(\pi/2, 0 \text{ or } \pi)$, and hence that

$$\bar{M}_{22}(\hat{\underline{R}}_o) = - \left[\frac{3}{2}\right]^{1/2} \bar{M}_{20}(\hat{\underline{R}}_o) \ , \tag{10}$$

$$\bar{M}_{21}(\hat{\underline{R}}_o) = 0 \ . \tag{11}$$

The last formula when used in eq. (6) just confirms the result (8), but eqs. (10) and (6) give something new i.e.

$$\bar{T}_{22}(\theta) = - \left[\frac{3}{2}\right]^{1/2} \bar{T}_{20}(\theta) \ . \tag{12}$$

This means that there is only one independent tensor analyzing component. The relationships implied by eqs. (8) and (12) are found to be quantitatively very well satisfied in exact quantum mechanical calculations with physically realistic T_R potentials.

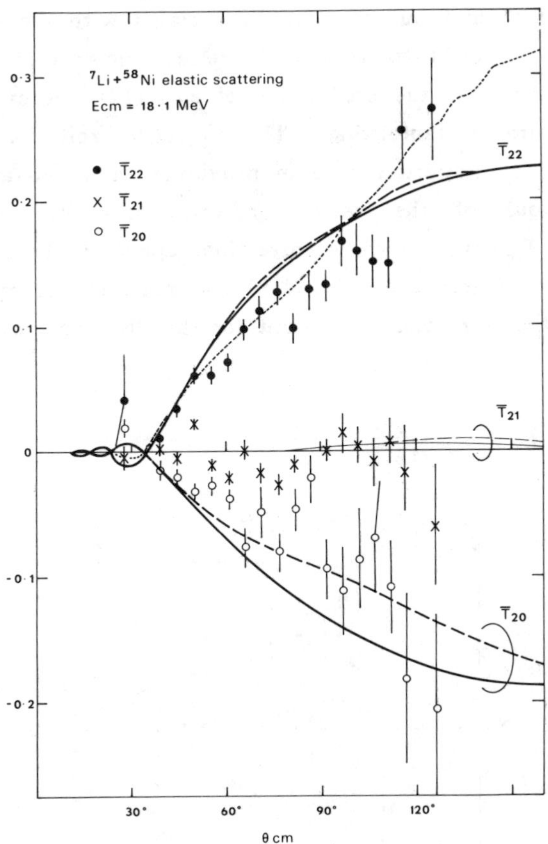

Fig. 7. Experimental data [10)] and calculations [12)] for the 'barred' components of the tensor analyzing powers for ^{7}Li + ^{58}Ni elastic at E_{cm}= 18.1 MeV. The solid curves are one-channel calculations with folding-model potentials and the dashed curves include a T_P potential. The dotted curve shows a 4-channel calculation including the ^{7}Li excited states given in Fig. 3 ($\bar{T}_{22}$ only shown).

The experimental results [10] are displayed in Fig. 7, where they are compared with calculations [5,15] based on a single-folding, cluster-model approach to the calculation of the complex central and T_R components of the ^{7}Li optical potential. It is clear that the main feature of all three components of the tensor analyzing powers can be accounted for in this way, i.e. in terms of effects associated with the non-spherical shape of the ^{7}Li matter distribution. There do, however, appear to be systematic deviations from the predictions of eq. (12), as can be seen in Fig. 8. The situation is tantalizing. The projectile excitation mechanism, which played such an important rôle in providing an understanding of the anomalous behaviour of the vector analyzing powers, also produces corrections to the $\bar{T}_{2q}(\theta)$; but these corrections appear to be predominantly of the T_R type [14,15] (see Figs. 7 and 8) and preserve the relationship of eq. (12). They therefore cannot account for the discrepancies.

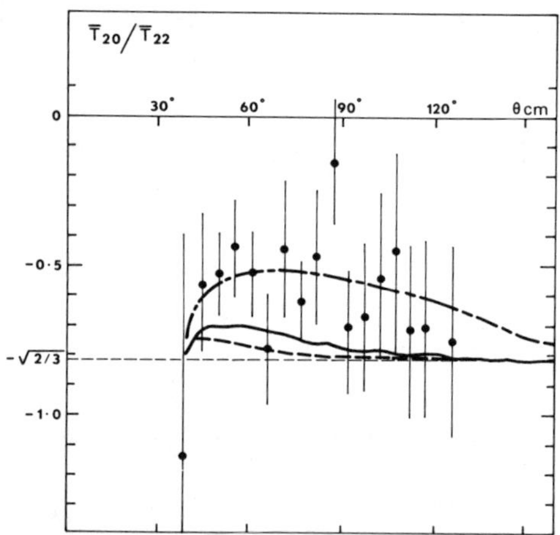

Fig. 8. The ratio $\bar{T}_{20}/\bar{T}_{22}$ for the same case as Fig. 7. Solid and dashed curves are one-channel and 4-channel calculations respectively. The dash-dotted curve is the result of a one-channel calculation including a T_P potential.

A possible explanation of the discrepancies has been given by Nishioka and Johnson [14,15] who point out that a tensor force of the T_P type is expected to produce relations between tensor observables which differ radically from eq. (12). For a T_P-type force, $V_{2q} = Y_{2q}(\underline{P})$ and hence, using the turning point model,

$$\bar{T}_{22} = 0 \,, \tag{13}$$

as well as the general result $\bar{T}_{21} = 0$.

Hence, one way to explain the observed discrepancies is to introduce a T_P-type potential of suitable strength. The dash-dotted curve in Fig. 8 shows the result of an exact quantum mechanical calculation with a T_P force of the type suggested by Ioannides and Johnson [16]. A good fit is obtained, but it is clear that more accurate data for other energies and target is required before a very firm conclusion can be made.

What physical mechanism can produce a T_P-type potential? Folding models can in principle produce such a term, but calculations [15] show that this is very small. (In a single folding model such a term would arise from a nucleon-nucleus $\underline{\ell}.\underline{s}$ potential with a momentum dependent form factor). Ioannides and Johnson [16] showed that such a term always arises when a non-spherical nucleus propagates through a sea of identical fermions. For example, the overlap of the momentum distribution of a non-spherical deuteron with the Fermi sea of a target nucleus depends on the orientation of the deuteron's shape (and hence its spin) relative to the momentum of its centre-of-mass. Pauli-blocking effects, which modify the deuteron optical model, will therefore also depend on the relative orientation of the deuteron's spin and momentum. No quantitative estimates of the analogous effect for ^{7}Li have been published. The T_P potential used to obtain the fit shown in Figs. 7 and 8 was simply carried over from the predictions for deuterons in ref. 16). Attempts to detect the effect in deuteron scattering have not so far been successful, mainly because of the large $\underline{L}.\underline{I}$ interaction experienced by a deuteron [15]. For heavy-ions the relative smallness of the $\underline{L}.\underline{I}$ interaction and the usefulness of the semi-classical turning point model mean that in principle it should be possible to detect such a term if its magnitude is comparable to that expected for deuterons.

3. ($\vec{d}$,α) REACTIONS AND THE α-PARTICLE D-STATE

Polarization phenomena in transfer reactions with light ions are known to show a strong sensitivity to the structure of the initial and final states involved. The classic example is the (d,p) reaction where the vector analyzing powers for two final states with the same neutron orbital angular momentum ℓ_n but different $j_n (= \ell_n \pm \frac{1}{2})$ have opposite sign. Tensor analyzing powers are extremely sensitive to the amplitude and phase of the deuteron D-state component. These results are well established experimentally and can be understood in a direct reaction framework [17].

New applications of these ideas have arisen from work on (d,α) reactions with vector and tensor polarized deuterons at Birmingham and TUNL. These studies [18-23] are providing important tests of the amplitude and phases of deuteron-like cluster components in shell-model wavefunctions as well as giving the clearest evidence to date for a D-state component of the alpha-particle.

In the simplest one-step direct reaction model, the A(d,α)B reaction amplitude involves two vertices: d + (np) → α and the associated overlap $\langle d,(np)|\alpha\rangle$, and A → (np) + B and its overlap integral $\langle B(np)|A\rangle$, where (np) denotes a cluster with the quantum numbers of the deuteron. The possible angular momentum structure of these overlaps is severely restricted by the spins of the nuclei involved. For example, the $\langle d,(np)|\alpha\rangle$ overlap has two components in which the deuteron clusters have relative angular momentum $\ell = 0$ and $\ell = 2$. The latter component arises mainly from non-spherical components in the α-particle wavefunction induced by the tensor force in the nucleon-nucleon interaction. There is considerable interest in checking the predictions of 4-nucleon calculations of these components. In the case of a $0^+ \rightarrow 1^+$ transition the $\langle A|(np),|B\rangle$ overlap also contains L = 0 and L = 2 components in general. Polarization studies are important, because for example, in the absence of a deuteron-nucleus spin-orbit potential, the tensor analyzing powers in a $0^+ \rightarrow 1^+$ transition are predicted to be zero unless L = 2 admixtures are present in at least one of the overlaps.

An important consideration here is that at the low energies studied so far (9 MeV < E_d < 16 MeV) the only property of the α-particle

D-state which enters the tensor analyzing powers is the number $D_2(d,\alpha)$ which is closely related to the asymptotic D/S ratio of the <d, (np) | α> overlap. This number is not related in any simple way to the D-state probability, but it is nevertheless an important property of the α-particle, as is the analogous quantity for the deuteron [24,25]. All the transitions studied so far involve this same quantity D_2, whereas of course they cover a range of completely unrelated <A | (np), B> overlaps. The relative phases of the components of the latter and the sign and magnitude of $D_2(d,\alpha)$ have a crucial influence on the tensor analyzing power data. The Birmingham-Surrey study [5] showed that the $^{40}Ca(d,\alpha)^{38}K$ data at 12.3 MeV could be fitted with $D_2 = 0$, but only if the relative amplitude of L = 2 and L = 0 (np) clusters in ^{40}Ca had the opposite sign to that predicted by shell-model calculations [26]. Consistency with the shell-model calculations could be obtained provided $D_2(d,\alpha)$ was given a value close to - 0.31 fm^2. This value is close to that obtained from other [19,20,23] (d,α) studies as well as simple estimates based on α-particle wavefunctions [18,20]. Refined calculations of $D_2(d,\alpha)$ values corresponding to more complete α-particle wavefunction would be of great value. Further $(\vec{d},\alpha)$ measurements involving targets and final states whose shell-model configuration have been tested independently are also needed.

The reactions (d,^{6}Li), (^{6}Li,d), (^{6}Li,α), (^{7}Li,α) hold out similar promise with regard to the structure of d and α-cluster states as well as the <d,α | ^{6}Li> and <α,t | ^{7}Li> vertices. The deuteron-nucleus spin-orbit force can by itself generate tensor analyzing powers in a transfer reaction and this can cause problems. The vector analyzing power can often give an important guide here as well as confirming the dominant orbital angular momentum transfer involved. The ^{6}Li and ^{7}Li spin-orbit forces are probably less important, and we can learn here from the insights gained from elastic scattering which I discussed in Section 2. The whole question of the rôle of multi-step processes does introduce some uncertainty. In some cases these can be ignored with some confidence [23], but the subject has been surrounded with considerable cosntroversy and a lot of open questions remain [27]. The (d,α) studies I have highlighted certainly ignore these contributions and obtain a consistent account of the polarization data.

4. ACKNOWLEDGEMENTS

I am grateful to Dr. P.K. Iyengar and his Organising Committee for inviting me to India and the Indian National Science Academy and the Royal Society for financial support. I am also grateful to my colleagues H. Nishioka, J.A. Tostevin, G. Windham, and F.D. Santos for the many discussions we have had about the physics discussed here.

REFERENCES

1. Kramer, D, Becker, K., Blatt K., Caplar R., Fick, D., Gemmeke, H., Haeberli, W., Jansch, H., Karban, O., Koenig, I., Luh, L., Mobius, K.-H., Necas, V., Ott, W., Tanaka, M., Tungate, G., Turkiewicz, I., Weller, A. and Steffeus, E., Nucl. Inst. and Methods 220, 123 (1984). An extensive list of references to the experimental results relevant to this talk can be found in ref 5) below.

2. Johnson, R.C., Lecture Notes of the 1983 RCNP-Kikuchi Summer School (Ed. by H. Ogata, M. Fujiwara and A. Shimizu, Osaka University, Osaka, 1983) p.194; Johnson, R.C., Genshikakukenkyu 23, 1 (1979).

3) Polarization Phenomena in Nuclear Reactions, (Edited by H.H. Barschall and W. Haeberli, U. of Wisconsin Press, Madison, 1971) p.143.

4. Rusek, K, Moroz, Z., Caplar, R., Egelhof, P., Mobius, K.-H., Steffens, E., Koenig, I., Weller, A. and Fick, D., Nucl. Phys. A407, 208 (1983).

5. Nishioka, H., Tostevin, J.A., Johnson, R.C. and Kubo, K.-I., Nucl. Phys. A415, 230 (1984).

6. Tungate, G., Bottger, R., Egelhof, P., Mobius, K.-H., Moroz, Z., Steffens, E., Dreves, W., Koenig, I. and Fick, D., Phys. Lett. 98B, 347 (1981).

7. Ohnishi, H., Tanifuji, M., Kamimura, M., Sakaguchi, Y. and Yahiro, M., Nucl. Phys. A415, 271 (1984).

8. Windham, G., Nishioka, H., Tostevin, J.A. and Johnson, R.C., Phys. Letts. 138B, 253 (1984).

9. Dreves, W., Zupranski, D., Egelhof, P., Kassen, D., Steffens, E., Weiss, W. and Fick, D., Phys. Letts. 78B, 36 (1978).

10. Moroz, Z., Zupranski, D., Bottger, R., Egelhof, P., Mobius, K.-H., Tungate, G., Steffens, E., Dreves, W., Koenig, I. and Fick, D., Nucl. Phys. A381, 294 (1982), and references therein.

11. Zupranski, D., Dreves, W., Egelhof, P., Mobius, K.-H., Steffens, E., Tungate, G. and Fick, D., Phys. Letts. 91B, 358 (1980).

12. Nishioka, H., Tostevin, J.A. and Johnson, R.C., Phys. Letts. 124B, 17 (1983).

13. Johnson, R.C., Nucl. Phys. A293, 92 (1977); Hooton, D.J. and Johnson, R.C., ibid, A175, 583 (1971).

14. Nishioka, H. and Johnson, R.C., Phys. Rev. Letts. 53, 1881 (1984).

15. Nishioka, H. and Johnson, R.C., "Tensor interactions in heavy-ion scattering I. The turning point model", accepted for publication in Nucl. Phys. A, Jan. 1985.

16. Ioannides, A.A. and Johnson R.C., Phys. Rev. C17, 1331 (1978).

17. Knutson, L.D. and Haeberli, W., in Prog. Part. and Nucl. Physics (Pergamon Press 1980) 3, p. 127.

18. Santos, F.D., Tonsfeldt, S.A., Clegg, T.B., Ludwig, E.J., Tagishi, Y. and Wilkerson, J.F., Phys. Rev. C25, 3243 (1982).

19. Tostevin, J.A., Phys. Rev., C28, 961 (1983).

20. Santos, F.D., Prog. Theor. Phys. 70, 1679 (1983).

21. Santos, F.D. and Eiro, A.M., Portugaliae Physica 15, 65 (1984).

22. Tostevin, J.A., Nelson, J.M., Karban, O., Basak, A.K. and Roman, S., Phys. Letts. 149B, 9 (1984).

23. Karp, B.C., Ludwig, E.J., Thompson, J. and Santos, F.D., Phys. Rev. Letts. 53, 1619 (1984).

24. Johnson, R.C. and Santos, F.D., Part. Nucl. 2, 285 (1971).

25. Ericson, T.E.O. and Rosa-Clot, M., Nucl. Phys. A405, 497 (1983); Ericson, T.E.O., Nuclear. Phys. A416, 281C (1984).

26. Chung, W., and Wildenthal (unpublished, quoted in ref. 22).

27. Pinkston, T., Comm. Nucl. Part. Phys. 12, 133 (1984).

ON THE CONSTRUCTION AND USE OF THE NORMALISABLE RESONANT STATES

A.V. Lagu
Department of Physics, Banaras Hindu University
Varanasi-221 005
INDIA

ABSTRACT

The concept of a resonance occurs widely in many areas of physics. As far as the structure and characteristics of transition amplitudes containing resonance poles are concerned these are well understood and widely used. The question of an appropriate wavefunction, however, is usually treated with much less emphasis. In this article we discuss the various formulations for a resonant state in eigen and wave packet approaches, their usefulness in the general study of decay of quantum unstable systems, implications for the decay law and other remifications. It is also demonstrated how analyticity and some plausible quantum requirements play an important role in the formulation of a viable description of a normalisable resonant state.

1. INTRODUCTION

Resonances occur very frequently in both classical and quantum physics. Their presence in classical electricity, acoustics and optics is well known. In quantum physics, Gamow's description[1)] of alpha decay of radioactive nuclei dates back to more than fifty years where solutions of the Schroedinger equation, which behave at large separation distances as pure outgoing waves and belong to complex eigenvalues, were employed. These solutions are popularly called Gamow functions[2] and have been widely used in the formal theories of nuclear resonance reactions[3] and in the extension to the continuum of the nuclear shell model[4]. More recently its applications have been extended to the nuclear cluster model of collisions and reactions of light nuclei[5].

A major difficulty in using these functions in practical applications is caused by the fact that the Gamow function represents a decaying state in position representation and is not confined to a finite volume of space. Hence it represents a wave of exponentially increasing amplitude which is not square integrable. Our purpose here is to discuss in what ways the Gamow function is inadequate and how viable alternatives can be constructed. We shall also try to demonstrate through applications the usefulness of the alternatives.

Our attention shall be focussed on a two body system of reduced mass μ in the center of mass frame interacting via a well behaved[6,7] potential V, usually a sharp cut-off one or rational separable one. We shall confine ourselves to only a particular partial wave. Of course, these shall not be strong restrictions but rather useful ones. We shall use this article to both review the subject matter and also to discuss some new prescription. Because of the limitations of space, many topics have to be necessarily omitted or glossed over. Non-relativistic dynamics is used even if $k \to \infty$.

2. RESONANCE AND UNSTABLE STATE

Let us consider, among many others, the following three topics:

i) Transfer to unbound states[8],

ii) $\triangle$ and N* components in nuclei[9,10], and

iii) Radioactive decay law.

The item which is common to all the three is the existence of a quantum unstable system. In usual terminology, an unstable system is closely identified with a resonance. It really depends on a particular situation whether the term unstable state; we shall, in the following, use them synonymously. Indeed when time development is being studied "unstable state" is more appropriate.

Coming now to the first case, if we consider the specific example of the stripping reaction[8]

$$A(d,p)\ B^* \rightarrow A+n \tag{1}$$

in which B* is a long lived resonant state, then a wavefunction for B* is needed during the evaluation of the angular distribution of the outgoing proton employing the DWBA[8]. Can we use the Gamow function for this purpose ?

In the second case[9], suppose we have a bound state $|B\rangle$ of three particles, of which 2 and 3 have a scattering resonance. The amplitude for this resonance in the three body bound state is $\langle \Psi_{23}, \vec{P}_1 \mid B \rangle$ where $|\Psi_{23}\rangle$ represents the resonance. Can we use the Gamow function for $|\Psi_{23}\rangle$? This is not a fictitious example; the π +d process, in which N* or Δ resonances are produced, is very much a real process.

In discussing the radioactive decay law, we note that the experimental observation

$$N(t) = N(0)\ e^{-\Gamma t} \tag{2}$$

holds good to an amazing degree of accuracy[11,12]. Here N(t) and N(0) are the number of unstable nuclei produced at times t and 0 respectively and $1/\Gamma$ represents the lifetime of the unstable nuclei.

Obviously, the general description of the law should read

$$N(t) = N(0)\ P(t) \tag{3}$$

where P(t) is the probability of finding, for a measurement at time t, the quantum system in the same physical situation i.e. in the same state in which it was with certainty at the initial time t = 0. More precisely, since a quantum state is uniquely associated with a precise physical situation, P(t) can be written as

$$P(t) = |\langle \Psi_{UN} | e^{-iHt} | \Psi_{UN} \rangle|^2 = |A(t)|^2 \tag{4}$$

where H is the hamiltonian governing the dynamical evolution of the quantum system under investigation. Clearly, we need to know the quantum state $|\Psi_{UN}\rangle$ for studying the time evolution of the system. Can we use Gamow function for this purpose ? Although the decay law is very nearly exponential as seen from eq.(2), it can be proved[13] very generally that there must arise deviations from it at very small and very large times. We shall also see that whereas 'causality' demands that a measuring counter should start recording events (flux) only after the time required for the decay products to traverse the distance upto the counter has elapsed, the time dependent Gamow function violates 'causality' in this sense.

We shall discuss a little later the various formulations for resonant states; however, it is quite instructive to have a quick glance at the various resonance reaction formulations because, in the process, some understanding about the resonant state is necessarily obtained.

3. RESONANCE REACTION THEORIES

Analogous to Robson[3], we can categorise the theories in four broad classes depending upon the use of a modified Hamiltonian and/or modified boundary condition. Obviously, there is considerable overlap of one approach with the other. These classes are (i) S-matrix theories based on analyticity[1,2,14] (ii) R-matrix approaches based on channel radii[15] (iii) Projection operator approaches[16] and (iv) Unified approaches trying to combine best features of all[17]

4. THE RESONANT STATE-EIGEN CHARACTER

The information about the resonant state that is obtained in the process of formulating resonance reaction theories can then be briefly classified into (i) Gamow-Siegert type[1,2], (ii) Kapur-Peierls type[15], (iii) Wigner-Eisenbud type[15], (iv) Feshbach type[16] and (v) Tobocmann-Nagarajan type[17].

This does not, of course, exhaust the list of the various kinds of resonant states used in practice. The earliest and most widely used is of course the Gamow function. Their use makes the problem non-self-adjoint and usual quantum mechanical rules for normalisation, orthogonality and completeness do not apply[18]. Although it is usual to call these as eigenfunctions of the hamiltonian belonging to the complex eigenvalues

$$H\,|\Psi_j\rangle = E_j\,|\Psi_j\rangle\,,\quad E_j = W - i\frac{\Gamma}{2} \tag{5}$$

the boundary condition (which is purely outgoing)

$$\Psi_j(r) \underset{r\to\infty}{\sim} \frac{e^{ik_j r}}{r} \tag{6}$$

with $$k_j^2 = 2\mu E_j\,,\; W = \frac{\alpha^2-\gamma^2}{2\mu}\,,\; k_j = \alpha - i\gamma\,,\; \Gamma = \frac{2\alpha\gamma}{\mu} \tag{7}$$

leads to a probability density

$$|\Psi_j(r)|^2 \underset{r\to\infty}{\sim} \frac{e^{2\gamma r}}{r^2} \tag{8}$$

and causes serious problems in many applications, some of which were mentioned in Section 2 above. Indeed many artifacts have been used to regularise these functions e.g. (i) use a Gaussian damping factor in $\psi_j(r)$ (Garcia-Calderon[14] and Berggren[14] i.e. replace $\psi_j(r)$ by $\lim_{\epsilon\to 0} \exp(-\epsilon r^2)\,\psi_j(r)$ which is a subtle problem of (ii) modifying the normalisation rule (Berggren[14])

$$\int_0^d dr\, r^2\, \psi_j^2(r) + i\, \psi_j^2(d)\, \frac{d^2}{2k_j} = 1 \qquad (9)$$

for the case of a sharp cut-off-potential of range d, or (iii) working entirely in momentum space if the potential admits a Fourier transform[19]. In this last prescription of Mondragon and Hernandez[19] the momentum representative of the Gamow function can be obtained as solutions of homogeneous Lippmann-Schwinger equation for outgoing particles corresponding to complex energy eigenvalues. This approach is close in spirit to Weinberg's Quasiparticle method[20] but greater care is needed. It is very elegant and is closer in spirit to the use of transition amplitudes for doing all dynamical calculations.

In fact, it may be asked that why worry about a resonant state, per se, whether in coordinate or momentum representation when almost all dynamical calculations can be done via transition amplitudes ? The answer is that there do exist certain situations where it is either easier to do calculations with the use of a state, or to obtain greater insight into the problem. The topics mentioned in Section 2 above fall in this category. Moreover, there is the nagging aesthetic question "What a resonant state is after all ?". Considerable misconception exists about a resonant state, from identifying it to some kind of a bound state to its close connection withscattering states with complex eigenvalues ! In fact, in the study of the decay theory of quantum unstable systems, the need for a state vector is in some sense crucial.

It must be mentioned that there exist some alternative approaches to the Gamow state e.g. boundary condition models[21], modified hamiltonian-perturbative methods[22], the powerful dilation analytic method[23] and the unconventional rigged Hilbert space formulations[24]. Despite their apparent differences what is common amongst all is the concept and use of some kind of eigencharacter for the resonant or unstable state. The advantages of this class are (1) the underlying problem is mathematically clear cut, and (2) a unique solution is usually obtained because an equation is normally solved.

The major disadvantage is that the state is usually non-normalisable leading to a loss of probabilistic interpretation. This has the effect of creating the genuine risk that the matrix elements involving these states may either blow up or converge slowly. An alternative exists via the wave packet approach.

5. THE WAVE PACKET MODELS

The wave packets signify concentrated bunches of waves with a mean position and spread in configuration space at any time. These have been used in as varied areas as in non-linear field theories and in the derivation of the uncertainty principle.

5.1 S-matrix and Analyticity

But before we discuss these models, we note from figure 1 the distribution[6] of the poles and zeros of the S-matrix because we would be employing analyticity arguments a great deal. We shall assume that a one to one correspondence exists between the complex poles of the S-matrix given by

$$S^{+}(k) = D^{-}(k)/D^{+}(k) = \exp\left(2i\,\delta(k)\right) \tag{10}$$

which are close to the real axis, unstable particles and state vectors. $D^{\pm}(k)$ are Jost functions (or Fredholm determinant).
Fonda et al[12] have discussed at some length the fact that it is not necessary to associate poles to the unstable systems i.e. it does not

follow from the general requirements to be imposed on the S-matrix. However, to get a calculable theory for the unstable state, it is a natural and convenient starting point.

We then note that the resonance poles occur in pairs if the pole at k_j represents a decaying state, then its reflection at $-k_j^*$ can be identified with a capture state. This has considerable significance which is often missed. In fact, unless a state has been formed through a production mechanism, it is inappropriate to discuss its decay[11]. We shall assume that the poles of S-matrix in the LHP are simple and their number may be finite or infinite depending respectively upon the potential being rational separable or sharp cut-off type.

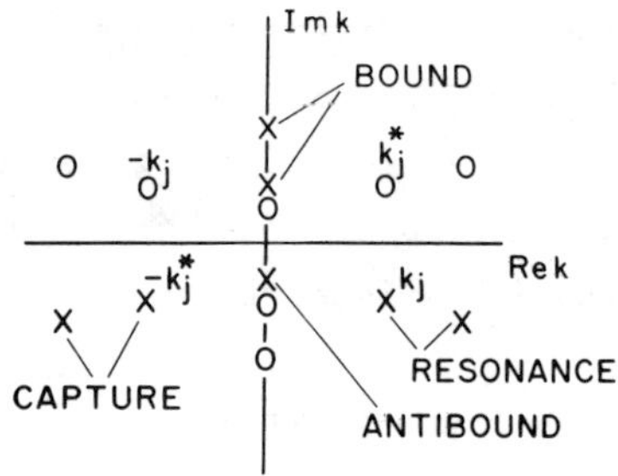

Figure 1 Poles (X) and zeros (O) of the S-matrix in the complex momentum plane.

We assume for physical potentials that every zero of $D^+(k)$ is simple. Interestingly, however, a pole of S at K_j is not only accompanied by another at $-k_j^*$, but also by zeros in the UHP at the conjugates of these two points, as shown in figure 1, denoted by k_j^* and $-k_j^*$ and $-k_j$. A pole on the positive imaginary axis is similarly accompanied by a conjugate zero in the LHP. It is the conjugate zero, equidistant from the real axis, which prevents S from violating the unitarity condition on the real axis when a pole occurs nearby.

We can also discuss the importance of these pole-zero pairs in the complex energy plane. An elegant and detailed exposition can be found in McVoy[6]. We would, however, be mostly concerned with k-plane analyticity and shall see, in what follows, that the zeros in UHP have an important role to play apart from the capture pole, a fact often missed by several workers involved in the construction of wave packet models for the resonant state.

5.2 The Wave Packet of Krylov-Fock[25)]

With these preliminaries, we can set to construct an isolated, elastic resonant state $|\xi_j\rangle$ through the superposition of a complete set of bound states $|\psi_b\rangle$ and scattering states $|\psi_k^\pm\rangle$ at t = 0. In the subspace spanned by $|\psi_k^\pm\rangle$ (suppressing other quantum numbers),

$$|\xi_j\rangle = \frac{1}{2\pi^2}\int_0^\infty dk\, k^2 |\psi_k^\pm\rangle \rho_j^\mp(k) + \sum_b |\psi_b\rangle \rho_j(b) \qquad (11)$$

We have deliberately chosen to use $|\xi_j\rangle$ to signify a resonant state in wave packet model to distinguish it from the Gamow function $|\psi_j\rangle$. $|\psi_k^+\rangle$ are outgoing scattering functions for the hamiltonian $H = H_o + V$.

The expansion coefficients ρ are unknown and their determination will be the major task in this approach. However, Krylov-Fock's[25)] interest was in the study of the general features of the time dependent decay process of quantum unstable systems and not determining the resonant state wavefunction as such. Subsequent workers e.g. Khalfin[26)], Coester and Schlessinger[27)], Fonda et al[12)] and many others[28)] have also been concerned mostly with establishing the validity of (or deviations from) the exponential law. Thus in most of these models, $|\xi_j\rangle$ has not been constructed with sufficient care. We shall come to this aspect several times in the following discussion. Nevertheless many general theorems about the survival probability P(t), see eq. (3), could be proved using only standard theorems of mathematical analysis and assuming the states to have only a finite norm and finite average energy. However, in all such proofs, it was necessary to assume $\rho_j(b) = 0$.

Inspite of the generality of the model (11) it has the following undesirable features:

(i) the dependence of ρ's on the magnetic projection m is superfluous because of rotational invariance (because we have suppressed m and l in eq.(11), these do not appear explicitly) a fact ignored by earlier workers,

(ii) although the choice $\rho_j(b) = 0$ is invariably made, the necessity for this choice is not discussed except in the sense of the previous paragraph, and

(iii) because of lack of guidelines for $\rho_j^-(k)$, the resulting state $|\xi_j\rangle$ has large ambiguities.

5.3 Khalfin's Model[26)]

It was shown by Khalfin that if the momentum spectral density $|\rho_j^-(k)|^2$ has the Lorentzian shape ($E=k^2/2\hbar$), a very obvious choice

$$|\rho_j^-(k)|_k^2 = \pi\Gamma/\{\hbar k((E-W)^2+\Gamma^2/4)\} \qquad (12a)$$

$$= 8\pi\alpha\gamma/\{k(k^2-k_j^2)(k^2-k_j^{*2})\} \qquad (12b)$$

then an approximate exponential decay law holds for narrow resonances. This was obviously an important advance. However this prescription still suffered from the drawback that the canonical average energy of the state is infinite, and the phase of $\rho_j^-(k)$ is left arbitrary.

It must be obvious by now that $|\xi_j\rangle$ is not an eigenstate of H, and in sharp contrast to Eigenstate theories, the knowledge of the hamiltonian H alone is not enough to fix the wave packet state. It is also clear that $|\xi_j\rangle$ cannot become an eigenstate of H unless $\rho_j^-(k)$ happens to be just delta function. Indeed, there is no completely self-contained unique method to generate ρs from any deep dynamical principles. We can only hope to construct a minimal form for $\rho(k)$ consistent with some mathematical requirements and plausibility considerations.

5.4 The Primitive Wave Packets with Poles

In an effort to overcome this great arbitrariness in $\bar{\rho}(k)$, some ansatz's were proposed by Coester-Schlessinger-Payne[27] and Fonda-Ghirardi-Rimini[12] by trying to incorporate two pieces of information viz.

i) $|\rho_j^-(k)|^2$ is expected to have poles at $E=E_j$, E_j^* in conformity with Lorentzian form, and

(ii) Gamow state $|\psi_j\rangle$ has an exact vertex function[29] $a_j(k) = \langle k|V|\psi_j\rangle$. $a_j(k_j)$ is the on-shell value of the vertex function and measures the coupling strength of the state $|\psi_j\rangle$ to its two constituents in a state of magnitude of relative momentum k. Alternatively $a_j^2(k_j)$ measures simply the residue of the on-shell amplitude $T^+(k)$ at $k = k_j$.

With these pieces of information some models for $\bar{\rho}(k)$ were constructed with the general form

$$\rho_j^-(k) = 2i\, a_j(k)\, \mathrm{Num}(k)/\mathrm{Den}(k) \qquad (13)$$

where Num (k) is essentially arbitrary but slowly varying over resonance peak and vanishing rapidly as E moves away from W, while Den (k), contains information about the dynamical points k_j, $-k_j^*$. The nice features of these ansatzs are (i) the state can be made normalizable, (ii) $|\rho_j^-|^2$ is sharply peaked around E=W and (iii) reaction matrix elements involving $|\xi_j>$ are expected to be well behaved. Specifically, these are

$$\rho_j^-(k)_{Coester} = 2\ \mu a_j(k)\ Num(k)/(k^2-k_j^2) \qquad (14)$$

$$\rho_j^-(k)_{Fonda} = 2\ \mu a_j(k)\ Num(k)/D^+(k) \qquad (15)$$

However, for lack of any guidelines for Num(k), the resulting state becomes highly ambiguous; indeed rather wild exponential and Lorentzian shapes with large powers have been tried by Schlessinger and Payne[27].

On the other hand injudicious choices for Den(k) can lead to serious conflict with analyticity, e.g. the use of $D^+(k)$ in (15) for Den(k) gives rise to double poles in the saturated T-matrix at all the dynamical points which is a bad feature.

5.5 Conformity with Analyticity

An ad-hoc choice by Menon & Lagu[30]

$$\rho_j^+(k)_{ML} = S^+(k)\ \rho_j^-(k) = \frac{2\ \mu\ a_j(k)}{(k-k_j)(k+k_j^*)} \qquad (16)$$

is somewhat superior as far as analyticity of the saturated T-matrix is concerned because of only simple poles at k_j and $-k_j^*$, but none of the properties of the corresponding state can be calculated in closed form.

At their face value the above models viz. (12), (14)-(16) look reasonable because the spectral density $|\rho_j^-(k)|^2$ has a Lorentzian shape near E=W, but as mentioned in above paragraphs all of them are unsatisfactory because of wrong analytic properties and/or absence of specific physical/mathematical constraints leading to a highly arbitrary state, whether in closed form or otherwise.

6. AN IMPROVED WAVE PACKET DESCRIPTION

Menon & Lagu[31)] showed that a satisfactory ρ can be constructed by using analyticity as a powerful tool andimposing upon the system a few plausible quantum mechanical requirements. By analyticity is meant that ρ should be made up of standard quantities like the dynamical points $\pm k_j$, $\pm k^*_j$, the functions $D^{\pm}(k)$ etc. while the quantum requirements shall ensure the normalisability of $|\xi_j\rangle$ and correct pole residues of the saturated T-matrix. We may caution that they have been unable to establish a rigid set of necessary and sufficient conditions for fixing one hundred percent. The resulting ρ is by no means unique, at best it can be viewed as a minimal form with the hope that it is able to demonstrate its viability and superiority in applications.

Menon & Lagu[31)] discuss in detail the reasons for setting $\rho_j(b) = 0$. Because the appearance of a resonance is ascribed to the continuum, the bound state need not play a role in this appearance. Also like in most eigenstate theories, resonance state is expected to be orthogonal to all bound states. At a deeper level, in a time dependent picture we would require that the survival probability should not contain constant terms coming from the bound state.

The general form of $\rho^+_j(k)$ as proposed by the authors[31)] is

$$\rho^+_j(k) = X_j(k)+(-)^1 S^+(k)X_j(-k) \tag{17a}$$

with $$X_j(k) = \frac{D^-(k)F(k)Y(k)}{(k^2-k_j^2)(k^2-k^{*2}_j)} \tag{17b}$$

The denominator tells that only the points corresponding to the jth resonance + capture pair and their reflections through the real axis are relevant so that X_j will be singularity free at other dynamical points. The use of $D^-(k)$ ensures that only the poles possessed by T^+- matrix survive in the resonance saturation limit. We shall discuss saturation at some length a little later.

The adjustable function F(k) is taken to be analytic and non-zero at $k = \pm k_j, \pm k_j^*$ and serves to compensate for essential singularities (as well as kinematic poles which appear due to the geometric shape of the potential) of the half shell amplitude; while other adjustable function Y(k) is a polynominal and its purpose is to guarantee correct residues in the resonance saturated amplitude. The addition of the second term on the rhs of (17a) is to ensure that maximal exploitation of Cauchy's residue theorem could be made, so that closed expressions could be obtained as much as possible. A packet vertex function $\Omega_j(p)$ is now defined by $\Omega_j(p) = \langle p | V | \xi_j \rangle$ in analogy with the Gamow vertex function $a_j(p)$ having the same physical significance. This function is needed later. The mathematical implication is that $\Omega_j(p)$ is directly expressible as a linear combination of $a_j(p)$ and its conjugate function. This would not have been possible had the ansatzs (14) or (15) been used.

The general behaviour of the radial wave function $\xi_j(r)$ cann now be obtained by considering in detail the structure of the various integrals involved. However, one has now to distinguish between a sharp cut off and rational separable potential case. In particular, for both cases, $\xi_j(r)$ behaves like $r^{-(\ell+1)}$ symptotically instead of exponential like. However, this is not really detrimental. The various integrals would still converge. In any case, this behaviour is definitely better than the asymptotic behaviour of the Gamow function (see eq.(6)). Interestingly, at small distances, for a sharp cut-off potential of range d, $\xi_j(r)$ is regular at the origin and becomes equal to a linear combination of the Gamow function and its conjugate within $r \leq d$. This is a very nice feature. Similarly $\xi_j(r)$ for a rational separable case can also be shown to be regular at the origin. For intermediate regions, without the specific knowledge of the potential or form factors and F & Y, much information cannot be obtained excepting some analytic properties.

At this point, some plausibility conditions have been imposed so that the form of ρ could be pinpointed as much as possible.

(i) Finiteness of $\langle \xi_j | H^n | \xi_j \rangle$, n = 0,1,2

We can obtain some more insight into $|\xi_j\rangle$ and its usefullness through the matrix elements $M^{(n)}_{j'j} = \langle \xi_{j'} | H^n | \xi_j \rangle$ for n=0,1,2 where j and j' label two wave packet states. Clearly $M^{(0)}_{jj}$ is the squared norm of the $|\xi_j\rangle$ itself while $M^{(1)}_{jj}$ and $M^{(2)}_{jj}$ are intimately linked with average canonical energy ξ_{jc} and mean canonical spread Δ_{jc} in the energy viz. (see also Lane[32] and Peres[28])

$$\xi_{jc} = M^{(1)}_{jj} / M^{(0)}_{jj} \quad \text{and} \quad \Delta^2_{jc} = M^{(2)}_{jj} / M^{(0)}_{jj} - \xi^2_{jc} \qquad (18)$$

Without going into details, it must be mentioned that finiteness of $M^{(n)}_{jj}$ imposes strong restrictions on the threshold behaviour of X_j and hence on the pole structure of ρ^+_j. It should also be pointed out that ξ_{jc} and Δ_{jc} need not be equal to W and $\Gamma/2$ respectively; it is enough that these are finite giving rise to a peak in $|\rho^+_j(k)|^2$ at $k \simeq \alpha$ (or $E \simeq W$).

It is necessary at this point to invoke the pole dominance hypothesis. The philosophy behind this is in our initial statement that resonant phenomena can be associated with a pole of the T-matrrix near the real axis. We reiterate that this pole actually dominates the said phenomenon if the resonance is narrow enough and the background is negligible. This viewpoint is the physical basis behind the unitary pole expansion of Weinberg[20] and Harms. In our context, it helps us enormously in estimating the various integrals by taking into account only the resonance poles in the LHP and ignoring all essential singularities of the integrand. It can be shown, in this limit, that $\langle \xi_{j'} | \xi_j \rangle = 0$

if $j' \neq j$, j(capture), $\langle \xi_j | \xi_j \rangle$ = finite number, $\langle \xi_j | H | \xi_j \rangle = W \langle \xi_j | \xi_j \rangle$, $\langle \xi_j | H^2 | \xi_j \rangle$ = finite but not well determined.

(ii) Approximate equality of $\Omega_j(p)$ and $a_j(p)$

The desirability of this imposition is quite obvious because it establishes a correspondence (not equality) between $|\xi_j\rangle$ and $|\psi_j\rangle$ the latter being known to be useful descriptions of the resonances. Mathematically, this condition results in certain restrictions resulting in the fact, e.g. in the case of sharp cut off potentials

$$\xi_j(r) \approx \mathrm{Re}\, \psi_j(r), \quad r < d \tag{19}$$

in the scheme in which $|\xi_j\rangle$ is made equal to its time reversed state $|\xi_j\rangle$ Of course one need not assume the equality of $|\tilde{\xi}_j\rangle$ & $|\xi_j\rangle$ and keep them linearly independent; but then (19) in a simple form cannot be obtained. This situation has also been investigated in detail.

(iii) Condition on resonance saturated amplitude

To answer how $|\xi_j\rangle$ and $|\underline{\xi}_j\rangle$ play a role in generating the scattering amplitude, we take a hint from the Low equation in which the unit operator was saturated (i.e. built up) exactly by the continuum and bound states. In a similar spirit a saturation operator $|\xi_J\rangle\langle\xi_J| = 1$ acting in the subspace of $|\xi_J\rangle$ and $|\underline{\xi}_J\rangle$ is constructed, and a resonance saturated half-shell amplitude

$$R_J^+(p,k) = \langle p|V|\xi_J\rangle\langle\xi_J|\psi_k^+\rangle = \Omega_J(p)\rho_J^+(k) \quad (20)$$

is defined in the scheme in which $|\xi_J\rangle = |\underline{\xi}_J\rangle$ and the requirement is imposed that R_j^+ possesses approximately the same pole residues at $k=k_j$, $-k_j^*$ as the exact half shell T-matrix and no poles at positions of other resonances and bound states. This condition ensures a close correspondence with standard Breit-Wigner amplitude. A general saturation operator has also been considered in which $|\xi_J\rangle \neq |\underline{\xi}_J\rangle$ but shall not be discussed here due to space limitation.

Thus Menon-Lagu[31)] arrive at a minimal choice for F(k), which for a sharp cut-off potential of range d is

$$F(k) = \exp\ -i(kd-\ell\pi/2)\ /ik \quad (21)$$

$$\text{and} \quad F(k) = \text{a rational function of } k/ik \quad (22)$$

for the separable interaction case. It is clear that exp(-ikd) in (21) is needed to suppress the essential singularities so that various integrations could be done in closed form.

Consistent with the above requirements, Y(k) can be optimally chosen to be of the form $Y(k)=C_o+iC_1k$ with the real constants C_o and C_1 obtained from the relation $C_o+iC_1\alpha = -2i\ell\alpha\gamma a_j(\alpha)/[F(\alpha)D^-(\alpha)]$.

7. TIME DEVELOPMENT AND DECAY LAW

Having constructed above a reasonable $|\xi_J\rangle$ at time t=0, a natural question arises about such a state having reasonable time evolution characteristics. It is expected that since the above

ρ's also possess peaking characteristics as the Lorentzian, the required properties of

$$\xi_{j}(r,t) = \langle r | e^{-iHt} | \xi_{j} \rangle \tag{23}$$

for variable times are likely to be reasonable. Explicit determination is mathematically quite subtle and physically very important because of its intimate connection with the outgoing yield of the decay products, given by

$$\text{Flux}(r,t) = \frac{r^2}{\mu} \text{Im}\left[\xi_{j}^{*}(r,t) \frac{\partial \xi_{j}(r,t)}{\partial r}\right] \tag{24}$$

In the Gamow-Siegert approach, it may be noted that

$$\psi_{j}(r,t) \underset{r\to\infty}{\sim} \frac{1}{r} \exp\left[i(k_j r - E_j t)\right] \text{ for all } t \tag{25}$$

so that $\text{Flux}\,(r,t)\big|_{\text{Gamow}} = \Gamma \exp\left[-\Gamma(t-t_\alpha)\right],\ t_\alpha = \mu r/\alpha$ (26)

for narrow resonances. Clearly, although the probability density $|\psi_j(r,t)|^2$ for fixed r, has a strict exponential time dependence $\exp\left[-\Gamma t\right]$, the decay is not really casual because the yield does not become zero for $t < t_\alpha$. Further $|\psi_j(t)\rangle$ has infinite norm as seen from (25).

Using the form for $|\xi_j\rangle$ as developed in the previous section, $\xi_j(r,t), A_j(t)$ and Flux (r,t) have been determined by the steepest descent technique assuming for convenience narrow resonances, fixed large distance and positive times. The algebra is quite detailed and needs great care. The concerned integrals for $A_j(t)$ are decomposed in three time regions viz. small times $t \ll \mu r^2$ intermediate times and large times $t \gg \mu r^2$. In the small time region one obtains

$$P_j(t) = |A_j(t)|^2 = 1 - t^2 \Delta^2_{jc} + O(t^{5/2}) \tag{27}$$

A similar result has recently been obtained by Lane[32] and demonstrates the direct physical significance of the coefficient of t^2. As t becomes large and tends to ∞, $P_j(t)$ decreases ;ole 1/t if

$\rho_j^+(k)$ behaves at threshold as 1/k, a result identical with that of Fonda et al but derived differently. It can be seen that for small and large time regions, there are clear deviations from the exponential decay law. For intermediate times, one obtains a complicated expression.

In a similar manner $\xi_j(r,t)$ and Flux (r,t) are evaluated in three time regions involving Fresnel functions. At intermediate times viz. between t_α and $t_\alpha + n_j/\Gamma$ & very narrow resonances $\xi_j(r,t)$ becomes identical with the Gamow function $\psi_j(r,t)$ and Flux (r,t) is given by (26). The number n_j has interesting connotation viz. it is the number of life times for which the exponential decay law for yield is valid.

A transcendental equation determines n_j depending upon the fact that flux (r,t) deviates from (26) by a fraction ϵ . The relation is $\exp(n_j)/n_j = \pi\epsilon^2 W/\Gamma$. This is a new estimate and a simple example shall demonstrate its usefulness.

Consider a radioactive heavy nucleus emitting alpha particles with the parameters (these are hypothetical values, but are reasonable and chosen for demonstration purpose only) W = 1 MeV, lifetime = 1 sec -.e. $\Gamma = 0.66\mathrm{x}10^{-21}$MeV, μ=3725 MeV and assume $\epsilon = 10^{-2}$ i.e. only a 1% deviation from exponential law. We get $n_j = 49$! Clearly any possible statistically significant deviation of the flux from the classical value within 1% is difficult to measure for at least 49 lifetimes because after the lapse of 49 lifetimes, the activity of the source drops down so low that statistical reliability is lost. It can also be seen that for a counter placed at a distance of, say, 1 metre, the time $\mu r^2 \approx 2$ yrs, which is large enough time.

It will be interesting to estimate n_j for some other process where the parameters are such that n_j turns out to be around 10 with ϵ =0.01. Unfortunately in direct reaction processes while this can happen, the lifetimes being of the order of 10^{-21} sec, measurements are not possible. Nevertheless this is an interesting estimate.

8. A DIRECT APPLICATION

Menon & Lagu[31,33] demonstrate the viability of their wave packet by calculating $\sigma_p(\theta)$ for the stripping reactions $^{16}O(d,p)\,^{17}O^*$ using the DWBA, a topic mentioned at the beginning of this article. Since Woods-Saxon potential is not analytically solvable they have replaced only the neutron-nucleus potential V_{nA} by a square well of depth V_0 and range d. This is reasonable for low energy resonances where the shape dependence of the potential is not crucial. They employ $(\Gamma_e/\Gamma)^{1/2}|\xi_j\rangle$ for neutron wavefunction in the nucleus where Γ_e and Γ are experimental and single particle widths. This scaling is natural and takes care of the fact that while $\xi_j(r)$ behaves like $\Gamma^{1/2}\eta_\ell(\alpha r)$ outside the potential range, for a given set of parameters reproducing α correctly, Γ_e is not necessarily obtained. Further this scaling also normalises the vertex function $a_j(\alpha)$ from the old value $\sqrt{\Gamma}$ to the new value $\sqrt{\Gamma_e}$.

In figure 2 are shown [33] the results for various calculations alongwith the experimental[34] points. For comparison a Gaussian damping calculation with the same square well potential parameters. after using a suitable spectroscopic factor of 0.87, has also been included. The other curves correspond to pseudobound[35], contour rotation[36] and primitive wave packet models[27] using a Woods-Saxon potential. While the Gaussian damping and Menon-Lagu wave packet show close correspondence, normalization near forward angles is guaranteed in the latter without a spectroscopic factor. Compared to other prescriptions, the Menon-Lagu fit is rather good. This feature is seen in a more dramatic manner in the case of $^{12}C(d,p)$ $^{13}C^*(6.864\text{ MeV}, 5/2^+)$ case[31]. It is not unreasonable to believe that this is due to the presence of factors arising from additional structures in the momentum plane resulting in a ρ which has a a denominator close to Breit-Wigner but nevertheless different.

9. OTHER APPLICATIONS AND EXTENSIONS

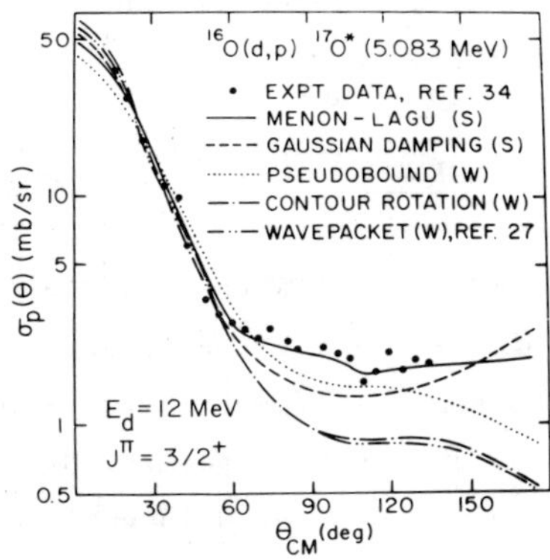

Figure 2 Proton angular distribution (ref.33)

Apart from the three topics mentioned in Section 2, tehe wave packet approach can be employed to compute the amplitude at the CBγ vertex in which unstable states C may be formed during the intermediate stage of A (n,γ)B process. Other applications could be in dealing with the excitation of isobaric analogue resonances by nucleon transfer, autoionizing states in electron-atom scattering or unbound states of composites produced in heavy ion collision[37)]. It must be noted that if additional complications are introduced e.g. spin, inelasticities, tailed potentials (say Coulomb or Woods-Saxon), the expansion coefficients ρ may or may not be calculable purely analytically and additional requirements may have to be imposed. However, the philosophy is well defined.

The most important aspect of a wave packet approach for a resonance, nevertheless, remains in the study of time-development of unstable quantum system. Indeed, in a dramatic work, Horwitz and Katznelson[38)] argue that the proton's existence is constantly being monitored through collisions with other nucleons in the nucleus, resulting in a continuous series of measurements which invalidate the usual analysis of a decaying state. But we shall not discuss it here. Fonda et al have also discussed this aspect of the environment having its effect on the measurement process.

Partial financial support from the DAE, Government of India is acknowledged.

10. REFERENCES

1. Gamow, G., Z.Phys. 51, 204 (1928)
2. Siegert, A.J.F., Phys.Rev.56, 750 (1939).
3. Robson, D., in Nuclear Spectroscopy and Reactions Part D, ed. J. Cerny (Academic, N.Y., 1975) p.179.
4. Mahaux, C., and Weidenmuller, H.A., Shell Model Approach to Nuclear Reactions (North-Holland, Amsterdam, 1969).
5. Giraud, B.G. et al, Ann.Phys. (N.Y.) 140, 29 (1982).
6. McVoy, K.W., in Fundamentals in Nuclear Theory, ed. A.de Shalit and C.Villi (IAEA, Vienna, 1967), chap.8.
7. Newton, R.G. Scattering Theory of Waves and Particles (McGraw-Hill, N.Y., 1966), Chap.12.
8. Satchler, G.R., Direct Nuclear Reaction (Clarendon, Oxford 1983), chap.17.
9. Amado, R.D., Phys. Rev. C19, 1095 (1979).
10. Green, A.M., Rep. Prog. Phys. 39, 1109 (1976).
11. Ref.7, chap.19
12. Fonda, L., Ghirardi, G.C. and Rimini A., Rep.Prog.Phys. 41, 587 (1978).
13. Ersak, I., Sov.J.Nucl.Phys.9, 263 (1969).
14. Humblet, J. and Rosenfeld, K., Nucl.Phys.26, 529 (1961); Garcia-Calderon, G. and Peierls, R., Nucl.Phys, A 265, 443 (1976) Berggren, T., Nucl.Phys. A169, 353 (1971); Romo, W., Nucl.Phys. A116, 618 (1968); Gyarmati, B. and Vertse, T., Nucl.Phys.A160, 523 (1971).
15. Kapur, P.L. and Peierls, R.E., Proc. Roy. Soc. (London) A166, 277(1938); Eisenbud, L. and Wigner E.P., Phys. Rev. 72 29 (1947); Pattanayak, D.N. and Wolf E., Phys.Rev. D13, 2287 (1976).

16. Feshbach, H., Ann.Phys. (N.Y.) 5, 357 (1958) and 19, 287 (1962) MacDonald, W.M. Nucle.Phys. 56, 647 (1964); Herzenberg, A. et al, Proc. Roy. Soc. (London) 84, 477 (1964).

17. Nagarajan, M.A., and Tobocman, W., Phy.Rev. 138, 1351B (1965); Lane, A.M. and Robson, D., Phy. Rev. 185, 1403 (1969).

18. Hart, C.F., and Girardeau, M.D., Phys. Rev. Lett. 51, 1725 (1983).

19. Hernandez,E. and Mondragon, A., Phy.Rev.C29, 722 (1984)

20. Weinberg,S., Phy.Rev.133, B232 (1963).

21. Bloch, C., Nuc.Phy. 4, 503 (1957)

22. Nicolaides, C.A. and Beck, D.R., Int.J.Quant.Chem.14, 457 (1978) Fano, U., Phy.Rev.124, 1866 (1961).

23. Balslev, E. and Combes, J.M., Comm.Math.Phys. 22, 280 (1971)

24. Parravicini, G., et.al, Texas Univ. Preprint (1979).

25. Krylov, N.S. and Fock, V.A., Zh.Eksp.Teor.Fiz. 17, 93 (1947).

26. Khalfin, L.A., Sov.Phys. JETP 6, 1053 (1958).

27. Coester, F. and Schlessinger, L. Ann Phy (N.Y.) 78, 90 (1973) Schlessinger, L. add Payne, G.L., Phys.Rev. C6,,2047 (1972).

28. Nussenzveig, H.M., Nuovo Cimento 20, 694 (1961); Kilian, V.H. and Petzold, J., Ann.d.Phys.24, 356 (1970); Sudarshan, E.C.G. et al, Phy.Rev.D18, 2914 (1978); Peres, A., Ann Phy (N.Y.) 129, 33 (1980).

29. Menon, V.J. and Lagu, A.V., Lett. Nuovo Cimento 24, 105 (1979).

30. Menon, V.J. and Lagu, A.V., Hadronic Jour. 2, 159 (1979); Proc. IX Few Body Conf. Eugene, Vol.I p.II 28-S54, ed. Moravcsik, M. and Levin, F.S. (1980).

31. Menon, V.J. and Lagu, A.V., Phys.Rev.Lett.51,1407 (1983); Menon, V.J. and Lagu, A.V., to be published; Menon, V.J., Ph.D. Thesis, Banaras Hindu University (1983), unpublished.

32. Lane, A.M., J.Phy.A.Math.Gen. 17, 1011 (1984).

33. Menon, V.J. and Lagu, A.V. in Few Body Problems in Physics, Vol.II, ed. Zeitnitz, B. (Elsevier, Amsterdam, 1984), p.473.

34. Alty, J.L. et al, Nucl.Phys. A97, 541 (1967).

35. Cole, B.J. et al, Phys. Lett. 33B, 320 (1970).

36. Vincent, C.M. and Fortune, H.T., Phys. Rev. C2, 782 (1970).
37. Tiereth, W. et al, Phys. Rev. C28, 735 (1983);
Dasgupta, S., Invited Talk - this conference.
38. Horwitz, L.P. and Katznelson, E., Phy.Rev.Lett.50, 1184 (1983).

INTERMEDIATE ENERGY NUCLEUS-NUCLEUS COLLISIONS

Daniel Sperber
Department of Physics, Rensselaer Polytechnic Institute
Troy, New York 12181
UNITED STATES

ABSTRACT

This brief qualitative overview focuses on the main features of nucleus-nucleus collisions at intermediate energies. The expected differences between the low and intermediate energy regimes are discussed. The question of how the dissipated energy is distributed in target and projectile is addressed.

1. INTRODUCTION

In this review we discuss the main features of nucleus nucleus collisions in the intermediate energy range. We consider the important physical phenomena to be investigated and understood, without going into considerable details discussing one theory or another. We will try, to present different and conflicting points of view. Also we will try to show that some theories may have a common denominator. So far there is no consensus as to which theory or model is most appropriate to describe the observed properties of nucleus nucleus collisions in the intermediate energy range. The lack of such a consensus is a great challenge to both theorists and experimentalists to explain in a consistent way the significance of experimental results.

There is no exact definition as to what is meant by intermediate energies. In the broadest sense one considers reaction in the range of 10 $\frac{\text{MeV}}{\text{AMU}}$ to 300 $\frac{\text{MeV}}{\text{AMU}}$ as intermediate. Clearly this is a very wide energy range at which new phenomena may play an important role. In this energy range the relative energy changes from about 1/4 of the Fermi energy to about seven times that energy. At higher energies the compression degree of freedom may play a significant role, to mention just two important sudden transitions.

For practical reasons first nucleus-nucleus collision for energies of few MeV/AMU above the barrier have been first investigated. Later taking advantages of the BEVRLAC the region of 800 MeV/AMU - 2 GeV/AMU for lighter ions became available.

The low energy regime has been successfully understood in terms of classical and semi-classical models on one hand or in terms of TDHF and the diabatic approximation on the other.

The high energy regime has been understood in terms of thermodynamical models such as the fire ball or fire streak.

It became clear that the exploration of nucleus-nucleus collisions over the entire energy is necessary. Accordingly a number of heavy ion accelerators have been constructed and are supplying rich amounts of information about nucleus-nucleus collision for the lower part of the intermediate energy regime. Clearly there is a need to span the entire energy regime and to go to ultrarelativistic energies.

On the theoretical side it would be very satisfying to have a unified model describing heavy ion reactions over the entire energy span from a few MeV/AMU up to the ultrarelativistic range. However, since in practice all models are approximations, it is clear that no such model can emerge in the near future since in different regimes different approximations are appropriate. For different energies one has to consider different normal modes and different degrees of freedom. For example in the lower energy regime one can ignore pionic degrees of freedom and Δ resonances, this is not the case for higher energies. Furthermore, it is believed that in the ultra-relativistic regime quark degrees of freedom may play a dominant role. Also there probably are a number of phase transitions. The intermediate energy range is a very good candidate in which phase transitions may occur.

2. GENERAL CONSIDERATIONS FOR THE INTERMEDIATE ENERGY RANGE

When nuclei collide at intermediate and high energies they become so highly excited that one finds oneself in a region of very broad overlapping levels. In this region it is not very useful to study individual quantum states. On the other hand in this region an equation of state is very useful in describing the gross properties of nuclei. Furthermore a "good" equation of state can also shed some light on the interesting question of phase transitions. As already mentioned before there are adequate methods to study nucleus-nucleus collision in both the low energy and BEVRLAC regime. An interesting question is how the intermediate energy regime is connected to both.

Current experimental information is available mainly for the lower part of this energy range. The main signatures of these reactions are the spectra and angular distributions of nucleons light systems and fragmentation. Attempts have been made to analyze some of the experimental data using appropriate models.

The main results from nucleon emission can be summarized by the two following observations (a) the angular distributions are highly forward peaked and (b) the energy spectra have a high energy component which cannot be accounted for by an evaporation spectrum.

These observations may contradict some of the observations in the lower energy regime. Original analysis of the neutron energy spectra in low energy nucleus-nucleus collision seems to indicate that these neutrons come from two completely energy equilibrated systems of equal temperatures. More recent results suggest that this equilibration may not be as complete and that the two fragments may have different temperatures.

In any event the difference between the low and intermediate regime is dramatic. Clearly the proton and neutron spectra suggest that there is a component in the spectra which originates in a non-equilibrated system. In other words there is some degree of localization.

The question whether the transition from an equilibrated to a non-equilibrated system occurs continuously or suddenly is still open. This is a very important question to settle, since a sudden change would suggest a phase transition. This question can only be resolved by studying the neutron spectra of the same system over a wide range of bombarding energies. At the present, no such data is available.

The next question is how to account for such a lack of equilibration. To date there are a number of models which have been used and are now reviewed briefly.

3. THE TIME DEPENDENT HARTREE-FOCK METHOD

The time dependent Hartree Fock wavefunction allows one, using Fourier transform, to determine the probability distribution in phasespace. The distribution has a contribution corresponding to a high momentum component spacially removed from both colliding nuclei.

4. PROMPTLY EMITTED PARTICLES

The promptly emitted particles or Fermi jets are due to the coupling to the Fermi motion for high nucleus nucleus velocities. Adding the two velocities may yield a velocity high enough to "spill" out nucleons with a high velocity. Estimates, using kinematic consideration for PEPS have been obtained. In the model Pauli blocking has been incorporated. According to the model, nucleons exceeding considerably the beam velocity are allowed. These nucleons energy in a narrow angle. The predicted distribution overestimates the low momentum, but decreases very fast for high momenta above the Fermi momentum. In some models allowance for second chance emission is made. The PEPS model predicts a very small cross section for the fastly emitted particles. The model leads to an exponential decay of the spectrum for high energies. The spectra of PEPS for protons and neutrons differ due to the Coulomb barrier. The effect of the Coulomb barrier has been incorporated using the Hill Wheeler approximation. To date there is no clear experiment which proves the validity of the PEPS model, although the importance of PEPS in some reactions cannot be ruled out. As will be discussed later in some reactions they must be incorporated.

5. HOT ZONES

The high energy component of emitted nucleons and their forward peaking suggests a non equilibrium process or a localization. This localization may be in the number of excitons, leading to the precompound model. On the other hand one may consider also a special localization, in other words the energy dissipated in a nucleus nucleus collision is not distributed evenly, but is restricted to a smaller volume of the dinuclear complex when the ions are at close proximity. This excitation is due to the dissipation which has to be determined from the dynamics. It is important to note that some experiments studying the correlation of emitted particles suggests directly that there is a hot excited region.

The existence of localized hot zones has not been established rigorously. If the hot spot model is equivalent to the precompound model the hot zone must correspond to a complicated superposition of particle hole states or excitons. Such a superposition is unlikely in the adiabatic spproximation. However, the intermediate energy regime may be just the energy regime in which the adiabatic approximation fails. The adiabatic approximation is particularly valid in the low energy regime. In this region the collective motion is very slow (small frequency) as compared to the single particle and particle hole motion (higher frequency). In this energy range the adiabatic approximation is valid. The system may be compared to Brownian motion where the collective degrees of freedom play the role of the Brownian particles and the particle hole states supply the heat bath corresponding to the molecular motion. However, the approximation fails when the two velocities frequencies are close to one another. Therefore, in the low energy regime the system can be studied using the Fokker-Planck equation. On the other hand in the higher energy regime one may get a hot spot consistent with a rather complicated superposition of particle hole states different from the superposition leading to the average behaviour described by the collective dynamics in the lower energy region.

We now focus on one description of hot zones. In these models an a priori existence of a localized hot zone is assumed. For very high energies in the relativistic regime one reaches the fire ball limit in which the hot zone moves within the medium. For lower energies one first studies the consequences of a stationary zone. In principle one considers a zone characterized by two parameters (i) the radius of the zone (ii) the density of the zone. Since for the lower limit of the intermediate range of nucleus-nucleus collision nuclear matter can still be considered incompressible one uses only one parameter, namely the radius of the zone. (This is definitely reasonable for reactions at energies of up to 30 MeV/AMU.) One assumes that the zone itself is in thermal equilibrium and is made of all the nucleons of the projectile and some of the target. For simplicity a spherical zone is assumed. Later improvements are discussed. The relative energy is equated to the excitation energy allowing one to determine a temperature. It is assumed that when a nucleon emerges from the hot region it can be emitted only if its kinetic energy of motion perpendicular to the wall is greater than the potential depth. Using some lengthy mathematical manipulation one obtains the spectra and angular distribution of the emitted nucleons. Comparing with experiment one concludes that at energies of 10 MeV/AMU to 17 MeV per nucleon compression plays no significant role.

The previous model is valid only for central collisions, using a spherical shape for the hot zone. Clearly the model has to be generalized for non central collisions and even peripheral collisions. To do that one first assumes that the nuclei stick together instantaneously. In this approach one needs to determine the energy which is used up to heat up the hot zone and the temperature of this zone. (It is assumed that the energy is distributed uniformly over the volume of the zone.) It is assumed that the zone consists of two caps near to where the colliding nuclei touch. The width of the caps being equivalent to one layer of nucleons. From the boundary of the hot regions nucleons are emitted in all directions with equal probability. One assumes that the mean free path through the cold matter is very large (Knudson limit, this assumption is discussed below in this paper). Particles from the plane boundaries of the caps reach the nuclear surface with no absorption. On the other hand nucleons emitted from spherical surfaces are completely absorbed. For the hot region one assumes a very short mean free path or complete absorption. At the nuclear surface the nucleons undergo refraction as only the radial components of the velocity are changing in overcoming the nuclear potential, but the tangential component remains unchanged. The condition for a nucleon being emitted is the same as previous. This allows one to calculate explicitly the total flux of emitted nucleus. However, the angular distribution is calculated with respect to an axis of cylindrical symmetry corresponding, to the axis determined by a line connecting the center of the sticking target and projectile. In order to compare with experiment, a geometric transformation is required to obtain the angular distribution with respect to the direction of the beam. Improved agreement is obtained for the excitation functions. However, the differential cross section in forward angles is highly underestimated.

The lack of complete agreement between theory and experiment suggest the need for improvements of the theory.

Clearly the assumption that the cold zone remains cold at all times and the hot zone remains hot at all times can be considered only as an approximation. The temperature of the hot zone is determined by the energy loss due to dissipative forces as previously discussed. The initial temperature of the cold zone vanishes. The rate of change of excitation of the hot zone is due to transfer of particles through the boundary between the hot and cold zone respectively. This leads to a differential equation for the temperature of the hot zone. It can be shown that for all practical purposes the temperature of the hot zone decays exponentially with a characteristic time of the order of magnitude of 10^{-22} sec. Since the spectrum and angular distribution depend explicitly on temperature both become time dependent. In order to compare with experiment one has to integrate the instantaneous spectrum and angular distribution over time. Performing such an integration one again gets improved agreement with experiment without using any adjustable parameters. Yet, the agreement is not completely satisfactory no matter how one changes some of the initial input. One has to conclude that the hot zone model by itself cannot account for the data. We already mentioned that PEPS alone do not account for the data. One therefore has to include the contribution from both PEPS and hot zones and determine if such composite model can account for the data. Comparison with experiment shows that indeed the combined contribution from hot zones and PEPS accounts for the data very well. At forward angles for high energies PEPS dominate, for backward angles and the same energies the contribution from the hot zone dominates.

6. PREEQUILIBRIUM MODELS

Preequilibrium models have been very useful to account for the high energy component in nucleon and alpha induced reaction. The preequilibrium models were designed to bridge the gap between direct reactions and totally statistical compound reactions. In a direct the projectiles interact with one nucleon and the reaction is characterized by almost complete memory. In the other limit the projectiles make many collisions with the constituents of the target so that the energy brought in by the projectile is shared among the constituents of the target leading to a complete thermalization and loss of memory. According to the precompound model a few collisions occur leading to the excitation of a number of particle hole states and then an ejectile with reduced energy is being emitted. An incident particle of an energy above the Fermi sea undergoes a collision with a nucleon below the Fermi surface creating one particle and one hole or 2 excitons. Such processes are a result of two body interaction. They are not existent in the extreme one body case. This is important since mean aspects of nucleons nucleus collisions are anlalyzed in terms of mean field theories. However, mean field theories are valid only when the mean free path is considerably larger than the dimension of the system. However, as will be discussed later the mean free path decreases dramatically with temperature so that in the intermediate energy regime two body collisions play a very significant role. One can write a master equation for the probability of having a specified number of excitons. The transition probabilities appearing in such an equation can be determined from the imaginary part of the optical potential for free nucleon nucleon scattering. Using preequilibrium models in nucleus reactions one has to assume an initial number of excitons. This has been done mainly for central collisions. The results are somewhat sensitive to the choice of the initial number of excitons, this being one of the weak points of a rather promising approach. The big success of the model is that it predicts very well the shape of the spectrum. There is a trade off between the choice of original number of excitons and normalization. The model does not yield information about the angular distribution. The shape of the spectra depend on the bombarding energy, which is reasonable. Good fits to the experimental data may be obtained by adding the contribution of evaporation and preequilibrium contributions. The preequilibrium model has not been investigated to its fullest, neither has there been made a consistent comparison with experiment. However, the model predicts the gradual onset of non equilibrium process which seems to be indicated by the data.

7. MEAN FREE PATH

It is clear from the above discussion that the mean free path plays a considerable role in the study of heavy ion reactions. This is true for both the low and intermediate energy ranges. In the low energy range a long mean free path suggests one body dissipation or that the dissipation is due to the time variation of the single particle potential. A short mean free path suggests two body dissipation or viscosity is the dissipative mechanism. In this case dissipation is due to the residual time dependent two body interaction. The mean free path is closely related to the transport coefficients, the static thermal conductivity and the sheer viscosity. A small thermal conductivity allows for the formation of a hot spot where large thermal conductivity results leads to a fast equilibration. Therefore we consider the dependence of the thermal conductivity on the nuclear temperature. Different approximations are required for different temperatures. For the liquid phase at very low temperatures one must use Landau theory. For higher temperatures of the liquid one can treat the nuclear fluid as a classical fluid. The gas phase also can be divided into two phases. For the lower regime one has to consider the gas as a Fermi gas whereas for higher energies, when Pauli blocking does not play a significant role one can consider the nuclear gas a classical gas. The question of phase transition (from fluid to gas is of great importance and has been recently addressed both theoretically and experimentally but not addressed here). For a temperature below the gap energy (about 2 MeV) the fluid has to be treated as a Landau Fluid. However, a two-fluid model is not applicable since the coherence length exceeds the size of nuclei. For temperatures between the gap and boiling the fluid behaves as a classical liquid, (the boiling temperature is approximately 8 MeV) i.e. effects of two body correlations and Pauli principle are very important for the calculation of the transport coefficients. For temperatures between the boiling temperature and a temperature the Sommerfeld temperature, one has to use the Fermi gas model [the Sommerfeld temperature can be estimated from the condition that thermal wave length should be smaller than the interparticle distance, this temperature is about 30 MeV.]

For a Landau liquid expression for the transport coefficients in terms of the Fermi momentum, the temperature, a nucleon reduced mass and scattering probability have been derived. It was found that the main contributions are due to the scattering from s, p and t waves and that higher waves contribute very little to the transport. Coefficients corrections for finite size were made. For a classical model below the boiling point the old Eyring equations have been used. For gases expressions for the transport coefficients have also been derived long ago. However, for a Fermi gas corrections for Pauli blocking have to be included. Only scattering from occupied states to empty states is allowed which are consistent with the conservation laws. The effects of Pauli blocking can be neglected for high temperatures where there is no difference between Fermi Dirac statistics and Boltzman statistics. The transport coefficients depend on the nucleon-nucleon cross section which is determined from experiment. The calculations show that there is a sudden reduction in the mean free path somewhere between a temperature of 3 MeV and 6 MeV. On the other hand for low temperatures the mean free path is very long. This suggests that for low temperatures one body dissipation prevails where for high temperatures two body dissipation prevails. Furthermore, after the two ions touch the immediate regions to where the nuclei touch is heated up. But due to the long mean free path in cold matter the original regions which were at zero temperature will remain so. Thus a hot zone is formed.

8. PROSPECTIVES

Considerable progress has been made in the understanting of nucleus-nucleus collisions at intermediate energies. Here we focussed on models explaining nucleon emission. Models for the study of the excitation function of complex nuclei and fragmentation are avilable but not discussed here. The up to date models for nucleon emission do not include the details of the dynamics. Such a dynamical model must include among the collective degrees of freedom the relative distance between the colliding nuclei, the mass asymmetry and a neck. A parameterization including the above degrees of freedom using the same analytical expression for one and two body configurations has been suggested by Blocki and Swiatecki. Using this parameterization and taking of the hydrodynamical models allows one to obtain a very adequate description of nucleus-nucleus collisions in the low energy regime. However, the model can be generalized to make it more powerful. These generalizations allow one to obtain (from the known dissipation) the distribution of the excitation energy. Thus the original temperature profile can be determined. This allows one to obtain the temperature profile at all times. Work in this direction is underway. Preliminary results indicate a higher temperature in the neck region for intermediate energy nucleus nucleus collisions. Thus if one couples hydrodynamics and the heat equation, one does not have to assume an arbitrary "hot zone" The formation of the hot zone is now included in the model from first principles. In previous models for nucleus-nucleus collision only the decay of the zone is included or the exit channel behaviour of the zone is studied. In the new approach both the entrance and exit channel time evolution of the zone are incorporated dynamically. This approach is very promising.

9. REMARKS

In this review no equations, figures or references, both to our work and to the work of many other colleagues were included. It was felt that the inclusion of some equations and figures could not do justice to the topic. The list of references is so extensive that it was felt it should be omitted. In a more detailed review, these details will be incorporated.

LINEAR AND ANGULAR MOMENTUM TRANSFER IN HEAVY ION REACTIONS*

R. K. Choudhury,** D. Fabris, K. Hagel, Z. Majka,
M. N. Namboodiri,*** J. B. Natowitz and G. Nebbia

Cyclotron Institute
Texas A&M University
College Station, Texas 77843
UNITED STATES

The study of the dominant reaction mechanisms induced by heavy ion projectiles of intermediate energy, e.g., 10 - 100 MeV/u, has been stimulated by the recent availability of accelerators capable of producing such beams. This energy region promises to be an interesting one since it is expected that as the energy increases through this region the interactions become increasingly more dominated by individual nucleon behavior and less sensitive to mean field effects. The projectile velocities encountered in this energy range bracket the velocity of sound in nuclear matter and the fermi velocities of the constituent nucleons, two characteristic velocities the surpassing of which should lead to qualitatively different behavior in the reactions between two heavy nuclei.

Measurements of linear momentum or angular momentum transfer in such heavy ion collisions can provide a great deal of information on the reaction mechanisms and the systematic trends in the intermediate energy region. Correlation of these quantities where possible offers the possibility of delineating the partial wave dependence of the mechanisms. The latter is important since one real utility of the mechanism studies will be found in the ability to produce nuclei with reasonably well defined but extreme conditions of excitation energy and angular momentum and to characterize their behavior under these extreme conditions.

The systematics of linear momentum transfer in the intermediate energy range is becoming well established using two basic experimental techniques, residue velocity measurements and fission fragment angular correlation measurements. A number of studies have been carried out at energies up to 84 MeV/u with a variety of incident projectiles. Most of the experiments have involved inclusive measurements. An example of the type of data which can be obtained in such measurements is shown in Figure 1.

Fig. 1 Residue Velocity Distributions (Ref. 1)

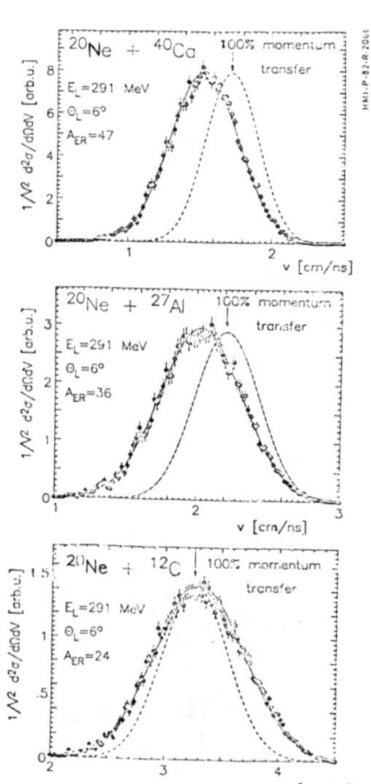

These residue velocity distributions have been measured by Morgenstern, et al[1)] for the reactions of 14.5 MeV/u ^{20}Ne projectiles with several targets. It is noted that for targets heavier than the projectile the distributions peak at velocities corresponding to less than full momentum transfer and are somewhat broader than the distributions calculated using the evaporation code CASCADE.[2)] For ^{20}Ne + ^{12}C the shift in velocity is not seen but the experimental distribution is much broader than the calculated distribution. At these and higher velocities, the typical experimental results are most probable velocities well below those corresponding to total momentum transfer.

In fission fragment angular correlation measurements with heavy targets, distributions such as that shown in Figure 2 for 27 MeV/u $^{40}Ar + ^{238}U$ are commonly seen for the two fragment angular correlation.

Fig. 2 Angular Correlation Distribution for Fission Fragments - 27 MeV/u $^{40}Ar + ^{238}U$ (Ref. 3)

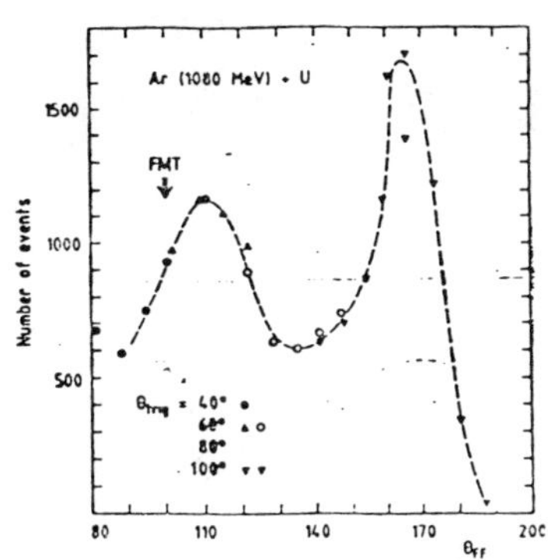

This measurement by Rivet and Borderie[3)] may be interpreted as reflecting a rough division into peripheral collisions with little momentum transfer (and thus folding angles close to 180°) and more central collisions with larger momentum transfers (and much smaller folding angles). Even for the central component, however, the most probable momentum transfer is characteristically less than full.

This qualitative decomposition into central and peripheral collisions is further supported by coincidence measurements in which fission fragments are detectable in coincidence with forward energetic projectilelike fragments. For example, measurements by Back, et al[4)] for the system 310 MeV ^{16}O with ^{238}U have shown that the large folding angle distribution is observed in coincidence with relatively heavy ejectiles of Z = 4 - 6 while the central component is observed only with light, Z=1,2, ejectiles, some of which appear to result from fast processes and others of which are evaporated from either the composite system or the fragments. Thus the lowest momentum transfers appear to be associated with more peripheral incomplete fusion or massive transfer reactions while the larger momentum transfers which are still less than total momentum transfer are associated with fast light particle emission.

If one focuses on the more central collisions, the systematics of momentum transfer with increasing projectile velocity for such collisions may be viewed in either relative or absolute terms. In Figure 3 a recent summary of the results of both residue velocity and folding

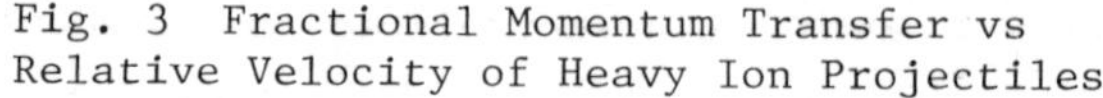

Fig. 3 Fractional Momentum Transfer vs Relative Velocity of Heavy Ion Projectiles

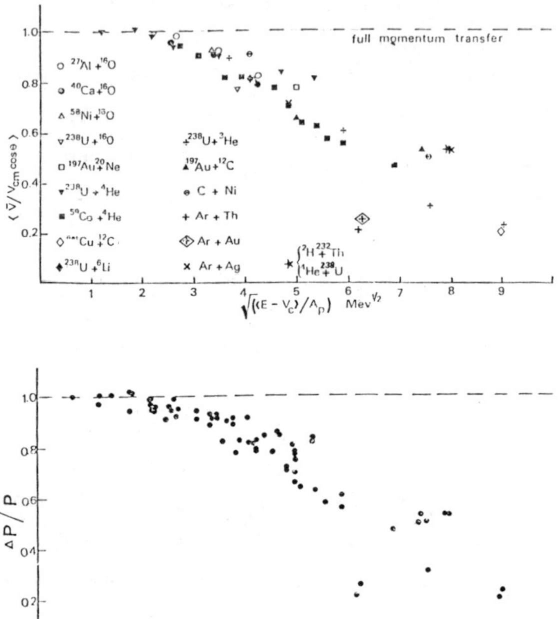

angle correlation measurements is presented.[5] The fraction of the projectile momentum which is transferred appears to fall linearly with the relative velocity of the projectile above projectile energies of 8-10 MeV/u. In absolute terms there may be some evidence for a saturation of the most probable momentum transfer near 0.19 Gev/c per nucleon but such a saturation is not clearly established.

One notable aspect of both types of measurements is that even though the difference between the most probable momentum transfer and full momentum transfer increases rapidly with projectile velocity, the observed distributions still encompass the momentum region corresponding to full momentum transfer. Whether this reflects the existence of fused systems with full momentum transfer and the full excitation energy and temperature characteristic of such systems, or whether it merely is a fluctuation reflecting the particle de-excitation of the system of lower p transfer and excitation energy is one of the interesting questions to be explored in these measurements.

Another interesting, if difficult question, is that of which partial waves contribute to the production of such systems. In this regard, it is useful to view schematically some possible scenarios for the correspondence between linear momentum transfer and angular momentum transfer to the heavy product of an intermediate energy collision. In Figure 4 several possible correlations between these quantities, corresponding to qualitatively different dependencies of the dominant reaction mechanisms on impact parameter, are presented. For example, Figure 4 (a) corresponds to a division into central fusing collisions with total momentum transfer and peripheral collisions with decreasing momentum transfer. The lowest J values in the second type of events correspond in fact to the highest partial waves for which most of the projectile escapes. The other diagrams in Figure 4 (b) show the possibility of a low ℓ window with less than full momentum transfer (c) no full momentum transfer for any partial waves and (d) full momentum transfer only for very central collisions. In a typical linear momentum transfer measurement, it is the projection of such distributions onto the p axis which is observed.

Fig. 4 Schematic Representation of Residue Angular Momentum vs Linear Momentum (see text)

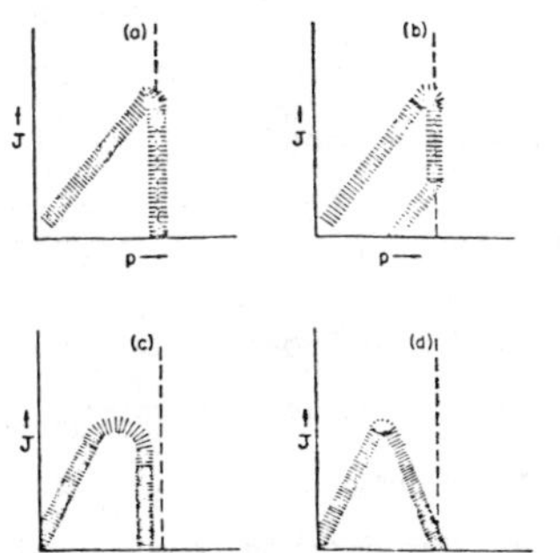

In order to explore the partial wave dependence of the dominant mechanisms, several techniques may be utilized. For example, the measurent of fission cross sections in concert with momentum transfer is one method which provides some information. This technique has been exploited by Huizenga and coworkers[6)] for the reactions of ^{20}Ne with a number of targets. By empirical decomposition of the folding angle distribution using information obtained from the out of plane, evaporation dominated, fragment correlations they have derived cross sections for total momentum transfer consistent with a dynamic model. Assuming these cross sections to represent only the lowest partial waves they have further derived some information on the range of partial waves contributing to TMT and to massive transfer events. Using both excitation function data and momentum transfer measurements at energies in the 20 MeV/u range, Gavron et al[7)] have also drawn some conclusions regarding the range of partial waves contributing to the dominant reactions. Again the arguments are based on cross section measurements alone.

While these results are interesting in their implications regarding the partial waves contributing to various incomplete fusion processes, they are not definitive since they rely on cross section arguments and on the packing of cross section into well defined ℓ windows with little or no overlap. Indeed, some impact parameter localization for such reactions is predicted by several models but the actual range of impact parameters contributing to a particular reaction may be quite different from that derived from cross section arguments.

Models for incomplete fusion reactions have been proposed by Wilczynski, et al[8)] by Udagawa and Tamura[9)] and by Harvey and Homeyer.[10)] While the model of Udagawa and Tamura appears to be the most complete theoretical treatment, it has not been extensively applied to heavy ion collisions such as those we consider in this paper. Thus we will restrict our present comparisions with the other two models.

It is probably fair to characterize the first model, that of Wilczynski et al as a low energy model where the range of partial waves contributing to a given mass transfer is determined by the nucleus nucleus potential. In contrast, the Harvey-Homeyer model is a geometric overlap model analagous to the high energy abrasion-ablation models. Determining the range of applicability of these models in the intermediate energy region is useful.

Both models have had some success. For example various γ-ray multiplicity studies in the 6-10 MeV/u range[8,11] have been interpreted as being in agreement with the Wilczynski model prediction of the average ℓ transfers. On the other hand, a comparision of the predictions of the Harvey-Homeyer model with excitation functions for ejectile cross sections for the reactions of ^{20}Ne with ^{197}Au to energies as high as 20 MeV/u shows very good agreement, supporting the application of this model.[10)] In particular this model successfully predicts the relative constancy of the cross sections at higher energies which is in contrast to the predicted decrease expected from the Wilczynski model.[8)]

We have undertaken a more detailed test of the models by attempting to determine the angular momentum transfer associated with incomplete fusion reactions in the reactions of 310 MeV ^{16}O with ^{154}Sm using an experimental set-up such as that illustrated in Figure 5.

Fig. 5 Schematic of Measurement Using the Total Energy Crystal

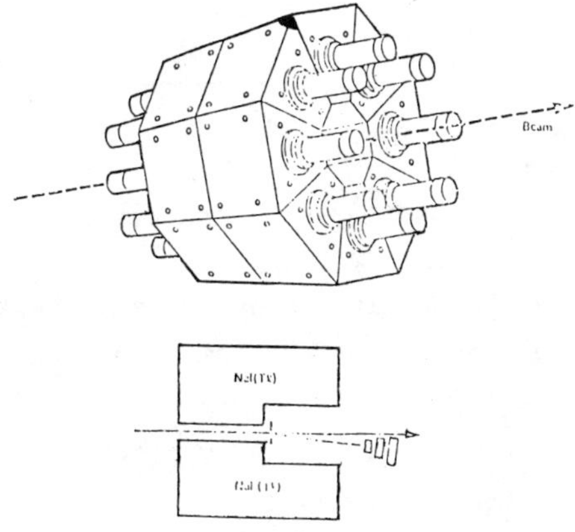

In that experiment the average total γ-ray energy measured in coincidence with mass and atomic number identified ejectiles observed at Θ_L=12°, close to the grazing angle, were obtained for ejectiles with Z=3 to 9.[12)] Energy spectra for the observed ejectiles shown in

Figure 6, are characterized by most probable energies close to the same energy per nucleon as that of the projectile for ejectile masses near that of the projectile. As the mass difference between the projectile and ejectile increases, the difference between the peak energy and that corresponding to a beam velocity ejectile increases. Similar spectra have been observed for 20 MeV/u ^{16}O incident on other targets.

Fig. 6 Ejectile Energy Spectra for Reactions of 180 MeV (dashed lines) and 310 MeV (solid lines) ^{16}O with ^{154}Sm. Energy Scale is for 180 Mev. For 310 MeV add 120 MeV to Scale Value

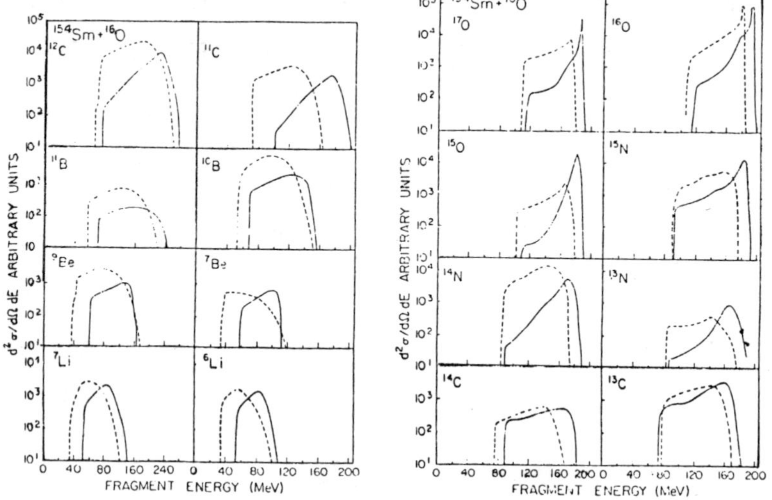

Figure 7 presents the results of our measurements of the average total γ-ray energy as a function of the energy of the identified ejectile for two different projectile energies 180 MeV and 310 MeV.

Conversion of the average total γ-ray energy to transferred angular momentum is, of course, a crucial point in the experiments. To accomplish this we have used conversions based on empirical systematics.[13] Such systematics lead to the calibration curve presented in Figure 8 and is the one employed in the work. While it is difficult to assign errors to such a determination comparison of our calibration with a similar effort by Sarantites, et al[14] leads to a

Fig. 7 Average Total Gamma Ray Energy vs Ejectile Energy for Reactions of ^{16}O with 180 MeV (solid circles) and 310 MeV (open circles) ^{16}O with ^{154}Sm

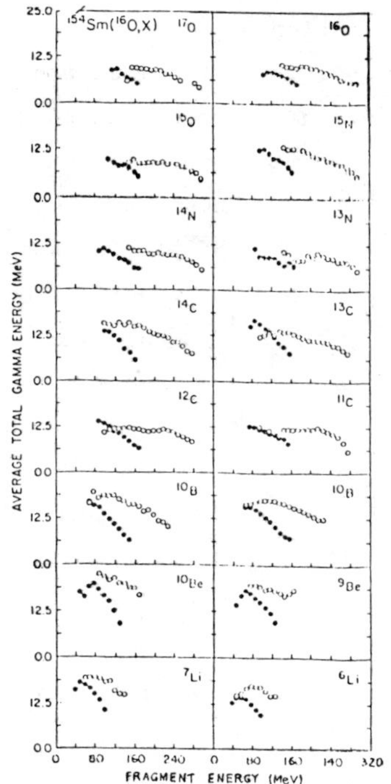

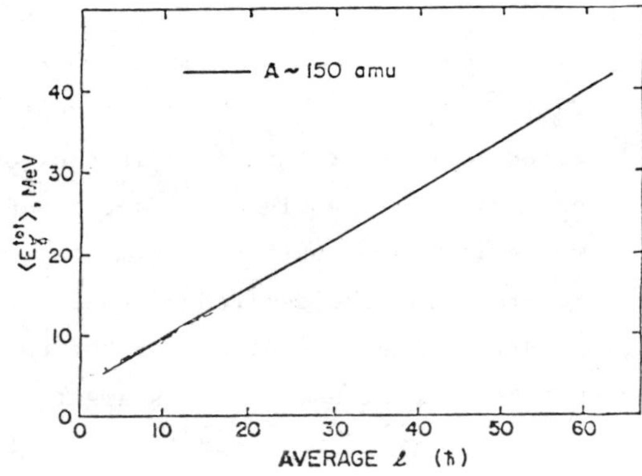

Fig. 8 Empirical Relations Between Average Total Gamma Energy and Average Angular Momentum for A ∿ 150 amu

consistency within 2-3 ħ which gives some further confidence in the average angular momenta.

In Figure 9 the observed distributions of angular momentum transfer (weighted by cross section at Θ_L=12°) are presented. In the following we confine ourselves to the most probable value of the angular momentum transfer and compare that to the predictions of the Wilczynski and the Harvey-Homeyer models.

Fig. 9 Distributions of Derived Angular Momenta Associated with Ejectiles Observed at Θ_L = 12°

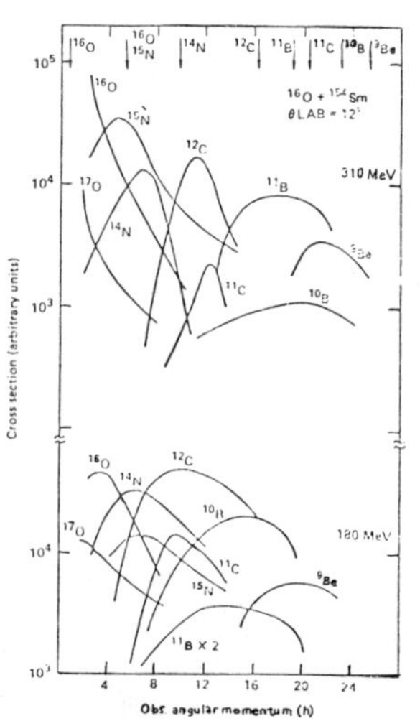

Some distributions representative of the predictions of these models for the reactions of 310 MeV ^{16}O with ^{154}Sm are shown in Figures 10 and 11. The predicted partial wave distributions for fusion and for reactions leading to ^{4}He, ^{12}C and ^{14}N ejectiles are depicted. While the partial waves contributing to fusion are similar in the two models, it will be noted that those contributing to specific incomplete fusion reactions are quite different. The predicted most probable waves contributing to a particular incomplete fusion reaction for the Harvey-Homeyer model are much larger than those of the Wilczynski model as a consequence of the different model assumptions.

In order to present the comparison of our results with the model predictions we follow in figures 12a-c a sequence of steps involving successive refinements to the experimental data and the models.

In Figure 12(a) the derived values of the product angular momenta are compared with the predictions of the two models. It should be recalled that the experimental points are those corresponding to the most probable energy of the identified ejectile. The models implicitly assume that the ejectiles are emitted with the beam velocity.

Fig. 10 Predicted ℓ Wave Distributions for Fusion and Three Ejectile Channels (Ref. 8)

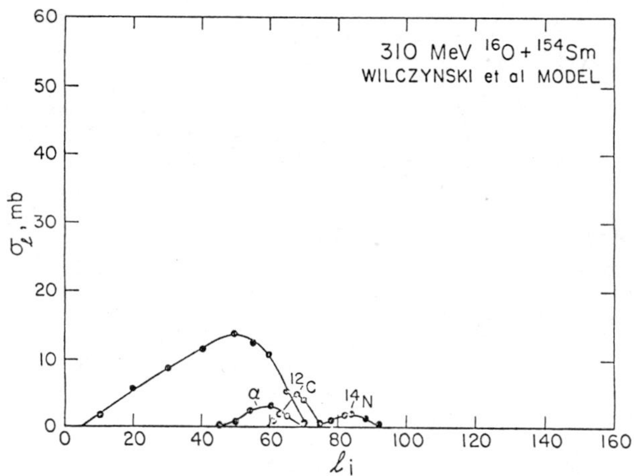

Fig. 11 Predicted ℓ Distributions for Fusion and Three Ejectile Channels (Ref. 9)

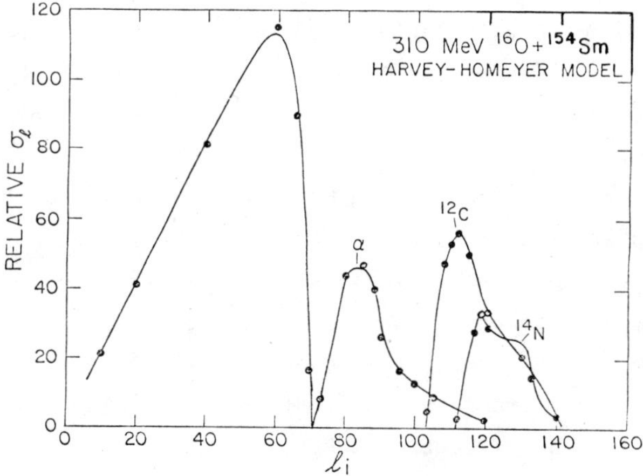

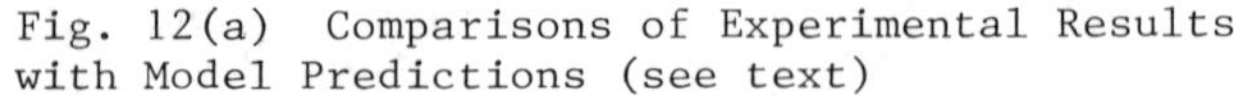

Fig. 12(a) Comparisons of Experimental Results with Model Predictions (see text)

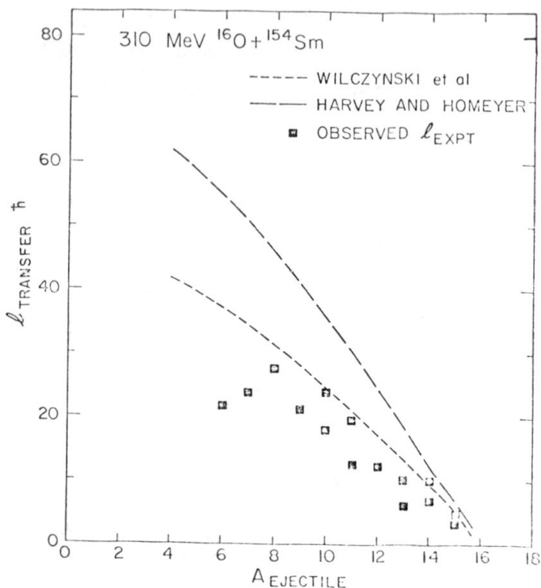

However, perhaps the most serious problem in the comparison is the fact that the model calculations are for the primary ejectiles while the experimental data are for observed ejectiles which do not necessarily correspond to all of the mass removed in the early stages of the collision.

In order to correct for missing mass we have employed the data of Back, et al[4)] obtained with 310 MeV ^{16}O on ^{238}U. Through coincidence measurements between observed ejectiles and fission fragments they were able to demonstrate an a discrepancy between the initial projectile momentum and the sum of the momenta of the ejectile and the heavy residue (derived from the fission fragment angular correlation). This was interpreted as reflecting undetected particles. From their data the missing momentum could be determined.

If we assume that the probabilities of missing momentum is the same in our experiment as in theirs, we can modify the experimental points so that they reflect the angular momentum which would have been observed if there were no missing mass. An alternative would be to scale down the model predictions for missing mass. The missing mass may result either from sequential emission from the ejectile or fast particle emission from the composite system. In the former case this alternative approach might be more logical but would not change the conclusions presented here.

In Figure 12(b) the "scaled" values of the measured angular momenta are compared with the model. Corrected for missing momentum the data are in much better agreement with the predictions of the Wilczynski model than with that of the Harvey-Homeyer model.

Fig. 12(b) Comparisons of Experimental Results with Predictions (see text)

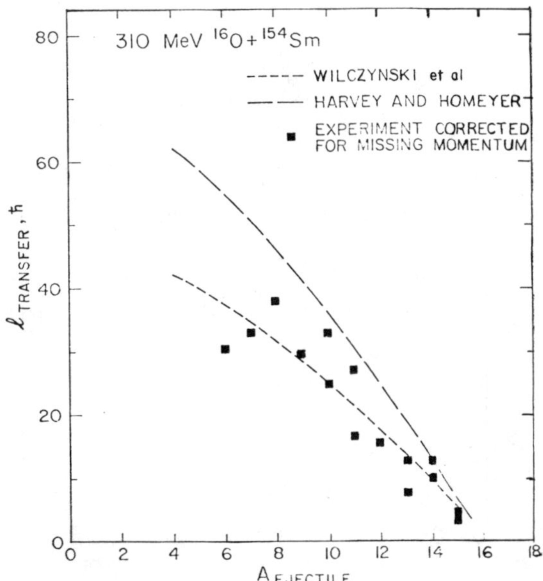

As a final refinement on the comparison, we note that for ejectiles far from the projectile the most probable velocities are well below the beam velocities indicating some energy loss and therefore greater momentum loss than predicted by the models which assume emission at beam velocity. It is possible to modify the experimental predictions by assuming that the same partial waves as those calculated are the important ones but scaling the angular momentum transfer for increased momentum loss using the experimental energy observations. We have done this and present in Figure 12(c) a comparison of the ratio of the experimental values, connected for missing momentum, to the model values scaled for increased momentum transfer with decreasing ejectile mass. This ratio of experimental and calculated transfers could also be viewed as a ratio of the experimentally determined partial waves to those predicted by the models. Here again the agreement with the predictons of the Wilczynski model is much better.

Fig. 12(c) Comparisons of Experimental Results with Model Predictions (see text)

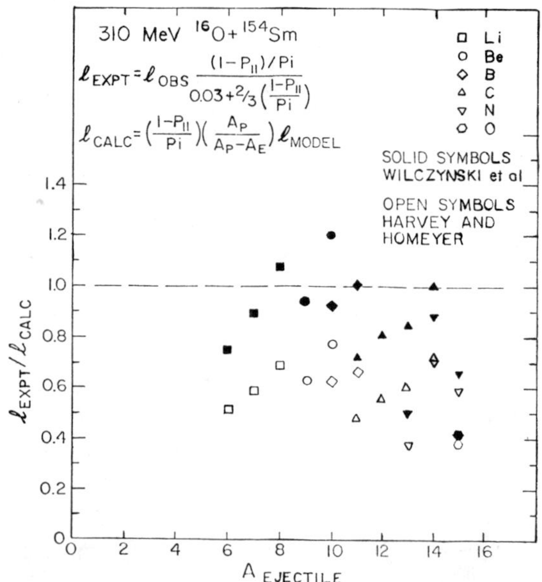

Given the success of the Harvey-Homeyer model in explaining ejectile excitation functions it appears that there are some further investigations to be done to clarify this result. Certainly a determination of the partial waves involved should be a critical test of any of the models proposed.

In this regard, M. N. Namboodiri has presented at this conference a preliminary report on an experimental collaboration between MPI Heidelberg, LLL, University of Geissen, BARC and Texas A&M in which the necessity to correct for missing momentum is avoided by direct measurements of the heavy residue velocity in coincidence with γ-rays and neutrons. This experiment with 248.9 MeV ^{19}F incident on ^{154}Sm was carried out in the MPI-GSI crystal ball. These data have been discussed by Dr. Namboodiri. The analysis of the data in terms of the partial wave dependence of the residue producing mechanisms is presently being carried out. Attempts are also being made to derive neutron multiplicity information from the data. It is hoped that it may prove possible to make an event by event reconstruction of the partial wave distribution through such techniques.

Similar attempts to explore the residue groups with higher energy particles, i.e., ^{14}N at 19 and 35 MeV/u are underway in a collaboration between TAMU, LLL, ANL, Hope College and MPI.[16] In those experiments a time-of-flight arm consisting of a channel plate start detector and a Si stop detector has been used to measure the velocity distributions of the residues. In Figure 13 the distribution $\frac{1}{V^2}\frac{d^2\sigma}{d\Omega dV}$ for residues for residues from the reactions of 261 MeV ^{14}N with ^{154}Sm is shown. The experimental distribution is compared in the figure to that calculated using the statistical model Code LILITA.[17] The widths of the experimental and calculated distributions are quite similar but the experimental distribution peaks at a significantly lower velocity consistent with earlier systematic observations of momentum transfer.[5]

Since there are, however, residues observed having velocities corresponding to those of total momentum transfer, it is possible that some fusion has occurred. To explore this in greater detail it would be useful to have mass distributions of the residues and to make detailed comparisons of the mass distributions for various velocity regions with those calculated using the equilibrium assumption. Unfortunately, for such low energy recoils, obtaining accurate mass information is very difficult, complicated by pulse height defects and plasma delay considerations.

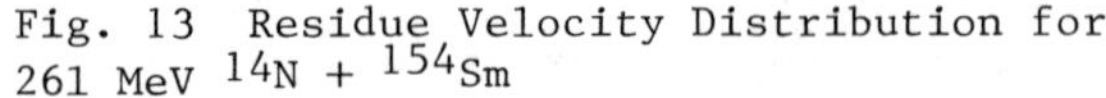
Fig. 13 Residue Velocity Distribution for 261 MeV ^{14}N + ^{154}Sm

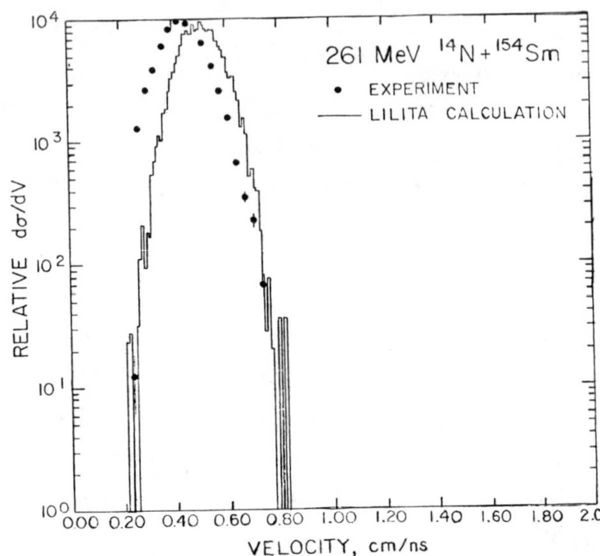

At somewhat higher energies, the situation is somewhat better. Thus in Figure 14 we show the distribution $\frac{1}{V^2}\frac{d^2\sigma}{d\,\Omega dV}$ for the reactions of 35 MeV/u ^{14}N with ^{154}Sm and in Figure 17 the mass distribution derived from the time and energy measurements. We have employed the scheme of Kaufman, et al[18] in correcting the observed energies for pulse height defect. As a test of our methods we compare the radiochemically determined mass distribution for the similar 420 MeV ^{12}C + ^{154}Sm reaction with our instrumentally derived mass distribution in Figure 15.

The broader distribution in our case probably reflects the difficulties in accurately ascertaining the pulse height defects which are a significant fraction of the total energy for such low energy recoils.

Recognizing that there are some uncertainties in the process one can nevertheless compare the mass distributions for different regions of residue velocity to observe trends in those distributions. In

Fig. 14 Residue Velocity Distribution for Reactions of 490 MeV $^{14}N + ^{154}Sm$. Velocity Corresponding to Total Momentum Transfer is Indicated

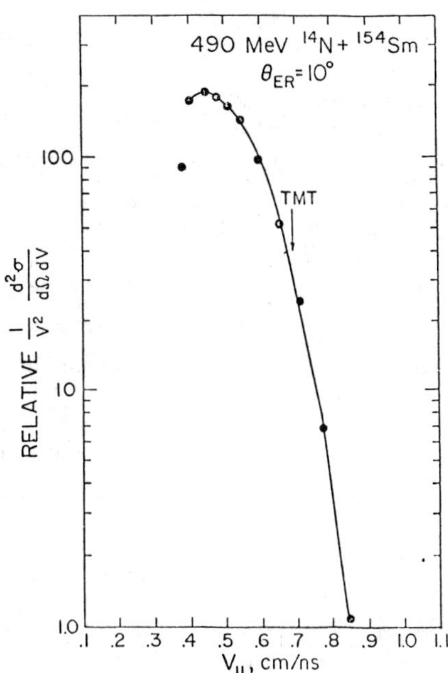

Fig. 15 Mass Distribution of Residues at Θ_L=10° for 490 MeV $^{14}N + ^{154}Sm$. Radiochemical Mass Distribution for 420 MeV $^{12}C + ^{154}Sm$ Shown for Comparison

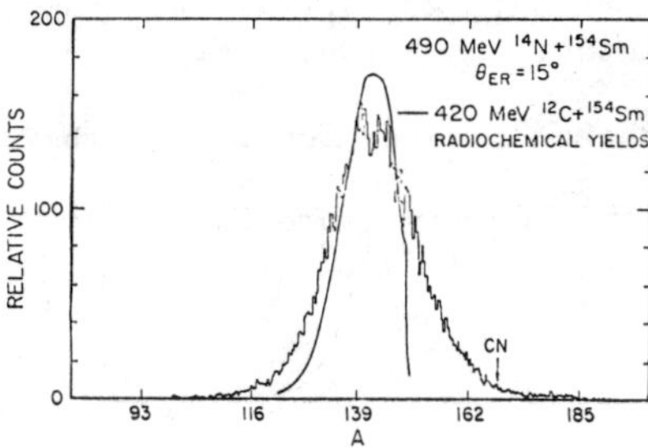

Figure 16 mass distributions derived for three different regions of momentum transfer ranging from full momentum transfer to 55% of that value are presented.

Fig. 16 Mass Distributions for Three Windows of Fractional Momentum Transfer

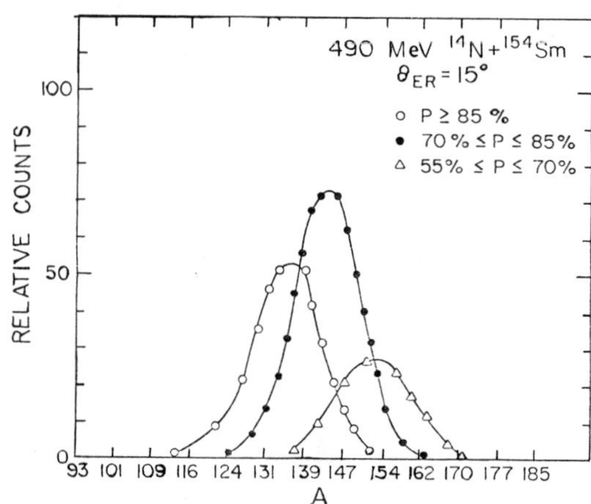

As the momentum transfer and therefore excitation energy increases, the number of evaporated particles increases. Statistical model calculations are currently underway to determine whether the experimental trend is consistent with de-excitation of compound systems having the temperatures even as high as those corresponding to total momentum transfer.

* This talk reports on works in which many different co-workers also participated. These collaborators who will appear as co-authors of various publications resulting from the work in progress which is described are L. Adler, J. Bronson, U. Garg, P. Gonthier, D. Haenni, Y. W. Lui, G. Mouchaty, T. Murakami, R. P. Schmitt, S. Simon, J. Sullivan, D. Youngblood, G. Berkowitz, H. Ho, M. Bantel, D. Habs, J. P. Wurm, A. Brucker, B. Lindl, R. Muffler, A. Pfoh, L. Shad, R. Fischer, W. Kuhn, V. Metag, and R. Novotny.

** Current address Bhaba Research Centre, Trombay, Bombay, India.

*** Current address Lawrence Livermore Laboratory, Livermore, California, USA

1. Morgenstern, H., et al, Phys. Lett. 112B, 463 (1982).

2. Pulhofer, F., et al, Phys. Rev. C16, 1010 (1977).

3. Rivet, M.F. and Borderie, B., Orsay Report 1PNO-DRE-84-32 (unpublished).

4. Back, B.B., et al, Phys. Rev. 22C, 1927 (1980).

5. Tamain, B., Universite de Caen Report LPC 84-03 (1984) unpublished.

6. Huizenga, J.R., et al, Phys. Rev C28, 1853 (1983). Tubbs, L.E., et al, Univ. of Rochester Report UR-NSRL-288 (1984) unpublished.

7. Gavron, A., et al, Phys. Rev. C30, 1550 (1984).

8. Wilczynski, U., et al, Nucl. Phys. A373, 109 (1982).

9. Harvey, B.G. and Homeyer, H., LBL Report 16882, unpublished.

10. Udagawa, T. and Tamura, T., Phys. Rev. Lett 16, 1311 (1980).

11. Geoffroy, K.A., et al, Phys. Rev. Lett 43, 1303 (1979).

12. Choudhury, R.K., et al, to be published.

13. Namboodiri, M.N., et al, Phys. Rev C20, 982 (1979).

14. Sarantites, D.G., et al, Phys. Rev. C18, 774 (1978).

15. Bantel, M., et al, contribution to this conference.

16. Nebbia, G., et al, work in progress.

17. Gomez del Campo, J. and Stokstad, R.G., ORNL Report ORNL TM 7295 (1981).

18. Kaufman, S.B., et al, Nucl. Inst. and Meth. 115, 47 (1974).

19. Aleklett, K., et al, LBL Report 1984 unpublished.

Subthreshold Pion Production in Nucleus - Nucleus Collisions What is the Mechanism*

R. Shyam and J. Knoll
GSI, Postfach 110541, Planckstrasse 1, D-6100 Darmstadt 11
Federal Republic of Germany

Abstract

Various reaction mechanisms proposed to explain the pion production at 'subthreshold' energies (below 290 MeV/A) are examined. They range from the nucleon-nucleon single collision mechanism to a co-operative multi-nucleon process. With a shell model prescription for the initial state energies the single collision picture can not explain the data. The participation of many nucleons in the pion production process appears to be necessary. We present a statistical model where the co-operative action of several of the target and projectile nucleons in the pion production process is invoked. We also consider the formation of the fragments in the final channel alongside the produced pion. Calculations performed within the model provide a good overall description of the experimental data over a wide range of beam energies and masses of the participating nuclei. Fragment formation in the final channel is seen to be vital to understand the experimental data within our model.

* Invited talk presented by R. Shyam in the International Conference on Nuclear Physics, Bhabha Atomic Research Centre, Bombay, Dec. 27-31 1984.

1. Introduction

One of the interesting aspects of heavy ion collisions at intermediate energies is the observation of a significant pion production cross section at the so called 'subthreshold beam energies' (i.e. below 290 MeV/A)[1-6]. These pions open up a kinematical domain which is not accessible in free nucleon-nucleon (NN) collisions. Theoretical attempts to understand their production mechanism range from the nucleon-nucleon single collision (NNSC) picture[7-8] (suggested even prior to the laboratory detection of the pion[9]) where the intrinsic (Fermi) motion of the nucleons in the initial states of the two colliding nuclei provides the necessary kinetic energy, to the collective processes like mean field approach[10], decay from the compound nucleus[11], and the pionic bremsstrahlung[12]. As will be shown latter on in this paper, the NNSC mechanism with a realistic shell model prescription to describe the intrisic nucleonic motion in the initial states of the colliding nuclei is insufficient to explain both the absolute yields and the shapes of the pion spectra[8]. In the mean field approach, the dynamics of the collision is described by the time dependent Hartree-Fock theory, and the pion production is treated in a first order approximation by the so called one nucleon or two nucleon mechanisms out of the microscopic time dependent orbits. The mechanism of pionic bremsstrahlung is even more collective in nature as it involves only the ground state densities of the two colliding nuclei. However, this depends sensitively on the assumed parameter related with the deceleration of the nuclei during the collision.

The collective processes mentioned above treat the one body part of the nucleus-nucleus interaction rather rigorously, i.e. they account for that part of all the interactions which can be included in a one-body field, while the residual two-body interaction part is treated at best only perturbatively. In this talk we present a model, where contrary to the collective picture one takes care of the residual interaction part of the nuclear collision dynamics and the more smooth one body part is ignored, an approach which becomes more and more valid with increasing beam energy. In the sense of a quantum multiple collision picture our model considers the genuine co-operation among several nucleons which allows the pooling of their energies to produce a pion. A further co-operative effect results from the formation of the fragments of various masses in the final channel alongside the produced pion. In order to simplify the calculations of the spectral forms we invoke the assumption of equal occupation of the available phase-space.

Before presenting the details of our model, we feel it to be worthwhile to understand clearly the bare kinematical implications of the NNSC and the multi-nucleon mechanisms. In this context we re-examine the NNSC mechanism rather carefully in the next section. The main aim of our discussions in this respect is to point out the implication of using an improved energy prescription for the initial state.

2. Nucleon - Nucleon Single Collision Mechanism

Besides the elementary production cross-sections the only information required for the NNSC model are the spectral functions $f_A(P,E)$ and $f_B(P,E)$ of the two colliding nuclei (A and B). These functions describe the distribution of momenta and energies available to a nucleon at the instant of impact (Fermi motion). Rather safe experimental information on these distributions up to momenta of about 500 MeV/c is available through (γ,p) and (γ,n) data[13] and for lower intrisic momenta through (e,e'p) and (e,e'n) coincidence measurements[14]. The Fermi-gas model implies a correlation corresponding to the motion of free particles

$$f_{FG}(P,E) = f(p)\delta(E-\varepsilon(P)) \tag{1}$$

Where

$$\varepsilon^2(P) = P^2 + (m_0 - V_0)^2$$

Here V_0 is the depth of the potential well and m_0 the nucleon mass. In the shell model on the other hand, particles are distributed in the orbits of discrete energies, and the one nucleon distribution function acquires the following form

$$f_{s.m.}(P,E) = \sum_i f_i(P)\delta(E-\varepsilon_i) \qquad (2)$$

which for five closed harmonic-oscillator shells is illustrated in fig. 1a. Here ε_i represents the binding energy of the ith orbit with a momentum distribution $f_i(P)$. Only in the limit of the large nuclei, both prescriptions merge in the quasi-classical sense, i.e. the most probable momentum components follow the Fermi gas prescription as illustrated by the dashed line in fig. 1a. The use of the smooth momentum distribution in the Fermi-gas picture (instead of a function with sharp cut-off at the

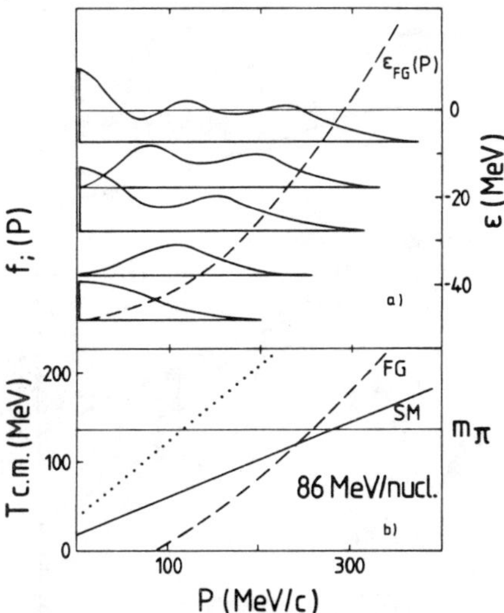

Fig. 1) (a) Display of the spectral function f(P,E) pertaining to the harmonic oscillator shell model with five filled shells. Dashed line shows the energy-momentum correlation implied by the Fermi-gas picture. (b) Total kinetic energy available for the production process at the beam energy of 86 MeV/A as a function of intrisic momentum P of the colliding nucleons relative to the respective nuclear rest frames for the shell model and Fermi-gas prescriptions. The dotted line shows the total kinetic energy available in a co-operative collision of two projectile nucleons with two target nucleons.

edge of the Fermi sea), however, brings in an inconsistency, particularly for those momentum components which are classically forbidden (the high momentum components). This allows the intrinsic energy of the nucleons to be greater than the nucleon separation energy, a defect which is not present in the shell model prescription, where all the nucleons remain bound in the initial state irrespective of the intrinsic momenta. All the calculations based on the NNSC model published so far, however, have used the Fermi gas prescription (in some cases even the binding potential V_o has been neglected), despite the fact that at subthreshold energies one is particularly sensitive to the classically forbidden momenta. Consequences of this on the total available c.m. kinetic

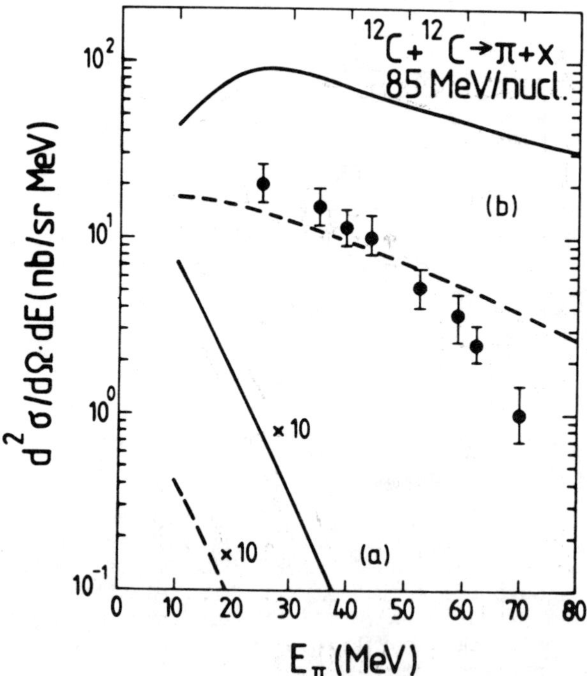

Fig. 2) Pion-production cross section at 90° laboratory angle (data from ref.[7]) calculated with the single N-N collision model. The group of curves (a) from a standard harmonic oscillator shell model fitting electron scattering data, and group (b) from the Fermi-gas model (V_o=0); solid and dashed curves correspond to the NN → d π and the NN → NN π channel, respectively.

energy ($T_{c.m.}$) can be seen in fig. 1b. Picking two nucleons out of the respective Fermi seas with equal and opposite momenta P_A=-P_B=P (which anyhow provides the maximum contribution to the total energy), one gets an available c.m. kinetic energy $T_{c.m.}$ which varies approximately linearly with P in the shell model picture, whereas an additional quadratic dependence follows for it in the Fermi gas prescription. This means that in the Fermi gas picture (in particular with V_o = 0) a pion can be produced already at far lower intrinsic momenta than in the shell model picture. Thus the calculations performed by the NNSC model with the Fermi gas prescription will result in larger yields and flatter slopes of the pion spectra in comparison to those performed with the more accurate shell model prescription, as illustrated in fig. 2. From fig. 2 we come to the following conclusion: the NNSC model with the improved energy prescription for the initial state is not able to reproduce the observed cross sections[2] in any respect. In fact the calculations miss the experimental yields by more than a factor of 100. As the calculations exploit intrinsic momenta up to just 500 MeV/c, we see no way that a more realistic momentum distribution can alter our above conclusion. Also the slopes of the calculated spectra are about a factor of 3 steeper than that of the experimental one. Note that the inclusion of Pauli blocking and re-absorption effects as considered in Ref.[7] would even further increase the discrepancy between experimental data and the calculations performed with the shell model prescription. Therefore, there is a need for a production mechanism other than the NNSC picture. One of the possibilities could be a co-operative action of several of the projectile and target nucleons, which is discussed subsequently.

Allowing for the participation of more than two nucleons in the pion production process, the kinematical situation in fact changes (fig.1b). As expected, the production threshold opens already at much lower intrinsic momenta in comparison to the NNSC picture, and the sensitivity of the process is shifted from the extreme tail region towards the clas-classically allowed components of the momentum distribution. It seems, therefore, that a multi-nucleon process is a more realistic picture for the subthreshold pion production.

In the next section we present the details of our model[15] where the pions are supposed to be produced via multiple collisions among a few of the projectile and the target nucleons.

3. The Multiple Collision Model

The starting point of our model is the general observation that in a quantum mechanical multiple collision picture, off-shell collisions allow a by far more flexible sharing of the available energy than it would be possible by sequential on-shell scattering (as in a cascade model). Such a genuine co-operative mechanism allows the pooling of energies of all those nucleons which are in the mutual interaction contact during the collision. Faced to such a picture the whole collision process can be visualised in terms of a grouping of all nucleons into what we like to call clusters, each cluster containing all those target and projectile nucleons which are in interaction contact with one another during the collision process. To be specific at this piont, we do not think about the pre-formed clusters in the sense of deuteron - or alpha particle correlations in the initial nuclear configurations, but rather the clusters to be formed dynamically through the interactions occurring during the collision process. As a consequence one body observables like the inclusive cross section for observing a particle γ can be expressed as an incoherent sum over the contributions arising from all the clusters[16,17)]

$$\varepsilon_\gamma \, (d^3\sigma/dp_\gamma^3) = \Sigma_{MN} \, \sigma_{AB}(M,N) \; F_{MN}^{\gamma}(p_\gamma) \qquad (3)$$

The labels M and N represent the numbers of the projectile and target nucleons respectively in each contributing cluster. In a diagramatic expansion of the whole collision dynamics a cluster contribution (M,N) would comprise the absolute square of the coherent sum of all those diagramms which contain a connected piece with M projectile and N target legs. Since different clusters reach different final channels we add them incoherently in eq.(3). The yield of each cluster is factorised into a formation cross-section $\sigma_{AB}(M,N)$, and a properly normalised probability distribution F_{MN}^{γ}. While the formation cross-sections specify the occurrens of a given cluster and thus carry the signatur of the incident channel, the spectra F^{γ}_{MN} give the partial flux that goes into the observed channel out of the cluster (M,N), i.e. to the observed particle γ with energy ε_{γ} and momentum p_{γ}. Note that eq. 3 as such is very general and accomodates quite a variety of multiple collision approaches. Different prescriptions to calculate σ_{AB} and F^{γ}_{MN} lead to different models like NNSC or the thermal models[18,19] for example. In the present study we employ the nuclear phase-space model as already explained in refs.[16,20]. However, we incorporate into it two essential extensions. One concerns the specification of the formation cross-section. Here we leave the straight-line communication limit (rows on rows[21]) which was quite successful at higher energies and turn to the communication as resulting from a 3-dimensional cascade[22]. The second extension concerns the final states, which here include also the possible formation of composite light nuclei besides free nucleons and the produced pion. Both extensions though less important at higher energies, become an essential part of the model due to the severe kinematical constraints imposed by the subthreshold condition.

4. Results and Discussion

What is the physics implemented in our model? Assuming that the violent interactions among the basic constituents, the nucleons, are the key mechanism to mediate the transport of energy and momentum, we consider the inclusive cross sections to arise from the incoherent contributions of all possible clusters. This assures a model that interpolates smoothly between the single collision picture and the limit of global interaction contact as in the fire-ball model. With regard to the final channels of the reaction, we allow, besides the production of the pion, also the formation of composite nuclei, in principle up to the fused nuclear system. This way our model interpolates up to the pionic fusion[23] limit. Two assumptions precise the model: a) the distribution of the cluster sizes is essentially determined from the collision geometry, and we calculate these distributions from our experience with the 3-dimensional cascade calculations; b) in order to allow for a sharing of energy in a less constrained manner than in the cascade models we consider the statistical limit for the cluster dynamics, i.e. we assume for each cluster all possible final states (these include the production of the pion and formation of the composite nuclei) with equal probability. The required level counting demands to specify the volume V_{MN} of the system. We take it proportional to the nucleon number of the cluster $V_{MN}\,\rho_c = M + N$ by means of a density parameter ρ_c. This parameter was fitted for one system. We found $\rho_c = \rho_0 = 0.17\ fm^{-3}$ optimal and kept it constant throughout. With a constant value for this parameter, the trends of the data with changes of both the bombarding energy as well as the nuclear masses are reproduced nicely. Let us see what the model can do.

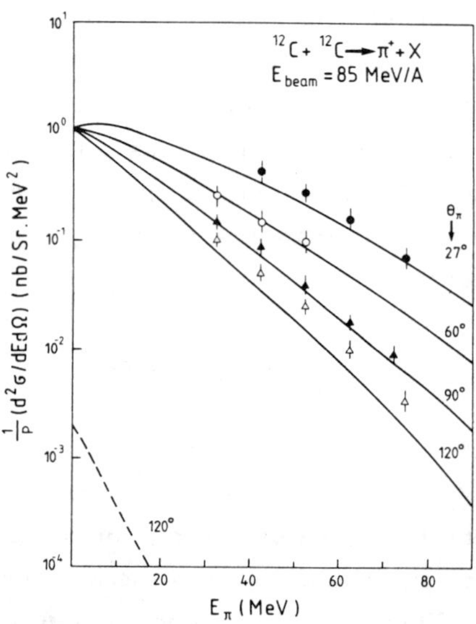

Fig. 3) Invariant cross section for the production of π^+ as a function of the pion kinetic energy for several pion angles. The colliding systems is ^{12}C + ^{12}C and the beam energy 85 MeV/A. Solid curves represent the results of our calculations where we have assumed the formation of the fragments of masses up to 12 in the final channel alongside the produced pion. The dotted curve represents the results of our calculations (for the pion angle of 120°) when only nucleons exist along with the pion in the final channel.

We begin by comparing our calculations with the experimental data for the charged pions corresponding to beam energies around 85 MeV/A. In figs. 3 we show the results of our calculations for the invariant cross section as a function of the pion kinetic energy for several pion angles. The experimental data have been taken from ref. 2. It is clear that our calculations reproduce the shapes as well as the absolute yields of the data reasonably well for all angles. In order to emphasize the importance of the formation of fragments in the final channel, we also show in this fig. the result of our calculations when final state comprises of only nucleons and the pion (dotted line) corresponding to the pion angle of 120°. In this case the absolute yield is down by nearly 3 orders of

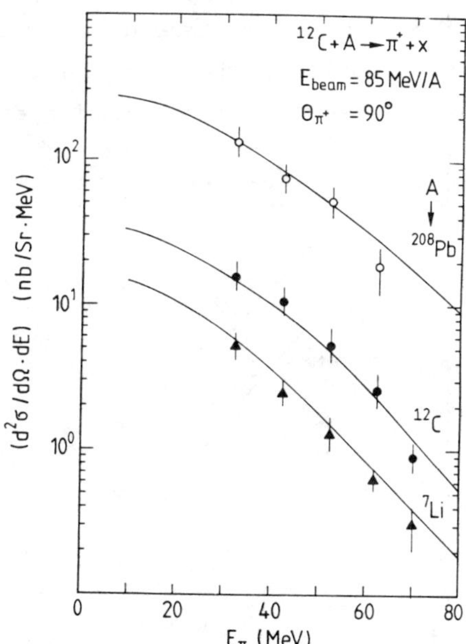

Fig. 4) Double differential cross section as function of the pion kinetic energy for the fixed pion angle of 90°. The projectile nucleus is ^{12}C whereas target nuclei vary from ^{7}Li to ^{208}Pb. The beam energy is 85 MeV/A. Solid curves have the same meaning as in fig. 3.

magnitude and also the slope is much steeper. This shows clearly that the formation of the fragments in the final channel is very important. One can easily understand this result by noting that the formation of fragments is tantamount to the freezing of the kinetic degrees of freedom in the final channel. Due to this, in our statistical picture the available total energy is shared less and hence more energy becomes available to the outgoing pion. This enhances the absolute yields and also the slopes of the spectra become flatter. First experimental studies have already been performed to check this prediction. The absolute yields of the fragments of various charge have been measured[24] recently in coincidence with pions in the collision of ^{12}C with ^{12}C at a beam energy of 85 MeV/A. There one sees relatively large yields of the light charged fragments. More experimental studies along these lines would be very welcome. In fig. 4 we show the comparison of the calculated double differential cross sections with the data for various target nuclei at a

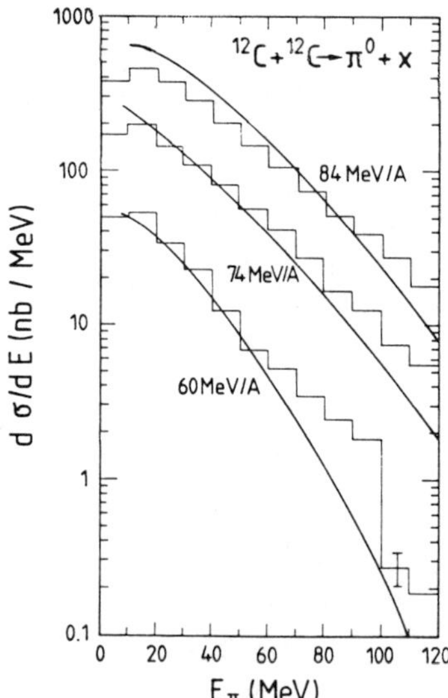

Fig. 5) Angle integrated cross sections for the production of π^0 as a function of the pion kinetic energy in the collision of ^{12}C with ^{12}C. The beam energies vary from 60 MeV/A to 84 MeV/A. The histograms represent the experimental data whereas the solid curves the results of our calculations.

fixed pion angle of 90°. we see that the target dependence of the double differential cross section is well reproduced by our calculations.

In fig. 5 we show the comparison of our calculations for the angle integrated cross sections with the data for the case of neutral pions for C + C collision at beam energies of 60 , 74 and 84 MeV/A. We see that the calculations are in a reasonable agreement with the experiments for all the three beam energies. We expect some differences between the calculated and the experimental cross sections in the extreme tail region of the energy spectra. There one approaches the limit of the available phase space. As the beam energy is lowered further, our calculations under predict the experimental cross sections more and more because of the increasing importance of the one body part of the interaction (mean field effects). This is apparent from figures 6 where we present a comparison of calculated and experimental angle integrated cross sections[6]

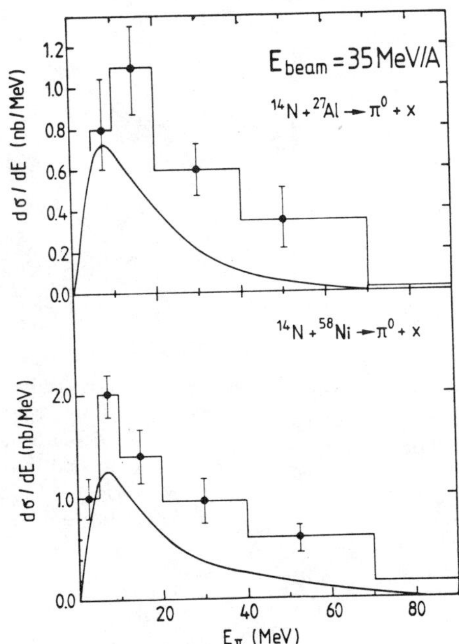

Fig. 6) Same as in Fig. 5 but for the systems ^{14}N + ^{27}Al and ^{14}N + ^{58}Ni at the beam energy of 35 MeV/A. The histograms and the solid curves have the same meaning as in fig. 5.

for the beam energy of 35 MeV/A. However, note that the background has not been substracted from the data[6] shown in this fig. This may account for some of the differences seen between the calculations and the experimental data particularly for larger pion kinetic energies. Nevertheless, collective processes are likely to be important for these lower energies. It would be interesting to see if our calculations combined with the mean field calculations in one nucleon model[10] provide a better description of the experimental cross sections.

In figs.7 and 8 we have shown the total pion production cross section as a function of beam energy and the target mass respectively. One notes

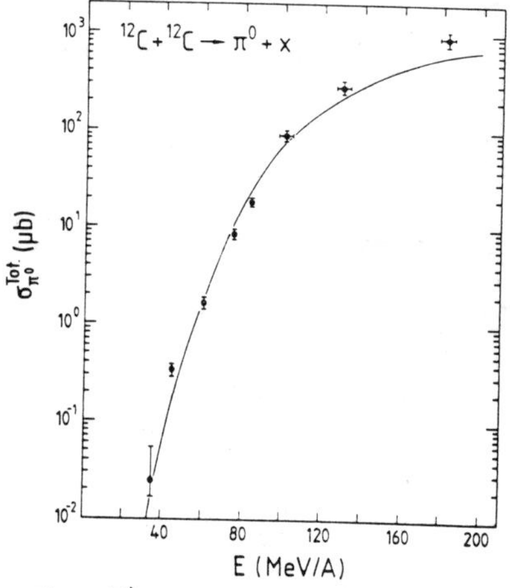

Fig. 7)

The total π^0 production cross section as a function of beam energy for the system ^{12}C + ^{12}C. The experimental data for the beam energies below 100 MeV/A have been taken from Refs. 4 and 6 and those above from Ref. 1. Where necessary the data have been translated to the ^{12}C + ^{12}C system by following an expression given in Ref. 6.

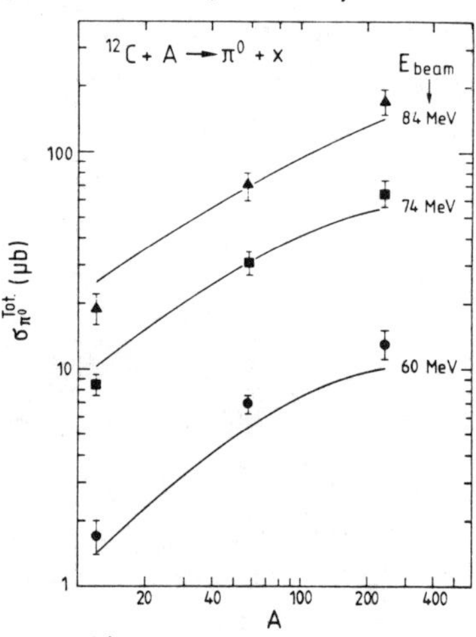

Fig. 8)

The target mass dependence of the total π^0 production cross sections for the beam energies of 60 MeV/A, 74 MeV/a and 84 MeV/A (the projectile being ^{12}C). The experimental points have been taken from Ref. 4.

from fig. 7 that the dependence of the cross section on the beam energy is rather well described by our calculations. The change of six orders of magnitudes which one sees in the experimental cross section in going from the beam energy of 35 MeV/A to 200 MeV/A, is reproduced by our calculations without requiring any re-adjustment of the density parameter. The experimental data show a dependence on the masses of the colliding partners in accordance with $(A_P A_T)^{0.68}$ where A_P and A_T are the masses of the projectile and the target nuclei respectively. Our calculations reproduce such a dependence for all the three beam energies.

The discussions presented above, make this quite clear that our model is able to describe almost all the systematics of the data accumulated so far on subthreshold pion production. The remarkable fact is that no readjustment of the density parameter is needed while changing from one beam energy to another or from one set of collision partners to another.

5. Conclusions

In conclusion we can say that the N-N single collision mechanism with an improved energy prescription (shell model) for the initial intrinsic nucleonic motion is unable to reproduce the absolute yields as well as the observed shapes of the experimental pion spectra at subthreshold beam energies. A mechanism which invokes genuine co-operation among nucleons which may arise through quantum off-shell processes appears to be more appropriate. We presented a model which is based on a picture of the multiple collision between several of the projectile and the target nucleons. Our multiple collision dynamics, however, is different from that incorporated in the intra nuclear cascade (INC) model. In the latter the collision between two nuclei is visualised as a sequence of classical binary on-shell nucleon-nucleon collisions. However, at these lower beam energies such a model would be no better than the NNSC model since due to the on-shell nature of the collisions only first nucleon nucleon collisions would be energetic enough to produce a pion. In our model, on the other hand, we take into account the quantum off-shell scattering between the nucleons. Thus a genuine co-operation among them is invoked in pooling their energies to produce a pion. The number of nucleons participating in the collision process has been estimated from a picture which in a way approches the 3 - dimensional cascade limit. The model involves a density parameter which we held fixed to a value equal to the normal nuclear matter density ρ_0 (= 0.17 fm^{-3}) for all the cases studied.

The calculations performed within the model provide a good overall description of the data over a wide range of beam energies and colliding nuclei. The dependence of the pion production cross section (which changes by 6 orders of magnitude between the beam energies 35 MeV/A to 200 MeV/A) has been reproduced without requiring any readjustment of the density parameter. The target mass dependence af the production cross sections is also well described by our calculations.

At lower beam energies the cross sections calculated by our model are somewhat smaller than the experimental data. This is due to the fact that at lower beam energies the mean field effects are expected to manifest themselves rather strongly which has been ignored by us.

The present study suggests that co-operative action of several nucleons in the pion production at subthreshold beam energies is indeed very important. In the model presented here this enters through multiple nucleon-nucleon collisions and the fragment formation in the final channel. The latter should reflect itself in a correlation between the produced pion and the mass distribution of the associated nuclear fragments. More exclusive measurments of the pions in coincidence with the associated fragments would be interesting in this context.

Useful conversations with P. Braun-Munzinger, W. Cassing, J. Cugnon, E. Grosse, C. Guet, B. Jakobsson, U. Mosel and B. Schürmann are gratefully acknowledged.

References

1) W. Benenson et al., Phys. Rev. Lett. **43** (1979) 683

2) T. Johansson et al., Phys. Rev. Lett. **48** (1982) 732; V. Bernard et al., Nucl. Phys. **A423** (1984) 511

3) S. Nagamiya et al., Phys. Rev. Lett. **48** (1982) 1780

4) H. Noll et al., Phys. Rev. Lett. **52** (1984) 1284; E. Grosse, Proc. Sixth High Energy Heavy ion Study and Second Workshop on Anomalons, Berkeley, June 28-July 1,1983, page 225

5) H. Heckwolf et al., Z. Physik **A315** (1984) 343

6) P. Braun-Munzinger et al., Phys. Rev. Lett. **52** (1984) 255

7) B. Jakobsson, Phys. Scri. **T5** (1983) 207

8) R. Shyam and J. Knoll, Phys. Lett. **136B** (1984) 221

9) W.G. McMilllan and E. Teller, Phys. Rev. **72** (1947) 1

10) M. Tohyama, R. Kaps, D. Masak and U. Mosel, Phys. Lett. **136B** (1984) 226; and Nucl. Phys. (to be published); W. Cassing, GSI pre-print GSI - 84 - 15

11) J. Aichelin and G.F. Bertsch, Phys. Lett. **138B** (1984) 350

12) D. Vasak, B. Müller and W. Greiner, Phys. Scr. **22** (1980) 25; D. Vasak, H. Stöcker, B. Müller and W. Greiner, Phys. Lett. **A360** (1980) 243

13) D.J.S. Findlay et al., Phys. Lett. **74B** (1978) 305

14) E.J. Moniz et al., Phys. Rev. Lett. **26** (1971) 445

15) R. Shyam and J. Knoll, Nucl. Phys. **A426** (1984) 606

16) J. Knoll, Phys. Rev. **C20** (1979) 779

17) J. Cugnon, J. Knoll and J. Randrup, Nuc. Phys. **A360** (1981) 444

18) J. Gosset et al.,Phys. Rev. **C16** (1977) 629

19) A.Z. Mekjian, Nucl. Phys. **A312** (1978) 491; Phys. Rev. **C17** (1978) 1051; S. Dasgupta and A.Z. Mekjian, Phys. Rep. **72** (1981) 131

20) J. Knoll, Nucl. Phys. **A343** (1980) 511

21) J. Hüfner and J. Knoll, Nucl. Phys. **A290** (1977) 460

22) J. Cugnon, T. Mizutani and J. Vandermeulen, Nucl. Phys. **A352** (1981) 505

23) K. Klingenbeck, M. Dillig and M.G. Huber, Phys. Rev. Lett. **47** (1981) 1654

24) M. Balore et al., Proc. Int. workshop on gross properties of nuclei and nuclear excitations XII, Hirschegg, Austria (1984) page 187

OPTICAL POTENTIAL AND MEAN FREE PATH IN SEMICLASSICAL APPROXIMATION

R.W. Hasse
Institut Laue-Langevin, Grenoble, France

P. Schuck
Institut des Sciences Nucléaires, Grenoble, France

ABSTRACT

By using a finite range force we calculate the polarisation and correlation contributions to the real and imaginary part of the optical potential in semiclassical approximation. The imaginary part, effective mass, single particle level density and mean free path - with and without temperature - are then derived from a Hartree-Fock potential plus correlations and, if possible, compared with experiments.

INTRODUCTION

The nucleon-nucleus optical potential or nucleon mass operator[1) is one of the fundamental quantities in nuclear physics. For positive energies its imaginary part, W, reflects two body scattering processes of the nucleon in the nucleus and it allows to study their energy dependence, the influence of the Pauli principle, the mean free path and many more. For negative energies, W accounts for the spreading widths of single particle states and, for instance, the width of the deep lying hole states is a very interesting quantity in this context. The real part is essentially given by the Hartree-Fock field but an energy dependent (dynamic) correction which corresponds to W can have an important influence on certain quantities as will be shown below.

Technically the dynamic part of the optical potential is quite difficult to handle since the first scattering process leads already to two particle - one hole (2p - 1h) intermediate states, i.e. with the possibility of two particles simultaneously lying in continuum scattering states which have to be summed over all possible momenta. We here will handle this problem with a novel Thomas Fermi theory for mp - nh states developed elsewhere[2] and which has turned out to be very accurate. This has the further advantage of being easily generalizable to finite temperatures.

2. THE OPTICAL POTENTIAL

To second order in the two body interaction the mass operator[1] which defines the optical potential at energy ω is given by

$$M^{\omega}_{11'} = \frac{1}{2}\sum_{234} \bar{v}_{1234}\left(\frac{n_2 \bar{n}_3 n_4}{\omega+\varepsilon_2-\varepsilon_3-\varepsilon_4+i\eta} - \frac{\bar{n}_2 n_3 n_4}{\omega+\varepsilon_2-\varepsilon_3-\varepsilon_4-i\eta}\right)\bar{v}_{4321'}\ , \tag{1}$$

where $\bar{n},n$ are the particle and hole occupation numbers, respectively, ε_i are particle and hole energies, and $\bar{v}_{1234}$ is the (antisymmetrized) matrix element of the effective two body interaction. The Feynman graph corresponding to (1) is represented in Fig. 1.

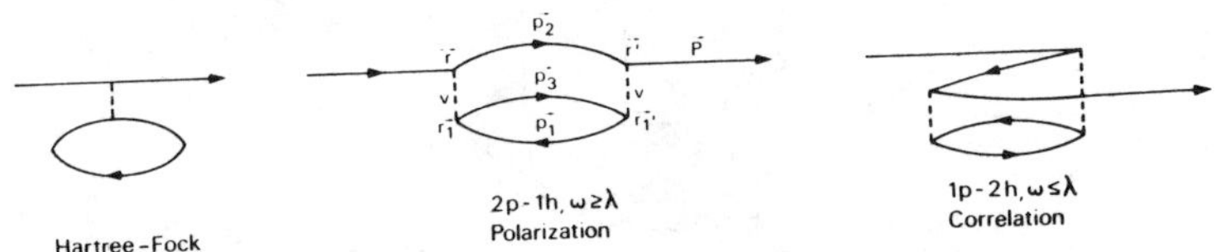

Fig. 1 : Feynman graphs corresponding to 2p-1h and 2h-1p intermediate states.

If the matrix elements in (1) were constant, then ImM would be proportional to the 2p-1h (2h-1p) level density. To explain the principle of our theory we first calculate this quantity in semiclassical approximation. To this end we write

$$g^{E}_{2p-1h} = \sum_{p_1 p_2 h} \delta(E+\varepsilon_h-\varepsilon_{p_1}-\varepsilon_{p_2}) = \mathrm{Tr}_1\,\mathrm{Tr}_2\,\mathrm{Tr}_3\,\Theta(\varepsilon_F-H_1)\Theta(H_2-\varepsilon_F)\Theta(H_3-\varepsilon_F)\,\delta(E+H_1-H_2-H_3)\ , \tag{2}$$

where H is the single particle Hamilton operator and the traces are to be taken of the complete set of single particle states. Semiclassical approximation means to replace the H_i by their classical counterparts $H_i^c = p_i^2/2m + V(R_i)$ and the traces go over into corresponding phase space integrals :

$$g_{2p-1h}^{TF} = c \int d\Gamma_1 d\Gamma_2 d\Gamma_3\, \theta(\varepsilon_F - H_1^c)\theta(H_2^c - \varepsilon_F)\theta(H_3^c - \varepsilon_F)\delta(E + H_1^c - H_2^c - H_3^c)$$
$$d\Gamma = \frac{d^3R\, d^3p}{(2\pi\hbar)^3} \,, \tag{3}$$

where c accounts for spin-isospin degeneracies. For a spherical harmonic potential all integrations can be done analytically[3] and the result is displayed in Fig. 2. We see that our approximation represents very well the average of the exact result. This is due to the fact

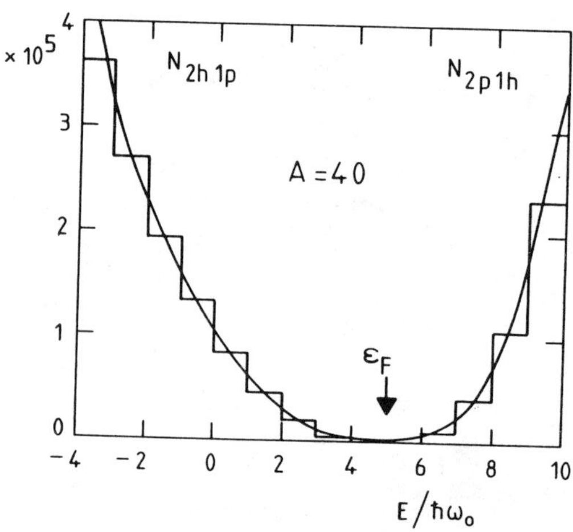

Fig. 2 : Exact versus semiclassical accumulated level density
$N(E) = \int dE' g(E')$

that the states are individually approximated by their T.F. expression. This means for scattering processes that in our T.F., particles (holes) can scatter at distance contrary to the local density approximation[4]. For finite range forces this is of importance in the nuclear surface.

Using (3) and a local effective Wigner force

$$\bar{v}(\vec{r}\vec{r}_1\vec{r}_2\vec{r}_1') = v(\vec{r} - \vec{r}_1)\delta(\vec{r} - \vec{r}_2)\,\delta(\vec{r}_1 - \vec{r}'), \tag{4}$$

the Wigner transform of the imaginary part of (1) can be written as

$$W(E,\vec{R},\vec{P})=\frac{2}{(2\pi)^5}\int d^3R'\,d^3P_1\,d^3P_2\,d^3P_3\left[\theta(P_F^2(R')-P_1^2)\theta(P_2^2-P_F^2(R))\theta(P_3^2-P_F^2(R'))+(-)\right]$$
$$\times\,\delta(\vec{P}+\vec{P}_1-\vec{P}_2-\vec{P}_3)\,\delta\left(E+\frac{P_1^2-P_3^2}{2m^*(R')}-\frac{P_2^2}{2m^*(R)}-V(R)\right) \tag{5}$$
$$\times\int d^3x\, e^{2i\vec{x}(\vec{P}-\vec{P}_2)}\,v(\vec{R}-\vec{R}'+\vec{x})\,v(\vec{R}-\vec{R}'-\vec{x})\,.$$

Here (-) indicates the same product of θ-functions as before but with arguments of opposite signs. As force we use an effective range approximation to the s-wave part of the Gogny force[5)] resulting in the following Gaussian :

$$v(r)=v_0e^{-r^2/b^2}\quad ,\ v_0=-26.5\,\text{MeV},\ b=2.25\,\text{fm}. \tag{6}$$

The effective mass $m^*(R)$ has also been calculated with the Gogny force in the local density approximation and for V we take the parameterization of [6)]. In Fig. 3 we show the radial dependence of W (on shell, i.e. $p^2 = 2\,m^*(R)(\omega - V(R))$ at various energies and in Fig. 4 is represented the depth of W as a function of E in the interior of the nucleus.

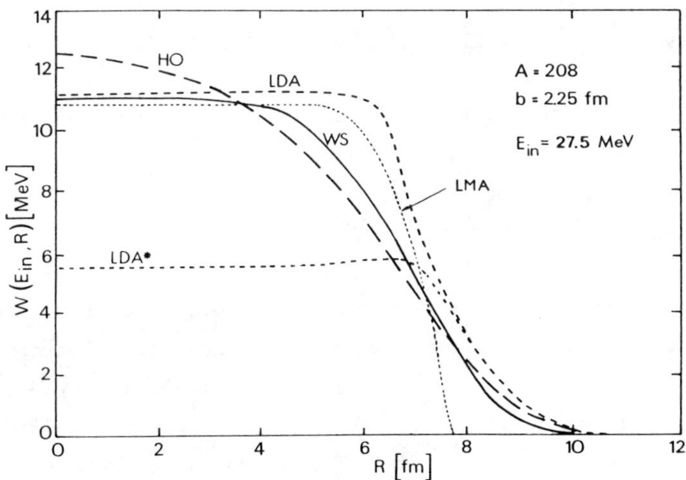

Fig. 3 : Radial dependence of W (on shell) at various energies.

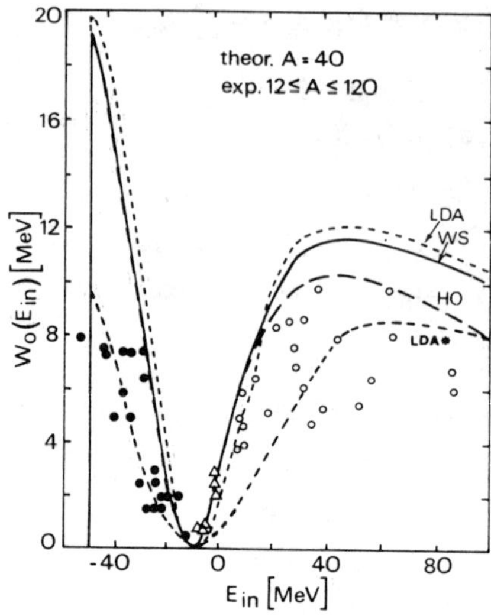

Fig. 4 : Dependence of the depth W(R=0,O.S.) on incident energy. Experimental points are taken from the compilation in [7].

The strong depression of W_o at the Fermi energy reflects the influence of the Pauli principle which suppresses scattering processes due to a restricted phase space. In this respect it is also interesting to write the on shell W in the following form

$$W^{\omega}(R) = \frac{\hbar}{2}\rho(R)\frac{p}{m}\frac{m^*}{m}\sigma_{NN}(\omega)\, f_p(R,\omega)\,, \tag{7}$$

where $\sigma_{NN}(\omega)$ is the free total nucleon-nucleon cross section calculated here with (6) using Fermi's golden rule

$$\sigma_{NN} = \frac{4\pi^2 v_o^2 m^2}{k_o^4}\,\frac{1-e^{-2p^2/k_o^2}}{p^2}\qquad, k_o = \frac{2}{b}\,. \tag{8}$$

The function $f_p(R,\omega)$ thus represents the influence of the Pauli principle as a function of energy and radius because the remaining expression in (7) represents just the same for a classical gas. In Fig. 5, f_p is shown as a function of ω and R for different ranges b of the force. It is interesting to see that only for a zero range force $f_p \to 1$ as $\omega \to \infty$. For finite range forces, f_p saturates at a value < 1 which comes from the restricted phase space available independent of energy due to the finite cut-off of the force. On the other hand, in the surface the

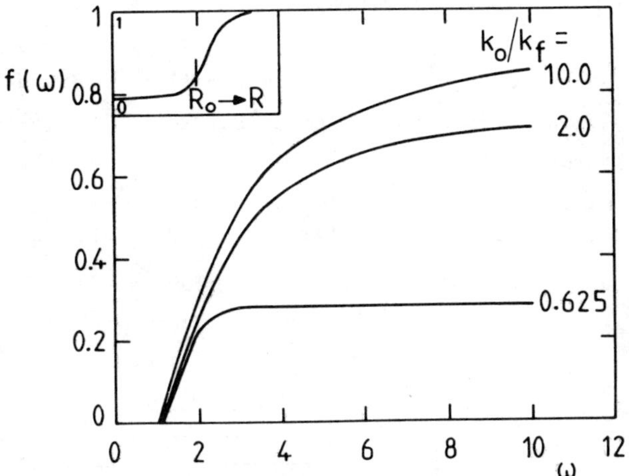

Fig. 5 : The Pauli blocking factor as a function of energy and radius. The lowest curve (k_o/k_F = 0.625) corresponds to our range b = 2.25 fm

influence of the Pauli principle vanishes.

3. THE REAL PART OF THE OPTICAL POTENTIAL

The major contribution to the real part of the optical potential is, of course, given by the Hartree Fock potential. There exists, however, a very important dynamical correction, $\tilde{V}$, which comes from the real part of (1) and which can be calculated from W via the usual dispersion relation. Since the dynamical correction at the Fermi surface is supposed to be included in the H.F. potential we have to use the subtracted dispersion relation [8)]

$$\tilde{V}(R,P,\omega) = \frac{\varepsilon_F - \omega}{\pi} \int_{-\infty}^{+\infty} ds \frac{W(s)}{(\varepsilon_F - s)(\omega - s)} \, . \tag{9}$$

Usually the nonlocality of the H.F. field and the momentum and energy dependence of $\tilde{V}$ are put into an energy and radius dependence of the effective mass. This goes via the relations

$$\frac{P^2}{2m} + V^{HF}(R,P) + \tilde{V}(R,P,E) \equiv \frac{P^2}{2m^*(R,P)} + V^{HF}(R,P=0) + \tilde{V}(R,P=0,E) \, , \tag{10a}$$

where E(R,p) is obtained from

$$E(R,P)=\frac{P^2}{2m}+V^{HF}(R,P)+\tilde{V}(R,P,E). \tag{10b}$$

In Fig. 6 we show the effective mass as a function of E and R, where we used the Perey and Buck potential[9] to calculate $V^{H.F.}(R,p)$ in the local density approximation. We see that $m^*/m \simeq 0.7$ in the interior of the nucleus as usual but close to the Fermi surface the influence of $\tilde{V}$ results for the R-dependence of m^*/m in a drastic increase in the surface region up to the value 1.4.

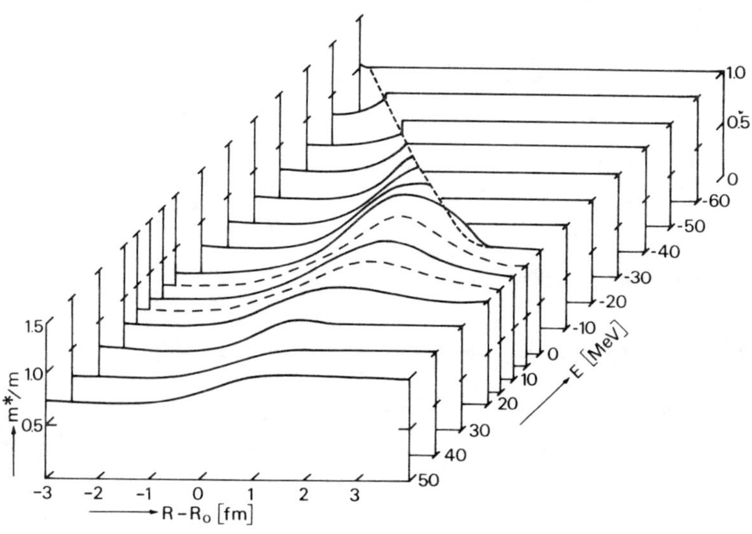

Fig. 6 : The effective mass as a function of E and R.

Intuitively this important enhancement of the effective mass close to both the real surface and the Fermi surface leads to an increase in the level density close to the Fermi energy since a heavier mass yields a denser single particle spectrum. More quantitatively, this can be calculated with the quasiparticle energies E(R,p) obtained from (10b) as follows :

$$g(\varepsilon_F)=2\int\frac{d^3R\,d^3P}{(2\pi\hbar)^3}\,\delta(\varepsilon_F-E(R,P))=\frac{4}{\hbar^2}\int dR\,R^2\,m^*(R,E)\,\mathcal{P}(R,E). \tag{11}$$

Performing the integration shows that the level density at the Fermi energy is enhanced by about 65 % percent (!) over the pure H.F. value thus resolving the old problem that in a H.F. calculation the level density at the Fermi surface always turned out to be roughly a factor of two too small.

4. MEAN FREE PATH

The mean free path of a nucleon in a nucleus is directly related to the imaginary part of the optical potential but also to the real part and for instance its nonlocality plays an important role. From

$$E = \frac{p^2}{2m} + V^{HF}(R,P) + \tilde{V}(R,P,E) - i\,W(R,P,E) \tag{12}$$

we obtain the complex momentum

$$p = p_r + i\,p_i \tag{13}$$

and therefore correspondingly a damped plane wave. The mean free path λ_{mf} is directly related to the damping as follows

$$\lambda_{mf}(R,E) = \frac{\hbar}{2p_i(R,E)} . \tag{14}$$

From our expressions we therefore can directly calculate λ_{mf} which is shown below in Fig. 8. The value $\lambda_{mf} \simeq 5$ fm in the interior for $E \geqslant 40$ MeV is in good agreement with experiment.

5. TEMPERATURE DEPENDENCE

It is straightforward to generalize our expressions to the case with finite temperature. The only modification we have to do is to replace the step functions in (5) by corresponding Fermi functions, e.g.

$$\theta(H-\varepsilon_F) \rightarrow \left(1 + \exp\frac{\varepsilon_F - H}{T}\right)^{-1} . \tag{15}$$

The resulting imaginary part for various temperatures is shown in Fig. 7.

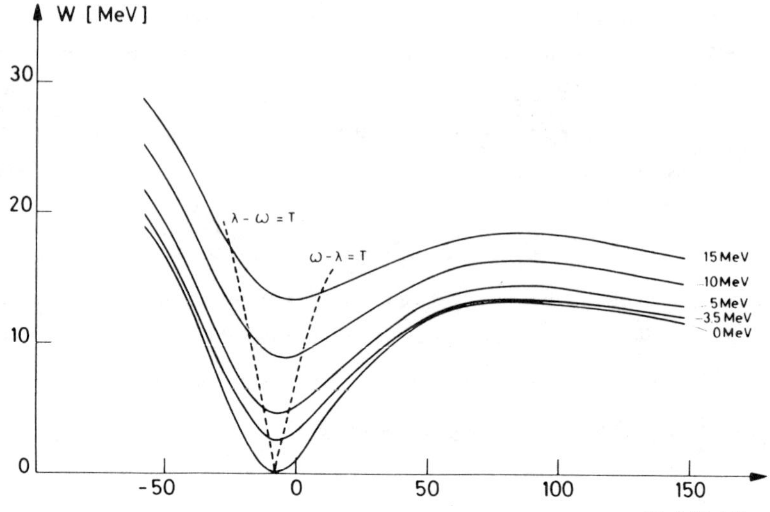

Fig. 7 : Temperature dependence of the imaginary part of the optical potential.

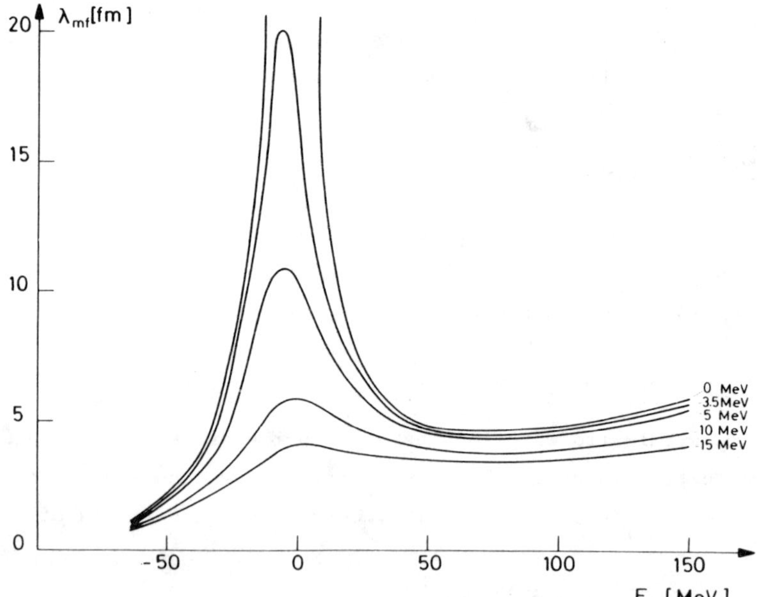

Fig. 8 : Influence of temperature on the mean free path of nucleons in nuclei (interior).

At finite temperatures the Pauli principle looses its influence and it is natural to see that the strong depression around the Fermi energy is gradually filled in.

This, of course, is also reflected in a decreasing mean free path which is shown in Fig. 8. It can be seen that only for very high temperatures ($T > 10$ MeV) the mean free path becomes sufficiently small that a local equilibrium description of nuclei may be justified.

6. CONCLUSIONS

We have presented the outline of a semiclassical theory for the calculation of the dynamic part of the optical potential of elastic nucleon-nucleus scattering. This formalism turns out to be simple and very accurate since we replace particles and holes individually by their Thomas Fermi approximation.

A number of results have been presented for the mass operator in second order approximation (2p-1h or 2h-1p). The imaginary part agrees well with phenomenological fits using the Gogny force. The influence of the Pauli principle has been discussed in detail. The unexpected finding is that for finite range forces its influence never vanishes even at extremely high bombarding energies. This is due to the restricted phase space exploited in the scattering due to the cut-off of the force.

The corresponding real part has been calculated via the usual dispersion relation. When transcribed into an effective mass the latter shows close to the Fermi energy an enhancement > 1 in the surface of the nucleus. This leads to the fact that the level density around the Fermi energy is increased by ~ 65 % over the H.F. value in almost quantitative agreement with experiment. The mean free path of nucleons in nuclei involves the imaginary but also the real part of the optical potential and our calculation again yields agreement with values extracted from phenomenological fits of the optical potential.

Finally we showed how to include finite temperatures into our formalism and displayed how the imaginary part increases and the mean free part decreases with raising temperature.

ACKNOWLEDGEMENTS

It is a pleasure to thank A. Blin and B. Hiller for many discussions and to allow us to present the mean free path as a function of temperature prior to a common publication[10).

REFERENCES

1. Bell, J.S. and Squires, E.J., Phys. Rev. Lett. 3, 96 (1959)
2. Ghosh, G., Hasse, R.W., Schuck, P., and Winter, J., Phys. Rev. Lett. 50, 1250 (1983)
3. Blin, A.H., Hiller B., Hasse, R.W., and Schuck, P., J. Physique 45, C6-231 (1984)
4. Brueckner, K.A., Eden, R.J. and Francis, N.C., Phys. Rev. 100, 891 (1955) ; Brueckner, K.A., Phys. Rev. 103, 172 (1956)
5. Déchargé, J. and Gogny, D., Phys. Rev. C 21, 1568 (1980)
6. Myers, W.D., Nucl. Phys. A 145, 387 (1970)
7. Mahaux, C. and Ngô, H., Nucl. Phys. A 378, 205 (1982) ; Nucl. Phys. A 410, 271 (1983)
8. Orland, H. and Schaeffer, R., Nucl. Phys. A 299, 442 (1978)
9. Perey, F. and Buck, B., Nucl. Phys. 32, 353 (1962)
10. Blin, A., Hiller, B., Hasse, R.W. and Schuck, P., to be published.

STUDY OF TRANSFER AND BREAKUP REACTIONS WITH THE PLASTIC BOX*

R.G. Stokstad[a], C.R. Albiston[a], M. Bantel[a], Y. Chan[a], P.J. Countryman[b], S. Gazes[a], B.G. Harvey[a], H. Homeyer[c], M.J. Murphy[a], I. Tserruya[a], K. Van Bibber[b], and S. Walda[a]

a) Nuclear Science Division
Lawrence Berkeley Laboratory
Berkeley, CA 94720

b) Department of Physics
Standford University
Stanford, CA 94305

c) Hahn-Meitner Institute
Berlin, Federal Republic of Germany.

*This work was supported by the Director, Office of Energy Research, Division of Nuclear Physics of the Office of High Energy and Nuclear Physics, and by the Nuclear Sciences of Basic Energy Sciences Program of the U.S. Department of Energy under Contracts DE-AC03-76SF00098 and DE-AM03-76-SF000326.

ABSTRACT

The study of transfer reactions with heavy-ion projectiles is complicated by the frequent presence of three or more nuclei in the final state. One prolific source of three-body reactions is the production of a primary ejectile in an excited state above a particle threshold. A subset of transfer reactions, viz., those producing ejectiles in bound states, can be identified experimentally. This has been accomplished with a 4π detector constructed of one-millimeter-thick scintillator paddles of dimension 20 cm x 20 cm. The paddles are arranged in the form of a cube centered around the target with small entrance and exit apertures for the beam and the projectile-like fragments, (PLF). The detection of a light particle (e.g., a proton or an alpha particle) in coincidence with a PLF indicates a breakup reaction. The absence of any such coincidence indicates a reaction in which all the charge lost by the projectile was transferred to the target. With this technique we have studied the transfer and breakup reactions induced by 220 and 341 MeV ^{20}Ne ions on a gold target. Ejectiles from Li to Ne have been measured at several scattering angles. The absolute cross sections angular distributions and energy spectra for the transfer and breakup reactions are presented. Relatively large cross sections are observed for the complete transfer of up to seven units of charge (i.e., a nitrogen nucleus). The relatively large probabilities for ejectiles to be produced in particle-bound states suggest that on the average, most of the excitation energy in a collision resides in the heavy fragment when mass is transferred from the lighter to the heavier fragment. The gross features and trends in the energy spectra for transfer and breakup reactions are similar.

I. INTRODUCTION

Heavy-ion reactions at bombarding energies from only a few MeV per nucleon above the barrier to the highest energies yet studied (~2 GeV/u) a exhibit a quasi-elastic peak. This peak consists of nuclei at forward or grazing angles with masses close to or smaller than the projectile mass and moving with approximately the beam velocity. While these products, when observed inclusively, exhibit some similarities at all energies[1)],the underlying reaction mechanisms are expected to be quite different at low and at very high bombarding energies. In the former case the transfer of nucleons from projectile to target yielding a two-body final state is predominant, whereas in the latter case the very high velocity of the projectile nucleons prevents their being captured by the target, even though the target might shear many nucleons from the projectile. Another mechanism that can produce a three-body (or more) final state containing a quasi-elastic fragment is the sequential decay of an excited projectile-like nucleus. Since the nucleus need have only ~10 MeV of excitation to be above threshold for particle decay, this process can occur in principle at all but the very lowest bombarding energies. The intermediate bombarding energy region, 10-100 MeV/nucleon, is possible host to all three mechanisms: transfer, sequential decay, and fragmentation.

It is thus important to design experiments that enable the separation of the various reaction mechanisms or, at least, that identify one particular component. Coincidence experiments address this problem and there have been many of these done at the lower bombarding energies, though relatively few at higher energies. Characteristic γ rays[2)] or x-rays[3)] emitted by the target-like fragment can identify transfer reactions; the number of neutrons emitted by the excited heavy partner can help separate transfer and breakup,[4)] and high resolution two-particle coincidence experiments are able to demonstrate sequential decay[5)]. A streamer chamber[6)], triggered by a projectile-like fragment (PLF), is able to detect the presence of any additional light charged particles and can effectively distinguish two-body from multibody reactions. The device described here, the "plastic box", is modeled after the streamer chamber in that it records the presence or absence of additional charged particles accompanying the PLF. It is a "4π detector" and can tell us whether a particular projectile fragment is one member of a two-body reaction. Unlike the streamer chamber, it is able to acquire data at relatively high rates and does not entail scanning a photograph of each event. It has been designed for use at intermediate energies.

The plastic box singles out the class of reactions in which charge is removed from the projectile, transferred to and captured by the target, and in which the emerging primary PLF is in a bound state. Operationally we refer to this class of events as transfer reactions. This class includes those cases in which the excited target-like fragment (TLF) subsequently decays by neutron emission, fission, or light-particle evaporation. Three-body reactions in which a fast light charged particle is emitted at forward angles along with the PLF will be referred to collectively as breakup reactions. This operational definition includes sequential decay of the primary excited PLF as well as all preequilibrium processes producing fast charged particles.

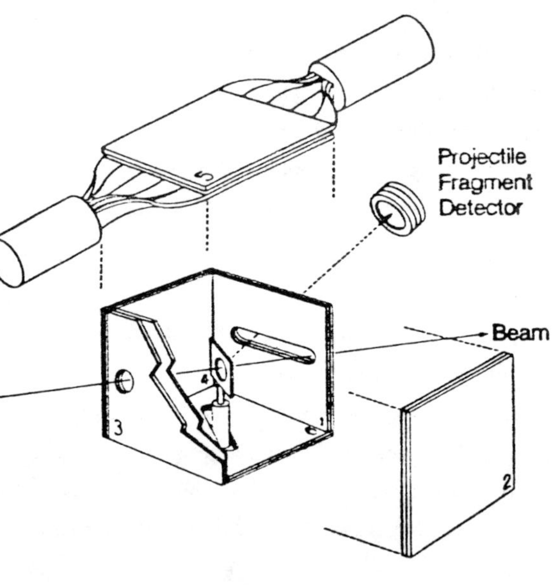

Fig. 1. Schematic diagram of the plastic box.

II. THE DETECTOR SYSTEM

The plastic box is an array of plastic scintillator paddles, light guides and phototubes arranged around the target in the fashion shown schematically in Fig.1. The light guides and phototubes are shown only for the top wall (wall 5) of the cube. Each wall consists of two separate scintillator paddles made of NE102A, each 20 cm high, 20 cm wide, and 1 mm thick. The beam and the projectile-like fragments emerge through a slot in wall 1. Two movable telescopes, each having three solid-state elements, can cover the angular range from 5° to 21°. Additional scintillator paddles (not shown) are located behind the telescopes and downstream close to 0° in order to detect most of the light particles that escape through the slot in wall 1. Each of the twelve phototubes on the cube plus the four on the additional paddles is serviced by a charge-sensitive ADC. All signals are handled by a CAMAC system.

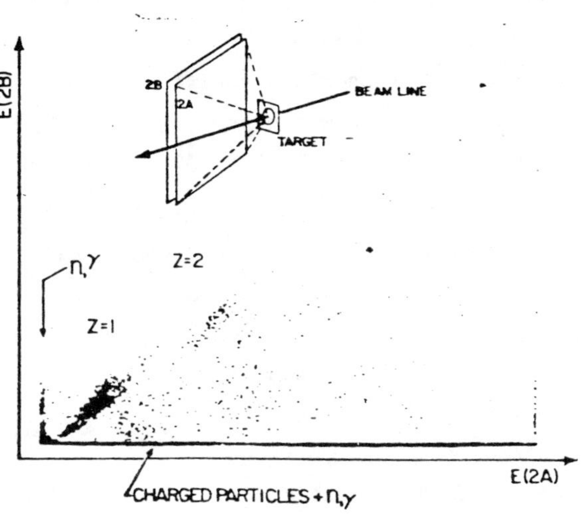

Fig. 2. A two-dimensional singles spectrum of the light output from the inner (A) and outer (B) paddles of the wall. Events appearing on the vertical axis ($\bar{A}B$) can only be due to neutral particles.

The response of a wall to various particles is illustrated in Fig. 2. The light output from the inner paddle (A) and outer paddle (B) can be used to identify neutral particles and to give some measure of identification for charged particles sufficiently energetic to penetrate the inner paddle. The type of event with a signal in paddle B and none in paddle A, denoted by ($\bar{A}B$), is a direct measure of the system's efficiency for detection of neutral particles. The paddles were made thin to minimize the efficiency for neutrals and chosen to be identical so that the number of those events ($A\bar{B}$) due to neutral particles is determined directly. Studies have been made of the procedures for (i) determining the energy thresholds for charged particles detection and (ii) correcting for neutral particles. A comparison was made of résults obtained with the plastic box for the reaction ^{16}O + Sn with results for ^{16}O + CsI from the streamer chamber, as a check on the whole technique. More information on the detection system is available in Ref. 7.

III. CHOICE OF THE Ne + Au REACTION

A number of factors influenced our choice of the projectile and target. A heavy target was desirable because neutron emission and fission are the favored modes of decay and the evaporation of charged particles is suppressed. The latter, in contrast to neutrons and fission fragments, would be detected by the scintillators. The projectile should be a "light" heavy ion because higher velocities for these are available from the cyclotron, and because the resolution of adjacent masses is still possible with a solid-state telescope. The ^{20}Ne + Au system satisfied these criteria and had the additional advantage that studies of the inclusive reactions had already been done at several energies at the Hahn-Meitner Institute.[8)]

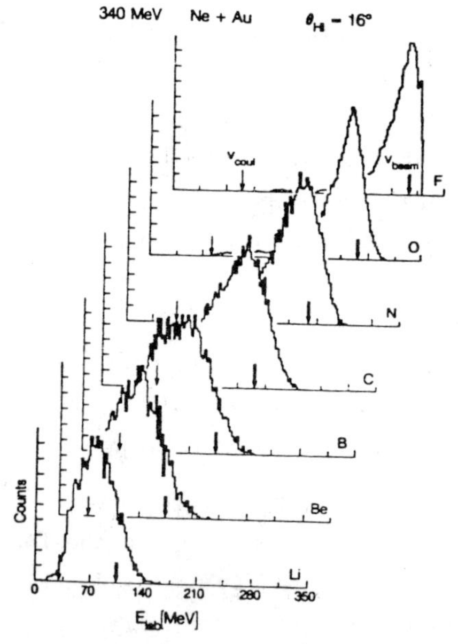

Fig. 3. Inclusive energy spectra for ejectiles near the grazing angle illustrate the quasi-elastic nature of the reaction products.

Measurements with the plastic box were made at Ne bombarding energies of 220 and 341 MeV, i.e., at 11 and 17 MeV per nucleon. The quasi-elastic peak is pronounced for all elements from fluorine down to lithium, as shown by the inclusive spectra in Fig. 3. These spectra were taken at an angle close to the grazing angle for 17 MeV/u ^{20}Ne+Au. The arrows labeled v_{beam} and v_{coul} denote, respectively, the energies corresponding to the beam velocity and the Coulomb repulsion energy. Kinematic models assuming linear and angular momentum matching for incoming and outgoing orbits reproduce the most probable velocities.[9)] The inclusive features of this reaction thus suggest that we are dealing with quasi-elastic (as opposed to deep-inelastic) reactions. The subject to be addressed by the plastic box concerns the mechanisms producing these fragments - how much is due to transfer, and how much to breakup.

IV. EXPERIMENTAL RESULTS

A. Yields

It is easiest to obtain an overview of the experimental results when individual isotopes are summed and elemental yields are considered. The telescopes did resolve the isotopes of the PLF's (at 341 MeV) and, therefore, most of the results presented below are also available by isotope as well as by element.

The first question to ask about any fragment observed in the particle telescope is how many of the six walls observed an associated charged particle. This number is denoted by S. The most straightforward case to interpret is S = O, i.e., no additional charged particles were observed. This implies that all the charge removed from the projectile must have been captured by the target and that the PLF was left in a bound state. F urthermore, the TLF must have emitted no light charged particle. The probability that a PLF of charge Z corresponded to S = 0, or to a large value of S (corrected for random coincudences) is given in Fig. 4 for 220 MeV Ne. One obtains immediately from this figure an overview of the transfer/breakup relationship, and its dependence on the ejectile charge. The same information for the higher bombarding energy, 341 MeV, is shown in Fig. 5.

The next question is which wall or walls registered particles. The symmetry of the cube and location of the beam axis along a diagonal gives us forward/backward and in-plane/out-of-plane information. A crude form of angular correlation for the light particles is

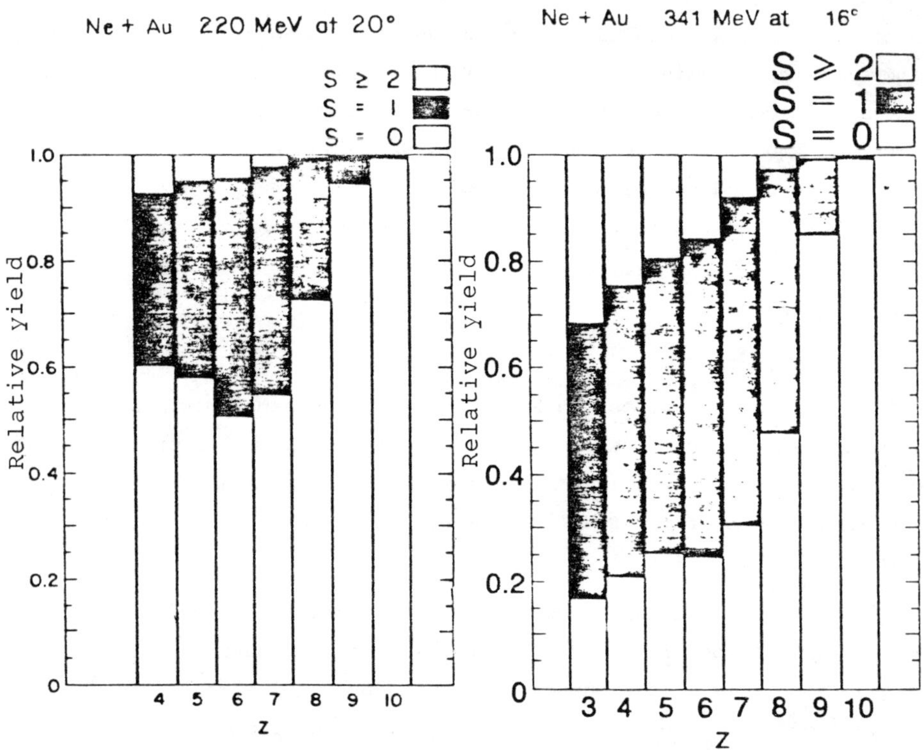

Fig. 4. The relative yield of ejectiles having none, one, or more than one scintillator walls in coincidence for an ejectile of charge Z, at a bombarding energy of 220 MeV.

Fig. 5. The same as Fig. 4 for the bombarding energy of 341 MeV.

given in Fig. 6. For each PLF, the count rate in the different walls is indicated. Recall from Fig. 1 that walls 1 and 2 are in the forward direction, 3 and 4 are back of 90° and walls 5 and 6 are up and down, respectively. In all cases, the forward walls are the most likely to observe a charged particle. The lighter PLF's (Li - C), however, are relatively more likely to be associated with particles in walls 3-6. It does not seem reasonable that these backward-going particles are the result of a projectile breakup reaction. The likely explanation is that they arise from the decay of the excited TLF by charged-particle evaporation. If this is indeed the case, these events belong in the category of a transfer reaction. The following consistency check was therefore made. Assuming a complete charge transfer occured, the excitation energy and the angular momentum of the TLF were estimated for each PLF. These estimates were then used in a statistical model calculation[10] to calculate the competition among neutron, charged particle, and fission decay. These predictions were found to be consistent with the observations. In addition, the most likely form of charged particle decay was predicted to be proton (rather than alpha-particle) emission. A separate experiment done with improved particle identification was performed to check this prediction and it was verified. Given this consistency, a correction based on the particles detected by the backward walls was therefore applied to all walls in order to include these events in the classification of transfer events.

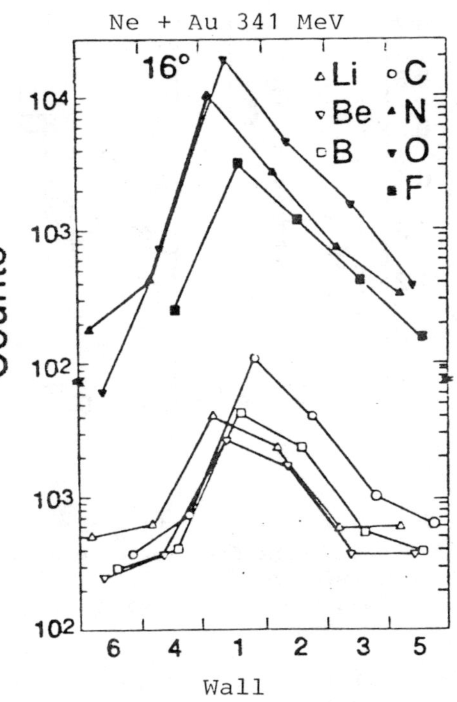

Fig. 6. The relative frequency with which each wall detects a light particle in coincidence with ejectiles from Li to F.

The dependence of the transfer/breakup relationship on the angle of the PLF is shown in Fig. 7. (The crosshatched area in each panel

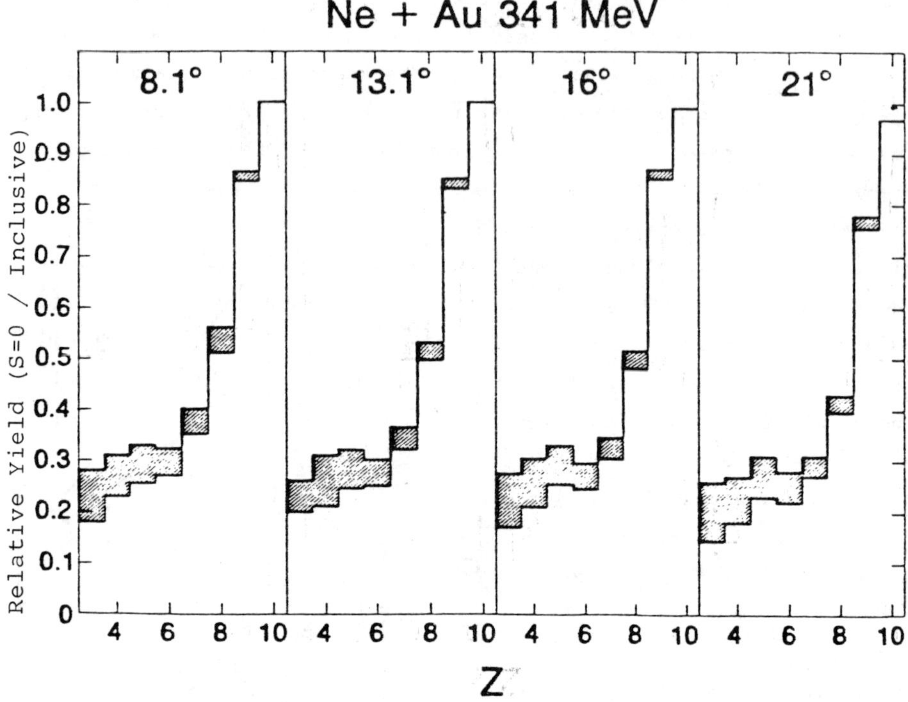

Fig. 7. The fraction of events that represent complete charge transfer for ejectiles at different angles. The shaded area gives the size of the correction for charged-particle evaporation by the target

indicates the size of the evaporation correction described above. Henceforth, all references to transfer or "S = 0" events include this correction). This dependence varies slowly with angle and thus can be extrapolated with confidence. This enables us to use inclusive measurements at other angles to establish the differential cross sections for transfer and breakup over a range of angles wide enough to integrate and obtain total cross sections. Fig. 8 gives the inclusive angular distributions and the transfer component for oxygen, carbon, and lithium ejectiles. The interpolated and extrapolated values of the transfer differential cross sections are given by the dashed lines. The angle-integrated cross sections are presented in Fig. 9. The absolute errors are typically ± 20%.

IV. B. Energy Spectra

One of the advantages of the plastic box over the streamer chamber is that it enables one not only to measure yields of various transfer reactions but also to acquire sufficient counting statistics to analyze energy spectra. The shapes of these spectra contain additional information on the reaction mechanism provided one is able to separate the contributing processes. The energy spectra for $^{16,17}O$ ions, produced by 341 MeV ^{20}Ne projectiles and observed at Θ_{lab} = 16°, are shown in Fig. 10. While the inclusive spectra for ^{16}O and ^{17}O exhibit some differences (the ^{17}O spectrum is broader), the shape differences are more marked when the spectra are separated into their transfer and breakup components. The differences between the S = 0 and S = 1 spectra for ^{17}O are particularly evident. The characteristic shape of a spectrum also depends on whether the PLF is close to or far from the projectile mass. In Fig. 11, the spectra of $^{6,7}Li$ ejectiles are shown for the same scattering angle as in Fig. 10. The spectra in Fig. 11 are much more symmetric and closer in form to a Gaussian. A comparison of transfer reaction spectra at two different laboratory angles is

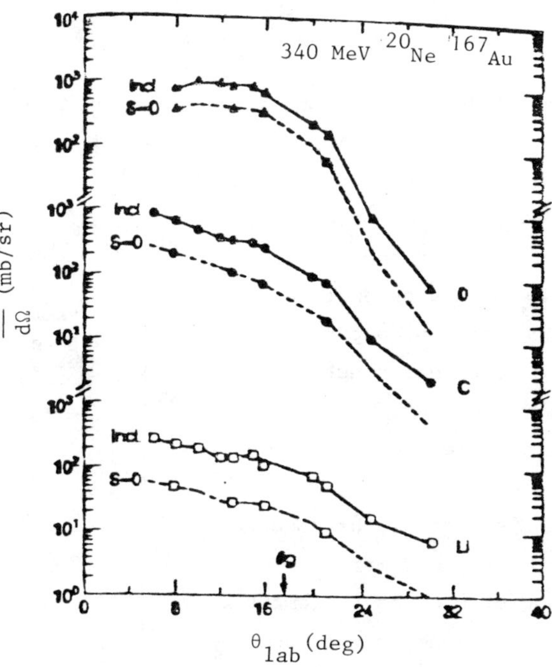

Fig. 8. Differential cross sections for inclusive and transfer reaction products.

shown in Fig.12. In this case the spectrum shape changes considerably in going to angles forward of the grazing angle.

One way of obtaining a concise overview of the energy spectra for the many different PLF's is to examine the moments of the spectra rather than the spectra themselves.

The moments - mean value E, standard deviation σ, skewness γ_1, and kurtosis β_2, defined by

$$\bar{E} \equiv \langle E \rangle \equiv \frac{\int \frac{d\sigma}{dE} E\, dE}{\int \frac{d\sigma}{dE}\, dE}, \quad \sigma \equiv (\langle E^2 \rangle - E^2)^{1/2}$$

$$\gamma_1 \equiv \langle (E-\bar{E})^3 \rangle / \sigma^3, \qquad \beta \equiv \langle (\bar{E}-E)^4 \rangle / \sigma^4,$$

have the advantage of an unambiguous, model-independent definition. In the case of the width, for example, use of the moment of the entire spectrum avoids the problem of choosing which portions of the spectrum to include or exclude when fitting a particular mathametical function to the spectrum. The moments are presented in Fig. 13 for the S = 0 and S = 1 components of the spectra for the most intense isotopes of each element. Each of the moments shows an overall trend as the charge and mass of the PLF vary. We shall come back to this in the discussion section.

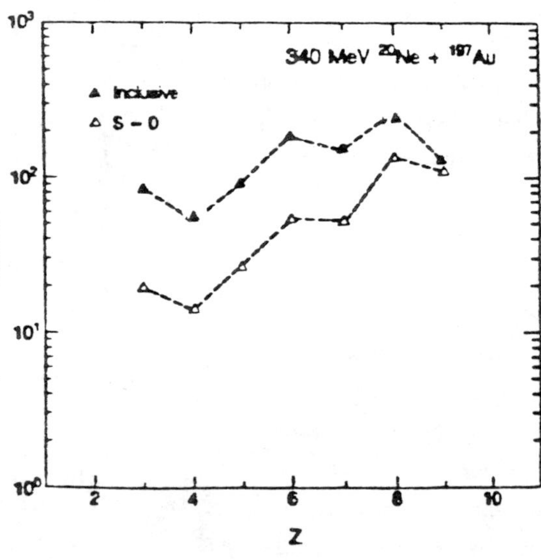

Fig. 9. Angle-integrated cross sections for inclusive and transfer reaction products.

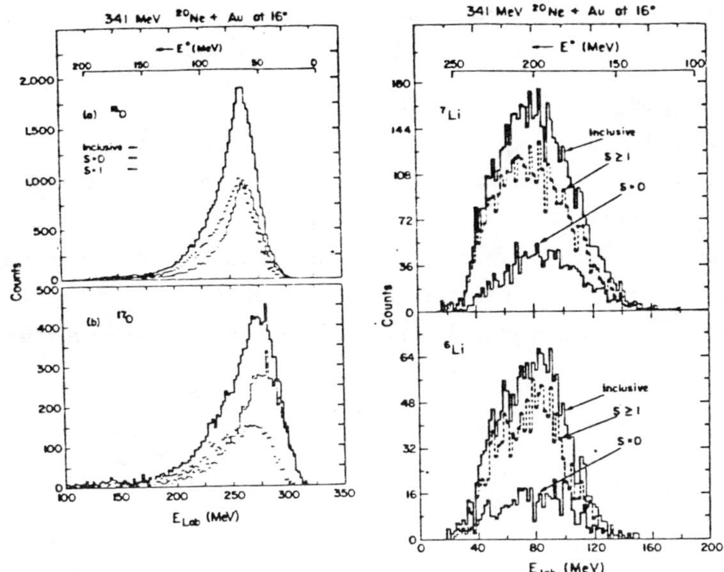

Fig. 10. Lab energy spectra of ^{16}O and ^{17}O ejectiles. The excitation energy E* is given for the reaction $^{197}Au(^{20}Ne, ^{16}O)^{201}Tl$, which corresponds to S=0.

Fig.11. Lab energy spectra of ^{7}Li and ^{6}Li ejectiles. The excitation energy E* is given for the reaction $^{197}Au(^{20}Ne,^{7}Li)^{210}Rn$, which corresponds to S=0.

IV. C. Other information and limitations of the method

In general, the multiplicity information from the plastic box for reactions at these energies may be fairly accurate even though the system is not highly segmented and even though there are only two walls covering the most forward angles. This is because the true multiplicities of reactions such as these (e.g., 16 MeV/u ^{16}O + CsI as measured by a streamer chamber[6]) are small, thus greatly reducing the probability of a multiple hit in a single wall.

Further information available from the plastic box consists of correlation patterns among the walls when two walls register particles.

Since S = 2 events are at most 30% of the total for a given PLF, and generally much less, we will not discuss those results here.

It would be very desirable to know which type of particles was detected in a S = 1 event. The large solid angle of a given wall and the thickness of the first scintillator paddle severely restrict the particle identification capability of the system. It is only possible to say that the energetic particles observed in the forward paddles are mainly protons and alpha particles. There does not seem to be much in the way of heavier fragments associated with any of the PLF's.

The light particles that are stopped in the cylindrical mounts holding the solid state telescopes are not recorded. Since the sequential decay process concentrates light particles in the general direction of the ejectile, the cross section for this effect amounts to a (maximum) 7% reduction to be made in the S = 0 cross sections.

The plastic box ignores not only neutrons evaporated by the target but also any neutrons associated with the projectile. An excited primary PLF my decay by neutron emission, of course. Thus, a detected ^{12}C might have originated as a ^{13}C, or an ^{16}O nucleus could be the product of an ^{17}O primary particle excited just above the neutron threshold. An examination of the separation

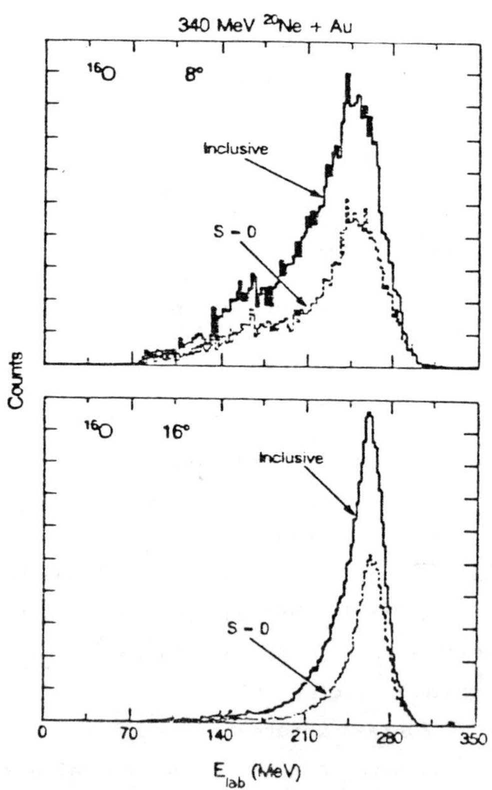

Fig.12. Lab energy spectra of ^{16}O ejectiles at 16° and at 8°.

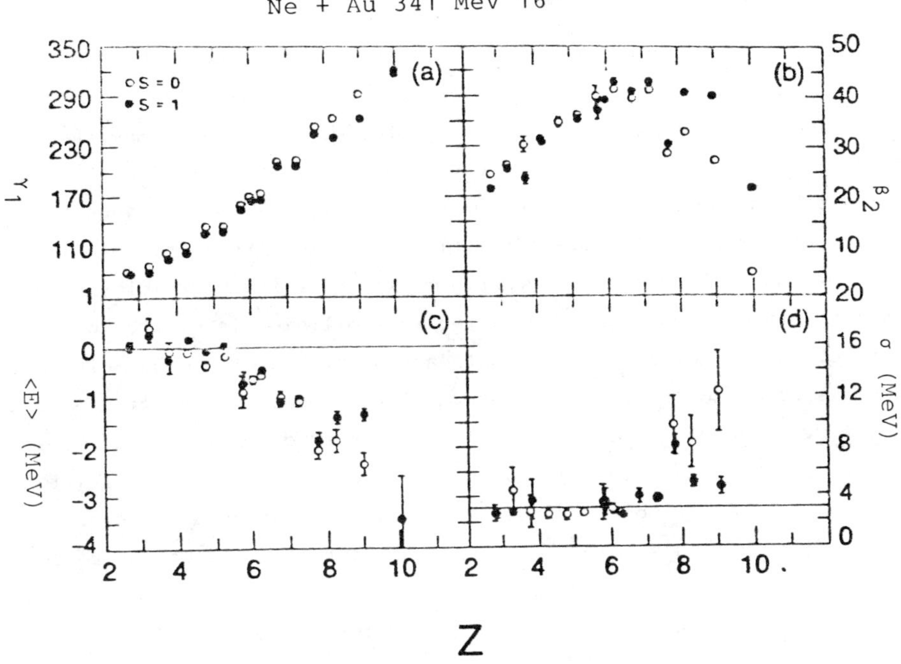

Fig. 13. The moments of the energy spectra. The PLF's are (left to right) ^{6,7}Li, ^{7,9}Be, ^{10,11}B, ^{11,12,13}C, ^{14,15}N, ^{16,17}O, ^{19}F, ^{20}Ne. Elastic scattering is excluded from the ^{20}Ne spectrum.

energies for n,p, and α emission by nuclei from F to Li indicates that ^{17}O, ^{14,13}C, and ^{10,9}Be are relevant cases in which the neutron threshold for charged particle emission. However, ^{9}Be* decays to α + α + n and is thus removed from consideration. In order for a collision involving neutron emission from the ejectile to be improperly labeled as a transfer (S = 0) reaction, it must consist of neutron without any other emission of a fast charged particle (whether preequilibrium or sequential). The case in which this could occur and have an effect would be the feeding of ^{12}C by neutron emission from ^{13}C or ^{14}C. In other cases, e.g., ^{17}O, inclusive measurements[8] suggest the primary population of the neutron emitter to be small. The important point is that the plastic box does measure complete charge transfer without ambiguity, and this should have a strong correspondence in most cases to mass transfer.

V. Discussion

It is useful to recall what the plastic box actually measures. A PLF observed without any associated charged particle is the result of a two-body reaction in which a transfer of nucleons to the target took place and the emerging PLF was produced in an excited state lying below the threshold for charged particle decay. This is a subset of a broader class of "transfer" reactions in which the light reaction partner (as well as the heavy partner) may emerge in particle-unbound excited states that decay long after the collision is over. This subject, nevertheless, is a significant portion of the quasi-elastic cross section for Ne on Au at intermediate bombarding energies - even for very large mass transfers. This latter fraction is of the order of 1/2 at 11 MeV/u (Fig. 4) and is still about 1/3 at 17 MeV/u (Figs.5 and 7). The first observation, therefore, is that this class of transfer reactions survives well into the intermediate energy region.

V. A. Reconstruction of primary cross sections

A second observation is that the events having $S \geqslant 1$, the breakup reactions, are on the increase as the bombarding energy is increased. This behavior is expected and might be explained in at least two ways. One possibility is an increase in the preequilibrium emission of light particles from the region of contact between target and projectile. However, it is known that the sequential decay of heavy ions is a dominant mechanism in the production of beam velocity ejectiles in coincidence with forward going light particles.[5,11] In this case the determining factor in the decay of an ejectile is simply whether it is an excited state above a particle decay threshold, i.e., it is the spectrum of excitation energy in the emerging fragment. (At high excitation energies the distinction between preequilibrium emission and sequential decay may blur.) It is a useful exercise to make the assumption that sequential decay is the dominant process and to see what information this permits us to gain from our measurements.

The assumption of sequential decay makes it possible to reconstruct the primary ejectile cross sections from the measured S = 0 and S = 1 cross sections. The information from the plastic scintillator walls suggests that mainly alpha particles or protons (as opposed to heavier charged particles) are in coincidence with the PLF. We will assume this to be the case. Thus, there are two possible charged-particle decay paths leading to the production of each observed ejectile and we will assume that the decay mode of each primary fragment is determined by its lowest threshold. In almost all cases, the alpha threshold of a primary fragment is lower than the proton threshold. The energies of the first alpha-, proton-, and neutron-decaying states of the most prominent ejectiles are indicated in Fig. 14. Therefore, in most cases the S = 1 events will be fed via alpha-decaying states (The assumption has been borne out by more recent coincidence experiments, which indicate a preponderance of alpha particles accompanying breakup.)

The low proton threshold of nitrogen provides an exception to this rule. As a result, the S = 1 carbon cross section could be expected to contain contributions from both oxygen and nitrogen breakup. Similarly, the S = 1 boron yield should be non-existent (sofar as our assumption that only the lowest thresholds contribute is valid). For these two cases, we have assumed that both proton and alpha sequential decays contribute to the observed breakup yield, and further assume that the relative contributions scale with the experimental S = 0 yields of the two possible primary nuclei.

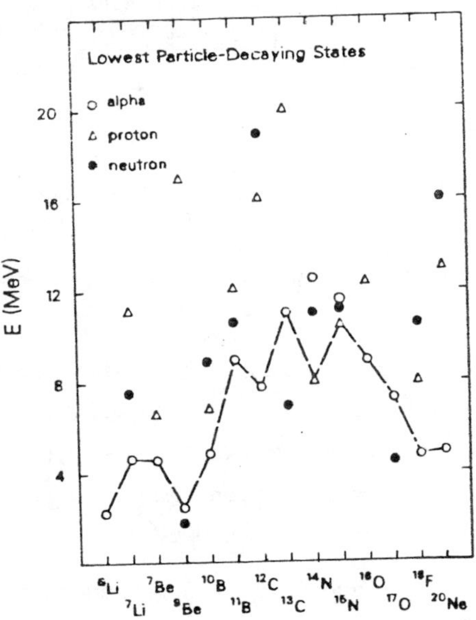

Fig. 14. The energy of the lowest alpha-, proton-, and neutron-decaying states for the most prominent ejectiles. The dashed line connects the lowest charged-particle decay thresholds.

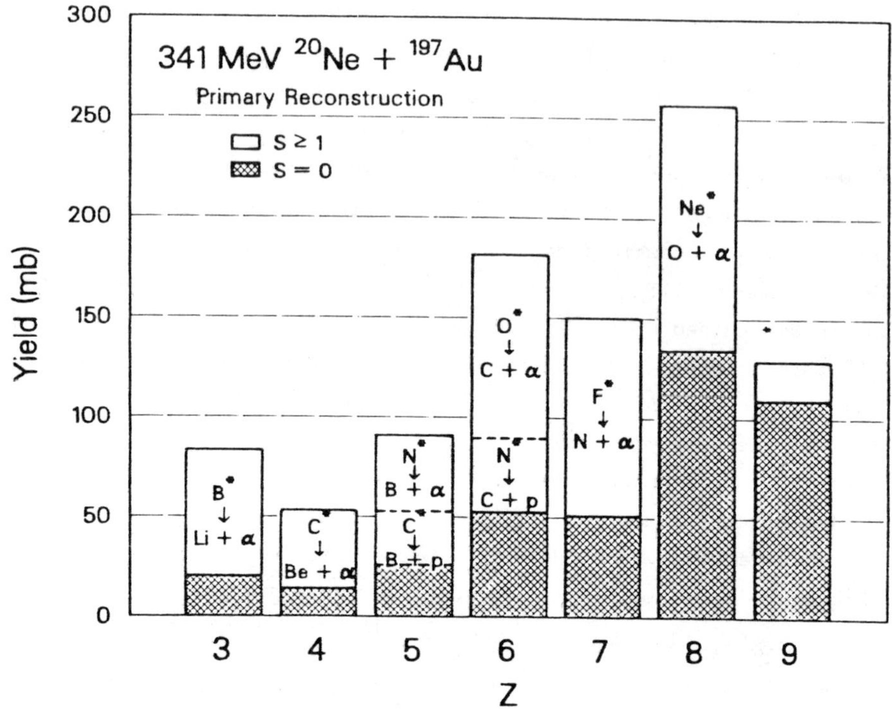

Fig.15. The reconstruction of primary ejectiles for the results at 341 MeV.

This provides us with a primary reconstruction as outlined schematically in Fig. 15.

The low-laying, neutron-decaying states of $^{14,13}C$ and ^{17}O have already been noted. While the presence of these states has an effect on the deduced breakup probabilities, they do not affect the accuracy of the reconstruction since the primary yields for a given z will be summed over all isotopes.

The reconstruction procedure just outlined generates primary cross sections over the range of primary charge Z = 5-9. The primary lithium and beryllium yields cannot be estimated since the reconstruction procedure would require S = 1 alpha and proton cross sections that, if experimentally available, could come from a multitude of primary fragments.

The results of the experimental reconstruction of the primary ejectile charge yields are shown in Fig. 16, at both 11 and 17 MeV/nucleon. The cross sections for the production of the heaviest ejectiles are remarkably similar at both bombarding energies. The higher beam energy is seen to enhance the yields of massive charge-transfer products. Given our assumptions, it is apparent that the large cross sections observed for the production of light ejectiles at higher beam energies are due to two effects: increased excitation energy of the primary fragment as well as greater charge transfer prior to breakup.

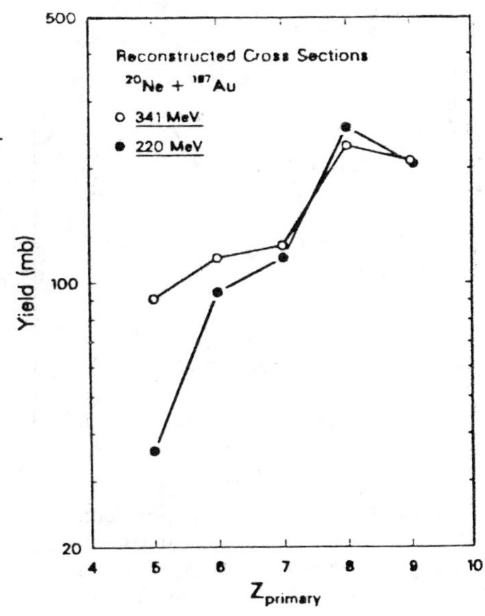

Fig. 16. The reconstructed primary cross sections plotted versus primary ejectile charge, as deduced from data at both bombarding energies.

The instability of ^{8}Be(g.s.) does not allow us to measure an S = 1 ^{8}Be cross section. Therefore, we miss a cross section that would have been added to the primary carbon yield in our reconstruction algorithm. For this reason, the reconstructed carbon yield will underestimate the abundance of primary carbon fragments.

V. B. Survival Fraction of the Primary Ejectiles, and the Division of Excitation Energy.

From the reconstructed primary cross sections we may calculate the probability that an ejectile will "survive" the transfer process without undergoing sequential decay. This is just the ratio of the S = 0 cross section to the total primary cross section and is of greater physical significance than the S = 0/inclusive ratio. The survival fractions calculated at 341 and 220 MeV are shown in Fig. 17.

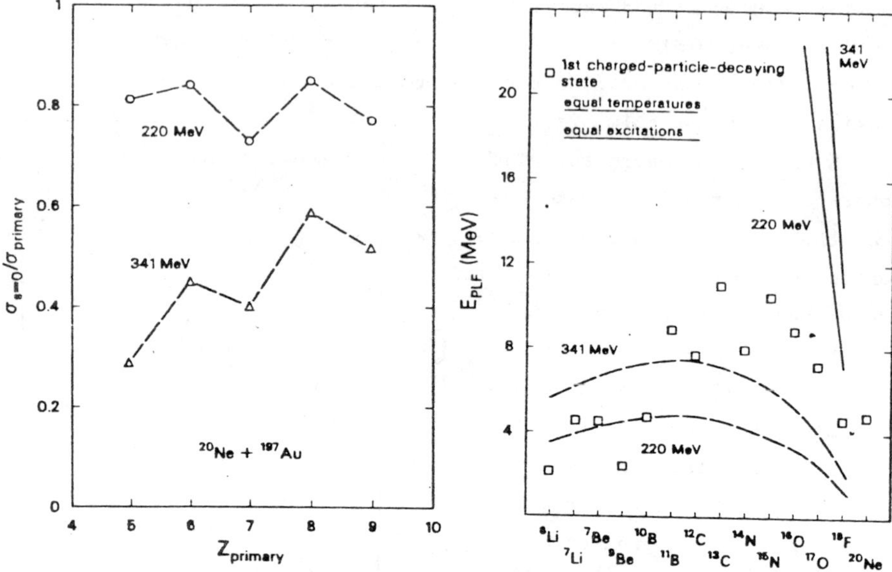

Fig. 17. The survival fractions are plotted as a function of primary ejectile charge at both bombarding energies. The larger values for carbon and oxygen may be associated with the unobserved removal of excitation energy by neutron decay of $^{14,13}C$ and ^{17}O primary fragments.

Fig. 18. The mean excitation energies of the primary fragments calculated under the extreme assumptions of equal excitation and division according to the masses. Also shown are the energies of the first charged-particle decaying states of the most prominent ejectiles.

As can be seen, the survival fractions are smaller at the higher bombarding energy. This can be understood in terms of the greater excitation associated with nucleon transfer at high energies. However, the striking aspect is that the survival fractions of the massive charge-transfer events are associated on the average with very large total excitation energies. This argues against an equal sharing of excitation energy between ejectile and target. However, if we assume that the excitation energy is divided unequally, for example as the ratio of the masses, then the values of the S = 0/primary ratios can be understood at least qualitatively: while the more massive transfers to the target will result in greater excitation energies in the primary system, the correspondingly lighter projectile-like fragments will, in turn, have a smaller fraction of the total excitation.

This is illustrated in Fig. 18, where the first particle-decaying states of the various ejectiles are compared with the average excitation energies deposited in the primary ejectiles assuming either an equal-excitation or mass-proportional division of excitation energy. As can be seen, the latter, asymmetric division of excitation results in ejectile energies that track roughly with the decay thresholds. While such an analysis is not quantative (lacking any information on the widths of the excitation-energy distributions), the observed survival fractions are more consistent with an excitation-energy division that is proportional to the mass asymmetry of the exit channel. We would like to note here that this asymmetric division of excitation energy does not necessarily imply that a high degree of equilibration has been reached in the collision. A direct transfer of several or many nucleons form light projectile to heavy traget leaves the heavy partner with the responsibility of converting the transferred nucleons' kinetic energy to excitation energy. The light partner need only have the transferred nucleons removed, and the energies associated with this are less than 11 or 17 MeV per nucleon.

V. C. Energy spectra

Consider next the shapes of the energy spectra. The mean energies, shown for both S = 0 and S = 1 reactions in Fig. 13a, are remarkably regular. They increase approximately linearly with the mass of the ejectile, as expected for PLF's with a beam-like velocity. The mean value for a two-body reactionis always higher than for a three-body reaction. This may be simply understood in terms of the additional excitation energy required to separate the third body. Note that for the isotopes of nitrogen and lighter PLF's, the trends in the mean energies for S = 0 and S = 1 are identical.

The standard deviations or widths of the spectra show a behavior with fragment mass that is (very roughly) reminiscent of the parabolic dependence found in high energy fragmentation.[12] Indeed, conversion of the energy widths for transfer or breakup to reduced momentum widths, $_{0}$, as in Ref. 12, yields a range of values (see Fig. 19) that fits into the existing systematics.[13)] Referring again to Fig. 13b, relatively large differences between S = 0 and S = 1 widths are found for ^{7}Br, ^{17}O, and ^{19}F. For isotopes of nitrogen and lighter PLF's, the trends exhibited by the S = 0 and S = 1 widths are very similar. As the PLF becomes further removed from the projectile, the shape of the spectrum becomes more symmetric and Gaussian-like, i.e., $\gamma_1 \rightarrow 0$ and $\beta_2 \rightarrow 3$. Again, for $Z \leqslant 7$, the behavior of γ_1 and β_2 are identical for S=0 and S=1.

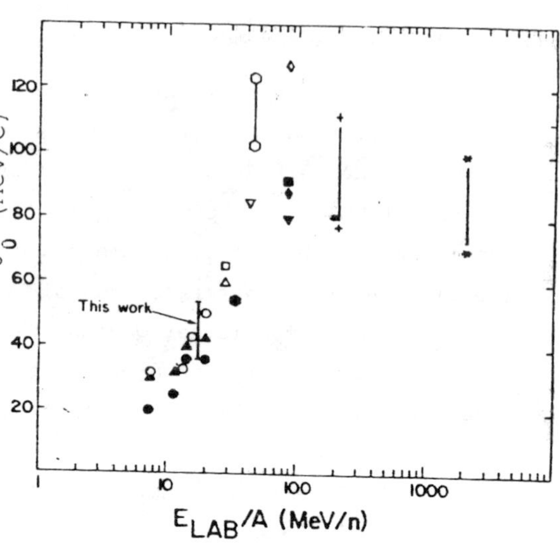

Fig. 19. The systematics of reduced momentum widths for PLF's from many reactions vs. lab energy. For references see Ref. 13.

The above observations on the moments of the spectra suggest that the transfer of three or more charge units is a natural dividing line. Smaller transfers, e.g., an alpha particle or less, are associated with spectra that may change rapidly in shape. The spectra for larger mass transfers change more slowly and regularly from one isotope and element to the next. (Note in Figs. 10 and 11 the similarity of the ^{6,7}Li spectra and the differences for ^{16,17}O.) This dividing line is also reflected in the ratio of transfer to inclusive yields (Figs. 5 and 7). The qualitative conclusion to draw from this is that the transfer of an alpha particle or less occurs in a very short time, i.e., is a fast "direct" reaction, whereas larger transfers are more characteristic of the fusion process. The fusion of heavy projectiles with heavy targets exhibits cross sections that vary smoothly with projectile mass provided one is well above the barrier, as is the case here. Few-nucleon transfer reactions, however, exhibit spectra that may change rapidly as the number of captured nucleons increases.

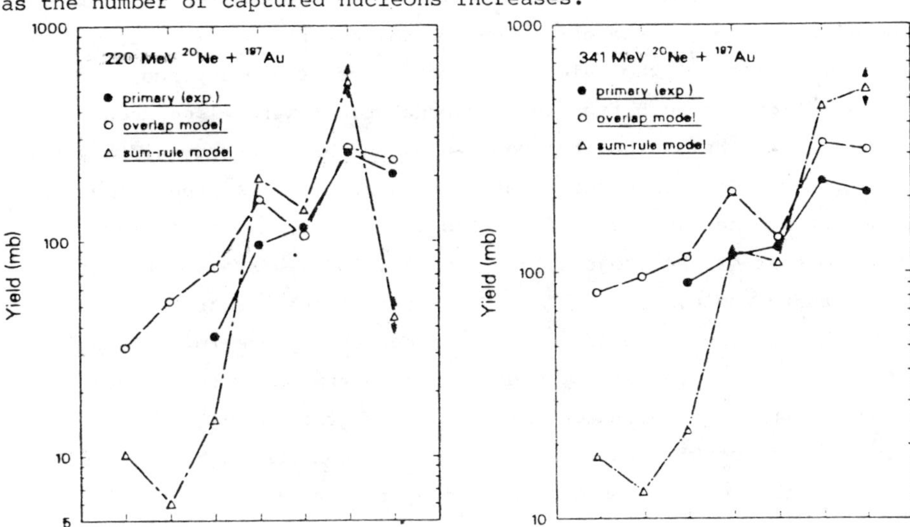

Fig. 20. The predictions of the sum-rule and overlap models are compared with reconstructed primary cross sections at 220 MeV and 341 MeV.

V. D. Comparison with Models

The qualitative remarks above could be sharpened or tested by quantitative comparisons with the predictions of different models provided the model and the experiment predict and measure the same quantity. As we shall see, the comparisons possible at present are few. Consider first the angle-integrated cross sections. The sum rule model of Wilczynska, et. al.,[14)] for incomplete fusion would be a natural starting point, and indeed it is straightforward to make a calculation based on the parameters of Ref. 14. As may be seen in Fig. 20 the agreement with the present experiment is poor.

The overlap model[15)] of Harvey and Homeyer calculates the yield of PLF's by assuming that the projectile can be factored into two portions, according to the fragmentation model of Friedman,[16)] and that there must be a geometrical overlap of the portion to be removed and the target. The calculation reproduces the main trends in the experimental results.

The shapes of the spectra clearly provide an important point of comparison. Some of the early theoretical effort in this area was by McVoy and nemes,[17)] who calculated energy spectra in a plane-wave approximation for both transfer and breakup. A general consequence of their assumptions is that the momentum-, and hence energy-width for breakup should always exceed that for transfer. This prediction is borne out by the experiment for PLF's at the grazing angle with Z very close to that of the projectile. However, the reverse is true at a more forward angle, Θ_{lab} = 80°. Here, the widths for transfer are (with the one exception of ^{19}F) all larger than those for breakup! The complexity of the energy widths is illustrated by the interesting change in the ^{16}O spectra in going from the grazing angle to the forward angle of 8° (see Fig. 12). The 8° spectrum is highly skewed - perhaps the more pronounced low energy tail reflects scattering from the far side of the target nucleus.

Other more elaborate models for which these energy spectra present an excellent opportunity for analysis are the direct-reaction models - the DWBA for the transfer of smaller masses,[18] and a multistep extension of the DWBA, the breakup-fusion model, for the larger mass transfers.[19] The S = 0 spectra are well suited for comparison with a DWBA calculation because one knows that the spectra are uncontaminated by breakup and that the excited states of the projectile that must be included in the calculation are limited to a relatively few bound states. The DWBA based on the diffraction model and extensively applied by Mermaz, et. al.,[18)] would seem well suited for the analysis of the present data for Z $\geqslant$8.

Models appropriate for more massive transfer reactions are based on a two-step process: the projectile is separated into two parts, one of which then fuses with the target. The overlap model[15] is an example involving geometrical overlap for the second step. A more sophisticated approach is the breakup-fusion model developed by Udagawa, et. al., which incorporates distorted waves.[19)] This model has been applied to a few cases at lower bombarding energies, e.g., at 6.8 MeV/u $^{14}N + ^{159}Tb$ for alpha-particle ejectiles.[19)] It would be of interest to see how well it does in describing a reaction at higher energies covering a wide range of PLF's.

VI. Summary and Outlook

Transfer and breakup reactions have been identified in the collisions of 11 and 17 MeV/u ^{20}Ne with Au. A 4π plastic scintillator system was used to separate the two types of reactions by detecting the presence or absence of additional light charged particles associated with the PLF. Angular distributions, total cross sections, and energy spectra were measured. The properties of the quasi-elastic reaction products are similar and vary smoothly for both transfer and breakup for PLF's with Z = 3 to 7, whereas for heavier ejectiles the behavior is less systematic. The assumption of sequential decay permits a reconstruction of the primary cross sections. A comparison of the absolute cross sections with the overlap model[15)] yields favorable results. The significant probabilities for producing an ejectile in a particle-bound state suggests that, on the average, most of the excitation energy is deposited in the heavy fragment, at least for the case of mass transferred from a light projectile to a heavy target.

At present, it appears that the charged particles accompanying a PLF are mainly protons and alpha particles rather than heavier ions. In the future it will be possible to identify the light particles and measure their angles with position-sensitive plastic walls that have been developed.[20] Such experiments should enable a more precise reconstruction of the primary yields and excitations from the observed sequential decay products.

REFERENCES

1) M. Buenerd et al., Phys. Rev. Lett. 37, 1191 (1976).

2) T. Inamura et al., Phys. Lett. 69B, 51 (1977).

3) H.W. Wiltschut et al., Phys. Lett. 123B, 173 (1983).

4) U. Jahnke et al., Phys. Rev. Lett. 50, 1246 (1983)

5) W. Rae et al., Phys. Lett. 105B, 417 (1981).

6) M.J. Murphy et al., Phys Lett. 120B, 319 (1983).

7) K. Van Bibber et al., IEEE Trans. Nucl. Sci. NS-31 35 (1984).

8) Ch. Egelhaaf et al., Nucl. Phys. A405, 397 (1983).

9) B.G. Harvey and M.J. Murphy, Phys. Lett. 130B, 373 (1983).

10) M. Blann, private communication.

11) H. Homeyer, et al., Phys. Rev. C26 1335 (1982).

12) A.S. Goldhaber, Phys. Lett. 53B 306 (1974).

13) M.J. Murphy and R.G. Stokstad, Phys. Rev. C28, 428 (1983).
R.G. Stokstad, Comm. on Nucl and Part. Phys. (1984).

14) J. Wilczynski et al., Nucl. Phys. A373, 109 (1982).

15) B. Harvey and H. Homeyer, LBL preprint 16882 (1983).

16) W. Friedman, Phys. Rev. C27, 569 (1983).

17) K.W. McVoy and M.C. Nemes, Z. Physik A295, 177 (1980).

18) M.C. Mermaz, Phys. Rev. C21, 2356 (1980), C28, 1587 (1983).

19) T. Udagawa, Proc. Int. Conf. Bad Honnef (1984);
Phys. Lett. 135B, 333(1984).

20) M. Bantel et al., Nucl. Inst. Meth. 226, 394 (1984).

DYNAMICAL CALCULATIONS OF NUCLEAR FISSION AND HEAVY-ION REACTIONS

J. Rayford Nix and Arnold J. Sierk
Theoretical Division, Los Alamos National Laboratory
Los Alamos, New Mexico 87545, USA

ABSTRACT

With the goal of determining the magnitude and mechanism of nuclear dissipation from comparisons of predictions with experimental data, we describe recent calculations in a unified macroscopic-microscopic approach to large-amplitude collective nuclear motion such as occurs in fission and heavy-ion reactions. We describe the time dependence of the distribution function in phase space of collective coordinates and momenta by a generalized Fokker-Planck equation. The nuclear potential energy of deformation is calculated as the sum of repulsive Coulomb and centrifugal energies and an attractive Yukawa-plus-exponential potential, the inertia tensor is calculated for a superposition of rigid-body rotation and incompressible, nearly irrotational flow by use of the Werner-Wheeler method, and the dissipation tensor that describes the conversion of collective energy into single-particle excitation energy is calculated for two prototype mechanisms that represent opposite extremes of large and small dissipation. We solve the generalized Hamilton equations of motion for the first moments of the distribution function to obtain the mean translational fission-fragment kinetic energy and mass of a third fragment that sometimes forms between the two end fragments, as well as dynamical thresholds, capture cross sections, and ternary events in heavy-ion reactions.

1. INTRODUCTION

Nuclear physicists have been struggling for years to determine the magnitude and mechanism of nuclear dissipation--to answer two elementary questions: Is a nucleus overdamped like a drop of honey, or underdamped like a drop of water? Does a nucleus dissipate its energy of collective motion primarily through interactions of nucleons with the mean field generated by the remaining nucleons, or do two-particle collisions play a substantial role? Despite numerous experimental clues provided by fission and heavy-ion reactions, the answers to such questions posed by this challenging many-body problem have proved elusive. This is because of the many complementary aspects displayed by the atomic nucleus. With its relatively small number of degrees of freedom, the nucleus is both microscopic and macroscopic on the one hand and both quantal and classical on the other, which gives it a rich dynamical behaviour ranging from elastic vibrations of solids to long-mean-free-path dissipative fluid flow with statistical fluctuations. On this occasion of the Golden Jubilee of the Indian National Science Academy, we would like to tell you about some of our recent calculations at Los Alamos directed toward answering these questions.

Our approach is *not* to explain the experimental data in terms of some model with adjustable parameters--since often several models with widely different physical bases are capable of doing this equally well--but instead to find and calculate physical observables that depend sensitively upon the magnitude and mechanism of nuclear dissipation. The difficulty arises because many of the gross experimental features of fission and heavy-ion reactions are determined primarily by a competition between the attractive nuclear force and the repulsive Coulomb and centrifugal forces, and any theoretical approach that includes correctly these relatively trivial forces reproduces the data with fair accuracy. Also, the final effects on observable quantities caused by dissipation are often very similar to the final effects caused by collective degrees of freedom.

In our studies here, we consider two prototype mechanisms that represent opposite extremes of large and small dissipation. For these two mechanisms we use a macroscopic-microscopic method to calculate observable quantities in fission and heavy-ion reactions and confront these predictions with experimental data in an attempt to determine the magnitude and mechanism of nuclear dissipation.

2. MACROSCOPIC-MICROSCOPIC METHOD

We focus from the outset on those few collective coordinates that are most relevant to the phenomena under consideration. In particular, for a system of A nucleons, we separate the 3A degrees of freedom representing their center-of-mass motion into N collective degrees of freedom that are treated explicitly and 3A - N internal degrees of freedom that are treated implicitly.

2.1 Collective Coordinates

In our earlier dynamical studies we have usually described the nuclear shape in terms of smoothly joined portions of three quadratic surfaces of revolution, with three symmetric and two independent asymmetric shape coordinates.[1-5] Although suitable for many purposes, this three-quadratic-surface parametrization breaks down in the later stages of many heavy-ion fusion calculations, is unable to describe division into more than two fragments, and leads to very complicated expressions for the forces involved.

Because of these disadvantages, we have switched[6] to a more suitable parametrization in which an axially symmetric nuclear shape is described in cylindrical coordinates by means of the Legendre-polynomial expansion[7]

$$\rho_s^2(z) = R_0^2 \sum_{n=0}^{N} q_n \, P_n[(z-\bar{z})/z_0] \quad . \tag{1}$$

In this expression, z is the coordinate along the symmetry axis, ρ_s is the value on the surface of the coordinate perpendicular to the symmetry axis, z_0 is one-half the distance between the two ends of the shape, $\bar{z}$ is the value of z at the midpoint between the two ends, R_0 is the radius of the spherical nucleus, P_n is a Legendre polynomial of degree n, and q_n for $n \neq 0$ and 1 are N - 1 shape coordinates. Since the nucleus is assumed to be incompressible, the quantity q_0 is not independent but is instead determined by volume conservation. Also, q_1 is determined by fixing the center of mass. In addition, we include an angular coordinate $\Theta \equiv q_{N+1}$ to describe the rotation of the nuclear symmetry axis in the reaction plane, which leads to a total of N collective coordinates $q = q_2, \ldots, q_{N+1}$ that are considered. Throughout this paper we use N = 11, corresponding to five independent symmetric and five independent asymmetric shape coordinates and one angular coordinate.

2.2 Potential Energy

We consider excitation energies that are sufficiently high that single-particle effects may be neglected and calculate the potential energy of deformation V(q) as the sum of repulsive Coulomb and centrifugal energies and an attractive Yukawa-plus-exponential potential,[8] with constants determined in a recent nuclear mass formula.[9] This generalized surface energy takes into account the reduction in energy arising from the nonzero range of the nuclear force in such a way that saturation is ensured when two semi-infinite slabs are brought into contact.

2.3 Kinetic Energy

The collective kinetic energy is given by

$$T = \frac{1}{2} M_{ij}(q)\, \dot{q}_i\, \dot{q}_j = \frac{1}{2}\, [M(q)^{-1}]_{ij}\, p_i\, p_j \quad , \tag{2}$$

where the collective momenta p are related to the collective velocities $\dot{q}$ by

$$p_i = M_{ij}(q)\, \dot{q}_j \quad . \tag{3}$$

In these equations and the remainder of this paper we use the convention that repeated indices are to be summed over from 2 to N + 1. At the high excitation energies and large deformations considered here, where pairing correlations have disappeared and near crossings of single-particle levels have become less frequent, the rotational moment of inertia is close to the rigid-body value[10] and the vibrational inertia is close to the incompressible, irrotational value.[11] We therefore calculate the inertia tensor M(q), which is a function of the shape of the system, for a superposition of rigid-body rotation and incompressible, nearly irrotational flow. For this purpose we use the Werner-Wheeler method, which determines the flow in terms of circular layers of fluid.[1-6]

2.4 Dissipation Mechanisms

The coupling between the collective and internal degrees of freedom gives rise to a dissipative force whose mean component in the i-th direction may be written as

$$F_i = - \eta_{ij}(q)\, \dot{q}_j = - \eta_{ij}(q)\, [M(q)^{-1}]_{jk}\, p_k \quad . \tag{4}$$

For the calculation of the shape-dependent dissipation tensor η(q) that describes the conversion of collective energy into single-particle excitation energy, we consider two prototype mechanisms that represent opposite extremes of large and small dissipation. The first mechanism is one-body dissipation,[4-6,12-15] which arises from collisions of nucleons with the moving nuclear surface and when the neck is smaller than a critical size also from the transfer of nucleons through it, with a magnitude that is completely specified by the model. The second mechanism is two-body viscosity,[2,4-6] which is responsible for dissipation in ordinary fluids. Because in nuclei the nucleon mean free path is long compared to the nuclear radius, the conventional result for this mechanism is *not* expected to apply. Nevertheless, with a coefficient of two-body viscosity that is adjusted to reproduce experimental results, it represents a tractable and useful phenomenological approach for describing small dissipation.

Compared to most of our previous calculations with one-body dissipation,[4,5,12] our present calculations incorporate three improvements. First, to describe the transition from the wall formula that applies to mononuclear shapes to the wall-and-window formula that applies to dinuclear shapes we now use the smooth interpolation[6]

$$\eta = \sin^2(\tfrac{\pi}{2}\alpha)\, \eta_{\mathrm{wall}} + \cos^2(\tfrac{\pi}{2}\alpha)\, \eta_{\mathrm{wall\text{-}and\text{-}window}} \quad , \tag{5}$$

where

$$\alpha = (r_{neck}/R_{min})^2 \tag{6}$$

is the square of the ratio of the neck radius r_{neck} to the transverse semi-axis R_{min} of the end fragment with the smaller value. Second, in determining the drift velocities of the end fragments relative to which velocities in the wall-and-window formula are measured, we now require the conservation of linear and angular momentum rather than using the velocities of the centers of mass.[6] However, the results calculated with both prescriptions for the drift velocity are nearly identical. Third, for asymmetric shapes we now also take into account the dissipation associated with a time rate of change of the mass asymmetry degree of freedom in the completed wall-and-window formula.[14,15]

2.5 Generalized Fokker-Planck Equation

In addition to the mean dissipative force, the coupling between the collective and internal degrees of freedom gives rise to a residual fluctuating force, which we treat under the Markovian assumption that it does not depend upon the system's previous history. At high excitation energies, where classical statistical mechanics is valid, we are led to the generalized Fokker-Planck equation

$$\frac{\partial f}{\partial t} + (M^{-1})_{ij}\, p_j \frac{\partial f}{\partial q_i} - \left[\frac{\partial V}{\partial q_i} + \frac{1}{2}\frac{\partial (M^{-1})_{jk}}{\partial q_i}\, p_j\, p_k\right]\frac{\partial f}{\partial p_i}$$

$$= \eta_{ij}\,(M^{-1})_{jk}\frac{\partial}{\partial p_i}(p_k f) + \tau\eta_{ij}\frac{\partial^2 f}{\partial p_i \partial p_j} \tag{7}$$

for the dependence upon time t of the distribution function f(q,p,t) in phase space of collective coordinates and momenta. The last term on the right-hand side of this equation describes the spreading of the distribution function in phase space, with a rate that is proportional to the dissipation strength and the nuclear temperature τ, which is measured here in energy units.

2.6 Generalized Hamilton Equations

Although a useful approximate solution of a two-dimensional

Fokker-Planck equation has been obtained recently,[16] it is still difficult in practice to solve the generalized Fokker-Planck equation except for special cases. Therefore, in some of our studies we use equations for the time rate of change of the first moments of the distribution function, with the neglect of higher moments. These are the generalized Hamilton equations

$$\dot{q}_i = (M^{-1})_{ij}\, p_j \tag{8}$$

and

$$\dot{p}_i = -\frac{\partial V}{\partial q_i} - \frac{1}{2}\frac{\partial (M^{-1})_{jk}}{\partial q_i}\, p_j\, p_k - \eta_{ij}\,(M^{-1})_{jk}\, p_k \quad , \tag{9}$$

which we solve numerically for each of the N generalized coordinates and momenta.

3. FISSION

As our first application, we calculate for the fission of nuclei throughout the periodic table their mean translational fission-fragment kinetic energies and compare with experimental values. Although similar to earlier studies,[2,4] our present calculations are performed, as discussed above, with a more flexible shape parametrization, with a more realistic set of constants, and with an improved treatment of one-body dissipation. Also, our initial conditions at the fission saddle point now incorporate the effect of dissipation on the fission direction[17] and are calculated for excited nuclei with nuclear temperature $\tau = 2$ MeV by determining the mean velocity of all nuclei that pass per unit time through the saddle point with positive velocity.[18] Because this procedure is no longer valid when the fission barrier is less than the nuclear temperature, in such cases we use the mean velocity of the nucleus whose barrier is 2 MeV high. The atomic number Z is related to the mass number A according to Green's approximation to the valley of beta stability.[19] Our calculations for two-body viscosity are performed with viscosity coefficient

$$\mu = 0.02 \text{ TP} = 1.25 \times 10^{-23} \text{ MeV s/fm}^3 \quad , \tag{10}$$

which as we see later is the value required to optimally reproduce experimental mean fission-fragment kinetic energies.

3.1 Dynamical Descent

In our fission calculations we specialize to reflection-symmetric shapes and zero angular momentum, so that only five coordinates are considered explicitly. The mean dynamical trajectories in deformation space for light nuclei correspond to short descents from dumbbell-like saddle-point shapes to compact scission shapes, whereas those for heavy nuclei correspond to long descents from cylinder-like saddle-point shapes to elongated scission shapes. Compared to the trajectories for nonviscous nuclei, those for one-body dissipation lead to more elongated scission shapes for light nuclei and to more compact scission shapes for heavy nuclei. In contrast, the trajectories for two-body viscosity always lead to more elongated scission shapes.

3.2 Ternary Division

An exciting new aspect of these dynamical calculations is the formation of a third fragment between the two end fragments for sufficiently heavy nuclei with either no dissipation or small two-body viscosity. As shown in Fig. 1, the mass of this third fragment increases with increasing $Z^2/A^{1/3}$ above a critical value that is slightly lower for two-body viscosity than for no dissipation. Since no third fragment is formed with one-body dissipation, accurate experimental information concerning such true ternary-fission processes should help decide the nuclear-dissipation issue. Further theoretical aspects of this problem are currently being studied at Los Alamos by Cârjan.[20]

3.3 Fission-Fragment Kinetic Energies

In calculating the mean fission-fragment translational kinetic energy at infinity, we treat the post-scission dynamical motion in terms of two spheroids, with initial conditions determined by keeping continuous the values of two moments and their time derivatives. When a small third fragment is formed in a realistic situation off the

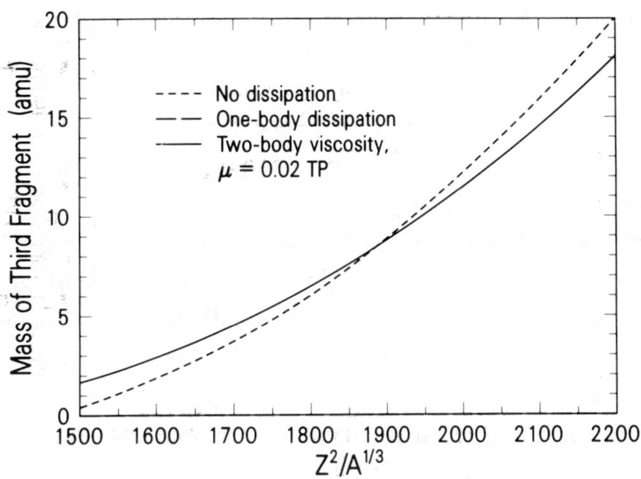

Figure 1
Effect of dissipation on the formation of a third fragment between the two end fragments.

symmetry axis and/or with some transverse velocity, it moves away and contributes less to the kinetic energy of the two larger end fragments than it would in our idealized calculation, where it remains stationary at its origin. In the presence of a third fragment, we obtain a lower limit to the fission-fragment kinetic energy by calculating the post-scission separation of the end fragments in the absence of the middle fragment. Also, we estimate an upper limit in terms of the kinetic energy at scission of the two end fragments plus the Coulomb interaction energy of three spherical nuclei positioned at their respective centers of charge.

As the nucleus descends dynamically from its fission saddle point, the repulsive Coulomb force can overcome the attractive nuclear force and rupture the neck prior to its reaching a zero radius, as is required in our calculations. Although such a neck rupture at a nonzero radius would increase the calculated kinetic energy slightly,[21] we neglect this effect here because of the difficulty of properly incorporating the nuclear compressibility energy, which plays a crucial role in the neck-rupture process.

We compare in Figs. 2 and 3 our mean kinetic energies calculated in this way with experimental values for the fission of nuclei at high excitation energy,[2,22,23] where single-particle effects have decreased in importance. As shown by the short-dashed curves in both figures, the results calculated with no dissipation are for heavy nuclei substantially higher than the experimental values. Dissipation of either type lowers the calculated kinetic energy. However, as shown by the long-dashed curve in Fig. 2, one-body dissipation with a magnitude that is specified by the theory predicts for heavy nuclei values that lie below the experimental data. This underprediction arises because the highly dissipative descent from the saddle point damps out much of the pre-scission kinetic energy, and our improved parametrization leads to moderately elongated scission shapes with lower Coulomb repulsion. We regard this discrepancy as experimentally

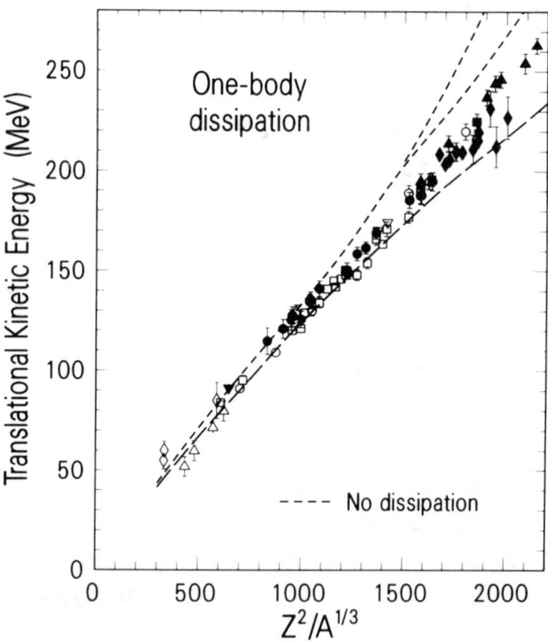

Figure 2
Reduction of mean fission-fragment kinetic energies by one-body dissipation, compared to experimental values.

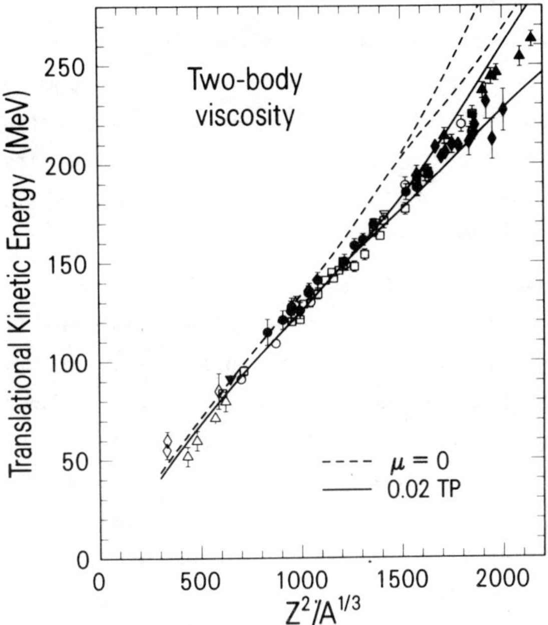

Figure 3
Reduction of mean fission-fragment kinetic energies by two-body viscosity, compared to experimental values.

demonstrating that one-body dissipation as presently formulated is *not* the complete dissipation mechanism in large-amplitude collective nuclear motion.

In contrast, as shown by the solid curves in Fig. 3, when the two-body viscosity coefficient is adjusted to the value μ = 0.02 TP, the experimental data for heavy nuclei lie between the calculated lower and upper limits and are adequately reproduced throughout the rest of the periodic table. For two-body viscosity, the dynamical trajectories lead to elongated scission shapes with less Coulomb repulsion, but this is supplemented by some pre-scission kinetic energy. These results calculated with several improvements demonstrate that mean fission-fragment kinetic energies are capable after all of distinguishing between dissipation mechanisms.

4. HEAVY-ION REACTIONS

Even better prospects for determing the dissipation mechanism reside with heavy-ion reactions, where we are able to choose the total mass of the combined system, the mass asymmetry of the entrance channel, and the bombarding energy with foresight. This permits us to select for study those dynamically interesting cases that involve large distances in deformation space.

4.1 Dynamical Thresholds for Fusion

A necessary condition for compound-nucleus formation is that the dynamical trajectory of the fusing system pass inside the fission saddle point in a multidimensional deformation space. For heavy nuclear systems and/or large impact parameters, the fission saddle point lies inside the contact point, and the center-of-mass bombarding energy must exceed the maximum in the one-dimensional zero-angular-momentum interaction barrier by an amount ΔE in order to form a compound nucleus.

This additional energy ΔE has been calculated for symmetric nuclear systems both by solving the generalized Hamilton equations numerically with the three-quadratic-surface shape parametrization and realistic forces[3,5,6] and approximately with the two-sphere-plus-conical-neck shape parametrization and schematic forces.[14,24,25] Such values calculated for symmetric nuclear systems have been compared with experimental values derived from asymmetric nuclear systems under various assumptions concerning the scaling of asymmetric systems into symmetric ones.[5,6,14,23-28] However, our recent calculations involving asymmetric systems indicate that none of these scaling assumptions are sufficiently accurate for detailed comparisons.

We therefore compare here our values of the additional energy ΔE calculated for five specific nuclear systems for which neutron-evaporation-residue cross sections have been recently measured[28,29] and analyzed to yield experimental thresholds.[28] As indicated in Table 1, as we progress through these systems the additional energy calculated with two-body viscosity increases from less than 1 MeV, representing only the energy that is dissipated during the approach stage,

Table 1

Comparison of calculated and experimental values of the additional energy ΔE required to form a compound nucleus, measured relative to the maximum in the calculated one-dimensional zero-angular-momentum interaction barrier. The calculated values of additional energy are for two-body viscosity with coefficient μ = 0.02 TP.

Reaction	Calculated one-dimensional barrier (MeV)	Calculated additional energy (MeV)	Experimental additional energy (MeV)	Note
$^{90}Zr + {}^{90}Zr \rightarrow {}^{180}Hg$	189.0	0.9	-7 ± 2	a
$^{86}Kr + {}^{123}Sb \rightarrow {}^{209}Fr$	209.0	0.9	0^{+4}_{-2}	
$^{124}Sn + {}^{96}Zr \rightarrow {}^{220}Th$	224.7	1.2	16^{+5}_{-3}	b
$^{124}Sn + {}^{94}Zr \rightarrow {}^{218}Th$	225.5	6.5	13^{+5}_{-3}	c
$^{124}Sn + {}^{92}Zr \rightarrow {}^{216}Th$	226.3	8.2	11^{+4}_{-3}	

[a]For this reaction involving nuclei lighter than those requiring an additional energy, the negative experimental value of ΔE suggests the importance of zero-point vibrations on the low-energy fusion cross section (Ref. 30).

[b]For this reaction involving a target for which the calculated value of Nilsson's spheroidal deformation coordinate $\varepsilon = 0.20$ (Refs. 9 and 31), the large experimental value of ΔE compared to the calculated value suggests the importance of static ground-state deformations on the additional energy.

[c]For this reaction involving a target for which the calculated value of Nilsson's spheroidal deformation coordinate $\varepsilon = -0.12$ (Refs. 9 and 31), the moderately large experimental value of ΔE compared to the calculated value suggests the importance of static ground-state deformations on the additional energy.

to several MeV, representing in addition the energy required to dynamically push the system inside its fission saddle point. The experimental values show a similar trend, but three large deviations from the calculated values suggest the important role played by zero-point vibrations and static ground-state deformations, as discussed in the footnotes to Table 1. Our analogous calculations with one-body dissipation are not yet completed; like the rest of you at this Conference, we are eagerly awaiting their outcome.

4.2 Capture Cross Section

For the reaction ^{208}Pb + ^{58}Fe that has been studied experimentally by Bock et al.,[26] we calculate the capture cross section corresponding to the transfer of 40 or more nucleons from the heavier ^{208}Pb nucleus to the lighter ^{58}Fe nucleus. Our calculated results are compared in Fig. 4 with experimental values resulting from a revised analysis in which the experimental capture cross section is defined in terms of reaction products with fully relaxed total kinetic energy and masses lying between the deep-inelastic peaks.[23] The cross section calculated with two-body viscosity is somewhat larger than the experimental points at all energies except near the threshold.

We have not yet finished our analogous calculations with one-body dissipation when the additional term in the completed wall-and-window formula is included because it requires the specification of a mass asymmetry, which is difficult for shapes with long necks. When this

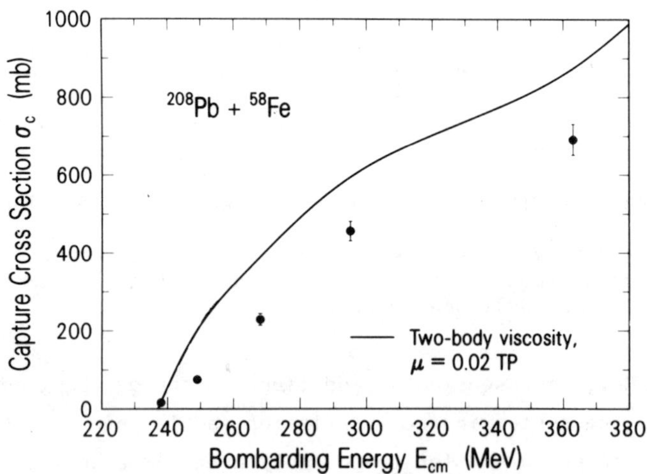

Figure 4
Capture cross section calculated with two-body viscosity, compared to revised experimental values (Ref. 23).

additional term is omitted, the capture cross section calculated with one-body dissipation is even larger at intermediate and high energies than that calculated with two-body viscosity, but the additional term is expected to reduce it. Similar calculations with a restricted shape parametrization where the present difficulties did not arise have been performed by Błocki,[32] who adjusted his interpolation procedure to reproduce the original *unrevised* experimental data for a comparable reaction.[26]

4.3 Ternary Events

We consider next the reaction $^{129}Xe + ^{122}Sn$ at a laboratory bombarding energy per nucleon of 12.5 MeV studied experimentally by Glässel et al.,[33] for which ternary events were observed approximately 10% of the time when the energy loss was large. Glässel et al. deduced that the time between successive scission events is approximately 1×10^{-21} s, during which the two primary fragments move only a few nuclear radii apart and perform only a fraction of a rotation. The ratio of mean fragment masses for the second scission event was determined to be approximately 1.5.

Figures 5 and 6 show sequences of shapes calculated for this reaction for angular momentum L = 250 and 350 $\hbar$, respectively. In these two figures our results with one-body dissipation are calculated for computational ease without the additional term in the wall-and-window formula, which has little effect since the system is nearly symmetric. With this dissipation mechanism, the process is essentially binary, with only extremely small third fragments forming between the two end fragments. In contrast, two-body viscosity with coefficient μ = 0.02 TP leads to true ternary events, with middle-fragment masses of 51.4 and 69.1 amu for L = 250 and 350 $\hbar$, respectively. The mass ratio of the forward-going fragment to the middle fragment is 1.93 for L = 250 $\hbar$ and 1.32 for L = 350 $\hbar$.

Although in our calculations with two-body viscosity, which refer to mean events, the two necks reach zero radius at essentially the same time, fluctuations could introduce some difference. Also, the scission-to-scission time of 1×10^{-21} s deduced by Glässel et al. was

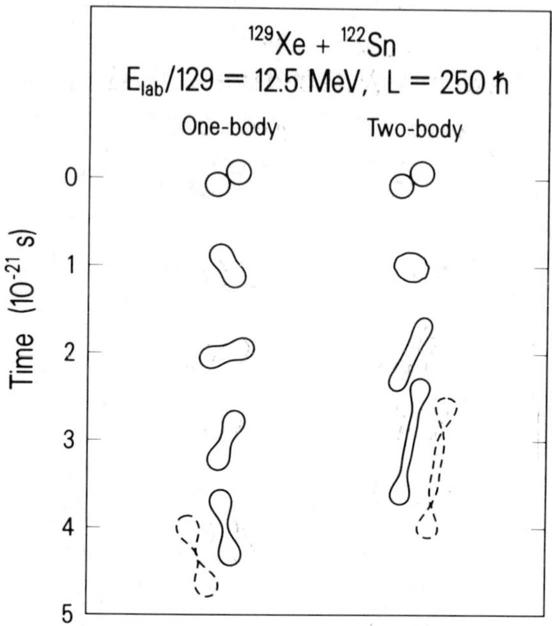

Figure 5
Effect of dissipation on ternary heavy-ion events for angular momentum $L = 250\,\hbar$. The ^{129}Xe projectile is incident from the right. For clarity, the dashed scission shapes are shown displaced from their proper horizontal positions.

based on certain assumptions concerning nuclear shapes that are very different from those calculated here. Although the probability for ternary events in our calculations with two-body viscosity is much larger than the approximately 10% observed by Glässel et al., the experimental arrangement could have missed events in which the middle fragment remained essentially at rest in the center-of-mass system and detected instead only those with some forward velocity resulting once again from fluctuations. Although several issues remain to be clarified, it is possible that the ternary events seen by Glässel et al. have a dynamical origin of the type calculated here for *small* two-body viscosity. If so, this could provide a convincing discrimination between the two extremes of dissipation that we are considering.

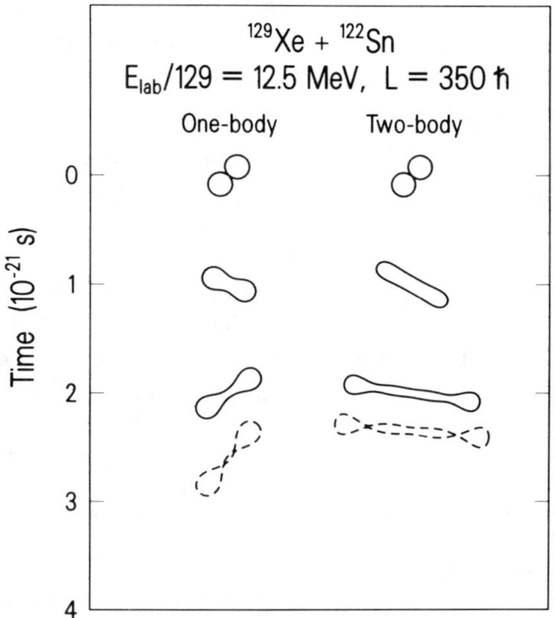

Figure 6
Effect of dissipation on ternary heavy-ion events for angular momentum L = 350 ħ. The ^{129}Xe projectile is incident from the right.

5. CONCLUSION

We are entering a new era in fission and heavy-ion reactions. Up to now theoretical approaches with vastly different pictures of the underlying nuclear dynamics have reproduced many of the gross experimental features of fission and heavy-ion reactions because they include correctly the dominant nuclear, Coulomb, and centrifugal forces. However, calculations are now being designed specifically to test the dissipation mechanism. When compared with mean fission-fragment kinetic energies, these calculations demonstrate that one-body dissipation is *not* the complete dissipation mechanism. The next step is to compute dynamical thresholds for fusion and capture cross sections with one-body dissipation and compare with experimental results.

Ternary heavy-ion events offer the most exciting prospect for finally determining the magnitude and mechanism of nuclear dissipation. If experimentally observed ternary events turn out to have a dynamical origin of the type calculated here with two-body viscosity, this would suggest small dissipation in nuclei. In this eventuality the theoretical challenge would be to understand the mechanism, since the long nucleon mean free path eliminates the conventional two-body mechanism that is present in ordinary fluids.

ACKNOWLEDGEMENTS

We are grateful to T. C. Awes, B. B. Back, S. Bjørnholm, N. Cârjan, A. Gavron, F. Plasil, J. Randrup, W. J. Swiatecki, R. Vandenbosch, and J. B. Wilhelmy for stimulating discussions. This work was supported by the U. S. Department of Energy.

REFERENCES

1. J. R. Nix, Nucl. Phys. A130, 241 (1969).
2. K. T. R. Davies, A. J. Sierk, and J. R. Nix, Phys. Rev. C13, 2385 (1976).
3. J. R. Nix and A. J. Sierk, Phys. Rev. C15, 2072 (1977).
4. A. J. Sierk and J. R. Nix, Phys. Rev. C21, 982 (1980).
5. K. T. R. Davies, A. J. Sierk, and J. R. Nix, Phys. Rev. C28, 679 (1983).
6. J. R. Nix and A. J. Sierk, Nucl. Phys. A428, 161c (1984).
7. S. Trentalange, S. E. Koonin, and A. J. Sierk, Phys. Rev. C22, 1159 (1980).
8. H. J. Krappe, J. R. Nix, and A. J. Sierk, Phys. Rev. C20, 992 (1979).
9. P. Möller and J. R. Nix, Nucl. Phys. A361, 117 (1981).
10. J. Kunz and U. Mosel, Nucl. Phys. A406, 269 (1983).
11. J. Kunz and J. R. Nix, Los Alamos National Laboratory Preprint LA-UR-84-2580 (1984).
12. J. Błocki, Y. Boneh, J. R. Nix, J. Randrup, M. Robel, A. J. Sierk, and W. J. Swiatecki, Ann. Phys. (N. Y.) 113, 330 (1978).
13. J. Randrup and W. J. Swiatecki, Ann. Phys. (N. Y.) 125, 193 (1980).
14. W. J. Swiatecki, Nucl. Phys. A428, 199c (1984).
15. J. Randrup and W. J. Swiatecki, Nucl. Phys. A429, 105 (1984).

16. F. Scheuter, C. Grégoire, H. Hofmann, and J. R. Nix, Grand Accélérateur National d'Ions Lourds Preprint GANIL-P.84.09 (1984).

17. K. T. R. Davies, J. R. Nix, and A. J. Sierk, Phys. Rev. C28, 1181 (1983).

18. J. R. Nix, A. J. Sierk, H. Hofmann, F. Scheuter, and D. Vautherin, Nucl. Phys. A424, 239 (1984).

19. A. E. S. Green, *Nuclear Physics* (McGraw-Hill, New York, 1955), pp. 185, 250.

20. N. Cârjan, private communication.

21. K. T. R. Davies, R. A. Managan, J. R. Nix, and A. J. Sierk, Phys. Rev. C16, 1890 (1977).

22. D. v. Harrach, P. Glässel, Y. Civelekoğlu, R. Männer, H. J. Specht, J. B. Wilhelmy, H. Freiesleben, and K. D. Hildebrand, in *Proc. Int. Symp. on Physics and Chemistry of Fission, Jülich, 1979* (International Atomic Energy Agency, Vienna, 1980), p. 575.

23. J. Tōke, R. Bock, G. X. Dai, S. Gralla, A. Gobbi, K. D. Hildenbrand, J. Kuźminski, W. F. J. Müller, A. Olmi, H. Stelzer, B. B. Back, and S. Bjørnholm, Gesellschaft für Schwerionenforschung Preprint GSI-84-51 (1984).

24. W. J. Swiatecki, Phys. Scr. 24, 113 (1981).

25. S. Bjørnholm and W. J. Swiatecki, Nucl. Phys. A391, 471 (1982).

26. R. Bock, Y. T. Chu, M. Dakowski, A. Gobbi, E. Grosse, A. Olmi, H. Sann, D. Schwalm, U. Lynen, W. Müller, S. Bjørnholm, H. Esbensen, W. Wölfli, and E. Morenzoni, Nucl. Phys. A388, 334 (1982).

27. R. Vandenbosch, University of Washington Preprint (1984).

28. C. C. Sahm, H. G. Clerc, K. H. Schmidt, W. Reisdorf, P. Armbruster, F. P. Hessberger, J. G. Keller, G. Münzenberg, and D. Vermeulen, Technische Hochschule Darmstadt Preprint (1984).

29. J. G. Keller, K. H. Schmidt, H. Stelzer, W. Reisdorf, Y. K. Agarwal, F. P. Hessberger, G. Münzenberg, H. G. Clerc, and C. C. Sahm, Phys. Rev. C29, 1569 (1984).

30. T. Kodama, R. A. M. S. Nazareth, P. Möller, and J. R. Nix, Phys. Ref. C17, 111 (1978).

31. P. Möller and J. R. Nix, Atomic Data Nucl. Data Tables 26, 165 (1981).

32. J. Błocki, J. Phys. (Paris) 45, C6-489 (1984).

33. P. Glässel, D. v. Harrach, H. J. Specht, and L. Grodzins, Z. Phys. A310, 189 (1983).

IMPLICATIONS OF INDEPENDENT PARTICLE MOTION FOR FISSION AND FUSION DYNAMICS

Sven Bjørnholm
Niels Bohr Institute, University of Copenhagen,
DK-2100 Copenhagen Ø, Denmark

ABSTRACT

Fission and the reverse process of heavy ion "fusion" are governed by the same type of dissipative dynamics and by conservative driving forces of identical, or similar, nature. The three experimentally observed "fusion" reaction types: deep inelastic scattering, quasi fission, and compound nucleus fission are discussed in terms of the topographical features of the relevant potential energy landscapes and in terms of the special temperature independent dissipation that appear to characterize the nuclear quantum fluid at moderate excitation energies.

1. INTRODUCTION

Bohr and Wheeler[1)] succeeded, as we know, already in 1939 in obtaining a good understanding of the fission process on the basis of the liquid drop model of the nucleus. The important feature was the interplay of the Coulomb repulsion and the cohesive surface tension, which determine the driving forces on the nucleus, also when it is far away from its equilibrium. In a droplet, one usually visualises the motion in terms of strongly interacting particles with a short mean free path, but this was not essential to their result. Nevertheless, when the shell model[2], with approximately independent particle motion as its central feature, made its appearance in 1949, it raised the question of the compatibility of the liquid drop model of fission with the shell model. This led Hill and Wheeler[3] to realize already in 1953 that nuclei could be expected to exhibit quite unusual dynamical properties. Here is a quotation[3] from their paper:

> "Nucleus as Quantum Fluid: It is considered to be completely transparent internally with respect to motion of the constituent particles, and to receive disturbances solely by way of surface deformations. Its near-incompressibility comes about, not by particle to particle push, as in an ordinary liquid, but by more subtle means. It is capable of collective oscillations, but it is the wall which organizes these disturbances, not nucleon to nucleon interactions. Oscillations experience damping, but the mechanism of damping is unlike that encountered in ordinary liquids. ...Altogether one is dealing with a most interesting new form of matter".

In the following, we shall try to review the progress made experimentally and theoretically in exploring the special dynamical behaviour of this new form of matter when it is far from equilibrium. We will restrict ourselves to moderately low relative velocities and excitation energies, where the mean field and independent particle concepts of the shell model are most likely to remain valid.

Important experimental insights have come from the study of heavy ion collisions. The discovery and exploration of the deep inelastic scattering process[4,5] and of quasi fission (or fast fission)[6] in addition to the process of complete fusion into a compound nucleus followed by ordinary fission (or particle evaporation), as well as the empirical rules[7,8,9] in terms of an extra or extra-extra push required to direct flux into one of these specific reaction categories, rcpresent major stages in the experimental development.

On the theoretical side, there has been considerable progress in describing the relevant conservative driving forces; both for the entrance channel[10,11,12,12,14,15] and for the subsequent motion, where a more or less complete relaxation of all shape degrees of freedom takes place. This phase in the development has many features in common with the long familiar fission process. The degree of freedom describing the initial step of the coalescense process in terms of formation of a neck between two colliding nuclei plays an especially crucial role. A peculiar feature of neck formation is worth mentioning here, because it implies a weakening of the nuclear attraction when the neck forms. Combined with a Coulomb repulsion of appropriate magnitude this can lead to a reversal of the sign of the net force between the two halves of the collision complex, from an initial net attraction (potential pocket) to a net repulsion (disappearance of the "pocket"). In the simplest formulation, the initial nuclear attraction between two <u>spherical</u> nuclear surfaces in contact is equal to $2\pi\gamma \cdot 2\bar{R}$, where γ is the surface tension and $\bar{R}$ is the reduced radius, $R_1 \cdot R_2/(R_1 + R_2)$.

Once a neck is formed the corresponding attraction is equal to γ times the circumference of the neck, or $2\pi\gamma\rho$, where ρ is the smallest radius of the neck. The force will decrease, because $\rho < 2\bar{R}$ (unless the neck is exceptionally thick). The deep inelastic process can be understood as a result of this effect.

A theory of the special damping mechanism, "not encountered in ordinary liquids" has also been developed. The formulation of the maximum damping (chaotic limit) that is to be expected in two limiting cases is available in form of the wall and the window formulae, respectively, thanks to W. J. Swiatecki et al.[16]. In the initial phase of a collision the relative momenta play a decisive role in determining the further course of the collision process. In this phase inertial terms play a crucial role. The dissipation expected on the basis of the wall and window formulae is on the other hand so strong that a rapid decelleration of the initial relative motion is to be expected. Once that has happened, the motion is almost entirely governed by the interplay of conservative driving forces and dissipative forces (proportional to velocity), whereas inertial forces tend to be neglegible.

In reviewing the situation we shall start discussing an experimental technique, which has played a particularly important role for the study of heavy ion collisions, together with some new results, section 2. A classification into the three reaction categories in terms of the topographical features of the relevant potential energy landscapes (conservative forces) is discussed in section 3. This is followed in section 4 by a general discussion of dissipation in fermion systems with constant density; and an attempt to emphasize the special features that serve to characterize the nuclear fermion system as "a most interesting new form of matter" will be made. The various dissipative phenomena encountered experimentally in nuclear collisions will be discussed with reference to these features. Finally, section 5 summarises the situation.

2. EXPERIMENTS

Our present knowledge of heavy ion dynamics owes much to the perfection of binary kinematic coincidence techniques[17)], especially in conjunction with the use of inverse reactions, i.e. bombarding a target of light nuclei with a beam of heavy particles. In this case all reaction products are projected into a forward cone. As a result, reasonably sized detectors can in principle record all types of binary events, irrespective of mass fractionation, energy loss, and center-of-mass scattering angle; although events very near to the beam direction remain difficult to detect. The realization of this scheme hinges on the development of efficient and fast position sensitive detectors. Here, the parallel plate, avalance detector developed at the GSI in Darmstadt, Germany, has played a central role[18)].

Figure 1 shows to the left the geometrical arrangement of four such detectors relative to the direction of the (pulsed) beam and the target position. To the right one sees the dot pattern of measured coincident events. For each event there are two dots. Together with flight time measurements, the laboratory velocity vectors of each fragment are determined in this way; and the reconstruction of full momentum transfer binaries in the center of mass system becomes straightforeward. With the particular arrangement shown, only scattering into a certain range of azimuthal angles φ will be recorded, but because there is symmetry around the beam direction it is straightforewardward, although tedious, to calculate the geometrical efficiency for each type of event and normalize to 2 geometry in the azimuthal range.

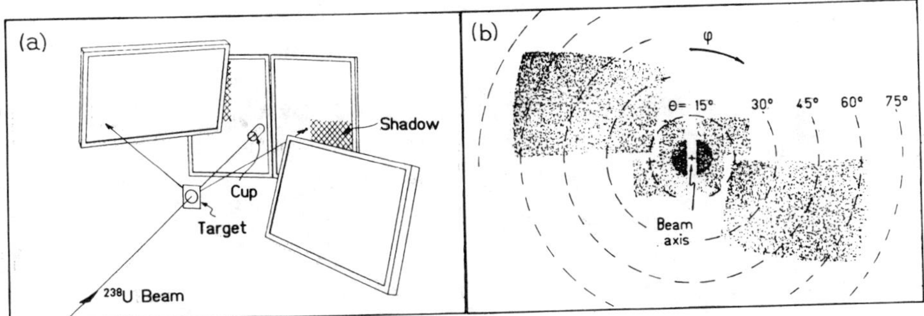

Fig. 1. (a) Arrangement of the fast, position sensitive, parallel plate detectors with respect to the pulsed beam. (b) Dot pattern of observed coincident events. (The two semicircles close to the beam are due to random coincidences of two ^{238}U-particles).

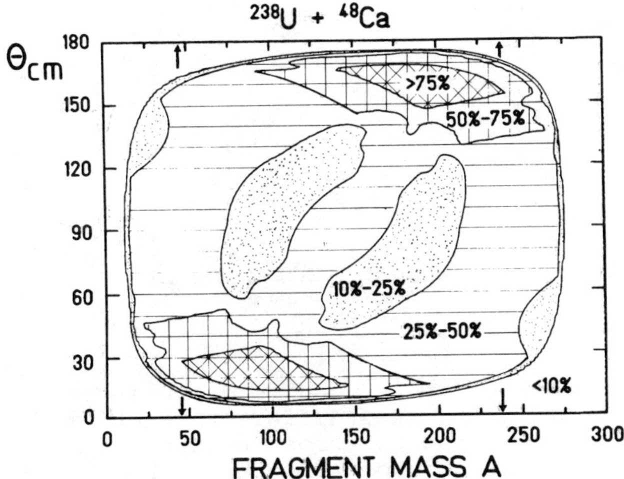

Fig. 2. Geometrical efficiency of the detector system for binary events with fully relaxed kinetic energies, as a function of the mass of one of the fragments and the scattering angle.

Figure 2 shows an example of the efficiency factors one has to apply. One notices the virtually complete coverage of this detection system. Only polar scattering angles near to the beam direction have efficiency zero, and therefore remain outside the measurable range.

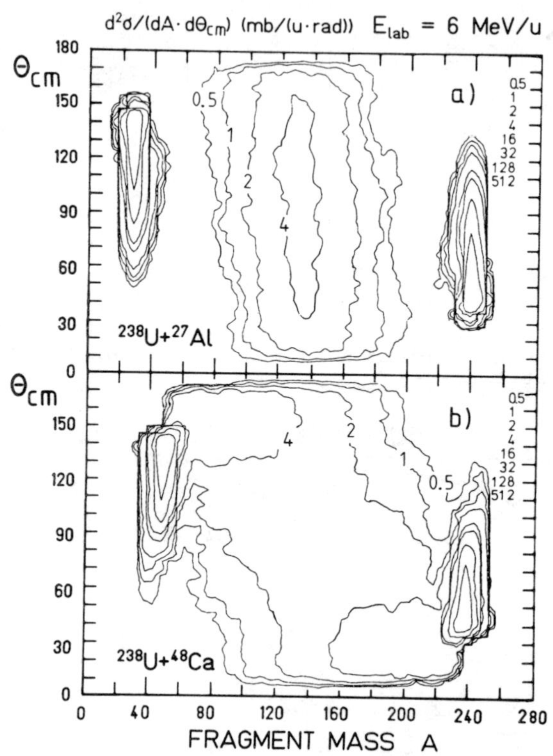

Fig. 3. Contour plots of the double differential cross section $d^2\sigma/dA \cdot d\Theta_{cm}$ for two different reactions. The contours are labelled with the cross section normalized to elastic events, in units of mb/u · rad. Elastic events at large scattering angles of target nuclei and small angles of ^{238}U-particles have been cut away.

Fig. 3 shows examples of the distributions in mass A and center of mass scattering Θ_{cm} for two different reactions. There is the usual concentration of elastic and inelastic (plus transfer) events near the target and projectile masses. Between these two masses there is also a considerable number of events. In the case of $^{238}U+^{27}Al$ the yield is concentrated near the symmetric masses, like in fission. One sees, however, that the angular distribution is squewed (to the left), and this is a clear signal that we are not dealing with compound nucleus fission. At least not exclusively. It can in fact be shown[19] that at most 30 per cent of the symmetric mass yield is ascribable to compound nucleus fission. The rest is quasi fission, a more direct reaction type, which has not gone through the complete equilibration that is typical of the compound nucleus state. This direct character becomes more pronounced in the case of the $^{238}U+^{48}Ca$ reaction, fig. 3b. Here the yield is uniformly spread over all intermediate masses, and at the same time concentrated at angles near zero and 180°, respectively. The picture is reminiscent of a Wilczynski plot (turned 90°), except that it is the mass drift that is plotted against angle, instead of the enery loss. In the present case one can interprete the observations analogously in terms of a definite turning angle $\Delta\Theta$ between the initial moment of contact and the binary decay of the reaction complex. This is illustrated more clearly in fig. 4. Assuming, as would seem reasonable, that the more central collisions (low ℓ-values) lead to the strongest mass drift, it becomes possible to estimate the reaction time as a function of mass drift. For details see ref.[20]. A preliminary summary of reaction time versus mass relaxation $\Delta A/\Delta A_{max}$ where ΔA_{max} represents a mass drift all the way to symmetry, is shown in fig. 5 for a large number of recently studied reactions[21]. Despite considerable experimental scatter of the points, one finds very strikingly that there seems to be a universal time

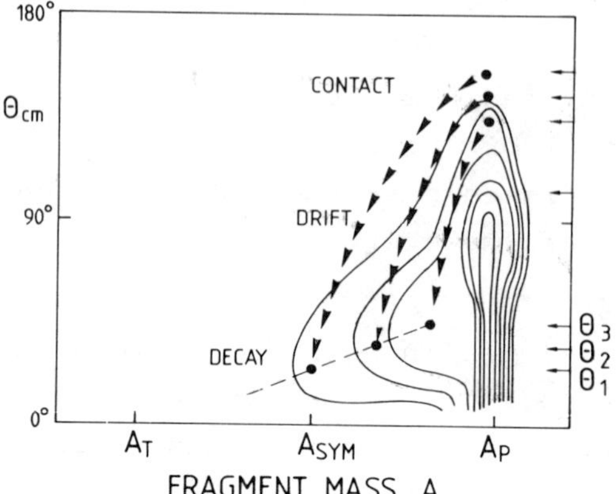

Fig. 4. Schematic picture of the angle versus mass distribution of projectile-like particles. Heavy arrows indicate the angle turned during contact with the target-like reaction partner. The reaction time, t, is estimated from $t = \Delta\Theta \cdot I/\langle \ell \rangle$, where $\langle \ell \rangle$ is found from the cross section, and I is the moment of inertia of the rotating reaction complex.

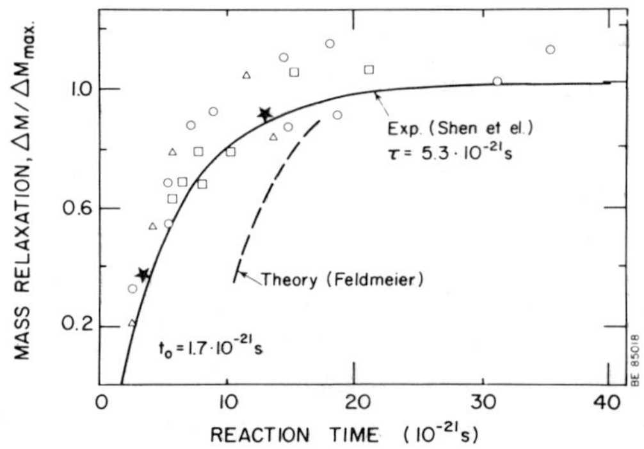

Fig. 5. Mass asymmetry relaxation versus reaction time for a large number of ^{238}U induced reactions[21]. For each reaction there are typically three points, cf. fig. 4.

constant of $5.3 \cdot 10^{-21}$s for the 1/e relaxation of the mass asymmetry; irrespective of bombarding energy and the nature of the target. (The empirical "universality" is so far restricted to ^{238}U-induced reactions). There also seems to be a common initial delay - before mass drift sets in - of $1.7 \cdot 10^{-21}$sec. We will return to these new results in connection with the discussion of dynamics, Sect. 4.

Before leaving the presentation of experimental results there is a particular aspect that should be stressed once more. It is the tendency for the reaction products to group themselves into distinct categories[4,5,6,7,20). There is (i) the deep inelastic process, where the net mass drift is essentially zero, and where the energy loss varies from a minimum value to the complete relaxation, characteristic of fission decay[22). There is (ii) quasi fission, where the energy is always completely relaxed, whereas the relaxation towards equality of the product masses varies. And there is (iii) compound nucleus formation, with complete relaxation of both mass and energy. Fig. 3 gives some evidence for this general phenomena, which is illustrated in fig. 6. In the next section we will discuss this tripartion of the reaction yield for strongly interacting collisions in terms of the relevant potential energy surfaces. (There is of course always in addition the elastic and quasi elastic (plus transfer) processes, but we will hardly be concerned with them here).

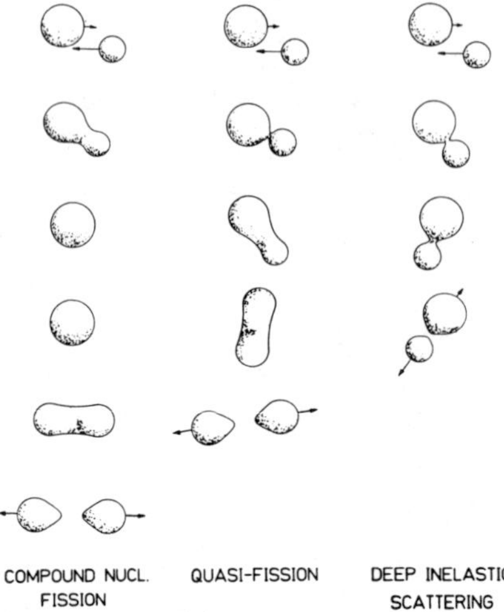

Fig. 6. Schematic illustration of the three distinct reaction types that occurs in strongly interacting heavy ion collision. They do not always occur simultaneously in a particular reaction.

3. POTENTIAL ENERGY SURFACES

The distinction between a potential barrier against fusion of two incoming heavy nuclei, one one side, and the saddle point that has to be overcome for a single compound nucleus to undergo fission on the other, dates back to discussions[23)] between Niels Bohr and Enrico Fermi as early as 1939. The difference between the barrier towards incoming motion and the saddle for the outgoing motion clearly lies in the number of shape degrees of freedom considered. The barrier V_B is the maximum energy in the relative ion-ion potential as a function of one single coordinate, the radial separation of the two spherical nuclei. The saddle, on the other hand is found after exploring the deformation energy of all possible shape degrees of freedom leading to binary decay. When we compare the configurations of the barrier and the saddle, respectively, a distinct difference emerges, fig. 7. The barrier configurations are always close to two spheres in contact, whether big or small, heavy or light. The saddle shape, on the other hand, will depend strongly on the total charge, or more precisely on Z^2/A. Guided by the empirical fact that deep inelastic collisions appear to proceed under excitation of deformation degrees of freedom (neck formation and elongation) but without change in mass asymmetry, one is naturally lead to explore saddle configurations as a function of mass asymmetry[24]. These conditional saddles will form ridges extending upward and outward from the true saddle point at symmetry, as shown by the hatched areas on fig.7. They will be more elongated than the symmetric saddle shape, because the effective value equivalent to $Z^2/A((Z^2/A)_{eff})$ becomes smaller as the asymmetry is increased. For systems considered here, this tendency prevails within a broad range of asymmetries, whereafter the saddle ridge turns back towards the barrier ridge. The turning point (Bussinaro-Gallone point) is a maximum of potential energy. In general, the barrier ridge will correspond to higher potential energies than those of the saddle ridge, because it results from variations within the more restricted configuration space of two charged spheres at varying distance from one another. The relative positions of the barrier ridge and the saddle ridge will change with the total charge of the reaction complex as shown schematically for three cases in fig. 7. In case a, a light system, the barriers lie inside the saddle. So, once the barrier is passed there is a net attraction both before and after shape relaxation. As a result, trajectories that pass inside the barrier will lead to complete fusion. The remaining reaction cross section corresponds to reflections before reaching the top of the barrier. In such cases the overlap remains relatively weak. This corresponds to the quasi elastic - and transfer - processes. In case b, the

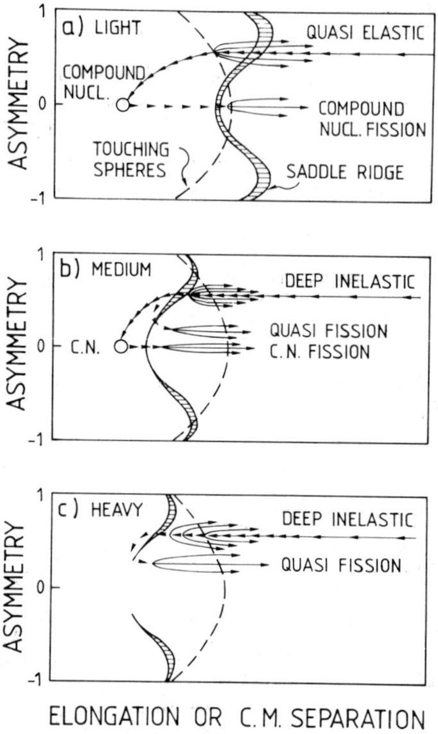

Fig. 7. Schematic illustrations of the relative positions of the ion-ion interaction barrier V_B (dahsed line) and the saddles towards symmetric and asymmetric fission (hatched areas) in a space spanned by an elongation coordinate and the mass asymmetry coordinate. (Neck formation and other deformation degrees of freedom are not shown). Different types of reactionsresulting from this kind of topography are indicated by the lines with arrows.

conditional saddle lies inside the barrier. To pass over the saddle, the two nuclei have to form a considerably more compact configuration than the two touching spheres at the barrier. Even though there might be a potential energy gain in doing so, a strong frictional force in the radial direction coupled with rapid neck formation may bring the system to a configuration outside the saddle ridge. If that happens the system will be driven outward again by the conservative forces. This description fits well to the deep inelastic scattering process and characterizes it as the fastest of the three "fusion" processes. But it should be stressed, that the description does not explain why the observed mass drift is negligibly small in deep inelastic scattering. One should also keep in mind that relaxation of the relative kinetic energy is only partial with this reaction. It presumably has to do with only partial relaxation of the shape degrees of freedom. Fig. 7b also illustrates the quasi fission process. If the barrier is passed with sufficient inward momentum (extra push) it is reasonable to expect a trajectory that passes inside the saddle ridge[7,25]. Now the net force will try to keep the system together and the reaction time will increase. There is on the other hand a driving force towards mass symmetry (except for extremely asymmetric systems). Depending on the dynamics, the system may evolve along a trajectory that runs inside the downsloping saddle ridge for a while and then crosses it. Such a course would explain the quasi fission reaction. The existence of this reaction suggest that dynamical trajectories of this kind actually occur. The main point is that the presence of the saddle ridge explains the (more or less) sharp separation of reaction trajectories into deep inelastic and quasi fission trajectories, respectively. The extra push concept also belongs naturally in this context. The same is true for the extra-extra push, required to bring complete fusion into a compound nucleus about. A trajectory with sufficient ingoing momentum may pass well inside the saddle ridge and stay inside the ridge during mass asymmetry relaxation. It will thus get behind the true saddle, which is the criterion for compound nucleus formation. The system depicted in fig. 7b will exhibit all three reaction types at the same time. In fig. 7c, a heavy system with a vanishing fission barrier is shown. In this case only deep inelastic and quasi fission processes are possible.

Before leaving the discussion of the schematic potential energy surfaces, fig. 7, it should be mentioned that the topographical changes that result from increasing the total charge of the colliding system are paralleled by similar changes when the angular momentum is increased in a system of constant charge. One may therefore observe all three reaction types, not only for a given system as a function of bombarding energy, but also at one single bombarding energy, with the different ℓ-values leading to population of the different reaction types.

The discussion in this section has focussed on the conservative forces. We have seen how even the most simple description in terms of Coulomb and surface energies gives rise to a relatively complicated topography, rich in distinctive qualitative features. They become enhanced by the fact that the shape degrees of freedom tend to be strongly constrained during the first part of the ingoing motion in contrast to the subsequent phases where the two nuclei interact strongly. These qualitative features are sufficient in order to understand the observed division of the reaction yield into three distinct categories. To see how the need for an extra or extra-extra push arises it is enough to invoke in addition some kind of dissipative force; but one does not have to be very specific.

On the other hand, for a real quantitative understanding of the dissipation so "unlike that, encountered in ordinary liquids[3)]" it becomes essential to be able to separate in an accurate way the effects due to conservative and inertial forces from the remaining dynamical response to these forces. The fact, that we are still searching for the answers to the questions raised[3] by the emergence of the independent particle model[2] more than thirty years ago, is to a considerable extent due to difficulties in isolating the dissipative aspects of the dynamics unequivocally from effects stemming from the complicated, multi-dimensional potential energy of the systems.

In addition to this, the treatment of irreversible, i.e. dissipative, processes in situations far from equilibrium has always been an especially challenging task. The challenge is so much greater in the nuclear case, as we have all reasons to expect new features to emerge that cannot be revealed by studying near-equilibrium situations.

4 DISSIPATIVE DYNAMICS

4.1 Fermion Matter

In order to appreciate the special features distinguishing dissipation in nuclei, of constant volume and under conditions where quasi-independent fermion motion within a mean field prevail, it may be instructive to consider the dissipative behavior of some other well known fermion systems.

The first example, shown in fig. 8.1 is the ohmic resistor. The current is carried by the electrons. In this case the dissipation comes about through collisions with fixed scattering centers, present in the form of the metal ions of the crystal lattice. The dissipation (resistance) is proportional to the (thermal- or Fermi-) velocity of the electrons and <u>inversely</u> proportional to the mean free path λ . This is a decreasing function of temperature because the ions oscilate stronger and scatter electrons more efficiently at higher temperatures. Thus the dissipation increases with temperature. The situation in liquid ^{3}He, fig. 8.2, is quite different. Here there are no fixed scattering centers and the dissipation (viscosity), being a question of upsetting a regular pattern of smoothly varying drift velocities through random particle motion, is directly proportional to the mean free path. The larger the difference in drift velocity at both ends of a particle path, the more effective is the randomizing momentum transport and the more power is required to maintain the drift pattern. This is true for any fluid or gaseous medium, consisting of weakly interacting particles. The viscosity equals $(1/3)\cdot\rho\cdot v_F\cdot\lambda(T)$, where v_F is the Fermi- or termal-velocity and ρ is the density[26]. The special features of a fermion system are two-fold. (i) The particle velocities are high, even at zero temperature, and (ii) the mean free path tends to be large, and it increases strongly with

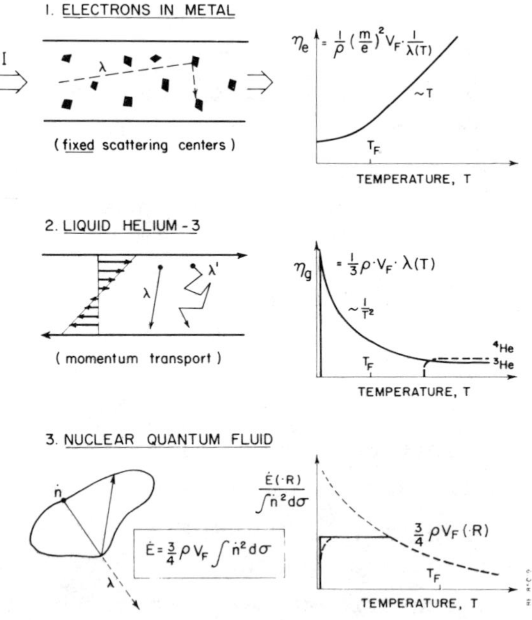

Fig. 8. Different kinds of dissipative processes in fermion matter. For details see text.

decreasing temperature. In liquid ^{3}He, λ varies as $1/T^2$; and the measured viscosity indeed increases upon cooling below the Fermi temperature as $1/T^2$, ref. [26].

The nuclear quantum fluid resembles ^{3}He. But nuclei are small in size. It means that the mean free path easily becomes longer than the dimensions R of the system, so that the relation between mean free path and dissipation coefficient breaks down. The relevant distance for randomizing momentum transport will be of order R, independent of temperature, instead of $\lambda(T)$ as in the previous example. In nuclei dissipation comes about primarily through momentum exchange at the surface of the matter distribution[3,16]. Because of the "subtle means" that ensure volume conservation, the effects of the first order term vanish. But there remains a second order term[16]. In the limit where it can be assumed that the particle will not return the extra (second order) momentum, which it picked up in an initial collision with the surface, as it later collides at the same place, this second order term will represent an irreversible dissipation. For small oscillations in a symmetric potential with closed classical orbits, the condition of irreversibility may not be fulfilled, but in situations far away from equilibrium one might expect it to hold. The wall formula for a single nucleus and the window formula for two colliding nuclei are the expressions derived under this (chaotic) assumption for two limiting cases, the mononucleus and the dinucleus, respectively[16]. The most unusual feature of these expressions is that the dissipation coefficient is independent of temperature. It is actually completely defined by the constant density ρ and the constant Fermi velocity v_F; see fig. 8.3.

4.2 Comparison With Experiment

Let us try to first ask for qualitative indications of the validity of the above picture, based on quasi-independent fermion motion, of a strong, temperature independent dissipation in nuclei.

4.2.1 Pauli Principle (Fermi Liquid). This principle implies high speeds, also at low temperature or excitation energy.

(i) The strong dissipation first revealed by the large energy loss in deep inelastic reactions[4)] can be seen as a direct reflection of the high particle velocities.

(ii) The absence of any noticeable mass drift in the deep inelastic reactions is also readily understood (Randrup[27)], Feldmeier[28)] as a result of the large Fermi velocity. It is because of this that the Fermi velocity differences behind the driving force towards mass symmetry tend to be minuscule compared to typical collision velocities. Therefore, particles are transferred either way with nearly equal probability, so that very little mass drift results. (It should be noted, though, that the driving forces towards charge equilibration are actually strong enough to relax this degree of freedom before the relative motion is damped away[29)]).

(iii) The Pauli principle tends to relegate the bulk of the particles to a passive role, leaving the least bound particles to "do all the work". In a random walk situation this will lead to large fluctuations; and indeed the mass fluctuations after deep inelastic collisions, $\sigma^2(A)$ vs. TKEL, are typically three times stronger than expected for a Boltzman gas; they are actually (v_F/v_{rel}) times stronger, ref.[30)].

4.2.1 Temperature Independent Dissipation. This is a consequence of the long mean free path and requires the chaotic limit to be reached, fig. 8.3.

(i) The almost complete insensitivity of the total kinetic energy release in fission to the excitation energy of the fissioning nucleus[22] is certainly consistent with temperature independent dissipation of the motion from saddle to scission. It can be seen as a very strong indication of temperature independence. It should of course be remembered that most of the 150-250 MeV of kinetic energy that is released in fission is due to post-scission Coulomb repulsion. It is only the drop in potential energy occuring between saddle and scission that may be converted to heat by a more or less temperature independent dissipation, leaving a bigger or smaller fraction of the potential energy release to pre-scission kinetic energy. In lighter nuclei, such as ^{230}Th, the saddle configuration is very close to the scission configuration in energy. In this case, the observed temperature independence should rather be interpreted to mean that the saddle and scission shapes do not change measurably with temperature. As a collorary, it should be noticed that superfluid, or cold, fission of ^{230}Th - as evidenced by a pronounced preference for divsion into even fragments - leads to the same kinetic energy release as ordinary fission at higher temperatures[31]. The situation is different with heavier systems. In the neutron induced fission of ^{235}U, the preference for even fragments is associated with higher than average kinetic energies, by about 5-10 MeV. This energy is a measure of the sensitive range for the dissipation. In still heavier systems the range becomes larger and therefore more suited to gauge the temperature dependence of the dissipation. So far, all evidence is consistent with no temperature dependence, because the observed total kinetic energy release is independent of excitation energy.

(ii) A temperature independent dissipation is implicitly assumed in a large number of good fits to measurements of the deep inelastic process, see refs.[32,33]. Most of these comparisons depend on window dissipation.

(iii) The recent experiments[21] on mass asymmetry relaxation, fig. 5, are carried out at different bombarding energies and correspond to excitation energies of the reaction complex that vary between 30 and 120 MeV. Nevertheless there are no discernable differences in relaxation time constant. (A more quantitative evaluation still needs to be carried out). This appears to be a strong indication of temperature independence, also for situations where wall damping is important.

(iv) In the analysis of extra push systematics[8,25,34] one assumes dynamical equivalence between pairs of reaction partners in terms of Coulomb and centrifugal effects. A central collision of two highly charged nuclei is expected to require the same extra push as a non-central collision of two somewhat less highly charged nuclei, if only the sum of centrifugal and Coulomb repulsion at contact is the same. Empirically such a relation holds, but only if it is assumed that no more than 50-70 per cent of the total angular momentum remains in the relative motion. The rest must then have been transferred to internal rotation of the two reaction partners by friction, and this necessarily implies heating them up. The non-central collision therefore proceeds with higher excitation energy than the equivalent central collisions, and yet the extra push is the same. (A more quantitative evaluation of the data with this aspect in mind remains to be carried out).

4.2.3 Absolute Magnitude of the Dissipation. These qualitative indication of the special type of dynamical behaviour encountered in nuclear systems far from equilibrium are on the whole very suggestive, but for a stringent test of the new dynamics it is of course necessary to ask for quantitative agreement between theory and experiment.

Here, the situation cannot be said to be entirely satisfactory, see for example ref.[42].

(i) The most satisfactory situation is found with the most intensely studied reaction, the deep inelastic scattering process. Here, there are several calculations[32,33] of energy loss versus deflection angle that are found to agree with experiments within a margin of accuracy of some 20 per cent. The principal problem is related to the question of uniqueness of these fits. Alternative descriptions, which include coupling to surface vibrations[35] or to incoherent particle-hole excitations[36] as important contributors to the damping process, also exhibit some measure of success. The absence of any noticeable mass drift in deep inelastic processes[37] seems to be an important clue in trying to distinguish between alternatives, and here the process of damping through unhindered particle exchange (windown friction) seems most successful[28].

(ii) There is now an increasing body of systematic experiments on the extra push[8,20,21,34] and on the extra-extra push[9]. We should first note that the data give no support whatsoever to the notion of a sudden onset of the extra-extra push (cliff), which was suggested on the basis of a simplified theoretical model[7] of the collisions processes[25,38,39]. After the model has been improved by incorporating the special damping term for motion along the mass asymmetry coordinate[40], the model itself actually fails to exhibit any cliff phenomenon[41]. Similarly, the originally suggested scaling in terms of the effective fissility, x_e, for the extra push and the mean fissility, x_m, for the extra-extra push appears not to be an accurately fulfilled law, neither by experiment[9,20] nor by more recent model calculations[41,42]. As far as a quantitative agreement between experimental and calculated values of the extra and extra-extra push energies is concerned, the situation seems to be that the calculated values are higher than the observed values by some 30-60 per cent[39,42,43]. The same applies to:

(iii) The relaxation times for the mass asymmetry. There is only one calculation[43] that can be compared to experiment, and as one sees from fig. 5 the calculated times are close to a factor of two longer than the observed relaxation times. (A more stringent comparison can be made by looking directly at the measured and the calculated angular distributions versus mass, cf. fig. 3. The conclusion remains roughly the same).

(iv) Finally, recent theoretical studies[44], of the angular momentum dissipation in deep inelastic scattering processes again seems to overestimate the damping strength as compared to experiment.

On the whole, therefore, real nuclei seem to be able to sometimes find their way through configuration space along less dissipative and faster trajectories than the model nuclei do. One of the most important outstanding problems is to better understand this. Is it due to the restricted configuration space that is used in the model calculations, with usually just three deformation parameters? Or does it really imply that the nuclear dissipation is weaker than assumed by the wall and window formulae?

5. CONCLUSION

Fission and heavy ion reactions allow us to explore the implications of the shell model of nearly independent particle motion in situations far removed from equilibrium. Both the shell model and the liquid drop model assume volume conservation, and the conservative driving forces towards equilibrium are therefore determined by the interplay between Coulomb (plus centrifugal) and surface effects. The topography of the resulting potential energy landscapes, in combination with different degrees of relaxation of the shape degrees of freedom, gives rise to a division of the strongly interacting heavy ion reactions into three distinct categories, as found experimentally, fig. 6.

The inertial terms tend to be important only during the initial stage of nuclear collisions, where the inertial mass can be assumed to be equal to the reduced mass. Otherwise the motion appears to be strongly dissipation dominated.

The observed dissipation in the nuclear quantum fluid has unique features. It is strong at all excitation energies; it is associated with large fluctuations; and it appears to be independent of temperature. The exact form and strength of the dissipation is given theoretically by the wall and window expressions; and model calculations of the energy dissipation based on these are found to be in agreement with experiment within a factor of two or better. At the same time, this picture of the dissipation mechanism has definite consequences for the evolution of several other observables during a reaction. For example the evolution of the correlated mass and charge distributions, of the partition of excitation energy between the two reaction partners, and for the dissipation, fluctuations and correlations of angular momentum.

On the whole it can be said that theoretical and experimental research during the last decade has brought very satisfactory progress towards unifying our understanding of nuclei in terms of independent particle motion in a mean field, not only in quasi-stationary situations near equilibrium but also in situations far from equilibrium.

Nevertheless, there remains a number of challenging problems before the subject can be said to be fully clarified. By way of example, one could mention:

(i) The expected temperature independence of the dissipation could be more clearly demonstrated experimentally.

(ii) There is a need for more extensive theoretical and experimental studies of the extra and extra-extra push in order to bring a possibly systematic discrepancy in evidence.

(iii) The theoretical models presently used are perhaps too restricted in terms of the shape degrees of freedom to permit accurate quantitative tests of the dissipation expressions, especially in strongly damped cases where wall dissipation is expected to dominate. Better criteria for making the transition from wall plus window to pure wall dissipation may also be needed[41].

(iv) Several studies of the gradual dissipation of the initial orbital angular momentum into alligned fragment rotation and into statistical spin modes, not least the tilting mode, remain to be made in order that this important testing ground for the dissipation theory[44] can be said to be covered in a satisfactory way.

(v) The magnitude of energy fluctuations associated with the extra and extra-extra push mean values needs to be clarified. This is of importance for:

(vi) The quantitative understanding of the mechanism by which superheavy compound nuclei with low excitation energy can be produced in heavy ion collisions. A better understanding may suggest new avenues to superheavy element synthesis.

(vii) The transition (cf. fig. 8.3) from superfluid, i.e. non-dissipative, dynamics encountered in connection with "cold" fission to a highly dissipative descent from saddle to scission needs to be better understood. The disappearance of superfluidity in heavy ion reactions could perhaps also be studied experimentally.

(viii) The transition from the observed, relatively weak damping of small amplitude collective oscillations to strong damping as expected in the chaotic limit, requires studies of the change in damping widths as a function of temperature[45]. The role of symmetry and shells in connection with this transition also remains unexplored.

(ix) The upper limit for a temperature independent, maximum dissipation also remains to be explored. The characteristic transition temperature cf. fig. 8.3, is not yet known.

(x) The role played by a diabatic[46] -i.e. essentially elastic, or weakly dissipative - collision mechanism should be clarified.

REFERENCES

1. Bohr, N. and Wheeler, J. A., Phys.Rev. 56, 426(1939).
2. Haxel, O., Jensen, J.H.D. and Suess, H.E., Phys.Rev. 75, 1766 (1949), and Mayer, M.G., Phys. Rev. 75, 1969 (1949).
3. Hill, D.L. and Wheeler, J.A., Phys. Rev. 89,1102(1953).
4. Hanappe, F., Ngo, C., Peter, J. and Tamain, B., Vol.II, 289, Proceedings of 3rd Int. Symp. on Phys. and Chem. of Fission, IAEA, Vienna, 1974.
5. Schröder, W.U. and Huizenga, J.R., Ann. Rev. Nucl. Sci. 27, 465 (1977) and Bock, R. ed., Heavy Ion Collisions, North Holland, Amsterdam 1980.
6. Heusch, B., Volant, C., Freiesleben, H., Chestnut, R.P., Hildenbrand, K.D., Pühlhofer, F., Schneider, W.F.W., Kohlmeyer, B. and Pfeffer, W., Z. Phys. A288, 391(1978) and Lebrun, C., Hanappe, F., Lecolley, J.F., Lefebvres, F., Ngo, C., Peter, J. and Tamain, B., Nucl. Phys.

A321, 207 (1979) and Borderie, B., Berlanger, M., Gardes, D., Hanappe, F., Nowicki, L., Peter, J., Tamain, B., Agrawall, S., Girard, J., Gregoire, C., Mutuszek, J. and Ngo, C., Z. Phys. A299, 263 (1981) and Gregoire, C., Ngo, C. and Remaud, B., Phys. Lett. 99B, 17 (1981).

7. Swiatecki, W.J., Proceedings of Nobel Symposium on Nuclei at Very High Spin, Physica Scripta 24,113(1981).
8. Bock,R., Chu, Y.T., Dakowski, M., Gobbi, A., Grosse,E., Olmi, A., Sann, H., Schwalm, D., Lynen, U., Müller, W., Bjørnholm, S., Esbensen, H., Wölfli, W. and Morenzoni, E., Nucl. Phys. A388, 334 (1982) and Sann, H., Bock,R., Chu, Y.T., Gobbi, A., Olmi, A., Lynen, U., Müller,W., Bjørnholm, S. and Esbensen, H., Phys. Rev. Lett. 47, 1248 (1981).
9. Back, B.B., Phys. Rev. C, (1985) to appear, and Sahm, C.-C., Clerc, H.-G., Schmidt, K.-H., Reisdorf, W., Armbruster, P., Hessberger, F.P., Keller, J.G., Münzenberg, G. and Vermeulen, D., Z. Phys. A319, 113(1984).
10. Möller, P. and Nix, J.R., Nucl. Phys. A281, 354(1977).
11. Blocki, J., Randrup, J., Swiatecki, W.J. and Tsang, C.F., Ann. of Phys. 105, 477 (1977).
12. Bass, R., Nucl. Phys. A231, 45 (1974).
13. Cohen, S., Plasil, F. and Swiatecki, W.J., Ann. of Phys. 82, 537 (1974).
14. Broglia, R.A., Dasso, C.H., Esbensen, H. and Winther, A., Phys. Lett. 104B, 11 (1981).
15. Randrup, J., Nucl. Phys. A327, 490 (1979) and Nucl. Phys. A383, 468 (1982).
16. Blocki, J., Boneh, Y., Nix, J.R., Randrup, J., Robel, M., Sierk, A.J. and Swiatecki, W.J., Ann. of Phys. 113, 330 (1978).

17. Lynen, U., Stelzer, H., Gobbi, A., Sann, H. and Olmi, A., in Detectors in Nuclear Science, ed. D.A. Bromley, special issue of Nucl. Instr. Meth., 162, 657 (1979).
18. Stelzer, H., Nucl. Phys. A354, 433 (1981).
19. Töke, J., Bock, R., Dai, G.X., Gobbi, A., Gralla, S., Hildenbrand, K.D., Kuzminski, J., Müller, W.F.J., Olmi, A., Reisdorf, W., Bjørnholm, S. and Back,B.B., Phys. Lett. 142B, 258 (1984).
20. Töke, J., Bock, R., Dai, G.X., Gralla, S., Hildenbrand, K.D., Kuzminski, J., Müller, W.,F.J., Olmi, A., Stelzer, H., Back, B.B. and Bjørnholm, S., Nucl. Phys. (1985) to appear.
21. Shen, W.Q.,Gobbi, A., Guarino, G., Hildenbrand, K.D., Stelzer, H., Back, B.B., Bjørnholm, S. and Sørensen, S.P., to be published.
22. Viola, V.E., Nucl. Data Sect. A1, 391 (1966).
23. Niel Bohr, Collected Works Vol. 9, Peirls, R., ed., North Holland, Amsterdam, to be published.
24. Swiatecki, W.J., Proc.Int.Conf. on Nuclear Reactions Induced by Heavy Ions, Heidelberg, (1969), and UCRL-19405.
25. Swiatecki, W.J., Nucl. Phys. A376, 275 (1982).
26. Wilks, J., The Properties of Liquid and Solid Helium, Clarendon Press, Oxford (1967).
27. Randrup, J., Ann. of Phys. 112, 356 (1978), and Nucl. Phys. A307, 319 (1978) and ref.[15].
28. Feldmeier, H. and Spangenberger, H., Nucl. Phys. A428, 223c (1984).
29. Freisleben, H. and Kratz, J.V., Phys. Rep. 106, 1(1984).
30. Schröder, W.U., Birkelund, J.R., Huizenga, J.R., Wilcke, W.W. and Randrup, J., Phys. Rev. Lett. 44, 308 (1980).
31. Armbruster, P., Quade, U., Rudolph, K., Clerc, H.G., Müller, M., Pannicke, J., Schmitt, D., Theobald, J.P.,

Engelhardt, W., Gönnenwein, F. and Schrader, H., CERN Conf. Proceeding 81/09, 675 (1981), and Signarbieux, C., Montoya, M., Ribrag, M., Mazur, C., Guet, C., Perrin, P. and Maurel, M.J., J. de Phys. 42, L437 (1981), and Bjørnholm, S., Phys. Scripta 10A, 110 (1974).

32. Gobbi, A. and Nörenberg, W., Heavy Ion Collisions, Vol. 2, Bock, R. ed., North Holland, Amsterdam (1980) and ref.[28].

33. Døssing, T. and Randrup, J., submitted to Phys.Lett. (1985).

34. Blann, M. and Akers, D., Phys. Rev. C26, 465 (1982) and Esterlund, R.A., Westmeier, W., Reus, U., Habbestad, A.M. and Patzelt, P., Nucl. Phys. (1985) to appear.

35. Broglia, R., Dasso, C. and Winther, A., Phys. Lett. 53B, 301 (1974).

36. Dakowski, M., Gobbi, A. and Nörenberg, W., Nucl. Phys. A378, 189 (1982).

37. Moretto, L.G., Nucl. Phys. A409, 115 (1983).

38. Bjørnholm, S., Nucl. Phys. A387, 51c (1982).

39. Bjørnholm, S. and Swiatecki, W.J., Nucl. Phys. A391, 471 (1982).

40. Randrup, J. and Swiatecki, W.J., Nucl. Phys. A429, 105 (1984).

41. Donangelo, R. and Canto, F., to be published.

42. Nix, R., contribution to this conference, and Blocki, J., private communication.

43. Feldmeier, H., unpublished results.

44. Døssing, T. and Randrup, J., Nucl. Phys. A433, 215 (1985) and Nucl. Phys. A433, 280 (1985).

45. Gaardhøje, J.J. et al., Phys. Rev. Lett., to appear.

46. Nörenberg, W., Nucl. Phys. A409, 191c (1984).

"FROM NONEQUILIBRIUM TO EQUILIBRIUM : A STUDY IN HEAVY ION COLLISIONS"

J.N. De
Nuclear Physics Division, BARC, VECC, Calcutta-700064

S.K. Samaddar
Saha Institute of Nuclear Physics, Calcutta - 700 009

and

K. Krishan
Variable Energy Cyclotron Centre, BARC, Calcutta-700064

ABSTRACT

In a dynamical Monte - Carlo calculation in the framework of the nucleon exchange model, the partitioning of the excitation energy between the target like fragment and the projectile like fragment is calculated for deep inelastic collisions. For small energy losses, the energy division is found to be nearly equal whereas for large energy losses, the energy is partitioned nearly according to mass. This establishes that there is a smooth transition from nonequilibrium to equilibrium, corroborating the recent experimental observations. The effects of shell gap in closed shell nuclei on the dynamical observables are also investigated and are found to be small.

I. INTRODUCTION

The idea of stochastic nucleon exchange was conceived by Ramanna[1)] in the context of fission phenomena. Since then, it has found wide application in the physics of deeply inelastic collisions between two heavy nuclei and has been found to be eminently successful in explaining various characteristics like the inclusive mass and charge distributions[2], isotopic and isobaric distributions[3], flow of total energy and angular momentum from ordered relative motion into disordered intrinsic motion[4] etc. These calculations have usually been done in the zero temperature approximation and in the cases where temperatures of the interacting nuclei have been accounted for, it has been assumed that they are in complete thermal equilibrium. Earlier experimental data[5] on the energy distribution of evaporated neutrons from the primary projectile-like and target-like fragments vindicate such an assumption though a later simplistic dynamical calculation[6] in the nucleon exchange model indicate on the contrary, for asymmetric systems and for incomplete energy damping.

Recently, experimental measurements[7] of the charge centroids for the reaction $^{58}Ni + ^{197}Au$ at 15.3 MeV/A incident energy were done. The observed values were found to be consistently smaller than the predicted values (where evaporation effects were taken into account) based on the assumption of thermal equilibrium. It was however shown that if the excitation energy was assumed to be shared equally between the reacting fragments (non-equilibrium division) rather than according to mass, the calculated results reasonably reproduced the measured values of the charge centroids. An analysis[7] of the neutron multiplicity data[8] for the reaction $^{56}Fe + ^{165}Ho$ at 465 MeV bombarding energy also indicated such a non-equilibrium partition of energy, at least for low energy losses. Such data being sensitive to evaporation effects, Vandenbosch et al[9] recently measured the relative yields of symmetric and asymmetric fission fragments (which is a sensitive function of excitation energy below 60 MeV) from the heavier partner in the reaction $^{56}Fe + ^{238}U$ at 476 MeV and concluded that the results precluded energy sharing based on thermal equilibrium. Energy partitioning in deep inelastic collisions thus seems to be still an open question.

If the driving force for nucleon transfer is neglected, nucleon exchange mechanism would lead one to expect similar flux of neutrons and protons in both nuclei. The excitation energy in both of them are therefore the same and so for asymmetric projectile-target combination, the temperature of the lighter one is expected to be higher. A temperature gradient would therefore be established towards the heavier nucleus, causing larger energy transport (and may be a little more particle transport) from the lighter to the heavier nucleus until equilibrium is reached. A proper accounting of temperature is therefore crucial for the calculation of energy division. The occupation numbers are temperature dependent and therefore the temperature gradient also regulates nucleon flow from one nucleus to the other. Driving force causes a further imbalance in nucleon flux. In this paper, we have made a quantitative investigation in a fully dynamical calculation on the role of these effects in energy division between colliding asymmetric nuclei and have studied its gradual evolution from non-equilibrium to equilibrium. The model of stochastic nucleon exchange has been employed. We also study the temperature dependence of the mass or charge flow and see how it is conditioned due to nuclear structure effect (the presence of a gap in the single-particle spectrum in magic or semi-magic nuclei). In the calculations, the effect of temperature dependent Pauli correlated exchange, the dynamics of the nuclei, the effect of driving force and the penetration of Coulomb and nuclear barriers are taken into account.

In Section II we briefly describe the model for accounting excitation energy in each nucleus and also other observables arising from nucleon exchange. In Section III A&B we compare the experimental data with calculated results for the reactions Fe + Ho and Fe + U at 465 MeV bombarding energy. The self-consistent friction arising from nucleon exchange is compared with the standard proximity friction in Section III C. Lastly, in Section III D the effect of the shell gap in the single particle spectrum on physical observables is examined. We summarise our results in Section IV.

II. MODEL

Each particle transfer generates a hole excitation in the donor nucleus and a particle excitation in the recipient nucleus. In the single particle picture, the hole excitation is given by

$$\Delta E_h = E_F - \tfrac{1}{2}mv_f^2 \qquad .. (1)$$

and the particle excitation is given by

$$\Delta E_p = \tfrac{1}{2}m(\vec{v}_f + \vec{v}_r)^2 - (E_F - w) \qquad .. (2)$$

Here $\vec{v}_f$ is the intrinsic Fermi velocity of the transferred nucleon in the donor frame of reference given by the finite temperature Fermi-Dirac distribution, $\vec{v}_r$ is the relative velocity of the nuclei, m is the nucleon mass, E_F is the Fermi energy of the donor nucleus and w is the macroscopic driving force towards particle transfer. The sum of the hole excitation and particle excitation energy is the elementary excitation energy in the dinuclear complex due to a single nucleon transfer. This excitation energy is generated at the expense of the energy in the relative motion which is therefore damped due to particle transfer. The total particle excitation energy in nucleus B due to transfer of nucleons from nucleus A in a certain reaction time t is thus given by

$$E_p^B = \frac{3}{4\pi v_F^3}\int dt \int d\mathcal{A} \int d\vec{v}_f \, f(\varepsilon_A, T_A)\left[1 - f(\varepsilon_B, T_B)\right] \eta(\vec{v}_f + \vec{v}_r)\, \mathcal{J} \, \Delta E_p \qquad .. (3)$$

In a similar way, an expression for the hole excitation energy can be obtained. In equation (3), the integrals are over the reaction time, the neck area $\mathcal{A}$ and the intrinsic velocities of the nucleons in the donor nucleus. This integral is evaluted in the Monte Carlo method. The energies of the transfered nucleons in system A and B are given by E_A and E_B, V_F is the Fermi Speed, $f(\varepsilon,T)$ is the occupancy at energy ε and temperature T, $\eta(\vec{v})$ is the particle flux and $\mathcal{J}$ is the barrier penetration probability. The change in the neck area, the temperatures, flux, barrier height etc are dynamically governed by the Euler-Lagrange equations which include the proximity[10)] and soft-Coulomb[11)] as the conservative forces and a self-consistent nucleon -exchange controlled friction as the dissipative force. Because of Pauli Correlations, temperature dependent occupation functions and the nuclear and Coulomb barrier, the nucleonic current and the energy loss arising from it may be different from that due to the standard proximity friction[12]. Therefore, in our calculations, whenever there is a particle transfer, the corresponding energy loss is taken into account and thus the dissipative effects on the radial and the tangential motions are accounted for self-consistently. The nuclear barrier is calculated in the proximity prescription[13]. The total barrier (nuclear + coulomb) is approximated to be parabolic and the penetration factor is calculated using the Hill-Wheeler formula. The maximum extension of the neck area at any moment is given by

$$A_{max} = 2\pi \bar{R} (D_{max} - S) \qquad .. (4)$$

where $\bar{R}$ is the reduced radius, S is the surface separation along the line joining the centres of the nuclei and D_{max} = 3.74 fm, a surface to surface distance beyond which no particle transfer takes place. In a given nucleus the sum of the particle and hole excitation energies gives the total excitation energy, part of which is thermal, and part of which goes into collective mode. The temperature of each fragment is then calculated as

$$T_{A,B} = \sqrt{10\ E^*_{A,B}/M_{A,B}} \qquad .. (5)$$

where $M_{A,B}$ is the mass number of either nucleus A or B and E* is the excitation energy except for the collective rotational energy which we calculate from the angular momentum poured in A or B from nucleon exchange.

III. RESULTS

A) Neutron and Proton Distributions

We have initially studied two asymmetric systems, namely $^{56}Fe + ^{165}Ho$ and $^{56}Fe + ^{238}U$, both at incident energies of 465 MeV[8]. The centroids of the charge and neutron distributions are plotted in Fig. 1. The driving force is taken as the liquid drop driving force. Due to the large asymmetry of the Coulomb forces, it is expected that the single-particle proton orbitals of the lighter nucleus would be pushed up more compared to those of the heavier nucleus. Consequently, protons would be driven more towards the heavier nucleus from the

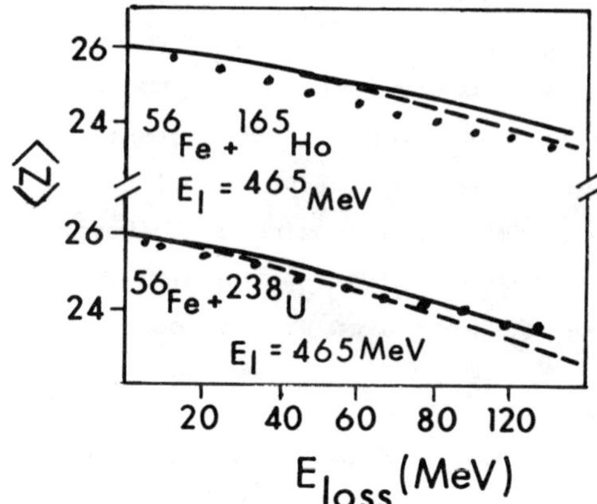

Fig.1(a). The centroid of the charge distribution as a function of energy loss. The full line corresponds to the calculated results with liquid drop(LD) driving force and the dashed line corresponds to those with the inclusion of shell-corrected (LD+SC) driving force. The dots refer to the experimental data.

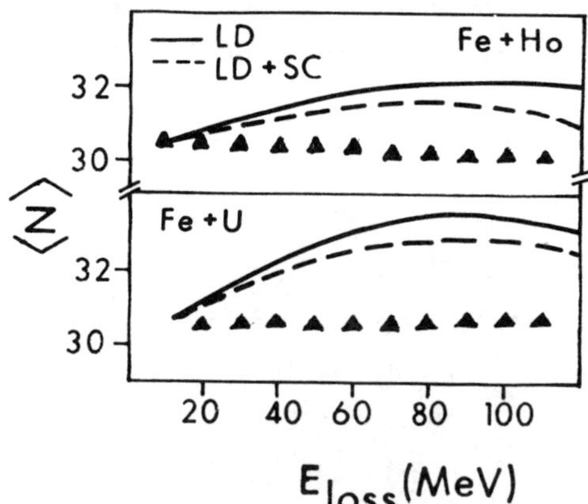

Fig.1(b). The centroid of the neutron distribution as a function of energy loss. The full triangles refer to the experimental data.

lighter nucleus. This effect is correctly reproduced. It is however found that for ^{56}Fe + ^{165}Ho system, the calculated values are around 0.5 units larger than the experimental data, particularly at higher energies. This may be a reflection of the inadequacy of the liquid drop driving force. The shell-corrected driving force[14)] is then incorporated and the fit to the experimental data is improved. It is however curious that the neutron centroids are larger than the experimental results by approximately 1.5 units at larger energy losses when liquid drop driving forces are taken. The gap between the experimental data and the theoretical results however narrows to some extent when the shell corrected driving forces are incorporated.

One cannot, however, escape a few questions at this juncture:
i) Should not the shell effect melt away at high excitations,
ii) Would not the evaporation corrected primary distribution obtained with the assumption of equilibrium division of energy change significantly if nonequilibrium partition is considered, or iii) Is the nucleon flow really kinetic driven rather than being potential driven?[15)]

In Fig. 2 the charge variances for the systems are plotted against energy loss. Variances obtained from the zero-temperature approximation very much underestimate the measured data. When temperature dependence is properly taken into account in the single -particle occupation functions, the experimental data are very well reproduced upto an energy loss of about 120 MeV. This is because of the effective dilution of the Pauli-blocking factors at finite temperature and also a larger opening of phase-space for particle exchange. We however note that the calculations done in the Monte-Carlo simulation method correspond to uncorrelated nucleon transfer; isospin correlation affects the proton or neutron variances very little[2)]. The charge distributions have also been calculated for few other systems. The general agreement is very good indicating the validity of the model.

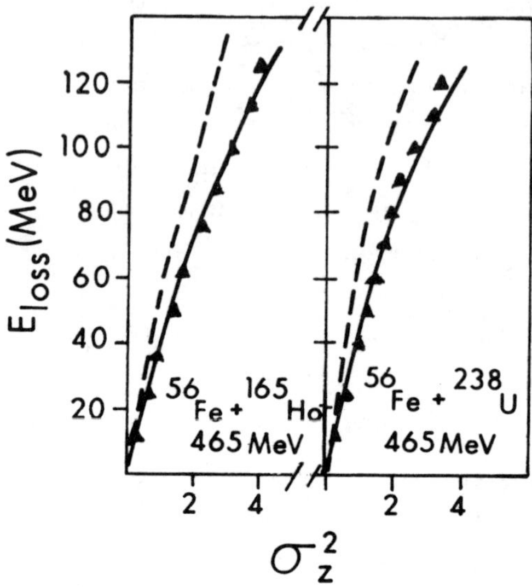

Fig. 2. The variance of the charge distribution as a function of energy loss. The dashed lines correspond to the calculated results at T = 0, the full lines correspond to those at proper finite temperature and the triangles refer to the experimental data.

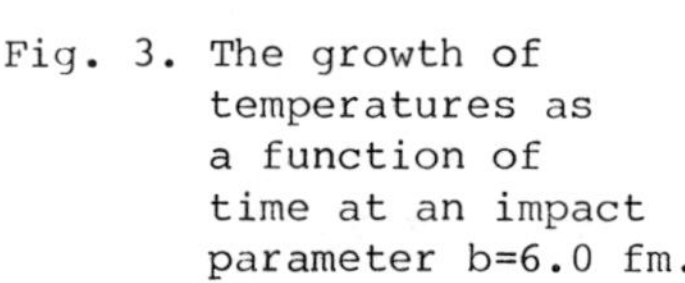
Fig. 3. The growth of temperatures as a function of time at an impact parameter b=6.0 fm.

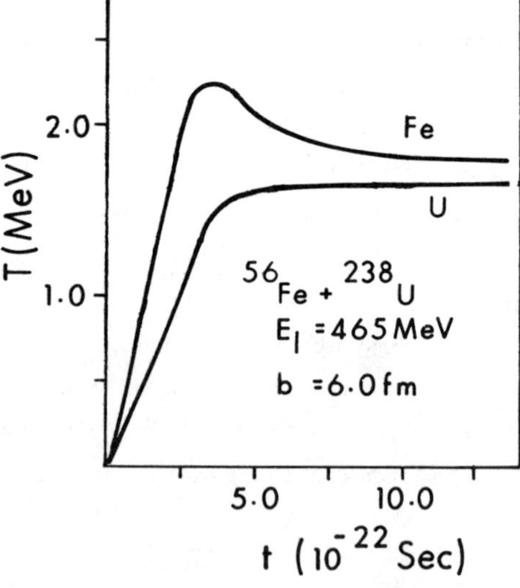

Recently Moretto[16)] has conjectured that the temperature gradient will drive an asymmetric system towards the direction of more mass or charge asymmetry. To test this conjecture, we repeated the calculations by forcing the temperatures of both the fragments to be the same. The thermal feedback was found to have no significant effect on both the drift and the variances.

B) Energy Partitioning

In Fig.3, we show the time evolution of the temperatures of the reacting fragments for the system Fe + U at an impact parameter of b = 6.0 fm which corresponds to a final energy loss of around 80 MeV. The temperatures of both the nuclei increase very quickly in the initial phase. This is, partly because the energy loss for particle transfer is very large at the beginning of the reaction and partly because the neck grows fast allowing for large number of nucleon exchange in a short time. The lighter fragment however gets hotter at a faster rate; this shows that energy is not divided in the ratio of the masses of the nuclei. After a time of $\sim 3.5 \times 10^{-22}$ sec., the temperature of the lighter fragment drops because it delivers more energy to the cooler system than it gets through particle diffusion. The de-excitation due to particle transfer from the tail of the temperature distribution seems to have a greater contribution towards equilibrium than the temperature-gradient-driven net nucleon loss[16)] from the hotter nucleus.

The ratio of the excitation energies for the two systems are displayed in Fig.4 against the total energy loss. For low energy loss, the lighter fragment gets a larger share of the excitation energy. Here the driving force drives more protons from the lighter nucleus to the heavier one due to Coulomb pressure; however this is counter weighted by more preferential transfer of neutrons from the heavier to the lighter nucleus. With increasing energy loss, the relative excitation of the heavier nucleus increases. The thermal gradient set in due to larger temperature of the lighter nucleus causes energy flow from the smaller to the heavier fragment. We see

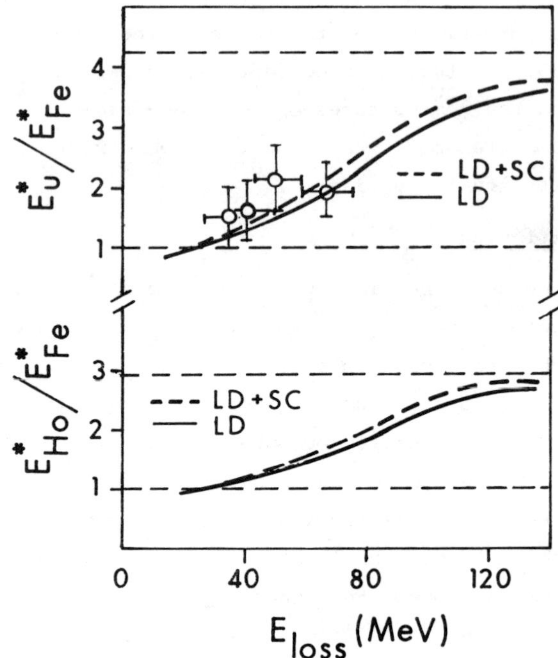

Fig. 4. The ratio of excitation energies as a function of energy loss. LD and LD + SC have the same meaning as in Fig.1(a). The crosses refer to the experimental data. The lower and upper horizontal lines refer to energy division based on equal energy and equal temperature.

that though the longer interaction time drives the system towards thermal equilibrium, complete equilibrium is not reached for the systems we have studied. The more asymmetric system Fe + U is more away from equilibrium as one would expect. The experimental data obtained by Vandenbosch et al[9)] for this particular case from the measurements of mass asymmetry for fission fragments from Uranium nucleus are displayed in the figure. They are in fair agreement with the calculated results. For the Fe + Ho case also, we find a gradual evolution from non-equilibrium energy division to equilibrium energy division in accordance with the experimental observations.

C) Asymmetric Nature of Friction

In all these calculations, the dynamical damping of the trajectory is not governed by the continuous proximity friction; to be self-consistent, whenever particle transfer occurs in the evolving trajectory, the associated energy loss is taken into account to control the radial and tangential velocity of the system. The change in energy loss as a function of time for a particular impact parameter (b=6.0 fm) for Fe + Ho case is plotted in Fig.5 for both the proximity prescription and our model. Though the total energy loss from the proximity friction agrees reasonably well with that obtained from our calculations for any impact parameter, there are some major differences. The rate of change of energy loss from our calculation is higher than that from Randrup's proximity friction in the entrance channel. In the exit channel the behaviour is completely different. Here we get a little energy gain in the relative motion in our calculation rather than energy loss as expected from proximity friction. We believe that this arises because Randrup uses full Fermi-momentum sphere model for nucleon exchange whereas the half Fermi-momentum sphere model[17)] has been used in this calculation.

In the full Fermi sphere model,

$$\frac{dE}{dt} = N'(\varepsilon_F) \langle \Delta E^2 \rangle_F \qquad .. (6)$$

whereas in the half-Fermi sphere model,

$$\frac{dE}{dt} = 2N'(\varepsilon_F)\langle \Delta E^2/(1-e^{-\Delta E/T})\rangle_H \qquad .. (7)$$

Here $N'(\varepsilon_F)$ is the differential one way current of particles at the Fermi surface, $\Delta E = \Delta E_p + \Delta E_h$ is the elementary excitation associated with a particle transfer and T is the temperature of the nuclei (assumed to be same for simplicity). The symbols F and H stand for full and half Fermi sphere flux averaging. For simplified cases, one can show[17)] that for head-on collision, in the entrance channel

$$\frac{dE}{dt}(H) = 2\frac{dE}{dt}(F) \qquad .. (8)$$

For the same condition in the exit channel, whereas dE/dt(F) is symmetric with the entrance channel, dE/dt(H) is of opposite sign and proportional to the temperature of the system provided the temperature is much higher than the elementary excitation energy. This behaviour is manifested in our calculation as is evident from figure 5. One can visualise this situation from the following analogy : when two boats full of stones approach each other with a certain velocity, the motion can be damped if the people on boats start throwing stones towards each other. Each throw gives a back kick and thus retards the motion. Similarly, if people on two receding boats exchange stones, because of the same back kick, the boats gain more relative velocity. This is a consequence of momentum conservation.

The energy gain in the exit channel may be understood in a more microscopic picture in the following way. In the exit channel (for illustration, let us assume head-on collision) the transfer of a particle above the Fermi Surface (the system has reached a finite temperature) goes predominantly to a hole-state below the Fermi Surface of the receipient nucleus. This causes killing of a particle excitation in the donor nucleus and killing of a hole excitation of the recipient nucleus (see figure 6). Thus the energy in the relative motion increases at the cost of the internal excitation energy. A recent study by Pi et al[18)] in the frozen density approximation also arrived at an asymmetric friction in the entrance and the exit channel.

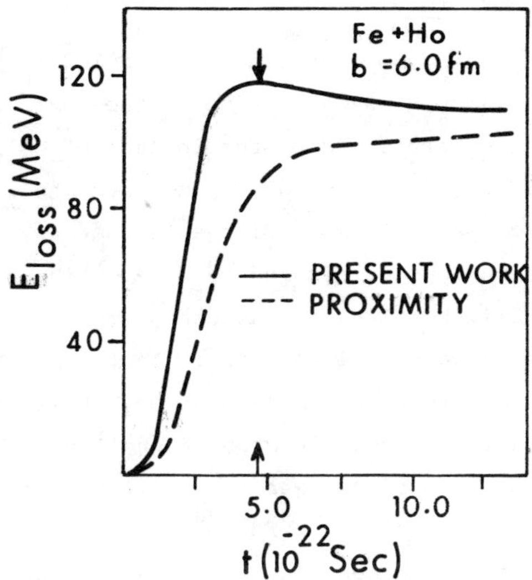

Fig. 5. Growth of energy loss as a function of time for the Fe + Ho system at 465 MeV. The full line corresponds to the present work in the half-Fermi sphere model whereas the dashed line refers to energy loss obtained from proximity friction.

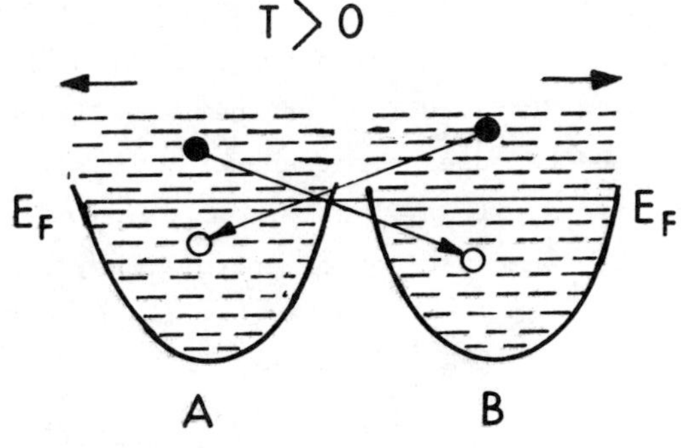

Fig. 6. Schematic representation for nucleon exchange at finite temperature in the exit channel. The open circle referes to hole state and the filled circle refers to particle state.

D) Effect of Shell Structure

The calculations being done in the Monte-Carlo simulation procedure allow for a simple investigation of the influence of the shell gap in the single particle spectrum in case of collisions between closed shell nuclei or closed and unclosed shell nuclei. As an example, we have chosen ^{56}Fe + ^{209}Bi reaction at 465 MeV bombarding energy. The Bismuth nucleus is assumed to behave like a doubly closed shell nucleus (Z = 82+1, N = 126, the effect of the extra proton is neglected) and Fe is an open shell nucleus. In the non-interacting Fermi gas approximation, there is no separation between the occupied states and the unoccupied states; in case of shell nuclei, the single particle spectrum shows a gap (Δ) near the Fermi Surface between the occupied and unoccupied states which for ^{209}Bi has been taken to be 3.5 MeV, for both protons and neutrons. The Fermi Surface is taken at the middle of the gap and the hole excitation and particle excitation energies are measured from this surface. Since there is no density of states in this gap, particles from the donor nucleus having energies corresponding to this gap can not be transferred to the shell nucleus. Similarly, because of the nonavailability of states in the gap, effective phase space for transfer from the shell nucleus is also reduced. We may therefore have a reduced flux of nucleons and a reduced dissipation, since both are interconnected.

In figure 7, we show the growth of the energy loss as a function of time without and with the inclusion of the gap for a peripheral collision (b = 7.0 fm) for the Fe + Bi system. A peripheral collision is chosen because here the energy loss is not very large and therefore, gap effects, if any, would not be washed away by temperature. We find that upto about 30 MeV of energy loss, the presence of the gap does not show itself up meaning that the effective truncation of the phase space is very small here because the relative velocity between the fragments is still large. Only in the last stage, when the fragments slow down, the energy loss is reduced approximately by 12 percent. We also plot the energy loss due to the proximity friction and the

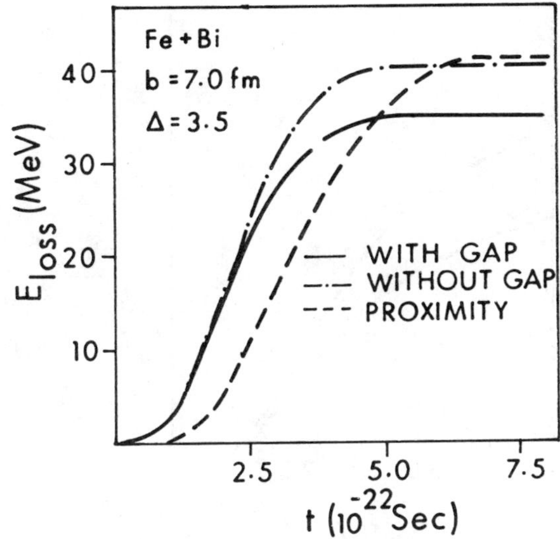

Fig. 7. Growth of energy loss as a function of time with and without the inclusion of the gap Δ in the single particle spectrum.

Fig. 8(a). Energy loss versus impact parameter with and without the inclusion of the gap parameter Δ.

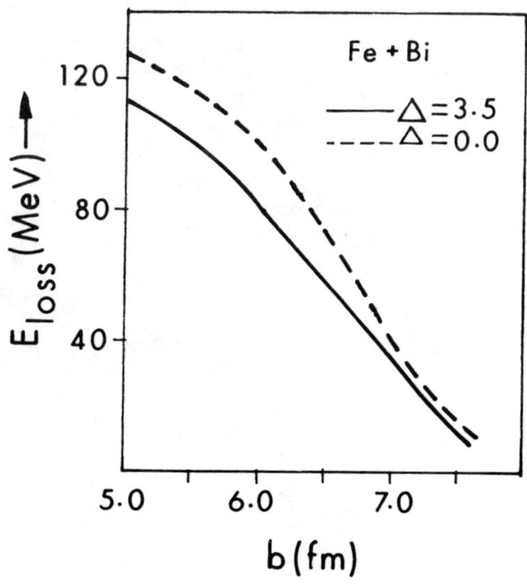

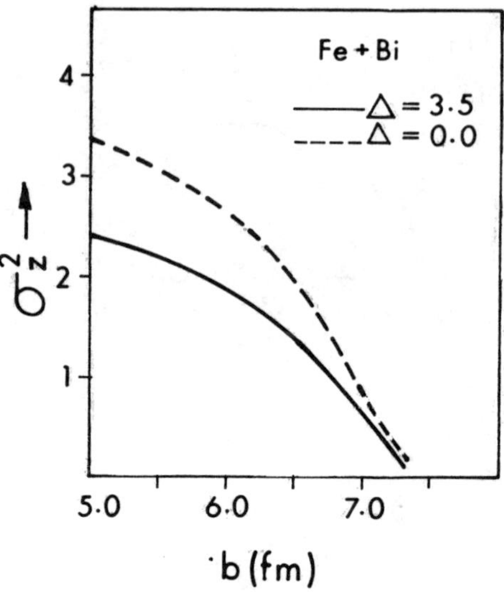

Fig. 8(b). Same as in Fig.8(a) for charge variance.

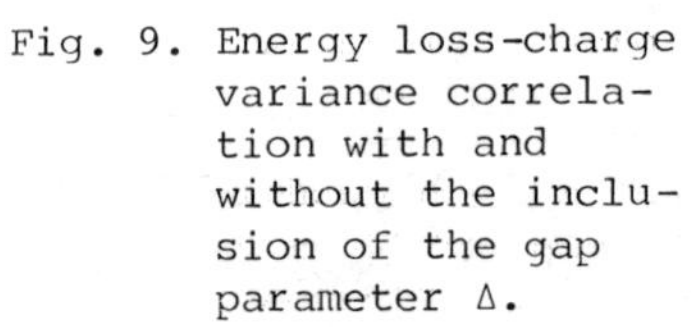

Fig. 9. Energy loss-charge variance correlation with and without the inclusion of the gap parameter Δ.

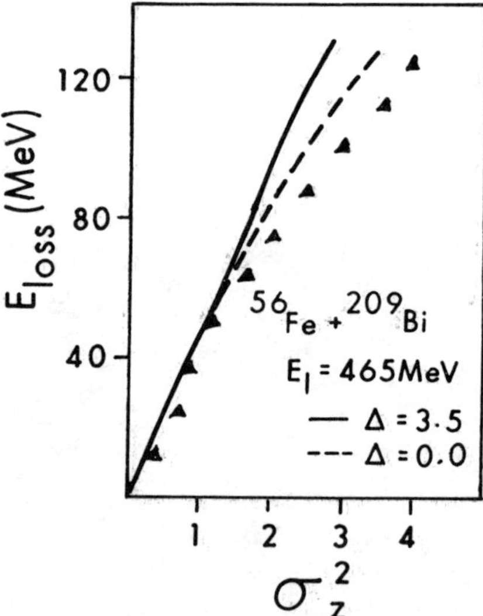

usual features already mentioned are again manifested.

In Fig. 8, we show energy loss and charge variances as a function of the impact parameter. For each b, with the inclusion of gap, the energy loss is reduced, but not much. Even for the lowest impact parameters, when the collisions are very close, the change in energy loss is not significant. For peripheral collisions, the flux is also not affected much. It is reduced for the deeper trajectories. This is expected; here the relative velocity is small for a longer interaction time and thus the particles may not get enough boost to cross over the gap. From these graphs, one can correlate the variance with the energy loss as shown in figure 9 and one sees that though σ_Z^2 and E_{loss} are both reduced due to the gap, E_{loss}- σ_Z^2 correlation is nearly unaffected till an energy loss $\sim$ 80 MeV. Beyond this σ_Z^2 is decreased due to the presence of the gap. On the other hand, in this region, probable deformation of the nuclei may possibly wash away any gap effects. From this study we may thus conclude that the influence of the shell gap on the energy loss-variance correlation is rather small. Such a conclusion was also reached in an earlier nondynamical study[19].

IV. SUMMARY

To summarise, we find that in the nucleon exchange model the colliding asymmetric nuclei are in thermal nonequilibrium for non-central collisions and gradually evolve towards thermal equilibrium for more central collisions[20]. This is supported by the recent experimental data. This observation coupled with the nice reproduction of charge centroids and variances as a function of energy loss brings out the dominant nature of the nucleon exchange mechanism in the deeply inelastic collisions between heavy nuclei. The dissipation function, arising from such nucleon exchanges is found to be highly asymmetric in the entrance and exit channels. This asymmetry is in agreement with recent study done by Pi et al in the frozen density approximation but is in contradiction with Randrup's proximity friction which is symmetric. Finally, from our study, we find that the reacting nuclei do not show any observable cut-out feature from where one can conclude that structure effects are important in the intimate collisions between nuclei.

REFERENCES

1. R. Ramanna, Phys. Lett. 10, 321(1964)

2. S.S. Kapoor and J.N. De, Phys. Rev. C 26, 172(1982)

3. J.N. De, Phys. Lett. 113B, 455(1982)
W.U. Schroder, J.R. Huizenga and J. Randrup, Phys. Lett. 98B, 355(1981)

4. J.N. De and D. Sperber, Phys. Lett. 72B, 293(1978)
H.C. Britt et al, Phys. Rev. C 26, 1999(1982)

5. D. Hilscher et al, Phys. Rev. C 20, 576(1979)
B. Tamain et al, Nucl. Phys. A 330, 253(1979)
Y. Eyal et al, Phys, **Rev.** C 21, 1377 (1980)

6. J. Randrup, Nucl. Phys. A 383, 468(1982)

7. T. Awes et al, Phys. Rev. Lett. 52, 251(1984)

8. H. Breuer et al, Phys. Rev. C 28, 1080(1983)

9. R. Vandenbosch et al, Phys. Rev. Lett. 52, 1964(1984)

10. J. Blocki et al, Ann. Phys. (N.Y.) 105, 427(1977)

11. J. P. Bondorf, M.I. Sobel and D. Sperber, Phys. Rep. 15C, 83(1974)

12. J. Randrup, Ann. Phys. (N.Y.) 112, 356(1978)

13. J. Randrup, Nucl. Phys. A 307, 319(1978)

14. W.D. Myers and W.J. Swiatecki, Ark. Fys. 36, 343(1967)

15. J.J. Griffin and W. Broniowski, Nucl. Phys. A 428, 145(1984)

16. L. Moretto, Z. Phys. A 310, 61(1983)

17. J.N. De, S.K. Kataria, S.S. Kapoor and V.S. Ramamurthy, BARC Preprint (1983)

18. M.Pi et al, Nucl. Phys. A 426, 163(1984)

19. J.N. De and S.S. Kapoor, Nucl. Phys. A (in Press)

20. S.K. Samaddar, J.N. De and K. Krishan, Phys. Rev. C (in Press)

PRE-EQUILIBRIUM FISSION: A NEW DECAY CHANNEL FOR COMPOSITE SYSTEMS FORMED IN FUSION REACTIONS

V.S.Ramamurthy
Bhabha Atomic Research Centre, Trombay
Bombay 400085
INDIA

ABSTRACT

In nuclear collisions, it is generally believed that if the target and the projectile can be brought to a configuration more compact than a critical configuration for fusion , they fuse. If however, the fused composite system has a vanishing fission barrier or a fission barrier which is more compact than the critical configuration for fusion, the system will undergo fission in a very short interval of time (fast and quasi fission) and no compound nucleus will be formed. We have recently shown from an analysis of fragment angular distributions in heavy ion induced fusion reactions that even in those cases when no fast or quasi fission events are expected, a fraction of the fission events take place , before the formation of the compound nucleus, in a time scale of the order of a few times 10^{-21} seconds. In analogy with the well known pre-equilibrium particle emission in similar time scales, we call these fission events pre-equilibrium fission. The importance of fission events following incomplete fusion in an analysis of fragment angular distributions measured in reactions involving highly fissile actinides as one of the collision partners is also pointed out.

1. INTRODUCTION

Complete fusion of the target and the projectile and the subsequent formation of a fully eguilibrated compound nucleus is an important reaction channel in nucleus-nucleus collisions. In the framework of the dynamical model of Swiatecki[1], these two processes are governed bythe existence of a few milestone configurations in a multidimensional potential energy surface . Figure 1 shows a schematic illustration of the different reaction channels which are obtained in this model in a typical nucleus-nucleus collision. For a given bombarding energy, if the collision trajectory does not reach the contact configuration, one gets elastic and quasi-elastic processes. If the target and the projectile make contact but do not reach the conditional saddle configuration of fixed mass asymmetry, we obtain the deep inelastic reactions. If however, the conditional saddle point is reached, there is fusion but not necessarily compound nucleus formation. For compound nucleus formation, the system should further cross the unconditional fission saddle npoint. It is generally believed that if the target and the projectile are brought to a configuration more compact than the unconditional fission saddle point, the composite system will automatically evolve towards a fully equilibrated compound nucleus. If, however, the system does not reach the unconditional fission saddle point or the composite system has a vanishing fission barrier, it will reseparate into two fragments without the formation of a compound nucleus. The names, quasi fission[2] and fast fission[3], have been suggested for these reaction channels. In the partial wave representation, the different l-windows associated with the different reaction channels are as shown in Fig.2. For a given bombarding energy, fusion is characterized by a critical angular momentum I_c, all partial waves beyond which do not lead to fusion. If I_c is smaller than the rotational liquid drop model limitinmg angular momentum I_F^{RLDM} for vanishing fission barriers, fusion results also in the formation of an equilibrated compound nucleus. If, on the otherhand, I_c is

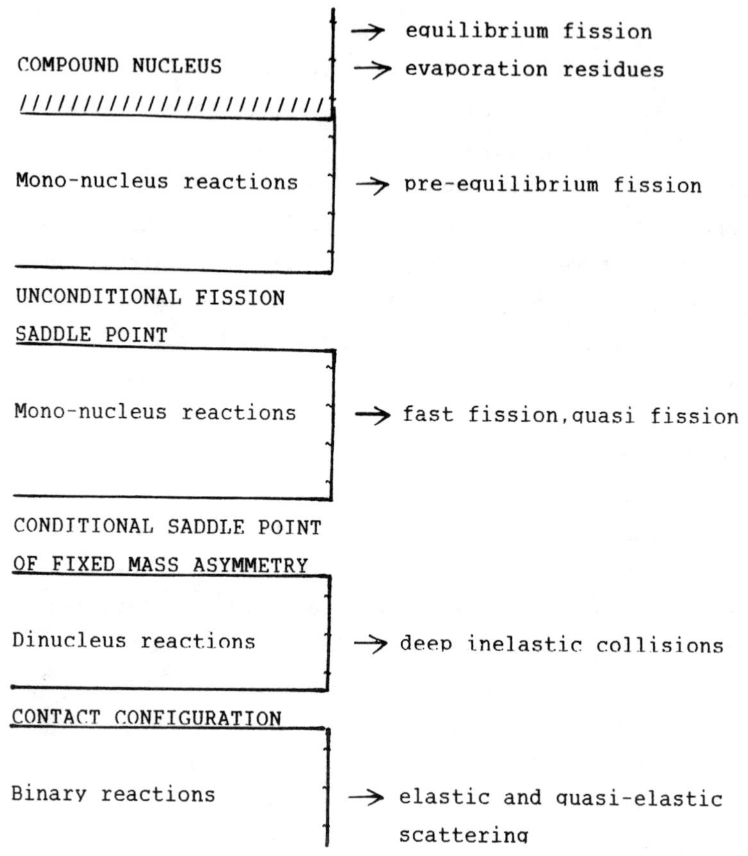

Fig.1. Schematic illustration of different reaction channels in heavy ion collisions in the framework of the dynamical model of Swiatecki[1].

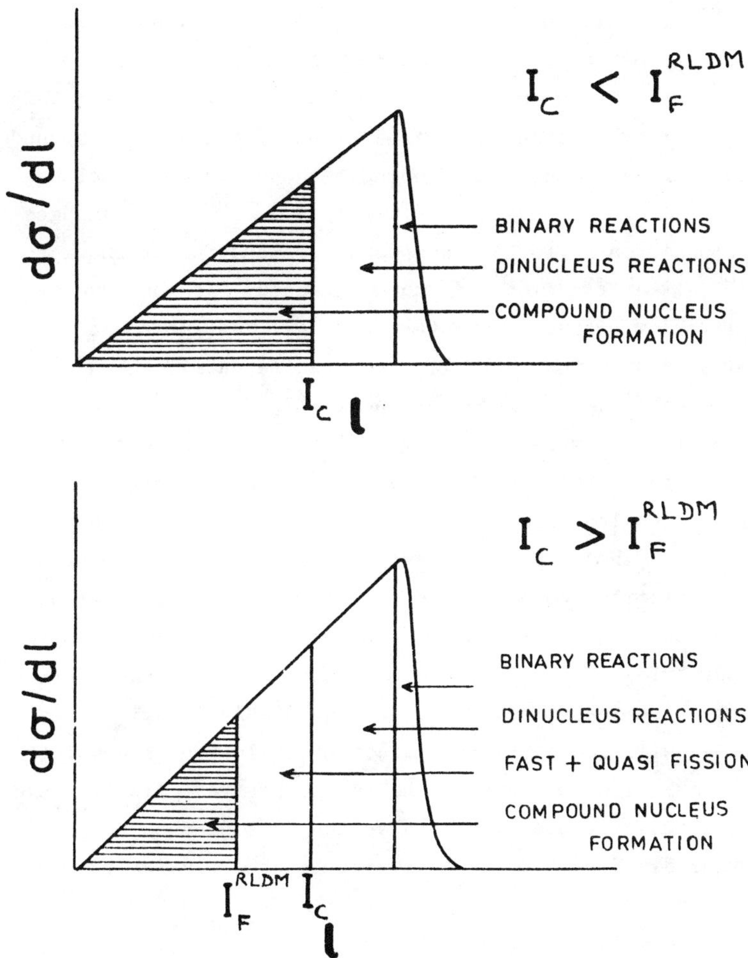

Fig.2. Schematic representation of the different l-windows associated with the reaction channels shown in Fig.1.

greater than I_F^{RLDM}, all partial waves in the l-window $I_F^{RLDM} \leqslant l \leqslant I_c$ result in fast fission and only partial waves with l less than I_F^{RLDM} result in the formation of a compound nucleus. If the fissility of the compound nucleus is large, fission dominates the decay of the compound nucleus. There is atpresent no experimental way of discriminating fission events following compound nucleus formation and those not going through compound nucleus formation. Therefore, in any experiment, if $I_c \leqslant I_F^{RLDM}$, one assumes that all fission events follow compound nucleus formation whereas if $I_c \geqslant I_F^{RLDM}$, a fraction of the detected fission events are non-equilibrium fission events such as fast fission and can be estimated only theoretically. We have recently shown[4-5] that even for those partial waves having $l \leqslant I_F^{RLDM}$, an appreciable fraction of the fused composite systems undergo fission in a very short time scale, comparable to the characteristic equilibration times of the compound nucleus and therefore before the formation of the compound nucleus. These fission events also have properties distincly different from those of fission events following compound nucleus formation. In analogy with the well known pre-equilibrium emission of particles, we call these events as pre-equilibrium fission events. In reactions involving thge highly fissile actinides as one of the collision partners, fission following incomplete momentum transfer also contribute to the observed fission events. The importance of these events in an analysis of the measured fragment angular distributions in heavy ion induced fission reactions is also demostrated.

2. FRAGMENT ANGULAR DISTRIBUTIONS IN HEAVY ION INDUCED FUSION-FISSION REACTIONS

Anomalous fragment angular distributions have recently been observed in a number of heavy ion induced fission reactions[6-12]. The data is usually analysed in terms of the standard theory[13-14] of fragment angular distributions in fission following compound nucleus formation. The discrepancy is often ascribed[15-16] to a possible breakdown of the standard theory or to the emergence[6] of the new reaction mechanisms such as fast or quasi fission. However, no single quantitative theory exists at present to explain all the measured distributions. We have recently proposed[4-5] a model based on the suggestion that in heavy ion induced fusion reactions, the observed fission events consist of an admixture of events of two types: (i) fission following the formation of a fully equilibrated compound nucleus and (ii) fission of a composite system which has equilibrated in all degrees of freedom except the K degree of freedom, where K is the projection of the angular momentum on the symmetry axis, which is also identified as the fission axis. Reaction mechanisms such as fast fission taking place for the case of composite systems with zero fission barriers and quasi-fission taking place for composite systems with fission barrier shapes more compact than the critical configuration for fusion are cases of reaction mechanisms of the type (ii) mentioned above. Another class of fission events which can contribute to non-compound nucleus forming fission events of type (ii) and which have not been considered earlier are those fission events occuring in a time scale comparable to the characteristic relaxation time in the K degree of freedom when the fission barrier heights are of the order of the intrinsic temperature of the composite system. The proposed model took into account the fission events of both types (i) and (ii) , each having its own characteristic angular distribution. It was shown that the data for a number of systems spanning a wide range of bombarding energies are well reproduced by the model.

For the fission events of type (i), we use the conventional theory[13-14] of the fragment angular distributions in fission following compound nucleus formation with the rotating liquid drop model extension[17] to high spins. We however include the modifications proposed recently by us[18] to take into account the dependence of the transition state energy and shape both on the magnitude of the spin I and its projection K on the symmetry axis. We have also made the reasonable assumption for heavy composite systems that all compound nucleus fission events correspond to first chance fission.

There exists at present no theoretical formulation for calculating the fragment angular distributions for the type(ii) events. However, based on some recent experiments[8] involving very high spins and large values of Z^2/A of the composite system where the compound nucleus formation probability is very small and highly anisotropic angular distributions have been measured, we had proposed[4-5] that the effective K distribution for all non-compound nucleus forming fission events may be represented by a narrow gaussian with a variance σ_K given by

$$\sigma_K{}^2 = I^2 \ \sigma_\theta{}^2$$

where σ_θ is the angular variance representing the misalignment of the symmetry axis of the fused composite system with respect to the K=0 plane. It was further assumed that σ_θ is nearly constant for all systems and bombarding energies considered in the analysis.

For a calculation of the relative probabilities of fission following compound nucleus formation and non-compound nucleus forming frission events, one first considers a fused composite system with an angular momentum I and temperature T. If I is larger than the rotational liquid drop model limit I_F^{RLDM} for vanishing fission barriers, no compound nucleus will be formed and the fused composite system will undergo fast fission with unit probability. If however, I is less than I_F^{RLDM}, one has to consider the dynamics between the fission saddle point and the compound nucleus. Let us assume that the system has been brought to a configuration more compact than the unconditional fission saddle point. This system is also initially formed predominantly in a K=0 configuration with a small width in the K distribution. Let $B_F(I,K)$ be the height of the barrier preventing this system from a fast binary split(fast fission). Left to itself, the system will then relax in the shape and orientation degrees of freedom to form a fully equilibrated compound nucleus. However, if the fission time is comparable to the characteristic time τ for K relaxation, the composite system has a finite probability of undergoing fission in a time shorter than the K equilibration time and the fragment angular distribution of such events will carry the memory of the entrance channel K distribution. For a given $B_F(I,K)$, one can calculate the fission probability per unit time making use of the Bohr-Wheeler transition state theory[19]. In principle, the evolution of the K distribution is continuous and the effective K distribution for fission events taking place at different times are different. However, for the sake of simplicity we assume that only those composite systems that survive fission for a time longer than τ will result in the formation of a fully equilibrated compound nucleus which subsequently undergoes fission while all fission events taking place in a time less than τ carry a memory of the entrance channel K distribution and have angular distributions similar those of the fast fission events. In anolagy with the emission of fast nucleons in nuclear reactions in time scales much shorter than the compound nuclear life times, we call these fission events competing with compound nucleus formation as pre-equilibrium fission events. We have fitted the experimental data on the fragment angular distributions in terms of the model described above for a number of systems. Figures 3 and 4 show a comparison of the calculated fragment angular distributions and the experimental data for a few typical systems involving different projectiles on the same target and same projectile on different targets. It was found that all the data shown in the figures can be fitted with a single set of the parameters corresponding to $\tau = 8\times10^{-21}$ seconds and $\sigma_\theta^2 = 0.06$. Figure 5 shows the different ial cross sections for fission following compound nucleus formation, pre-equilibrium fission and fast fission versus the entrance channel angular momentum l for the two typical cases of $^{19}F+^{208}Pb$ and $^{32}S+^{208}Pb$ as deduced from the above analysis. It can be seen that true compound nucleus formation is considerably reduced over a significant range of l values even below I_F^{RLDM}.

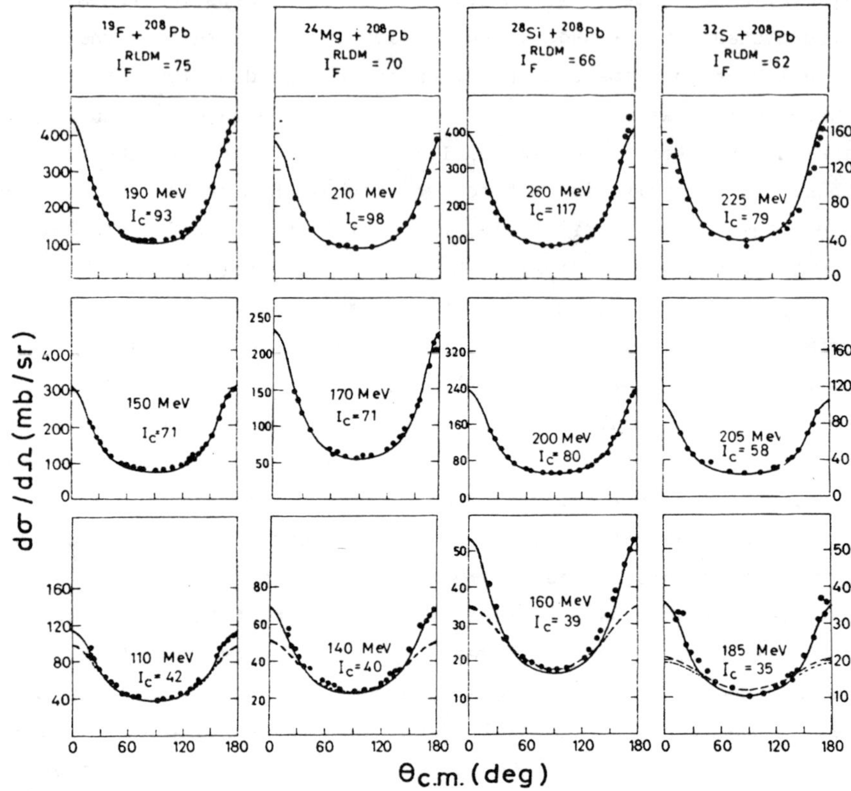

Fig.3. Fission fragment angular distributions for the $^{19}F+^{208}Pb$, $^{24}Mg+^{208}Pb$, $^{28}Si+^{208}Pb$ and $^{32}S+^{208}Pb$ reactions.The experimental points are from Ref.1 and 2. The continuous curves are the results of the present calculations. The dashed lines are the calculated fragment angular distributions based on the statistical theory for fission following compound nucleus formation based on the flexible rotor model[18]. The dotted line for the case of $^{32}S+^{208}Pb$ (185 MeV) is the result of the calculation based on the K-independent RLDM fission transition state shape.

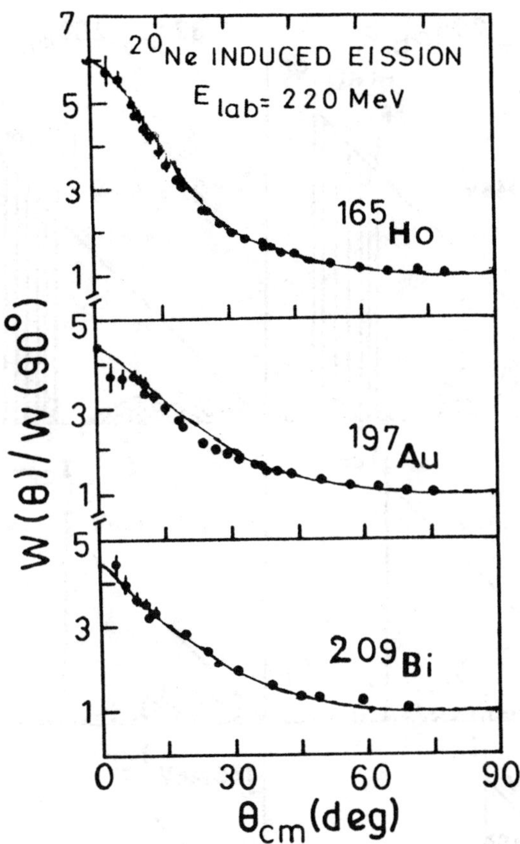

Fig.4. Fission fragment angular distributions for the $^{20}Ne+^{165}Ho$, $^{20}Ne+^{197}Au$ and $^{20}Ne+^{209}Bi$ reactions. The experimental points are from Ref.10. The continuous curves are the results of the present calculations.

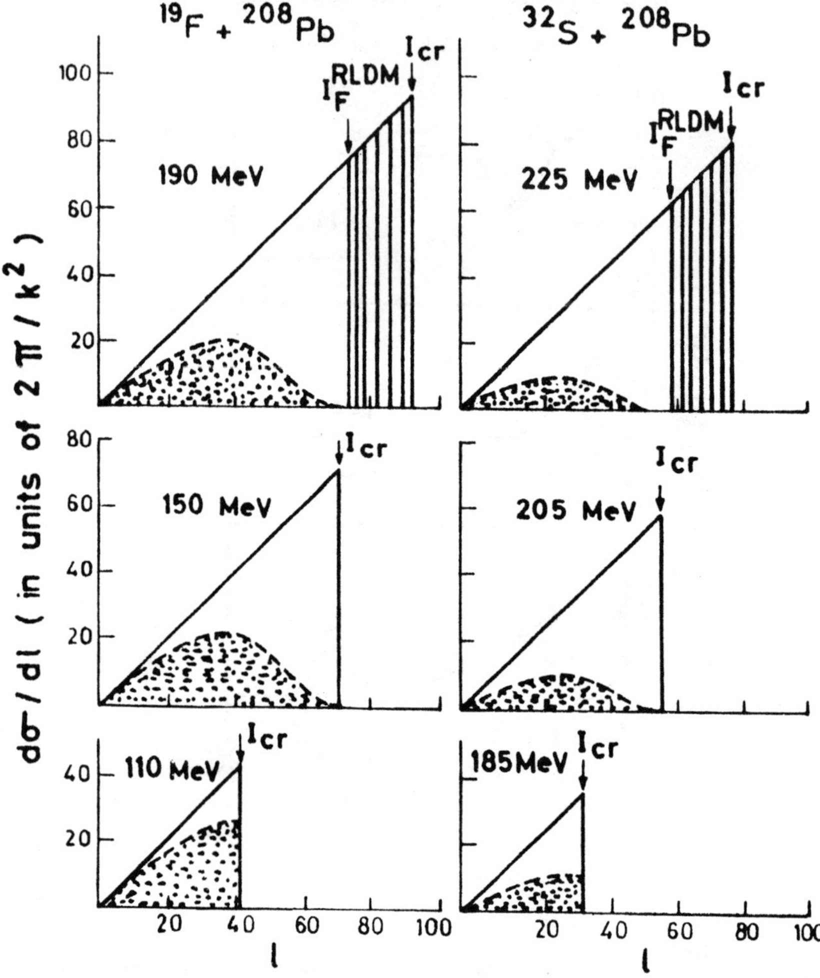

Fig.5. Differential cross sections for fission versus l in units of $2\pi/k^2$ where k is the entrance channel momentum. The shaded area represents fission following compound nucleus formation as deduced from the present analysis. The vertical lines define the angular momentum window for fast fission.

3. FISSION FOLLOWING INCOMPLETE FUSION

We have so far considered only systems inwhich both the target and the projectile are relatively nonfissile. However, there exist at present a number of measurenents of fission fragment angular distributions for systems inwhich one of the colliding partners is an actinide having a high fissility. As has been pointed out by Vaz and Alexander[20], in such systems, fission following incomplete fusion might contribute substantially to the observed fission events, unless special efforts are made to eliminate these by ensuring full momentum transfer. For reasons of experimental difficulties, most experiments do not incorporate this discrimination or do not provide full discrimination. In a theoretical analysis of the experimental fission fragment angular distributions in such cases, it is therefore necessary to take into account the presence of fissions following incomplete fusion.

There is at present no quantitative theory for calculating the yield and the fragment angular distribution of fission events following incomplete fusion. We have therefore adopted the following procedure. For a given target, projectile and bombarding energy, the complete fusion cross section is calculated using thestandard one dimensional potential barrier model[21]. This model is expected to provide a adequate description of the fusion cross sections systematics as long as the bombarding energies are close to the interactioin barriers. The parameters of the interaction barrier are obtained from the energy density model of Ngo and Ngo[22] fitted to satisfactorily reproduce the experimental interaction barrier heights and radii deduced from experimental fusion cross sections using the analysis of Vaz and Alexander[23]. Fission cross sections measdured in exess of the calculated fusion cross sections are then ascribed to fission following incomplete fusion. There are a number of factors which seem to indicate that the fragment angular distributions in fission following incomplete fusion are nealy isotropic. For example, if fission follows complete equilibration of the heavy reaction product in an incomplete fusion event, the fission fragments are expected to exhibit only a small anisotropy since the fissioning nucleus is highly fissile and therefore has a compact fission transition state shape. On the other hand, if fission occurs before equilibration of the heavy reaction product, because of the three body nature of the incomplete fusion event, the polarisation of spin I and its projection K on the symmetry axis of the fissioning nucleus is expected to be quite weak which again results in a weak anisotropy of the fragments. In the absence of a more formal theory, we make the extreme but reasonable asasumption that the fragment angular distributions in fission following incomplete fusion are isotropic. The calculated angular distributions are then comparecd with the experiments. Figures 6 and 7 show comparisons of the calculated fragment angular distributions with the experimental data for a few typical cases involving highly fissile actinides either as the target or as the projectile. It should however be mentioned that no free parameters are involved in the present comparisons with data. The very good agreement seen in the figures clearly demonstrate the conjecture of Vaz and Alexander that fission following incomplete fusion contributes substantially to the observed fission events and should be considered in any meaningful analysis of the angular distribution data.

4. SUMMARY

In summary, we have shown that, with the inclusion of pre-equilibrium fission and fission following incomplete fusion in addition to the normal equilibrium fusion-fission events, we are in a position to understand in a quantitative way the host of apparently anomalous fragment angular distributions measured in a number of heavy ion induced fission reactions.The analysis also brings out important new information on the dynamics of compound nucleus formation, namely, the equilibratrion time in the K degree of freedom for the fused composite system.

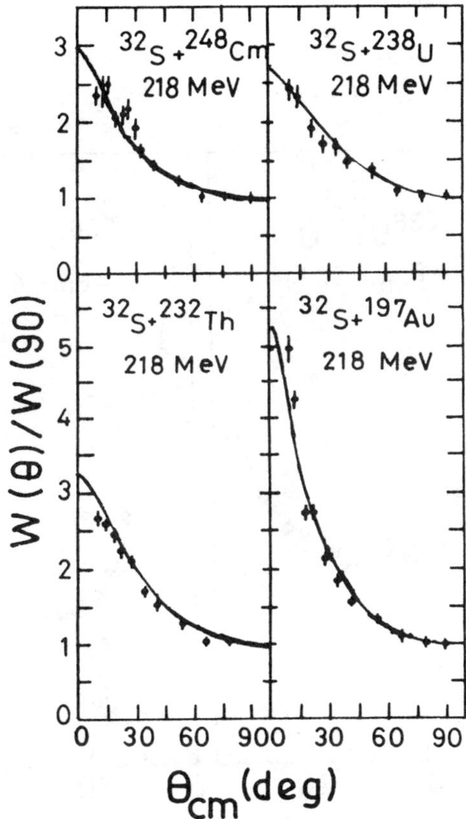

Fig.6. Fission fragment angular distributions for the $^{32}S+^{197}Au$, $^{32}S+^{232}Th$, $^{32}S+^{238}U$ and $^{32}S+^{248}Cm$ reactions. The experimental points are from Ref.9. The continuous curves are the results of the present calculations.

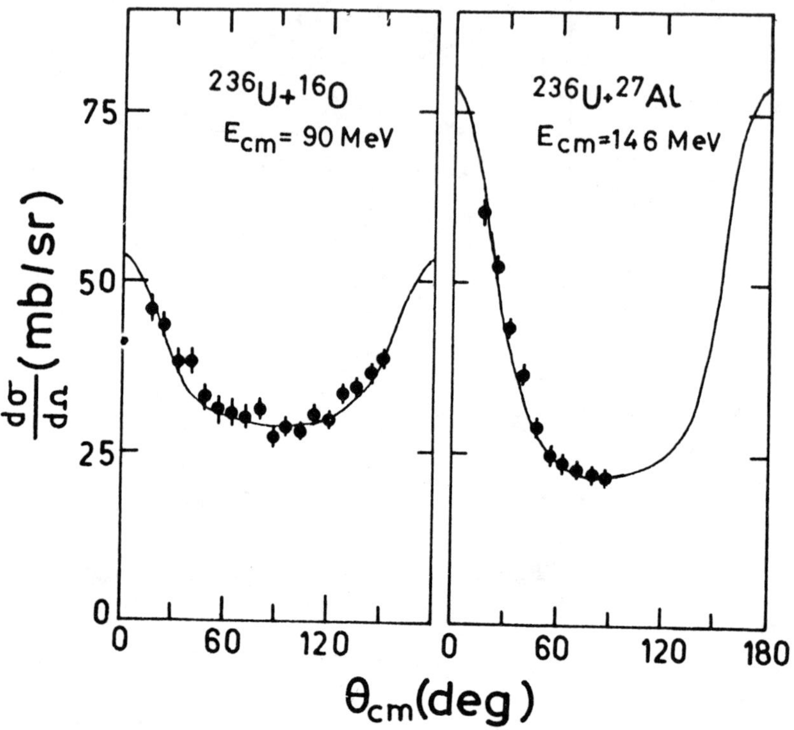

Fig.7. Fission fragment angular distributions for the ^{238}U+^{16}O and ^{238}U+^{27}Al reactions. The experimental points are from Ref.24. The continuous curves are the results of the present calculations.

I am thankful to Drs. R.Ramanna and S.S.Kapoor in collaboration with whom the investigations presented here were carried out.

REFERENCES

1. W.J.Swiatecki,Nucl.Phys.,A376(1982)275.
   S.Bjornholm and W.J.Swiatecki,Nucl.Phys.,A391(1982)471.
2. W.J.Swiatecki,Phys.Scr.,24(1981)113.
3. C.Gregoire,C.Ngo and V.Remaud,Phys.Lett.,99B(1981)17.
4. S.S.Kapoor,V.S.Ramamurthy and R.Ramanna,Pramana22(1984)275.
5. V.S.Ramamurthy and S.S.Kapoor,Phys.Rev.Lett.,54(1985)178.
6. B.B.Back et al.,Phys.Rev.Lett.,50(1983)818.
7. M.B.Tsang et al.,Phys.Lett.,129B(1983)18.
8. K.T.Lesko et al.,Phys.Rev.,C27(1983)2999.
9. B.B.Back et al.,Phys.Rev.Lett.,46(1981)1068.
10. H.Rossner et al.,Phys.Rev.,C27(1983)2666.
11. M.B.Tsang et al.,Phys.Rev.,C28(1983)747.
12. A.Gavron et al.,Phys.Rev.Lett.,52(1984)589.
13. A.Bohr,Proc. of the Int. Conf. on the Peaceful Uses of Atomic Energy,Geneva,1955(United Nations), Vol.2,page 131.
14. I.Halpern,V.M.Strutinski,Proc. of the Int. Conf. on the Peaceful uses of Atomic Energy,Geneva,1958(United Nations). Vol. 15,page 408.
15. P.D.Bond,Phys.Rev.Lett.,52(1984)414.
16. H.Rossner et al.,Phys.Rev.Lett.,53(1984)38.
17. S.Cohen,F.Plasil and W.J.Swiatecki,Ann.phys.(NY),82(1974)557.
18. M.Prakash et al.,Phys.Rev.Lett.,52(1984)990.
19. N.Bohr and J.A.Wheeler,Phys.Rev.56(1939)426.
20. L.C.Vaz and J.M.Alexander,Phys.Rev.Lett.,52(1984)396.
21. W.U.Schroder and J.R.Huizenga,Ann.Rev.Nucl.Sci.,27(1977)465.
    M.Lefort and C.Ngo,Ann.Phys.(Paris)3(1978)5.
22. H.Ngo and C.Ngo,Nucl.Phys.,A348(1980)140.
23. L.C.Vaz and J.M.Alexander,Phys.Rep.,5(1981)373.
24. J.Toke et.al.,Phys.Lett.,142B(1984)258.

RECENT ADVANCES IN HEAVY-ION-INDUCED FISSION

F. Plasil

Oak Ridge National Laboratory,* Oak Ridge, Tennessee 37831, U.S.A.

1. INTRODUCTION

It is an honor for me to be able to participate in the commemorative session on the occasion of the 60th birthday of Dr. Raja Ramanna, Chairman of the Indian Atomic Energy Commission, and a pleasure to know that we share a long-standing interest in the problem of nuclear fission. In this paper, I would like to cover three topics. The first deals with results that have been published recently on angular-momentum-dependent fission barriers.[1,2] I discuss them here because of the significance that we attach to them. We feel that, after a decade of study and controversy, we have arrived at a quantitative understanding of the competition between heavy-ion-induced fission and particle emission from compound nuclei at relatively low bombarding energies. The second topic concerns the extension of our heavy-ion-induced fission studies to higher energies.[3] It is clear that in this regime the effects, both of fission following incomplete fusion and of "extra-push" requirements, need to be considered. Finally, in Section 4, I will discuss our recent conclusions concerning the fissionlike decay of products from reactions between two ^{58}Ni nuclei at an incident energy, E/A, of 15.3 MeV, as well as the impact of our findings on the conclusions drawn from previous, similar measurements.[4-7]

2. ANGULAR-MOMENTUM-DEPENDENT FISSION BARRIERS

During the course of the last twenty years, a large number of heavy-ion-induced fission excitation functions have been measured.[8-19] The main purpose of most of these studies was to obtain quantitative

---

*Operated by Martin Marietta Energy Systems, Inc., under contract DE-AC05-84OR21400 with the U. S. Department of Energy.

information on the fission barriers of the compound nuclei formed in the various reactions. Usually, this information was extracted from the data within the framework of the statistical model by means of two-parameter fits. One of these parameters was related to the fission barrier, B_f, and the other was the ratio of the Fermi-gas level density parameter for fission to that for particle emission, a_f/a_n. The values of these extracted parameters, however, were determined in relatively narrow regions of angular momentum, where the fission barriers are similar to the binding energies of the evaporated particles. It was not possible to extract the angular-momentum dependence of the fission barriers directly from the data, and this dependence was usually introduced in the form of a theoretical model, such as the rotating-liquid-drop model (RLDM).[20] Typically, RLDM fission barriers, $B_f^{LD}(J)$, were incorporated in the statistical model code, and the parameter adjustment was then made according to some arbitrary prescription, such as $B_f(J) = k_f B_f^{LD}(J)$, where k_f is the adjustable constant. In almost all cases, for an adequate description of the excitation functions, it was found that $B_f(J) < B_f^{LD}(J)$, and k_f was found to vary from about 0.55 to 0.85.[11,12,14-19] Since for nonrotating nuclei, more realistic calculations, which take into account the diffuseness of the nuclear surface and the finite range of the nuclear force,[21,22] result in lower fission barriers than those of the liquid drop model, it was suspected[11,12] that if such effects were included in the calculated angular-momentum-dependent fission barriers, quantitative agreement between experiment and theory might result.

Recently, the required rotating-finite-range model (RFRM) has been developed independently by Mustafa et al.[23] and by Sierk.[24] In addition to corrections to the surface energy, the rotational energy has been corrected for effects of the diffuseness of the matter distribution, and the Coulomb energy has been corrected for effects of the diffuseness of the charge distribution. Furthermore, the shapes in the RFRM are not constrained to axial symmetry as was the case in the RLDM calculations. The results are illustrated in Fig. 1. Calculated $B_f(J)$ values are shown for ^{153}Tb, both for RLDM and for RFRM barriers. As expected, the RFRM values lie significantly below the RLDM values.

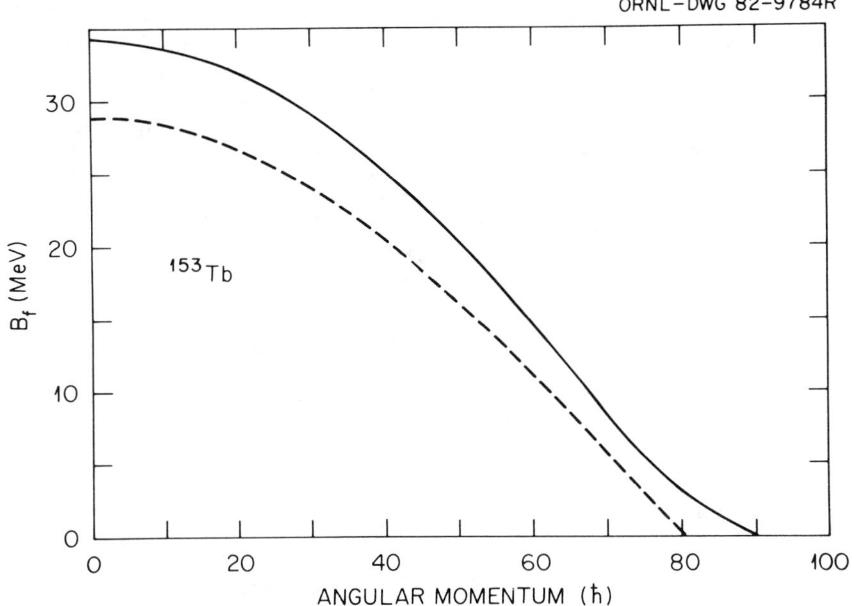

Fig. 1. Calculated fission barrier of ^{153}Tb as a function of angular momentum. The solid curve is from the rotating-liquid-drop model, and the dashed curve from the rotating-finite-range model.

We have made use of the new RFRM $B_f(J)$ results in the analysis of a large number of fission excitation functions which we have obtained during the course of a systematic investigation of heavy-ion-induced fission over a broad range of excitation energy, angular momentum, and compound nucleus mass.[1,2] The systems studied were ^{153}Tb, ^{158}Er, ^{181}Re, ^{186}Os, and ^{204,206,208,210}Po. For the statistical model analysis of the data, evaporation residue excitation functions are needed in addition to fission cross sections. These were measured in the case of the ^{153}Tb and ^{181}Re (Ref. 2) systems, and were found to be consistent with fusion cross sections calculated with the Bass model.[25] For the ^{158}Er, ^{186}Os, and the Po systems, we used evaporation residue cross sections deduced from the Bass model.[25] With the exception of some of the Po systems, two to five different reactions were used to produce the same compound nucleus. The projectiles ranged from ^{9}Be to ^{64}Ni. The various reactions are indicated in Fig. 2

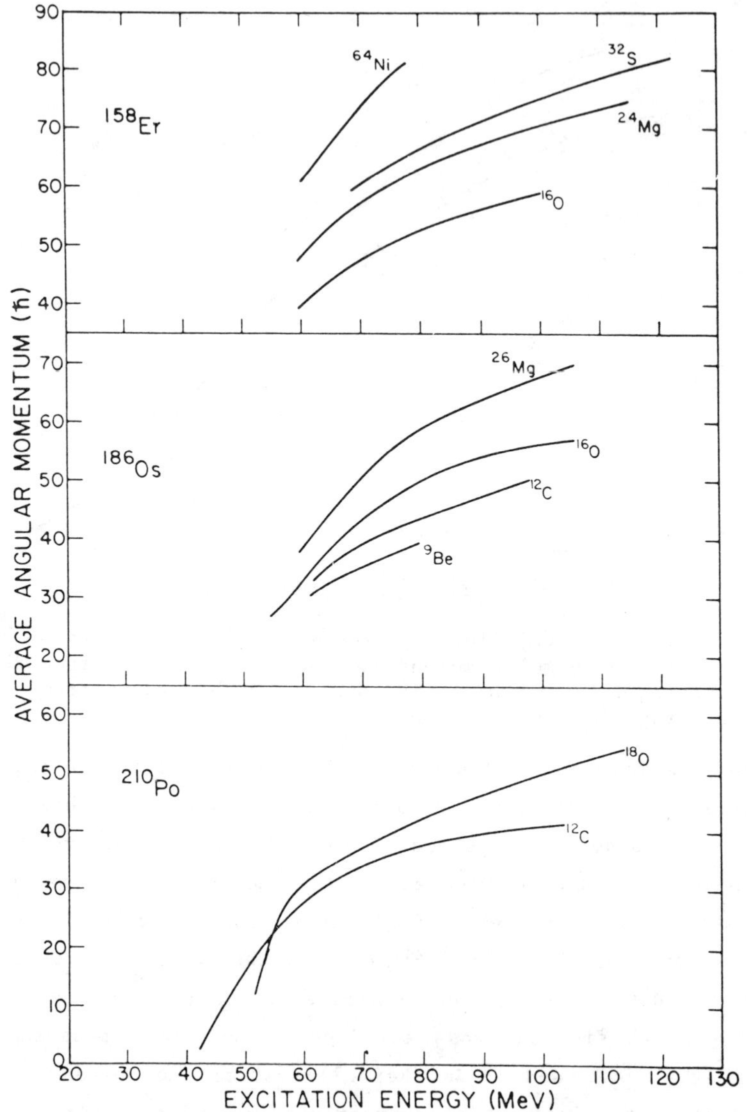

Fig. 2. Average angular momenta (from statistical model calculations) leading to fission as a function of the compound nucleus excitation energy for the ^{158}Er, ^{186}Os, and ^{210}Po systems. The curves indicate the range of excitation energy that our measurements covered, and the labels refer to the ions that were used to produce the compound nuclei.

for the ^{158}Er, ^{186}Os, and ^{210}Po systems, together with the relevant ranges of excitation energy and of the average angular momentum leading to fission.

The RFRM $B_f(J)$ values obtained from Ref. 24 were inserted into the statistical model analysis calculations and were not adjusted in any way. This reduced the fitting procedure to one that involved only a single adjustable parameter, a_f/a_n, whose value was constrained to be constant for any given compound nucleus. A typical set of measured excitation functions, together with calculated values, is shown in Fig. 3 for the ^{158}Er system. The adjusted value of a_f/a_n turned out to be 1.00 for this case. The quality of the fits in all other cases was similar to that of Fig. 3. It can be seen that the calculated

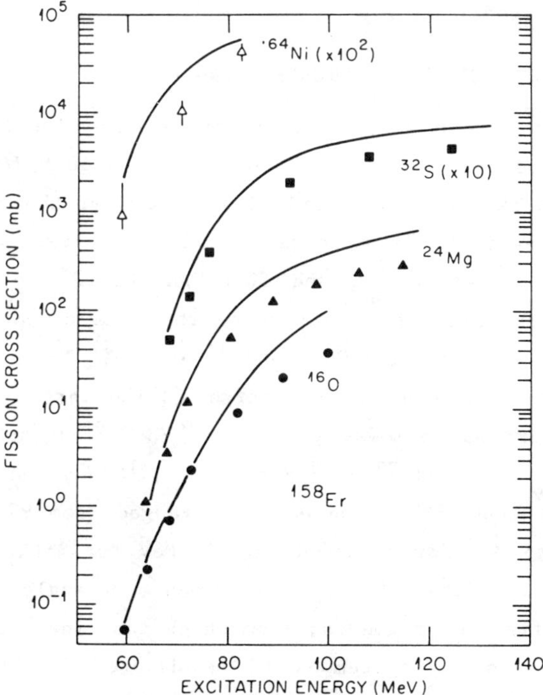

Fig. 3. Measured (points) and calculated (curves) fission excitation functions for the compound nucleus ^{158}Er formed in reactions with various projectiles.

curves represent the measurements very well, particularly in the steep portions of the excitation functions, which are most sensitive to the variations in the fission barrier.[12] We can conclude that calculated RFRM fission barriers are consistent with measurements, at least in the ranges of excitation energy (between 40 MeV and about 100 MeV) and of angular momentum (between 20 $\hbar$ and 80 $\hbar$) covered by our study. It is clear that the time has come for the RFRM[23,24] to replace the widely used RLDM in making estimates of various angular momentum effects in heavy-ion-induced reactions.

It can also been seen in Fig. 3 that the calculated curves and measurements tend to diverge at the highest excitation energies. This effect is also observed in the case of the other systems and may be due to incomplete fusion. This question will be addressed further in the next section.

3. FISSION CROSS SECTIONS AT HIGHER ENERGIES

Encouraged by our ability to describe heavy-ion-induced fission of a large number of systems at energies up to 10 MeV/nucleon in the combined framework of the statistical model and of the calculated Bass fusion cross sections,[25] we have extended our fission studies into the energy range between 10 and 20 MeV/nucleon. To the extent to which it was possible, we have studied the same compound (or composite) systems, i.e., ^{158}Er, ^{186}Os, and ^{208,210}Po. In addition, we have also made measurements on a ^{238}U target. The target and projectile combinations were as follows: ^{12}C on ^{174}Yb, ^{198}Pt, and ^{238}U; ^{16}O on ^{142}Nd, ^{170}Er, ^{192}Os, and ^{238}U; ^{32}S on ^{126}Te, ^{144}Nd, and ^{238}U; and ^{58}Ni on ^{96}Zr, ^{116}Cd, and ^{238}U. The energies ranged from 95 to 291 MeV for ^{12}C, 140 to 315 MeV for ^{16}O, 350 to 700 MeV for ^{32}S, and 352 to 875 MeV for ^{58}Ni. The data included measurements of angles and velocities of coincident fission fragments, from which the linear momentum transfer to the fissioning system could be deduced. Fission following incomplete momentum transfer was found to be substantial only in reactions on ^{238}U targets. In our cross section analysis, we have, as far as possible, included only events with complete and nearly complete momentum transfer. Experimental details are given elsewhere.[3]

We had expected at these higher energies to observe limitations resulting from various dynamical processes such as incomplete fusion[26,27] and/or extra push effects.[28,29] This, indeed, turned out to be the case, since it was no longer possible to describe the measured cross sections in terms of statistical model calculations with RFRM fission barriers and with Bass model fusion cross sections. It was expected that in reactions with ^{12}C and ^{16}O ions, incomplete fusion would result in lower observed fission cross sections, except in the case of the ^{238}U target, where even partially fused systems are highly fissile. Results of ^{12}C and ^{16}O bombardments of ^{238}U are shown in Fig. 4a for the near-full momentum transfer component, together with calculations using the Bass model. On the basis of the good agreement between theory and experiment, we conclude that the Bass model constitutes a reasonable parametrization of fusion (complete and incomplete) in this energy range. This situation did not prevail in the case of reactions of ^{32}S and ^{58}Ni with ^{238}U. The discrepancy between the Bass model results and the measured fission cross sections for the near-full momentum transfer component can be attributed to higher Coulomb repulsion between target and projectile nuclei, resulting in the need for an additional energy (extra push) to achieve fusion.[28,29]

In Fig. 4b the fission excitation functions for the $^{16}O + ^{142}Nd$ and the $^{12}C + ^{174}Yb$ systems are shown, together with statistical model calculations in which the Bass model fusion cross sections and the RFRM fission barriers have been incorporated. It can be seen that, at the highest energies, calculated cross sections are three to five times higher than measured cross sections, which saturate at a level consistent with a maximum angular momentum of the fissioning system of about 65 $\hbar$. This observed saturation in the fission excitation function may, presumably, be attributed to incomplete fusion, and the implication is that collisions with partial waves beyond ~65 $\hbar$ involve processes that result in residual nuclei with reduced angular momenta. While incomplete fusion is the most likely explanation, our results cannot be accounted for quantitatively on the basis of either version

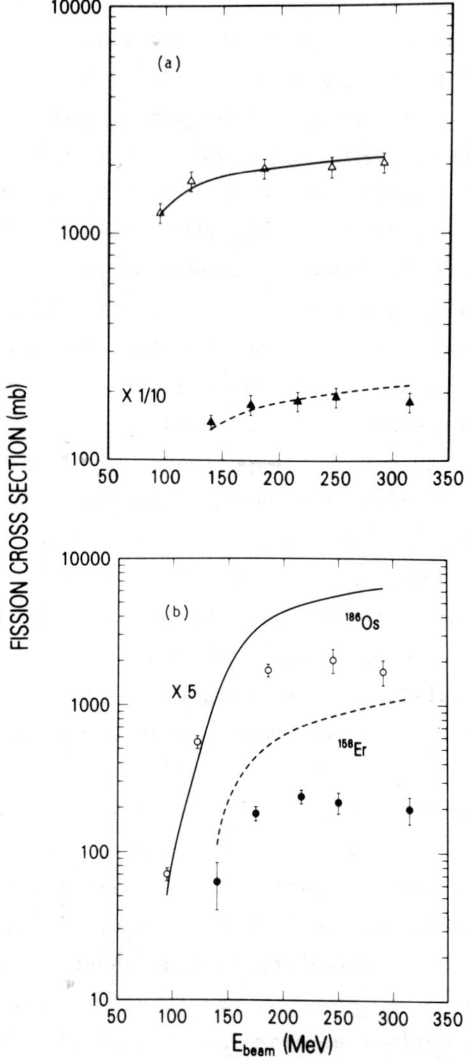

Fig. 4. Fission cross sections for reactions of ^{238}U with ^{12}C (open triangles) and ^{16}O (closed triangles) are shown in (a). In (b), cross sections are shown for the $^{16}O + ^{142}Nd$ (open circles) and the $^{12}C + ^{174}Yb$ (closed circles) systems. The solid and dashed lines depict Bass fusion cross sections in (a) and statistical model calculations using the Bass model in (b).

of the Wilczynski model.[26,27] This is because partial waves of only up to 60 $\hbar$ (for $^{16}O + ^{142}Nd$ at 315 MeV) are expected to contribute to complete fusion,[27] which would account for only 10% of our observed fission cross section. If, as appears to be the case, fission following incomplete fusion does not contribute significantly to the fission cross sections in the ^{158}Er and ^{186}Os cases, then we may conclude that fusion with nearly full momentum transfer must be taking place with partial waves above the limit predicted by the Wilczynski sum-rule model.[27]

For systems with high values of the Z_pZ_t product, it is expected that an additional energy (extra push)[28,29] is needed for fusion to take place. Even for reactions with ^{16}O ions, at energies above about 120 MeV, the highest partial waves involve angular momenta that are sufficient to increase the effective fissility to the point where extra push[29] effects are expected. Calculations with the parameters of Bjornholm and Swiatecki[29] (with the exception of our use of a = 10, which is within their limits of uncertainty) are compared with measurements in Fig. 5. For the ^{16}O-induced reactions (Fig. 5a), agreement between experiment and theory at the higher energies is much better than that which was obtained with the Bass model, except for the case of ^{238}U. We conclude that, even with ^{16}O ions, there is some evidence for a dynamical limitation to the fusion process and that the extra push model[29] determines the highest partial waves for which complete fusion takes place.

In Figs. 5b and 5c similar comparisons are shown for reactions with ^{32}S and ^{58}Ni. There does not appear to be a consistent quantitative agreement between experiment and theory, although it is clear that some kind of dynamical limitation plays a role in the cases that we have studied. Best agreement was found in the case of ^{58}Ni-induced reactions at the highest energy (875 MeV). In addition, a clear cut illustration of the extra push threshold effect is provided by the 352-MeV $^{58}Ni + ^{238}U$ result, where our measured fission cross section is 0^{+50}_{-0} mb and where a threshold of 460 MeV is predicted by the extra push model.

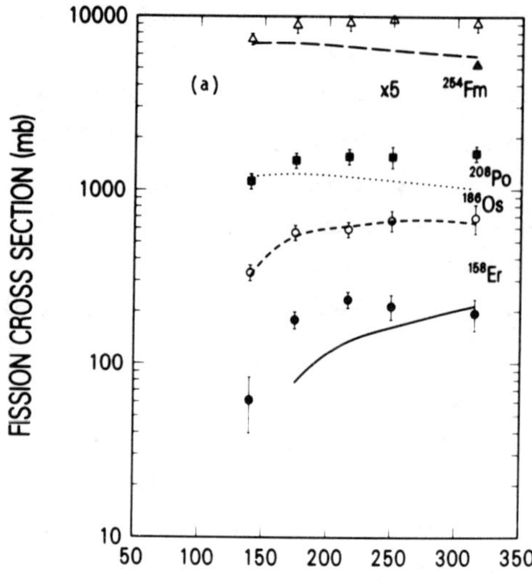

Fig. 5. Comparison of measured fission cross sections with extra push model cross sections (curves). In (a), reactions with ^{16}O are shown, and the labels indicate the composite systems. In (b), ^{32}S-induced reactions on ^{126}Te (circles, solid curve), ^{144}Nd (squares, short-dashed line), and ^{238}U (triangles, long-dashed line) are shown. Reactions of ^{58}Ni with ^{96}Zr (circles, solid line), with ^{116}Cd (squares, short-dashed line), and ^{238}U (triangles, long-dashed line) are shown in (c).

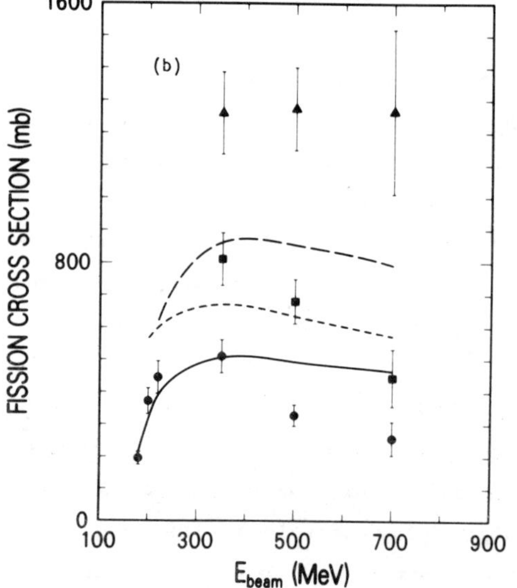

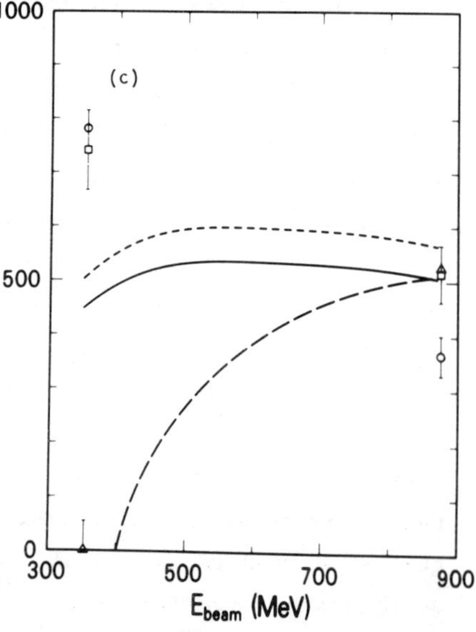

4. FISSION DECAY OF REACTION PRODUCTS WITH A $\lesssim$ 150

In this section I will describe the recent conclusions of Awes et al.[30] concerning the fissionlike decay of reaction products with A $\lesssim$ 150. In damped reactions between heavy ions at high incident energy, the strong dissipation of relative kinetic energy and angular momentum into intrinsic degrees of freedom may result in the fission decay of one of the reaction partners. Recent measurements at incident energies above E/A = 10 MeV have suggested a nonequilibrium fission process[4-7] occurring with unexpectedly large probability.[4,6,7] The fission has been shown to occur on a short time scale, by the observation of proximity effects.[5-7] In Oak Ridge we have studied the fissionlike decay of reaction products resulting from interactions of ^{58}Ni with ^{58}Ni at 15.3 MeV/nucleon.[30,31] Inclusive and two-fragment exclusive measurements of products were made using two large solid angle gas ionization chambers. These chambers were operated at angles that favored the observation of coincident fragments from the decay of one of the primary reaction products for all values of energy loss in the first step of the reaction.

In Fig. 6 the total charge, $Z_1 + Z_2$, of the coincident fragments is plotted against TKE, the total kinetic energy in the center of mass for the first step of the reaction assuming that the two observed fragments arise from sequential decay of a composite nucleus. Thus, TKE = $E_{12} + E_3$, where E_{12} is the kinetic energy of the 1-2 composite system and E_3 is the kinetic energy of the assumed third body which completes energy, momentum, and mass conservation. For a pure two-body reaction, TKE is identically zero (For a purely binary reaction the composite system would correspond to the compound nucleus which has no kinetic energy in the center of mass.); however, in the presence of particle evaporation this will be only approximately true. Thus, the cluster of events seen with $Z_1 + Z_2 \simeq 40$ and TKE ≈ 0 clearly arises from binary reactions accompanied by a large amount of particle evaporation. Those events with $Z_1 + Z_2$ less than about 30 are mainly due to three-body events. They are observed to occur for energy losses

(TKEL = $E_{c.m.}$ - TKE) ranging from TKEL ≈ 100 MeV to complete damping in the first step of the reaction.

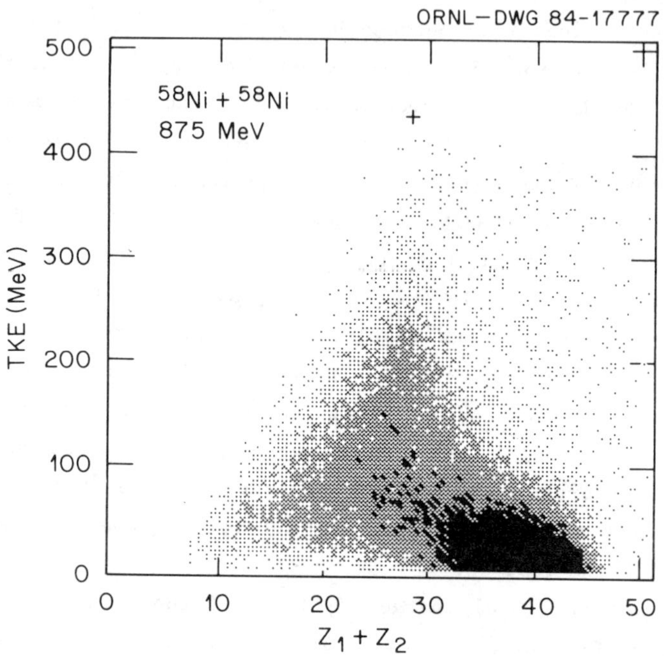

Fig. 6. Summed charge versus total c.m. kinetic energy during the first step of the reaction for all coincident products in reactions of $^{58}Ni + ^{58}Ni$ at 875 MeV.

Because the decaying fragments produced in this reaction have fissilities, [$X(^{58}Ni) = 0.27$], well below the Businaro-Gallone[32] point, they are expected to be unstable against asymmetric distortions. Thus, based on liquid-drop considerations, one expects a maximum in the potential and therefore a minimum in the yield for symmetric decay.[33] This behavior has recently been observed[34] for the decay of the $^{9}Be + ^{74}Ge$ system. From Fig. 7 it can be seen that asymmetric decay is the predominant mode of heavy fragment emission for the primary fragments ($Z_1 + Z_2 \approx 28$), with a strong contribution from $Z_2 = 6$ events. (The kinematics and asymmetry of our experimental geometry favor the observation of the heavy fragment at forward angles in the large-area ionization chamber.)

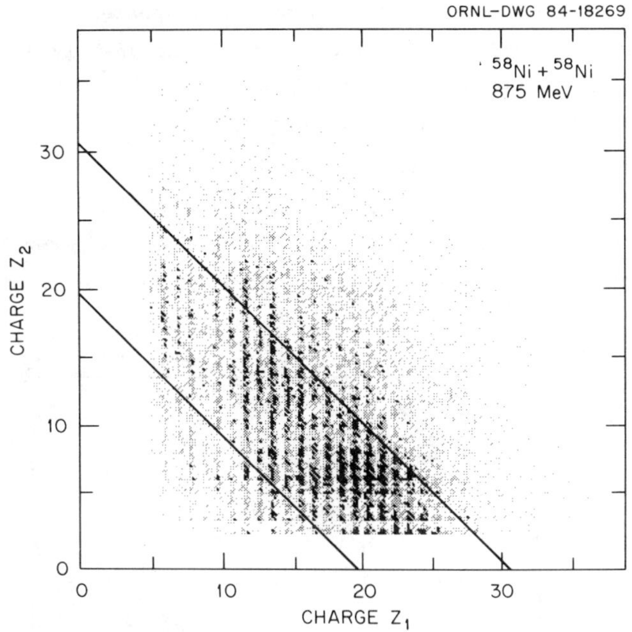

Fig. 7. Charge, Z_1, observed in one ionization detector versus charge, Z_2, observed on the opposite side of the beam for coincident products with TKE > 60 MeV (TKEL < 377.5 MeV) resulting from reactions of ^{58}Ni + ^{58}Ni at 875 MeV. Events in the region $19.5 < Z_1 + Z_2 < 30.5$ result mainly from decay of a nickellike primary fragment (see Fig. 6).

The observation of a strong contribution of coincidences with carbon immediately raises the question of possible light contaminants on the target which would undergo deep-inelastic or fusion-fission reactions yielding coincident distributions similar to those from projectile decay. To rule out this possibility, measurements were also made on a ^{12}C target under identical experimental conditions. Due to the inverse kinematics, light contaminants give apparent energy losses which are much smaller, and overlap very little, with those observed with the ^{58}Ni target. From this difference we can conclude that possible contributions from light target contaminants are less than 5% and are confined to the region of small TKEL. Although the experimental geometries and systems studied were different, there is a similarity between our results on carbon[31] and the ^{86}Kr + ^{89}Y coincidence measurements of Ref. 4 in the region of small apparent energy loss. A possible light contaminant would explain the apparently large projectile fission probability at small energy loss observed in Ref. 4 but not seen in our case or in that of Ref. 7.

The experimental fission probabilities are shown in Fig. 8 for all fissionlike events with Z_1 and $Z_2 > 3.5$ and also for symmetric fission events only (Z_1 and $Z_2 > 8.5$). In order to determine whether the observed probabilities for fissionlike decay are consistent with

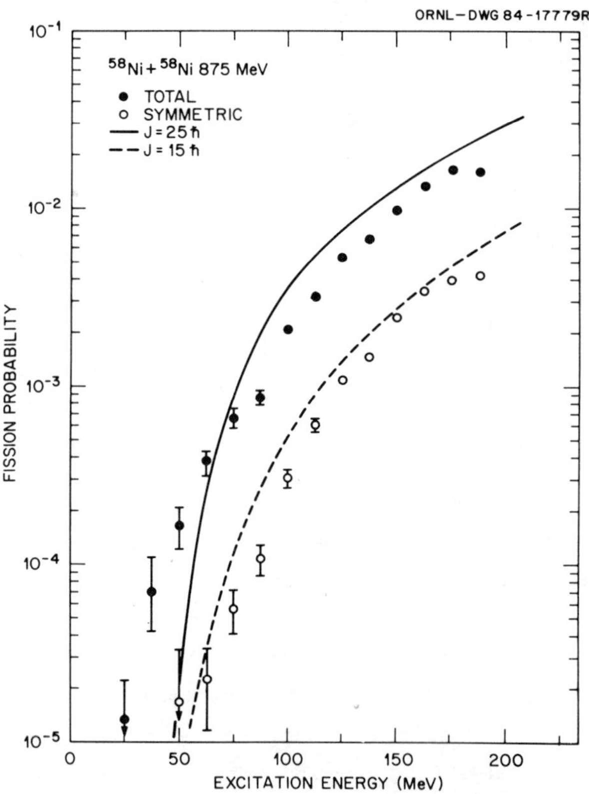

Fig. 8. Fission probability for all heavy fragment decays with $19.5 < Z_1 + Z_2 < 30.5$ for (Z_1 and Z_2) > 3.5 (solid points) and symmetric decay only (Z_1 and Z_2) > 8.5 (open points). Fission evaporation calculations for ^{58}Ni are shown for $a_f/a_n = 1.0$, $a = A/7.5$, and $J = 25\ \hbar$ (solid curve) or $J = 15\ \hbar$ (dashed curve).

an equilibrium process or instead indicate nonequilibrium effects, we have made statistical model calculations using the same evaporation code that was used to analyze the fission excitation functions described in Section 2. There are two parameters of importance for the calculation, the ratio of saddle point to ground state level densities, a_f/a_n, which we take to be 1.0, and the spin of the decaying nucleus. A distribution of contributing spins is expected, ranging from sticking at the critical angular momentum to sticking at the grazing angular momentum. This gives a range of spins from $J \simeq 10\ \hbar$ to $J \approx 35\ \hbar$. The calculated result is shown in Fig. 8 for the decay of ^{58}Ni with $J = 25\ \hbar$ (solid curve), which corresponds to two-thirds of the grazing value and is probably a realistic estimate of the average. An additional result is shown for $J = 15\ \hbar$ (dashed curve) to demonstrate the spin dependence.

From Fig. 8 it can be seen that the statistical model is able to account for the measured fissionlike decay probability quite well, and we may conclude that the observed decay is consistent with an equilibrated process.

This is in contrast to conclusions reached for the $^{129}Xe + {}^{122}Sn$ reaction,[6,7] which suggested a nonequilibrium fission process, based in part on measured fission probabilities which were found to be one to two orders of magnitude larger than those estimated using the statistical model[35] with reasonable parameter values.[6,7] However, we have used the statistical model computer code PACE[36] with the same parameters of Refs. 6 and 7 ($a_f/a_n = 1.08$ and $J = 40\ \hbar$), and with RFRM barriers, and obtained adequate agreement with experiment. For example, at 250-MeV excitation energy a fission probability, of $P_f = 6.2\%$ was obtained for fission of ^{129}Xe, which compares with a measured probability of $P_f = 9\%$ for fission of nuclei with mass 110 to 150 amu. (Complete agreement could be obtained by slightly increasing the average spin.) Also consistent with observation,[7] the calculated fission probability was found to vary by about a factor of only two over the mass range of 110 to 150 amu, for fixed values of excitation energy and spin. These calculations imply that, in this case, the observed fission is also consistent with a statistical process and not necessarily the result of a nonequilibrium dynamical mechanism.[37] In particular, the results imply that the fissioning nucleus is not deformed beyond the saddle point when it is formed in the first step of the reaction, as was suggested in Ref. 7. As a consequence, the scission-to-scission time of 0.75-1. x 10^{-21} sec extracted[7] from the observation of proximity effects provides an upper bound on the saddle-to-scission time of the fissioning xenonlike nucleus.

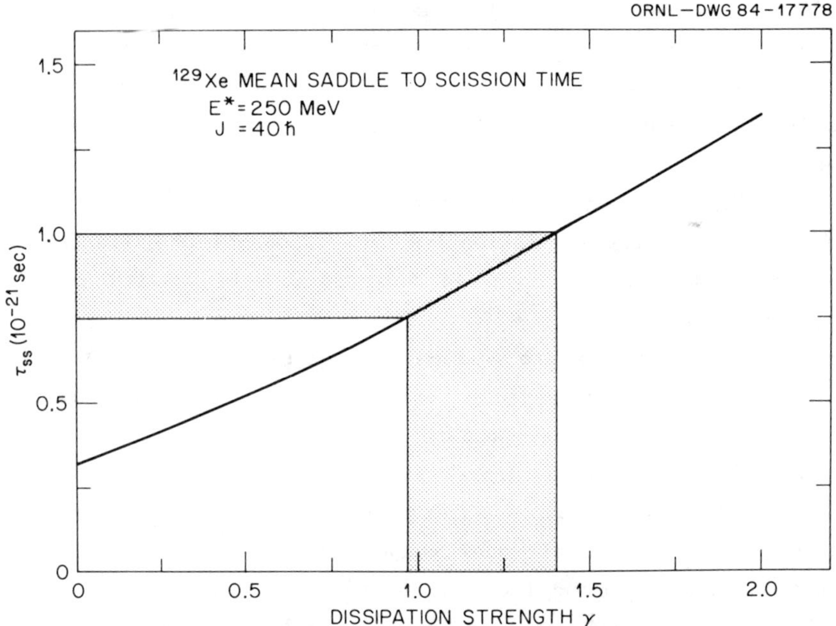

Fig. 9. Calculated saddle-to-scission time, τ_{ss}, for fission of ^{129}Xe as a function of the nuclear dissipation strength, γ (solid curve). The calculation was performed[38] for an excitation energy $E^* = 250$ MeV, spin J = 40 $\hbar$, and level density parameter a = A/7.5. Also shown by the shaded region is the experimentally deduced[7] scission-to-scission time of 0.75-1. x 10^{-21} sec, which provides an upper limit on the calculated saddle-to-scission time. Note: $\gamma = 1.0$ corresponds to critically damped motion.

The above limit, when compared to theoretical calculations of the saddle-to-scission time in fission,[39] can provide unique information regarding the nuclear dissipation strength, γ. (For a definition of γ, see Ref. 39.) Within the framework of Kramer's stationary solution of the Fokker-Planck equation for an inverted oscillator,[39] a limit on the saddle-to-scission time of $0.75\text{-}1.0 \times 10^{-21}$ sec for ^{129}Xe corresponds to an upper limit for the nuclear dissipation strength of $\gamma = 1.0\text{-}1.4$ (Ref. 38) (see Fig. 9), ruling out strongly overdamped dissipation strengths. Furthermore, if allowance is made for the time to reach the saddle point after the initial separation of the damped reaction products, which, although highly uncertain, might be expected to be of the order of the lifetime of the xenonlike product nucleus at 250-MeV excitation, $\hbar/\Gamma_{tot} \approx 5 \times 10^{-22}$ sec, then dissipation strengths of the order of $\gamma \approx 0.5$, corresponding to underdamped motion, are obtained.

5. ACKNOWLEDGMENTS

I would like to thank Drs. T. C. Awes, L. Grodzins, S. S. Kapoor, J. R. Nix, and H. J. Specht for fruitful discussions, and also S. J. Ball for help with the preparation of the manuscript.

REFERENCES

1. Van der Plicht, J., Britt, H. C., Fowler, M. W., Fraenkel, Z., Gavron, A., Wilhelmy, J. B., Plasil, F., Awes, T. C., and Young, G. R., Phys. Rev. C 28, 2022 (1983).

2. Plasil, F., Awes, T. C., Cheynis, B., Drain, D., Ferguson, R. L., Obenshain, F. E., Sierk, A. J., Steadman, S. G., and Young, G. R., Phys. Rev. C 29, 1145 (1984).

3. Gavron, A., Boissevain, J., Britt, H. C., Eskola, K., Eskola, P., Fowler, M. M., Ohm, H., Wilhelmy, J. B., Awes, T. C., Ferguson, R. L., Obenshain, F. E., Plasil, F., Young, G. R., and Wald, S., submitted to Physical Review C.

4. Olmi, A., Lynen, U., Natowitz, J. B., Dakowski, M., Doll, P., Gobbi, A., Sann, H., Stelzer, H., Bock, R., and Pelte, D., Phys. Rev. Lett. 44, 383 (1980).

5. Glässel, P., Harrach, D. v., Grodzins, L., and Specht, H. J., Phys. Rev. Lett. 48, 1089 (1982).

6. Harrach D. v., Glässel, P., Grodzins, L., Kapoor, S. S., and Specht, H. J., Phys. Rev. Lett. 48, 1093 (1982).

7. Glässel, P., Harrach, D. v., Specht, H. J., and Grodzins, L., Z. Phys. A310, 189 (1983).

8. Sikkeland, T., Phys. Rev. 135, B669 (1964).

9. Sikkeland, T., Clarkson, J. E., Steiger-Shafrir, N. H., and Viola, V. E., Phys. Rev. C 3, 329 (1971).

10. Videbaek, F., Goldstein, R. B., Grodzins, L., and Steadman, S. G., Phys. Rev. C 15, 954 (1977).

11. Beckerman, M., and Blann, M., Phys. Rev. Lett. 38, 272 (1977); Phys. Lett. 68B, 31 (1977); Phys. Rev. C 17, 1615 (1978).

12. Plasil, F., Ferguson, R. L., Hahn, R. L., Obenshain, F. E., Pleasonton, F., and Young, G. R., Phys. Rev. Lett. 45, 333 (1980).

13. Back, B. B., Betts, R. R., Henning, W., Wolf, K. L., Mignerey, A. C., and Lebowitz, J. M., Phys. Rev. Lett. 45, 1230 (1980).

14. Cabot, C., Gauvin, H., LeBeyec, Y., Delagrange, H., Dufour, J. P., Fleury, A., Llabador, Y., and Alexander, J. M., J. Phys. (Paris) C10, 234 (1980); Cabot, C. (D.es.Sc. Thesis), Institut de Physique Nucléaire, Orsay, France, Report No. IPN-T-83.02, 1983 (unpublished).

15. Leigh, J. R., Hinde, D. J., Newton, J. O., Galster, W., and Sie, S. H., Phys. Rev. Lett. 48, 527 (1982).

16. Sikora, B., Scobel, W., Beckerman, M., Bisplinghoff, J., and Blann, M., Phys. Rev. C 25, 1446 (1982).

17. Guillaume, G., Coffin, J. P., Rami, F., Engelstein, P., Heusch, B., Wagner, P., Fintz, P., Barrette, J., and Wegner, H. E., Phys. Rev. C 26, 2458 (1982).

18. Hinde, D. J., Leigh, J. R., Newton, J. O., Galster, W., and Sie, S., Nucl. Phys. A385, 109 (1982).

19. Hinde, D. J., Newton, J. O., Leigh, J. R., and Charity, R. J., Nucl. Phys. A398, 308 (1983).

20. Cohen, S., Plasil, F., and Swiatecki, W. J., Ann. Phys. (N.Y.) 82, 557 (1974).

21. Krappe, H. J., and Nix, J. R., in Proceedings of the Third International Atomic Energy Symposium on the Physics and Chemistry of Fission, Rochester, New York, August 1973 (International Atomic Energy Agency, Vienna, 1974), Vol. I, p. 159; Krappe, H. J., Nix, J. R., and Sierk, A. J., Phys. Rev. C 20, 992 (1979).

22. Möller, P., and Nix, J. R., in Proceedings of the Workshop on Nuclear Dynamics, Granlibakken, California, 1980, Lawrence Berkeley Laboratory Report No. LBL-10688, 1980 (unpublished), pp. 131-134.

23. Mustafa, M. G., Baisden, P. A., and Chandra, H., Phys. Rev. C 25, 2524 (1982).

24. Sierk, A. J., unpublished results.

25. Bass, R., Phys. Rev. Lett. 39, 265 (1977).

26. Siwek-Wilczynska, K., du Marchie van Voorthuysen, E. H., van Popta, J., Siemssen, R. H., and Wilczynski, J., Nucl. Phys. A330, 150 (1979).

27. Wilczynski, J., Siwek-Wilczynska, K., van Driel, J., Gonggrijp, S., Hageman, D.C.J.M., Janssens, R.V.F., Lukasiak, J., and Siemssen, R. H., Phys. Rev. Lett. 45, 606 (1980).

28. Nix, J. R., Annu. Rev. Nucl. Sci. 22, 65 (1972).

29. Swiatecki, W. J., Nucl. Phys. A376, 275 (1982); Bjornholm, S., and Swiatecki, W. J., Nucl. Phys. A391, 471 (1982).

30. Awes, T. C., Ferguson, R. L., Novotny, R., Obenshain, F. E., Plasil, F., Rauch, V., Young, G. R., and Sann, H., to be published.

31. Awes, T. C., Ferguson, R. L., Novotny, R., Obenshain, F. E., Plasil, F., Rauch, V., Young, G. R., and Sann, H., in Proceedings of the 11th International Winter Meeting on Nuclear Physics, Bormio, Italy, January 24-29, 1983, University of Milano Press, Italy (1983), p. 425.

32. Businaro, U. L., and Gallone, S., Nuovo Cimento 1, 629, 1277 (1955).

33. Moretto, Luciano G., Nucl. Phys. A247, 211 (1975).

34. Sobotka, L. G., McMahan, M. A., McDonald, R. J., Signarbieux, C., Wozniak, G. J., Padgett, M. L., Gu, J. H., Liu, Z. H., Yao, Z. Q., and Moretto, L. G., Phys. Rev. Lett. 53, 2004 (1984).

35. Vandenbosch, R., and Huizenga, J. R., Nuclear Fission, Academic Press, New York, 1983.

36. Gavron, A., Phys. Rev. C 21, 230 (1980).

37. Preliminary results for the $^{129}Xe + ^{122}Sn$ reaction at 18 MeV/nucleon indicate that the fission probability is not determined by the incident energy, but rather by the absolute energy loss in the collision, as would be expected for an equilibrium process. (H. J. Specht, private communication.)

38. Nix, J. R., and Sierk, A. J., private communication.

39. Hofmann, Helmut, and Nix, J. Rayford, Phys. Lett. 122B, 117 (1983).

FISSION ALONG THE MASS ASYMMETRY COORDINATE AN EXPERIMENTAL EVALUATION OF THE CONDITIONAL SADDLE MASSES AND OF THE BUSINARO-GALLONE POINT*

L.G. Moretto, M.A. McMahan, L.G. Sobotka and G.J. Wozniak

Nuclear Science Division
Lawrence Berkeley Laboratory
University of California
Berkeley, California 94720

ABSTRACT

The importance of empirically determining the ridge line of conditional saddle points is discussed in view of the recent liquid drop model refinements, like diffuseness and finite range. Two series of experiments are presented. The first series involves the complex fragment emission from compound nuclei resulting from the ^{3}He + Ag reaction. Kinetic energy distributions and excitation functions are shown, and the conditional barriers are obtained over a range of atomic numbers. In the second series, reverse kinematic reactions like Ge, Nb, La + Be, C are studied. The fragments emitted cover the entire Z range, from Z = 1 to symmetric splitting. Their origin from a full momentum transfer intermediate is shown. From the complete charge distributions it is possible to conclude that the Businaro-Gallone transition is observed.

1. INTRODUCTION

The understanding of the fission process and the development of the liquid drop model have followed a common history. The liquid drop fission saddle point determines, in heavier systems, the boundaries of nuclear stability and the dominance of symmetric splitting, while, in

*This work was supported by the Director, Office of Energy Research, Division of Nuclear Physics of the Office of High Energy and Nuclear Physics of the U.S. Department of Energy under Contract DE-AC03-76SF00098.

lighter systems, with the loss of stability of the mass asymmetry degree of freedom, it determines the disappearance of fission as a process separate from evaporation.

In a way, the mass asymmetry degree of freedom appears to be the most interesting normal mode at the saddle point. As mentioned above, its ever increasing stability above the Businaro-Gallone point defines fission as a separate process, whose mass distribution progressively sharpens as one moves toward the upper end of the periodic table. Below the Businaro Gallone point, the mass asymmetry mode loses stability; thus the associated fission mass distribution should portray a minimum at symmetry and should connect monotonically with the mass distribution arising from evaporation.

An experimental study of the mass asymmetry degree of freedom must rely on the measurement of the exit channel mass/charge distribution which, unfortunately, is a characteristic property of the scission point, rather than of the saddle point. Such a difficulty is mitigated by the fact that, for lighter nuclei, saddle and scission points are almost degenerate, so that little or no evolution can be expected from saddle to scission point. For heavier nuclei, saddle and scission points may be quite far apart for symmetric splitting, so that the observed mass distribution results from a combination of the initial asymmetry distribution at the saddle, and of the dynamical evolution from saddle to scission. Yet, even for the heavier systems, there is a range of mass asymmetries, near the most asymmetric limit, where the saddle point and the scission point are, in fact, nearly degenerate.

These general ideas have formed the basis of a theoretical paper that discusses in a unified way the compound nucleus emission of fragments with masses ranging from protons or neutrons to one half that of the compound nucleus.[1)] The key concepts in this development are the "conditional" saddle points at fixed mass asymmetry and their locus called "ridge line" in analogy to "saddle point."

Experimentally, information has been limited to the measurement of the fission barriers of heavy nuclei.[2)] These data were instrumental in removing the ambiguity arising from the Coulomb term and the surface term in the liquid drop model.[3] From the experimental barriers shell effects were obtained for both ground state and saddle point.

Similar information could be obtained by performing measurements of the conditional barriers as a function of mass asymmetry. This has been prevented so far by the very small decay rates that one encounters for lighter systems. Recently, however, we have been able to overcome some of the difficulties associated with the relevant experiments. Two classes of experiments will be reported here. The first deals with the emission of complex fragments from compound nuclei[4] and their excitation functions leading to the determination of the associated conditional barriers. The second is the measurement of the entire charge distribution from alpha particles to symmetric fragments and the verification of the Bùsinaro-Gallone transition.[5]

2. COMPOUND NUCLEUS EMISSION OF COMPLEX FRAGMENTS AND THEIR EXCITATION FUNCTIONS

Complex fragments have been observed in high-energy proton reactions.[6-9] However, the origin of these fragments cannot be easily or unequivocally traced to a well-characterized compound nucleus because precompound processes dominate and multiple fragmentation of the target is possible. Heavy-ion reactions at low bombarding energies[10] have been shown to produce equilibrated compound nuclei (e.g., ^{26}Al) from which Li or Be nuclei are emitted with low probability. At bombarding energies above ~8 MeV/nucleon,[11] however, the fragments of interest are also produced by projectile breakup, thus complicating the study of the compound nucleus decay.

We have obtained experimental evidence for the emission of complex nuclei from helium through fluorine by compound nuclei produced in the reaction 90 MeV ^{3}He + ^{nat}Ag.[4] The specific choice of ^{3}He as projectile was dictated by two reasons. On one hand it is desirable to have a relatively low velocity projectile in order to minimize preequilibrium losses, but massive enough to bring in sufficient energy. On the other, the mass of the projectile should be sufficiently smaller than those of the complex fragments of interest in order to rule out the ambiguity of projectile fragmentation or multinucleon transfer.

The ^{3}He beam was produced by the Lawrence Berkeley Laboratory 88-Inch Cyclotron. The atomic number and the energy spectra of the intermediate mass fragments were measured in a set of three standard gas, solid-state (300 μm, $d\Omega \sim 1.2$ msr) ΔE-E telescopes operated at a pressure of 100 torr. Alpha particles were detected in a separate solid-state telescope (40 μm, 5 mm, $d\Omega = 0.35$ msr). Angular distributions were obtained from 20° to 170° in the laboratory.

In order to determine the existence of an isotropically emitting source and its velocity, the laboratory energy spectra were transformed into invariant cross-section plots in velocity space which are presented in Fig. 1. The peak cross section for a heavy complex fragment, such as carbon, has a constant value and occurs at the same c.m. velocity from 170° to 40° (as indicated by the position of the X's relative to the circular arc). At the most forward angle the peak cross section occurs at a slightly increased velocity. Similarly, the higher velocity region (the region near the arc with the larger radius) shows no significant change in the backward hemisphere, but does stretch out at forward angles. For a light complex fragment such as Li, the peak of the cross section occurs at a constant c.m. velocity for a smaller backward angle region (170° to 120°). Forward of 120° the peak increases both in cross section and in velocity. The slope of the high-energy tail does not change significantly for the three most backward angles, but the intensity of the tail increases as the scattering angle decreases. The ^{9}Be and B fragments show a behavior intermediate between that of Li and C. In general, the heavier ejectiles show patterns more consistent with the emission from a single source.

Two conclusions can be drawn from these invariant cross-section plots. First, for all elements there is an angular region in the backward hemisphere where only a single component is observed, which can be characterized by the c.m. emission. This angular region

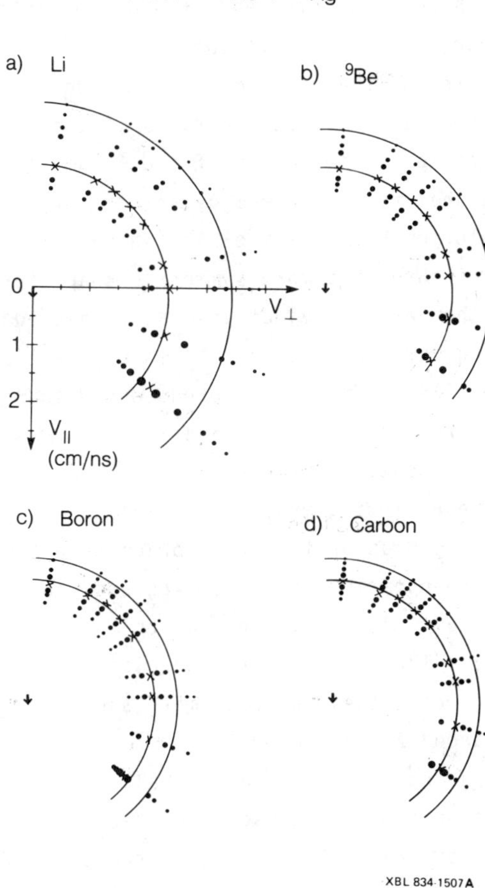

Fig. 1. Invariant cross section plots $\left(\propto \frac{1}{V^2}\frac{d^2\sigma}{d\ dV}\right)$ for representative ejectiles (Li, ^{9}Be, B, and C). The diameter of the dots is proportional to the logarithm of the cross section and the X's indicate the peak of velocity distribution. The two large arcs are sections of circles centered on the c.m. velocity (center arrow) appropriate for complete fusion. The beam direction (0°) is indicated by the c.m. velocity vector.

increases and extends to more forward angles as the ejectile mass increases. Second, there is a component of non-c.m. emissions that results in harder energy (or velocity) spectra at forward angles.

The energy spectra of the equilibrium component in the c.m. system are shown in Fig. 2. The mean energies of the spectra are Coulomb-like and increase as the charge of the fragment increases. The most interesting feature in the energy spectra of the equilibrium component is the evolution from a Maxwellian shape for α-particles (not shown) or Li ions through a more symmetric shape for B or C to a symmetric shape for the heaviest ejectiles, as predicted in Ref. 1. In previous high energy proton studies,[8-9] an exponential tail was observed for all ejectiles. This tail, produced by sources other than equilibrium emission from the center-of-mass system, masks the shape of the equilibrium component. At forward angles, our data also show the presence of a nonequilibrium exponential tail.

The experimental yields of the equilibrium component are shown in Fig. 3. To minimize contributions from sources other than the compound nucleus, we have plotted the yields only for the most backward angle (171°). These yields drop precipitously in going from Z = 2 to Z = 3, after which they decrease more slowly. The one exception is the enhanced Z = 6 yield.

The yield from an equilibrium statistical emission process should be roughly proportional to a factor $\exp[-B_Z/T_Z]$, where B_Z is the emission barrier for fragment Z and T_Z is the temperature at the barrier. More quantitatively, the decay width is given by

$$\Gamma_Z \propto T_Z[E/(E - B_Z)]^2 \exp[2\sqrt{a(E - B_Z)} - 2\sqrt{aE}] \tag{1}$$

$$\tilde{\propto} \exp[-B_Z/T_Z] \quad . \tag{2}$$

To calculate the theoretical yields, the following expression for the barrier was used

$$B_Z = U_1 + U_2 + \frac{Z_1 Z_2 e^2}{d} + U_{prox} - U_{CN} \quad , \tag{3}$$

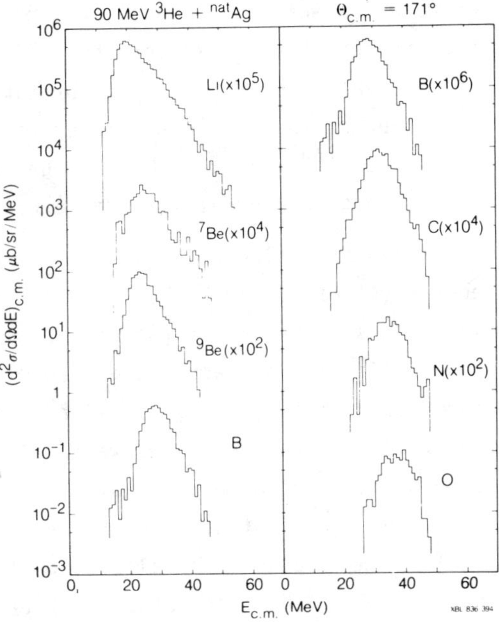

Fig. 2. Energy spectra in the c.m. system for various ejectiles detected at $\Theta_{c.m.}$ = 171°. Before correction for carbon contamination the lower level threshold varied from 4 to 12 MeV in the c.m. for Li to O ejectiles, respectively.

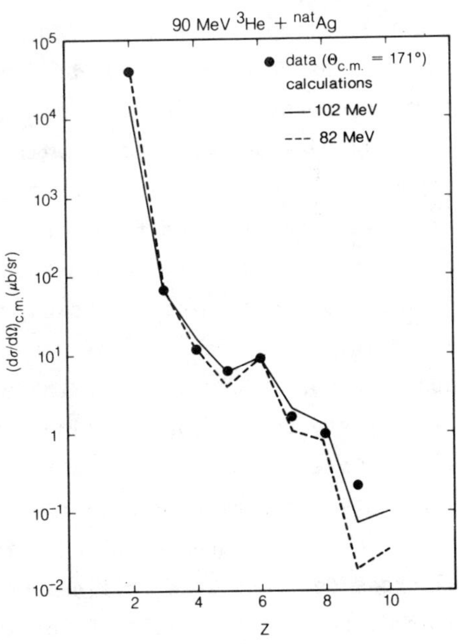

Fig. 3. Experimental (circles) and theoretical yields versus ejectile atomic number (Z).

where U_1 is the experimental mass of the light fragment, U_2, U_{CN} are the droplet model masses of the residual and compound nucleus, respectively, and U_{prox} is the proximity potential. The center-to-center distance d in the interfragment Coulomb term was taken to be $d = 1.225(A_1^{1/3} + A_2^{1/3}) + 2$ fm. The addition of 2 fm was done to obtain rough agreement with the energy spectra. The temperature (T_Z) was evaluated using $E - B_Z = aT_Z^2$. A compound nucleus excitation energy (E) of 102 MeV (the value for full momentum transfer) and a level density parameter (a) of A/8 were assumed. The calculated yields (Eq. 1) for each isotope were multiplied by 2I + 1 (where I is the ground state spin of the light fragment) and then summed. The theoretical ejectile yields were calculated as a ratio Γ_Z/Γ_6 and have been normalized to the data at Z = 6 in Fig. 3.

The agreement between the data (circles) and this simple equilibrium statistical calculation (solid line) is exceptionally good for Z = 3-9. The calculation underpredicts the α-particle yield because it only takes into account first chance emission, whereas substantial amounts of higher chance α-emission occur. Precompound emission is expected to leave the compound nucleus with a broad excitation energy distribution with a most probable value of ~85 MeV. A calculation (not shown) with this lower excitation energy also reproduces the relative yields of the heavy fragments quite well but overpredicts the yield of first chance α-emission. More detailed comparisons between the data and theory require calculations that include precompound emission; however, the substantial agreement depicted in Fig. 3 does indicate that an equilibrated process is responsible for the emission of these complex fragments.

Complete excitation functions obtained from the $^3He + {}^{nat}Ag$ measurements are shown in Fig. 4 for a series of decay products. The measurements were restricted to the backward angles (120° - 160°) in order to insure measurement of only the equilibrium component. The total cross sections were obtained by integrating over angle using the angular distribution formula of Moretto.[1)]

With increasing bombarding energy, the cross sections (see Fig.

4) rise rapidly and then flatten at higher energies. This is a characteristic signature of compound nucleus emission, and reinforces the assignment of compound nucleus decay that was made previously on the basis of data obtained at 90 MeV. The cross section for Z = 3 is a factor of 1000 lower than that for Z = 2, and for the heavier fragments it is even lower. In spite of these low cross sections, we were able to measure an excitation function over 2-3 orders of magnitude up to Z = 11, with a detection limit of about 50 nb.

The experimental excitation function data have been fitted using a transition state formalism, analogous to that used to fit fission excitation functions.[1] As shown in Ref. 1, the decay width for first-chance emission of a fragment of charge Z can be written as

$$\Gamma_Z = \frac{1}{2\pi\rho(E)} \int_0^{E-B_Z} \rho_Z^*(E - B_Z - \varepsilon)d\varepsilon \qquad (4)$$

where $\rho(E)$ is the compound nucleus level density, B_Z is the conditional barrier height, and $\rho_Z^*(E - B_Z - \varepsilon)$ is the level density at the conditional saddle with a kinetic energy ε in the decay mode. The neutron width Γ_n can be written as

$$\Gamma_n = \frac{2mR^2g}{2\pi\rho(E)} \int_0^{E-B_n} \varepsilon\rho(E - B_n - \varepsilon)d\varepsilon \qquad (5)$$

We make the assumption that the ratio of the decay widths, Γ_Z/Γ_n, is proportional to the ratio of the cross section for complex fragment emission, σ_Z, to that for complete fusion, σ_f, i.e.,

$$\Gamma_Z/\Gamma_n \simeq \frac{\Gamma_Z}{\Gamma_t} \simeq \frac{\sigma_Z}{\sigma_t} \simeq \frac{\sigma_Z}{\sigma_R} \qquad (6)$$

This is reasonable in this mass region because $\Gamma_n >> \sum_{Z\geq 1}^{\infty} \Gamma_Z$. One can then calculate $\Gamma_Z/\Gamma_n(E)$ using an appropriate choice for the level density expressions. A Fermi gas level density was used because it gives an analytical expression for Γ_f/Γ_n. A simple angular momentum dependence has been included by adding to the barriers the rotational energies appropriate to the ground and saddle point deformations.

Using the above expression for Γ_Z/Γ_n, the barriers B_Z, and the ratio a_Z/a_n, of the level density parameters were extracted from fits to the experimental data σ_Z/σ_R. These fits are shown by the solid lines in Fig. 4. The agreement between the data and the fits is remarkably good for all Z-values and confirms that these products originate from compound nuclear decay.

The barriers and values of a_Z/a_n extracted from the fits are shown by the circles in Fig. 5 as a function of Z/Z_{CN}. The extracted barriers increase dramatically as the exit channel becomes more symmetric. Some evidence of shell effects in the exit channel is visible in the barrier for carbon emission, Z = 6, which is lower than those of the neighboring elements. The values of a_Z/a_n extracted from the fits tend to oscillate in the range of 0.98-1.02, or 1.07-1.12 if preequilibrium emission is included. The values of a_f/a_n obtained in fitting fission excitation functions have been shown to reflect shell effects in the saddle point configuration.[2)] One might hope in the future to relate the variation of a_Z/a_n with exit channel charge and mass to saddle point shell effects. Future measurements of A as well as Z should give detailed information on saddle point shell effects.

The extracted barriers and level density parameters are subject to several uncertainties due to the assumptions made in this simple analysis. Not included in the fits are the effects of preequilibrium emission on the initial excitation energy. This can be estimated using the geometry dependent hybrid model from the statistical code ALICE[12] to calculate the excitation energy distribution after the fast precompound stage. Preequilibrium emission is negligible at the lowest bombarding energies, but significantly reduces the average excitation energy of the compound nucleus at bombarding energies of 90 and 130 MeV. However, when at each bombarding energy, the calculated excitation energy distribution after precompound emission was used to

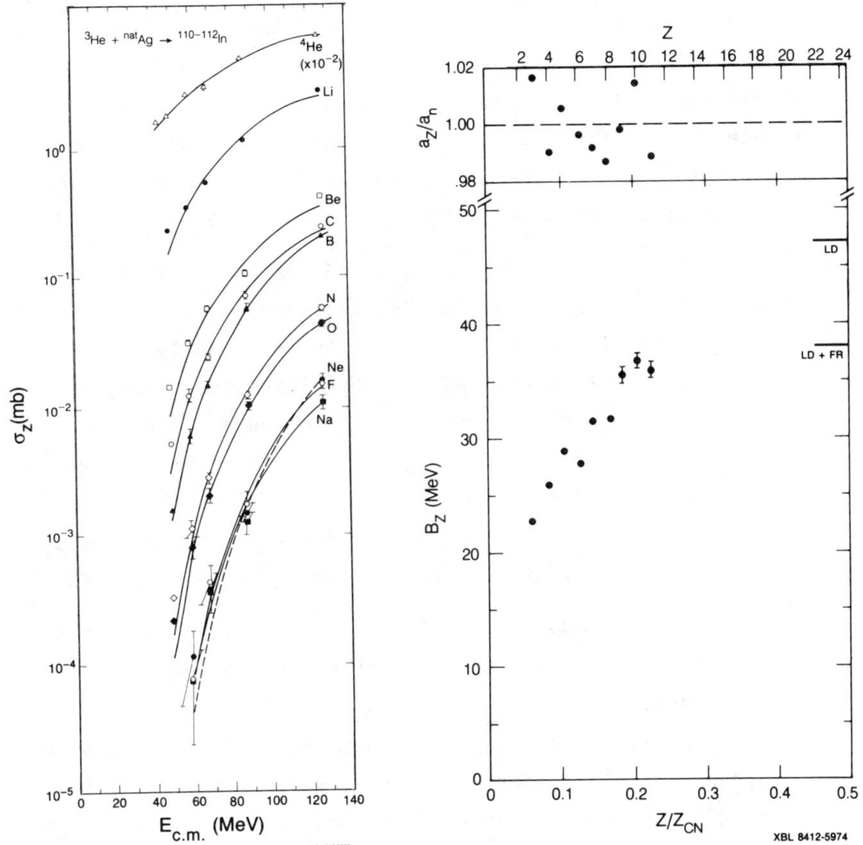

Fig. 4. Dependence of the total integrated cross sections for emission of complex fragments on the center-of-mass energy, $E_{c.m.}$ in the reaction ^{3}He + ^{nat}Ag. The points and error bars correspond to the experimental cross sections. The curves are fits with the parameters of Fig. 5.

Fig. 5. The emission barriers, B_Z, and ratio of level density parameters, a_Z/a_n, extracted in fitting the excitation functions as a function of asymmetry, Z/Z_{CN}. (Z/Z_{CN} = 0.5 corresponds to symmetric splitting.) The points are the extracted parameters, excluding pre-equilibrium emission.

fit the data, the values of the barriers are lowered only slightly, less than 1 MeV, and the values of a_Z/a_n are raised by ~7%. This is understandable since the barriers are most sensitive to the steep part of the excitation function at lower energies and a_Z/a_n is sensitive to the curvature at higher energies. First-chance emission from a fully equilibrated system was assumed, with the fusion cross section $\sigma_f(\ell)$ given by the sharp cutoff model with ℓ_{crit} = 16, as given by by the Bass Model.[13] This assumption should be valid for Z > 2, but is clearly invalid for Z = 2. Thus in the following discussion, the fitted parameters for Z = 2 are ignored. The sensitivity of the fit to the various input parameters has also been examined. Varying each of these parameters within reasonable limits makes a difference in the extracted barriers and values of a_Z/a_n, of at most 5% for all values of Z. In addition, the formalism was tested for "fission" by fitting fission excitation functions. The values of the fission barriers and a_f/a_n extracted with our fitting procedure agree quite well with values quoted in the literature.[2] This agreement indicates that our simple formalism gives a reasonable approximation to more sophisticated codes.

The sensitivity tests described above show that the extracted emission barriers have an uncertainty of ~2 MeV. These results are ideal for testing the finite-range correction to the liquid drop model.

3. COMPLETE CHARGE DISTRIBUTIONS AND THE ROLE OF THE BUSINARO-GALLONE POINT

The sharp distinction between evaporation and fission in relatively heavy compound nuclei is a result of a specific topological feature of the liquid drop model potential energy surface V(Z) as a function of mass asymmetry Z. This feature is a deep minimum at symmetry (fission region) flanked at greater asymmetries by the Businaro-Gallone mountains which in turn descend at even larger asymmetries ("evaporation" region). The corresponding mass distribution from compound nucleus decay is approximately proportional to $\exp[-V(Z)/T_Z]$ and shows a peak at symmetry (fission peak) and two wings at the extreme asymmetries (evaporation wings). The qualitative dependence of the potential energy and of the mass yield vs. asymmetry is shown in Fig. 6a for a heavy nucleus.

With decreasing total mass the potential energy surface undergoes a topological change when the fissility parameter x crosses the so-called Businaro-Gallone point.[14] At this point (x_{BG} = 0.396 for ℓ = 0 and decreases for larger ℓ values) the second derivative of the potential energy with respect to the mass asymmetry coordinate evaluated at symmetry vanishes.[1,14,15] Thus below the Businaro-Gallone point there is no longer a traditional fission saddle point, and the monotonically increasing potential energy towards symmetry implies the disappearance of fission as a process distinct from evaporation. Thus the mass distribution should show the two evaporation wings extending as far as symmetry where a minimum should be observed. This is illustrated in Fig. 6b.

Such a transition has never been observed, as it requires the measurement of the entire mass distribution from symmetry to the extreme asymmetry of α,p evaporation for a series of systems straddling the Businaro-Gallone point. This measurement is made very difficult by the low yield for symmetric decay of the compound nucleus in this general mass region, and by the need to verify that the fragments were produced by a compound nucleus mechanism.[4]

We have measured complete charge distributions from protons to symmetric splitting for a variety of nuclei and we have observed the Businaro-Gallone transition. Such a transition is inferred from the disappearance of the fission peak in the mass yield as the compound nucleus mass was decreased from ^{148}Eu, ^{102}Rh to ^{83}Kr.

The use of reverse kinematics (projectile heavier than the target) was crucial in performing these measurements. This technique virtually eliminates the problems associated with low cross section measurements due to the presence of light element target contaminants. Furthermore, reverse kinematics provides a large center-of-mass (c.m.) velocity which facilitates the verification of full momentum transfer and allows for easy identification of the fragment's atomic number at the higher lab energies. Finally the high energy solution at forward angles corresponds to very backward angles in ordinary kinematics. This enhances the observation of compound nucleus decay and virtually eliminates any possible deep-inelastic contamination.

The experiments were carried out at the Lawrence Berkeley Laboratory SuperHILAC utilizing beams of 550-Mev ^{74}Ge, 782-MeV ^{93}Nb and 1157-MeV ^{139}La, to bombard targets of 0.54 mg/cm^2 ^{12}C and 1.0 mg/cm^2 ^{9}Be. The detection system consisted of four solid state ΔE - E silicon telescopes (40-70 μm, 3-5 mm) situated at 7.5°, 15°, 25° and 35° from the beam with solid angles of approximately 1.0 msr. For the heavier ^{139}La beam these detectors were supplemented by two gas ΔE-solid state E telescopes at -7.5° and 22.5°.

The observed laboratory energies represent only the higher energy kinematic solution. In general, the lower solution is not observed because of the energy threshold due to the thickness of our ΔE detectors. Thus the laboratory energy of the upper solution, the measured atomic number (Z), and angle permitted us to verify that the recorded events both originated from a system with full momentum transfer and had a c.m. energy independent of angle. This is shown for two representative elements in Fig. 7.

The mean laboratory energies for each Z-value were converted to velocities with two different assumptions for the relationship between the Z and the mass of the detected fragments. These velocities are then decomposed into two components. One component, along the beam direction, is assigned an arbitrary value; the other component is that required to reconstitute the original velocity. (For convenience this second component is shown as the c.m. energy in Fig. 7.) In this way, for each laboratory angle we can draw a curve representing the dependence of the c.m. energy upon the source velocity. This procedure is followed for each angle that is smaller than the kinematically allowed maximum angle. The intersection of these lines determines, in a model independent way, both the momentum transfer and the energy in the center of mass. The error bars shown on the lines in Fig. 7 reflect the uncertainty in the mean laboratory energies.

The results from this type of analysis for the ^{93}Nb + ^{12}C

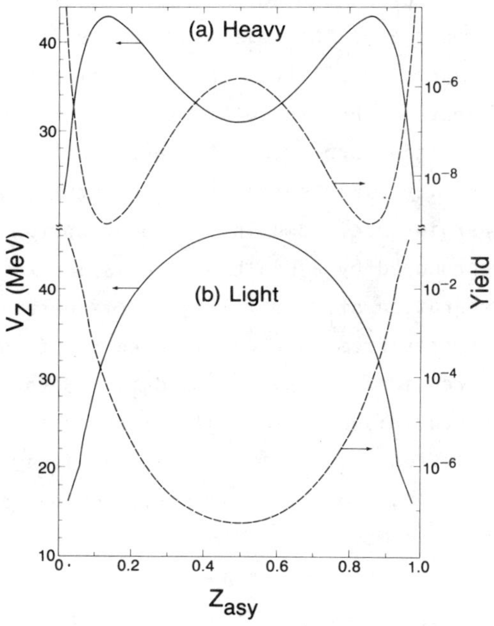

Fig. 6. Comparison of the potential energy surfaces (solid curve) and expected yields (dashed curve) for a) a heavy CN (Au at $\ell = 0$ and $E^* = 97$ MeV) and b) a light CN (Ge at $\ell = 0$ and $E^* = 72$ MeV).

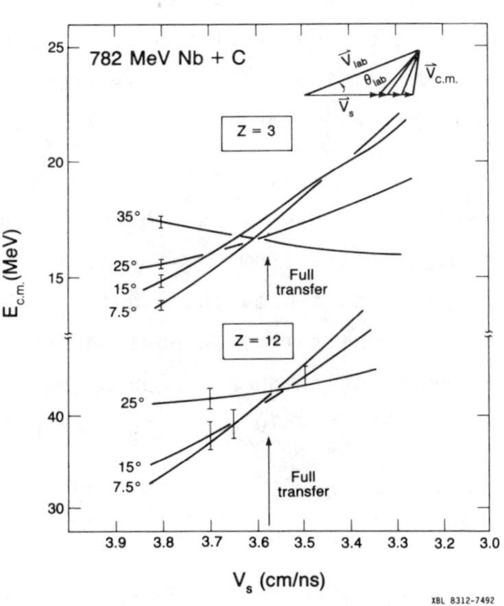

Fig. 7. The line for each angle gives the locus of solutions for both $E_{c.m.}$ and V_s. The intersection of the various lines fixes these quantities. The velocity corresponding to complete linear momentum transfer is indicated. This figure was drawn assuming that the masses follow the line of β-stability.

system are shown in Fig. 8. The upper part of this figure demonstrates that with either mass assumption all of the measured products result from the decay of a system with full momentum transfer. For the other systems studied, the extracted source velocities are also independent of Z within a few percent of the velocity expected for full momentum transfer. The deduced c.m. energies are shown in the lower portion of Fig. 8. These energies are reproduced by a Coulomb calculation for two spheres with a surface separation of 2 fm. This same separation also reproduces the c.m. energies from the ^{74}Ge induced reactions; however a larger separation is required for the ^{139}La data. Both the full momentum transfer and the invariance with angle of the c.m. energies seen above are consistent with compound nucleus decay.

The experimental cross sections for 530-MeV ^{74}Ge, 782-MeV ^{93}Nb and 1157-MeV ^{139}La + ^{9}Be systems are shown in Fig. 9. The cross sections are plotted as a function of charge asymmetry ($Z_{asy} = Z_{detected}/Z_{total}$). The lack of enhancement in yield near the target Z supports the compound nucleus origin of the products rather than a deep-inelastic origin. The yield from the ^{74}Ge + ^{9}Be system, with a fissility parameter of x = 0.31, decreases steadily as one moves towards symmetry. The yields from the ^{93}Nb + ^{9}Be system (x = 0.40) are essentially constant from Z_{asy} = 0.2 to 0.4 while the yields from the ^{139}La + ^{9}Be system (x = 0.50) show the characteristic fission peak at symmetry. These three systems clearly exhibit the qualitative trends expected from the topological changes in the potential energy surface predicted by the liquid drop model (see Fig. 6).

A quantitative comparison between these data and a compound nucleus calculation based upon the liquid drop model is also shown in Fig. 9. The absolute yields were calculated from the expression

$$\sigma_Z = \pi \lambda^2 \sum_{\ell=0}^{\ell_m} (2\ell + 1) \frac{\Gamma_Z(\ell)}{\Gamma_n + \Gamma_p + \Gamma_\alpha} \tag{7}$$

where $\frac{\Gamma_Z}{\Gamma_n}$ is given by Eq. (1).

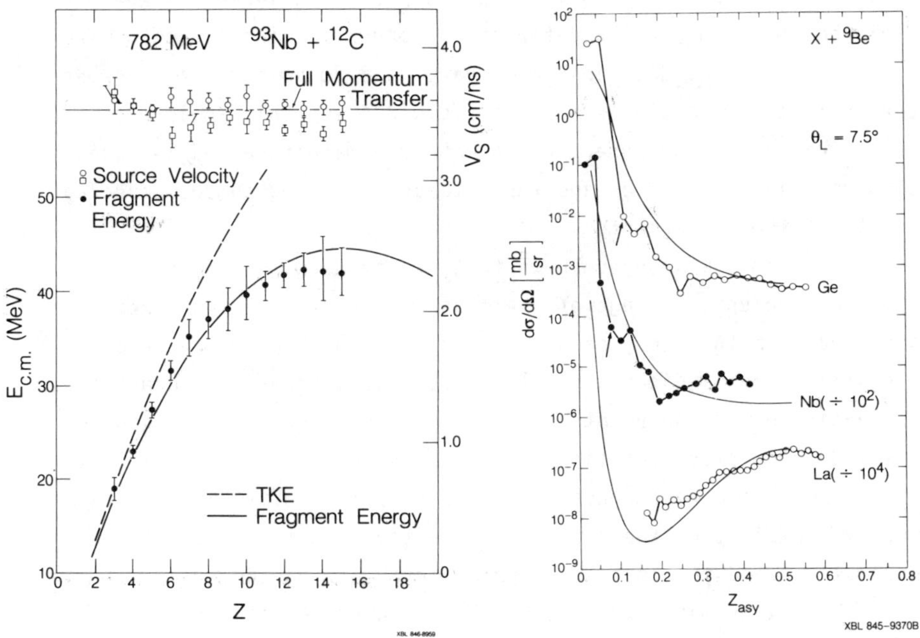

Fig. 8. The deduced c.m. energies (filled circles) and source velocities (open symbols) for the ^{93}Nb + ^{12}C system. Source velocities were determined assuming that the product mass followed the line of β-stability (open circles) or the charge equilibration line (open squares). A Coulomb calculation for two spheres is shown both for the c.m. energy of the light fragment (solid line) and the total kinetic energy (dashed line). The value of the source velocity expected for full momentum transfer is indicated by the horizontal line. The error bars indicate the uncertainty of the intersection point shown in Fig. 7.

Fig. 9. Center-of-mass cross sections for products from the ^{74}Ge, ^{93}Nb and ^{139}La + ^{9}Be systems detected at Θ_{Lab} = 7.5°. The solid line is a liquid drop model calculation of the fragment yield at $\Theta_{c.m.}$ = 30°. The arrows indicate the entrance channel asymmetry. See text. Data below z_{asy} = 0.15 were not obtained for the La + Be system, due to a limited dynamic range of the telescope.

The angular distribution expressions given in Ref. 1 were employed to calculate the differential cross section ($d\sigma/d\Omega$). The c.m. angles of the data in Fig. 9 vary as a function of Z. However, the average c.m. angle is approximately 30°, so this angle was chosen for comparison. The agreement in absolute magnitude and in trend between this calculation and the data confirms the compound nuclear origin of these fragments.

In summary, we have shown that fragments with atomic numbers covering the entire range of the mass asymmetry coordinate are produced from the decay of an excited compound nucleus. The observed Z distributions indicate that the topological transition expected at the Businaro-Gallone point does indeed take place in the region of $A \sim 100$. The exact position of the Businaro-Gallone point and its angular momentum dependence can in principle be established by a systematic study of the Z or A distributions as the fissility parameter x and the rotational parameter y are varied.

This work was supported by the Director, Office of Energy Research, Division of Nuclear Physics of the Office of High Energy and Nuclear Physics of the U.S. Department of Energy under Contract DE-AC03-76SF00098.

4. REFERENCES

1. Moretto, L.G., Nucl. Phys. A247, 211 (1975).
2. Moretto, L.G., in "Physics and Chemistry of Fission," Rochester, N.Y., 1973, pg 329.
3. Myers, W.D., "Droplet Model of Atomic Nuclei," IFI/Plenum Data Company, N.Y., 1977.
4. Sobotka, L.G., Padgett, M.A., Wozniak, G.J., Guarino, G., Pacheco, A.J., Moretto, L.G., Chan, Y., Stokstad, R.G., Tserruya, I. and Wald, S., Phys. Rev. Lett. 51, 2187 (1983).
5. Sobotka, L.G., McMahan, M.A., McDonald, R.J., Signarbieux, C., Wozniak, G.J., Padgett, M.L., Gu, J.H., Liu, Z.H., Yao Z.O. and L.G. Moretto, Phys. Rev. Lett.
6. Baker, E. et al., Phys. Rev. 112, 1319 (1958).
7. Dostrovsky, I. et al., Phys. Rev. 118, 781 (1960).
8. Westfall, G.D. et al., Phys. Rev. C17, 1368 (1978).
9. Green, R. and Korteling, R., Phys. Rev. C22, 1594 (1980).

10. Hanson, D.L. al., Phys. Rev. C9, 929 (1974).

11. Stokstad, R.G. et al., Phys. Rev. C16, 2249 (1977).

12. Blann, M., Lawrence Livermore National Laboratory Report UCID-20169 (1984).

13. Bass, R., Nucl. Phys. A231, 45 (1974).

14. Businaro, U.L. and Gallone, S., Nuovo Cim. 1, 629, 1277 (1955).

15. Nix, J.R., Nucl. Phys. A130, 241 (1969) and references therein.

GAMMA RAYS FROM EVAPORATION NUCLEI FORMED IN HEAVY-ION FUSION REACTIONS

L. Grodzins, R. Rohe, A. Smith, S. Steadman, and E. Vulgaris
Laboratory for Nuclear Science, M.I.T. Cambridge, MA

ABSTRACT

Gamma ray multiplicity distributions have been measured in coincidence with evaporation residues following the fusion of heavy ion. A recoil mass separator selected the evaporation residues, a NaI hodoscope detected the gamma rays. Studies have been carried out on compound systems from A=80 to 224 over a range of energies; in some cases down to far below the Coulomb barrier, in other cases to well above the energy where the spin saturates. Among the results which will be reported are: 1) The measurement of the maximum spin the evaporation residues can hold as a function of A and N. 2) Direct measurements of the spin carried per photon at the entry line of the gamma ray cascade. 3) The measurement of the broadening of the spin distribution with decreasing projectile energy, and the quantitative deterioration of the sharp cut-off approximation. 4) The low-energy asymptotic (statistical limit) to the mean multiplicity of gamma rays. 5) The observation of reaction-dependent trends in below - barrier multiplicity distributions.

I. INTRODUCTION

This paper is concerned with understanding the physics of the fusion of heavy ions through the study of the gamma rays which deexcite the evaporation residues. The experiments were carried out during the past two years with the Double Emperor Tandem Facility at Brookhaven National Laboratory. Many people were involved but the principal efforts were those of two PhD candidates, Andrew Smith[1] and Ronald Rohe[2]. The former studied the dissipation of angular momentum at the highest values of excitation energy and spin, while the latter was concerned with the lowest values which could be brought into the compound system. Smith's thesis was thus concerned with the decay channels while Rohe's was concerned with the entrance channels. All studies used the same basic apparatus; a recoil mass separator (RMS) to select the evaporation residues and a hodoscope of gamma ray detectors to measure the moments of the gamma ray distribution spectra. The experiments involved projectiles ranging from ^{16}O to ^{64}Ni and compound nuclei from A=80 to 224, with several tests of entrance channel effects. There has been one publication on the studies of high excitation and high angular momentum[3] and one preliminary report[4] on the below barrier work which is still in progress; the conclusions of the latter are necessarily tentative and fragmentary. Following a description of the experimental approach, we consider in Section III the results of measurments from just above the Coulomb barrier to well into the saturation region where the fused systems have as much angular momenta as they can retain without fission or significant -particle emission. Section IV discusses the results of gamma ray studies following fusion at energies below the Coulomb barrier; the bulk of the discussion will be on isotopes of Ni. The below-barrier experiments appears go give unique information about spin distributions which can shed light on the poorly understood below-barrier fusion cross sections.

II. EXPERIMENTAL OVERVIEW

There have number of studies of gamma ray deexcitation following fusion of heavy ions[5]. These studies have given us most of the deep insights we have of the bound states of fused systems, especially those at high angular momentum and excitation energy. With but a few important exceptions, these investigations used the selection of known gamma rays in the decay chain to identify the fusion. They were, therefore, limited, for the most part, to studies with known low-lying states, and to dominant channels. In most of the interactions reported on here, the decay schemes are unknown and there can be strong competition from fission or alpha decay. For these studies the RMS has proved invaluable, allowing the clean separation of fusion leading to evaporation.

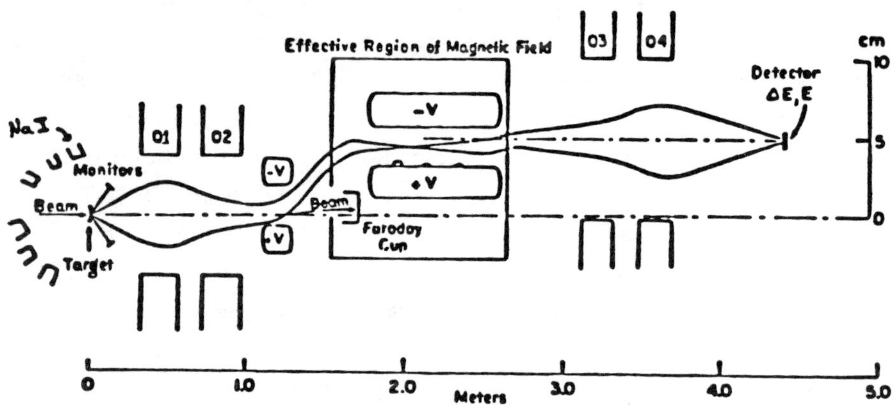

Fig. 1. Schematic of RMS-Multiplicity Apparatus

A schematic drawing of the experimental arrangement is shown in Fig. 1.[6] The target chamber was followed by a Wien-type velocity selector about 5 meters along[5] having an effective opening angle of about 16 mr in the horizontal and 32 mr in the vertical direction. The velocity bite of the RMS depends on the selected velocity but is typically $\Delta v/v = 0.55$. The RMS, set at 0°, has an excellent efficiency, generally grater than 5%, for detecting evaporation residues formed in heavy ion reactions. Only about 1 in 10^{12} of the beam particles reach the detector through the filter and the bulk of these are particles scattered by slits; they have the correct velocity but A_1/A lower energy than the fused nuclei A and so can be readily rejected by the ΔE(gas-E(Si) detector at the focus of the RMS. Surrounding the low mass target chamber were 10 to 14,3'x3' NaI detectors arranged so as to allow the determination of angular distributions of the deexcitation gamma rays and lead-shielded from neighbors; each detector was covered with appropriate absorbers to minimize the energy dependence of their efficiency. Two surface barrier detectors were placed at about ±22° to the beam to monitor the elastic scattered particles from the 100-150 $\mu g/cm^2$ targets. For each appropriate ΔE-E event, the energies and time relations of all detected gamma rays and particles were recorded through a CAMAC data collection system.

The coincidence resolving time was insufficient to discriminate against neutrons, so that appropriate (generally less than 5%) corrections for their presence were made to the data. The measured patterns of gamma rays detected in coincidence with the evaporation residues, were converted into a multiplicity distribution by techniques developed by Smith[1] and Rohe[3], which are refinements and augmentations of standard methods.

It is worth emphasizing that the RMS at 0° is not only insensitive to fission events but also discriminates against evaporation residues which follow alpha emission, since these ERs are recoiled out of the phase space accepted by the RMS. (Monti-Carlo simulations of the efficiency of the apparatus shows that, compared to xn reactions, the α emission is rejected by a factor of about 20. Thus, the gamma ray studies reported below are mainly from ERs formed after xn and yp evaporation, with some bias against the latter.

III. γ-DECAY FOLLOWING FUSION ABOVE THE COULOMB BARRIER

Fifteen reactions, ranging from $^{40}Ca + ^{40}Ca$ to $^{16}O + ^{208}Pb$ were studied as a function of bombarding energy.[1,3] Fig. 2A, showing the energy dependence of the first and second moments of the gamma ray distributions for the fusion reaction $^{17}Cl + ^{107}Ag$ is typical; a companion reaction, $^{56}Fe + ^{88}Sr$, was used for illustration in Ref 3. The mean <M> rises rapidly, reaching a plateau of <M> = 22. The root mean square value, RMSD or σ, on the other hand, plateaus much earlier. (We do not ascribe significance to the 'apparent' dip at 180 MeV.) The slope of the <M> vs E curve can, in principle, give the energy per gamma ray emitted at the entry state following particle emission; in practice, however, the measurement is swamped by the energy component from the particle emission. Much more informative is the multiplicity versus mean angular momentum. Fig 2B shows that dependence for the data of Fig 2A; standard methods were used for converting the beam energy into mean angular momentum.[1] Three distinct properties are worth noting: a linear initial slope; a relatively flat plateau; a reasonably sharp intersection.

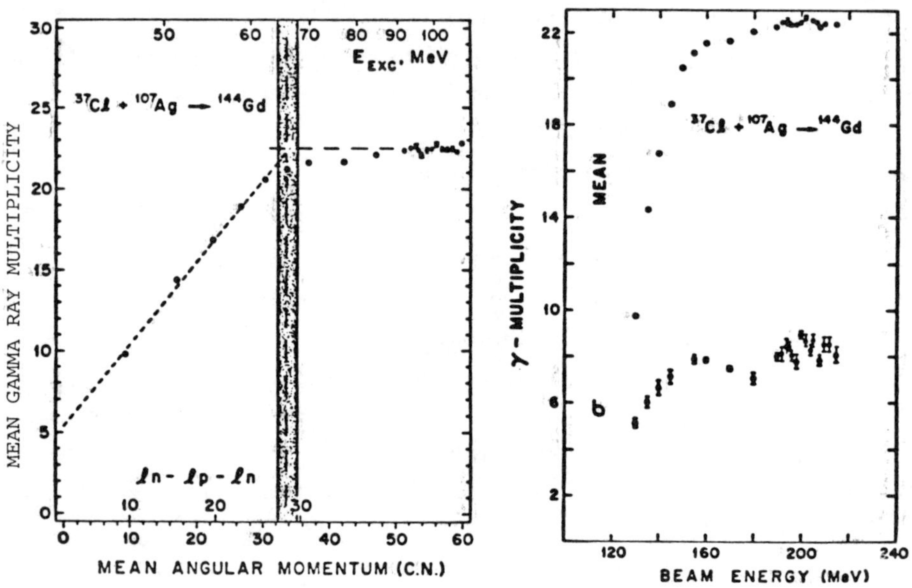

Fig. 2. Multiplicity moments versus energy and angular momentum.

Prior to saturation, the mean multiplicity rises linearly with input angular momentum; this behavior is typical of all reactions studied. There is therefore a definite slope, $\Delta\langle M\rangle/\Delta I$ representing increments at the entry level of the gamma ray cascade. The inverse of the slope, $_{\Delta}I/_{\Delta}\langle M\rangle$, is a direct measure of the total angular momentum carried away per emitted gamma ray. The abscissa values of I represents a sum of components including those from the particles emitted following the fusion. Since the magnitudes of the angular momentum carried by the neutrons and protons are small, they can be determined with sufficient accuracy using the statistical model code

Julian-Pace; recall that the RMS does not detect α-emitting ERs. The resulting mean angular momentum carried by the gamma rays is shown along the upper part of the abscissa scale of Fig 2B. The corresponding slope, $\Delta I(\gamma)/\Delta M$ yields the angular momentum carried by each gamma ray emitted at the entry point following the evaporation of particles. As far as we are aware, the slope method described here gives the most direct measure of this critical quantity. The measured value of $1.7 \pm .2$ is quite close to what might be expected for the principal evaporation nucleus ^{142}Eu. (The stated uncertainty is many times the statistical uncertainty; it is mainly from the estimated error in converting energy to angular momentum. The intercept, $\langle M \rangle = 4.6 \pm 1$, is a measure of the number of statistically emitted gamma rays; i.e. those which carry away no net angular momentum. The values of the slope and intercept depend on the final state; they are different, for example, for the two reactions, ^{37}Cl + ^{107}Ag and Fe + ^{88}Sr, which lead to ^{144}Gd, because the former reaction has almost 20 MeV more excitation energy (and therefore 1-2 more emitted particles) than does the latter reaction at the same input angular momentum.[3] The slope values can thus be used to determine the degree of collectivity of the evaporation nuclei. An example of such as study is reported in Refs 1 and 3. The effect of decreasing the neutron number from 82 to 72 for compound systems with Z=64, is dramatic. At the closed shell the angular momentum carried per photon is about 1.5. The value increases rapidly reaching 2K at N=76. In all 15 reactions the intercept ranged from 3 to 5K and the slopes ranged from 1.3 to 2K; these results are in good agreement with emperical formulas for converting multiplicity to spin.

The plateau of multiplicity versus energy of Fig 2B shows some structure. The mean value has about 1K smaller spin over the excitation range from 65 to 80 MeV. This 5% difference is readily accounted for by the fact that at the lower excitation energies the dominant ER has one fewer neutron than those at higher energies. Thus the structure in the plateau and at the intercept are both due to exit-channel effects as a result of the changing number of particles being evaporated as the excitation energy changes. The saturation can be experimentally correlated with the onset of alpha emissions which take increasing amounts of angular momenta and, at somewhat higher energy, by fission.[3] There is no evidence in any of our data for a rise in the reduced cross section following the initial saturation. Thus the distribution of angular momenta leading to evaporation residues appears to have a relatively sharp maximum with no long tails extending under the fission component of the cross section. Our overall view of the dissipation of the incoming angular momentum through various channels is shown in Fig 3 which is based on the energy depen-

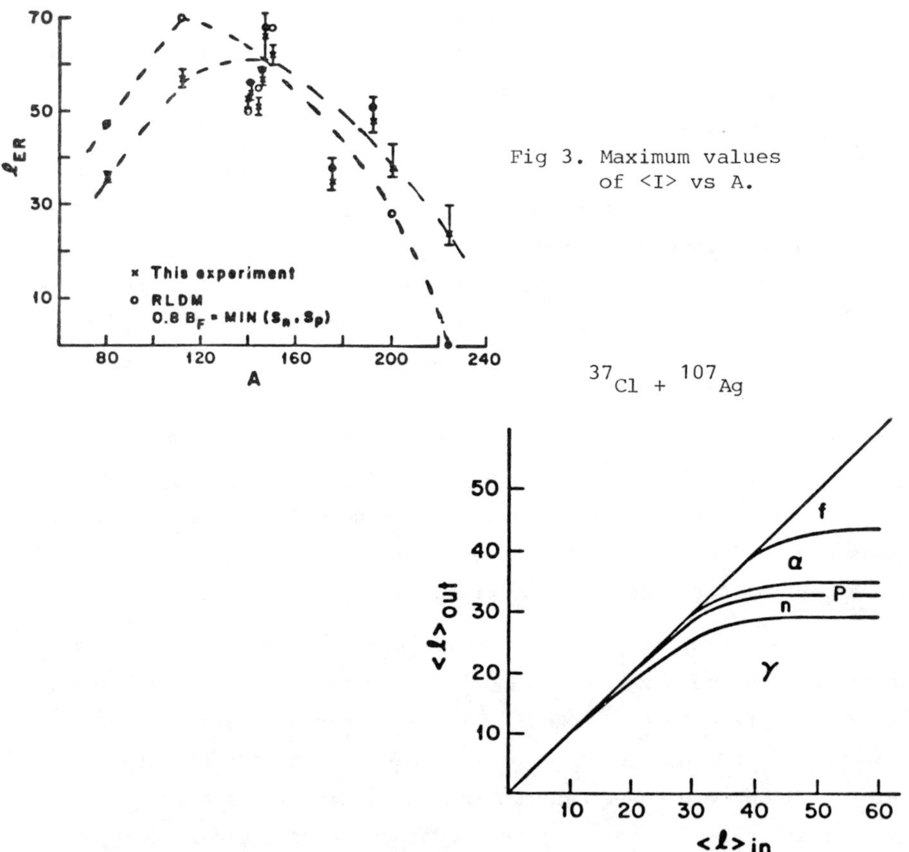

Fig 3. Maximum values of <I> vs A.

Fig. 4. Output components of angular momentum.

dence of gamma rays in coincidence with evaporation residues, α particles and fission.[1]

The intercept of the slope with the plateau region gives a direct measure of the maximum angular momentum which can be held by compound nuclei which decay by xn,γp emission. The saturation values are given in Fig 4,[3] together with the values expected from the rotating liquid drop model[7]. The RLDM model calculations were done for the specific reactions reported here, with fission barriers arbitralily lowered by 20%. The calculated values are in excellent agreement with the data for nuclei heavier than about A = 140. For lighter nuclei, the predictions are far higher than experiment, a not unexpected result since the RLDM model ignores a emission which is the dominant channel for dissipating the high angular momentum components. Overall, the maximum angular momentum that nuclei can sustain is about 70K.

FUSION BELOW THE COULOMB BARRIER

The mechanisms by which heavy ions fuse at energies below the classical Coulomb barriers are still not well understood despite the attention given to the problem in recent years.[8] The central puzzle, evident in the earliest experiments, and now observed in a wide variety of heavy-ion fusion reactions, is that generally the cross sections do not fall as rapidly as simple theory predicts. The problem is illustrated by the seminal study of oxygen fusing with isotopes of Sm.[9] The heavier isotopes, ^{152}Sm and ^{154}Sm have far larger below barrier fusion cross sections than do the lighter, ^{144}Sm and ^{150}Sm. The difference is presumably due to the large static deformations of the N $\geq$ 90 nuclei which have lower Coulomb barriers when the fusion takes place against the long ends of the ellipsoids. But the large below-barrier cross sections for the lighter Sm isotopes are still not well-understood though recent coupled channel calculations look promising.[8] The only data against which the models are tested are the energy dependences of the absolute cross sections for

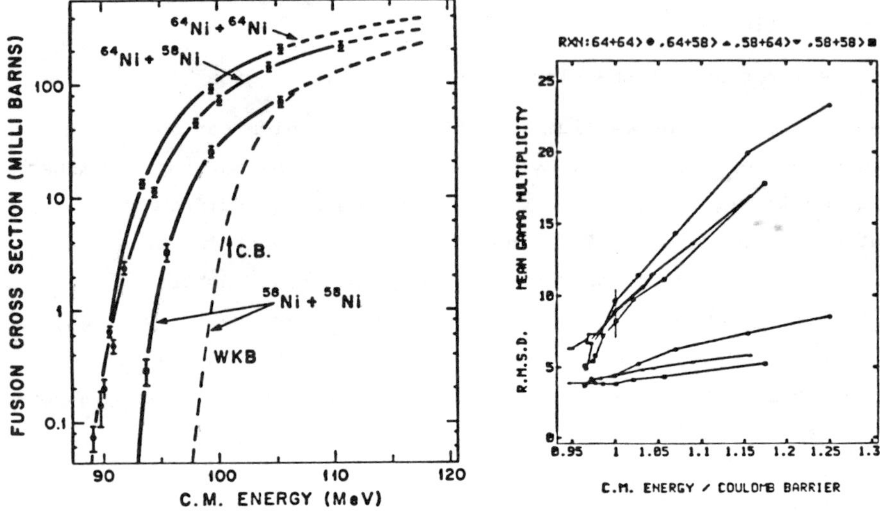

different pairs of interacting nuclei. When, as in the O-Sm case, the pairs include variations of static shapes one has the opportunity of making a decisive conclusion, though even in the case of the Sm isotopes there are other important variables besides deformation. When the interacting nuclei are not so distinctly different then the ability of the excitation function to discriminate among the models is limited. Such is the case with the provocative results of Beckeman who studied the fusion of the isotopes of nickel.[10] The energy dependence of the fusion cross sections are shown in Fig. 5 (These are recent data[2], carried out with some improvements in the technique. While the general results are similar to those of Ref. 10, there are important differences, particularly with the cross channel data.) All of the measured fusion cross sections are enhanced over that expected from simple Hill-Wheeler barrier penetration. And each of the combinations, ^{58}Ni + ^{58}Ni, ^{58}Ni + ^{64}Ni, ^{64}Ni + ^{64}Ni, shows a somewhat different behavior. Various phenomena have been invoked to explain these data, including the effects of nucleon-transfer channels, coupling to inelastic channels, giant dipole strengths, and zero point fluctuations. One would like to find additional parameters, intrinsic to the fusion reaction itself, which can give additional information about the process. The study which is potentially capable of illuminating the process is that of the gamma ray decay of the compound system after particle evaporation. One such study has been reported, on some of the ^{16}O + Sm fusion reactions, in which the mean multiplicity of the gamma distribution was measured as a function of the excitation energy. As we shall show, the width of the gamma ray multiplicity distribution and the ratio of mean to that width, are also important.

It might appear that the RMS is unessential to the studies of fusion below the barrier and this is indeed the case for studies with lighter projectiles such as ^{16}O which result in only one or two dominant channels. Vandenbosch et al[11] were able to study the reaction $^{154}Sm(^{16}O,4n)^{166}Yb$, down to about a mbarn by gating on the (4 - 2) transition. However, with heavier projectiles, such as Ni, the Coulomb excitation is so dominant, and the final decay products are so abundant and complex, that such small cross sections are very difficult to observe without gating on the fused recoil.

Fig 6 presents the multiplicity as a function of center of mass energy for the various Ni-Ni combinations. Several points are worthy of note: 1. The mean multiplicity rises rapidly from around 5 to values close to 20. 2. The energy dependences of the mean multiplicities reflect the cross section dependence in that the 64-58 values lie between the 58-58 and 64-64 values; the latter being the highest.3. The RMSD values rise slowly with energy; to first approximation the widths are constant.

In Fig 6 it appears that the mean multiplicities will continue to drop as the center of mass energy is lowered but we doubt if that will happen. On general grounds we expect that the minimum in $\langle M \rangle$ should be that of a cascade of statistical photons. To sec this effect, the mean multiplicities have been plotted as a function of the fusion cross section, Figs 7A,7B, and 7C. By stretching the abscissa one sees that the values of <M> asymptote at values close to 5, as expected.

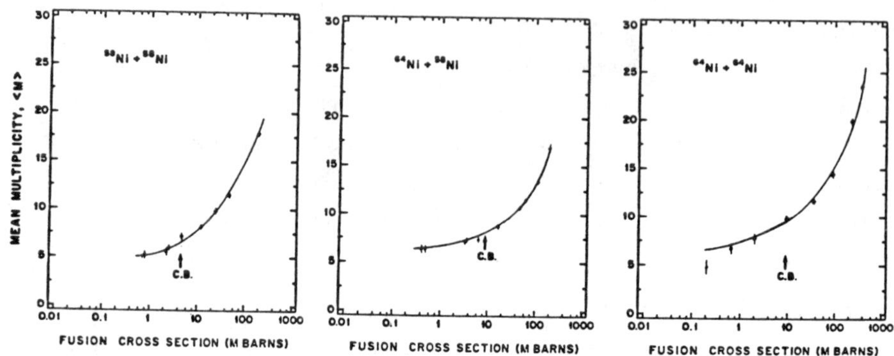

Fig 7. <I> versus fusion cross section for the Ni-Ni reactions.

The values of <M> are dropping far more slowly than expected from the sharp cutt-off model, a point made in some detail in Ref.11. This expected behavior is observed directly by plotting the ratio of the mean to the width of the multiplicity distribution. Fig 8A shows this parameter for the three Ni interactions. They are placed on an energy scale normalized to the Coulomb barrier energy. No points or errors are given in order to bring out the general features, but we emphasize that, as is clear from Fig 5, there are few points and each has a significant error, thus the asymptote is far from a flat line. The figure is meant to bring out what is clear from the data: the three reactions show distinct behaviors with the ^{58}Ni-^{58}Ni rising rapidly to the sharp cut-off value of 2.83 while the ^{58}Ni-^{64}Ni rises slowly. These data support the conclusions from the cross section measurements that the cross-channel is 'softer' than either of the others, with the ^{58}Ni-^{58}Ni being the most rigid.

It is worth noting that the energy dependence of the ratio of <M>/σ for the ^{17}Cl + ^{107}Ag reaction, shown in fig 8B, is essentially the same as that for the ^{64}Ni data of Fig 8A; that is, the onset of that deviation begins at about 15% above the barrier and the deterioration from the sharp cut-off approximation is slow.

With quantitative results for both the fusion cross sections and the multiplicity distributions, it should be possible to determine the correct explanation of the belowbarrier phenomena for Ni-Ni in a single global fit. Rohe has tried to do so starting from the 'natural'

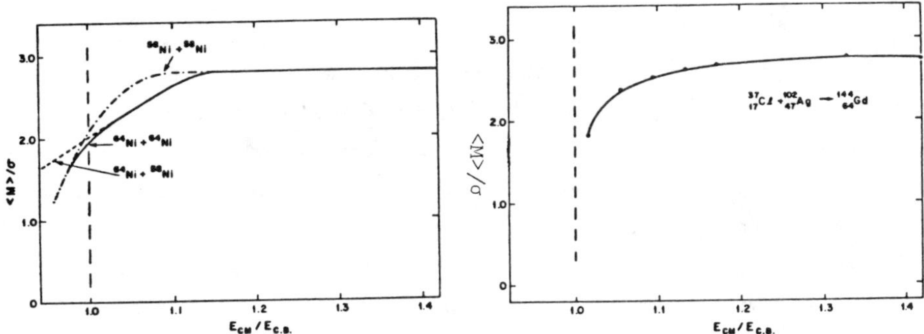

Fig 8. <I>/(RMSD) for Ni-Ni and for Cl-Ag reactions versus normalized energy.

explanation of zero point fluctuations based on B(E2) values from the ground to the first excited state; Esbensen suggested this process as the explanation for the cross section anomalies.[12] Both the spin distributions and the cross sections were calculated using the method of equilavalent spheres, assuming zero point fluxuations in both the projectile and target.[2] The calculations depend, of couse, on knowing the conversion factor from spin to multiplicity. Rohe has so far been unable to account satisfactorally for all the data assuming reasonable parameters for the zero point fluctuations and the spin to M conversion; the magnitudes agree reasonable well but a single set of parameters cannot account for the trands with energy.

We close this section by noting that the gamma ray deexitation contains much more information than we have presented above. In principle, the skewness of the multiplicity distribution can be obtained though in practice this parameter has proved elusive. On the other hand, the angular distribution of the gamma rays is relatively easy to obtain and gives information on the alignment of the fused system. Moreover, the energy distribution of the gamma rays can give insight into the collectivity of the transitions and, of course, the study of the discrete gamma ray transitions is a deep mine of detailed information about the bound states and their origin.

IV. SUMMATION

Gamma ray distributions obtained in coincidence with evaporation residues is a powerful tool for the study of fusion reactions. At reaction energies above the Coulomb barrier the method gives such information as the maximum angular momentum an evaporation residue can hold, and the collectivity of nuclear states at high excitation and high angular momentum. At reaction energies below the barrier, the studies can give unique information about the spin distributions leading to fusion and may thus provide the necessary measures to discriminate among the different explanations for enhanced fusion cross sections.

The authors acknowledge the support of the Department of Energy which supported this work through contract DE-AC02-76DR03069.

REFERENCES

1. A.P. Smith, PhD. Thesis, Mass Inst of Tech, 1983, unpublished.

2. R.C. Rohe, PhD. Thesis, Mass Inst of Tech, 1984, unpublished.

3. L. Grodzins, S. Gazes, A. Smith, S. Steadman, E. Vulgaris and J. Wiggins, Workshop on Electromagnetic Properties of High Spin Nuclear Levels, Israel, Jan. 1984.

4. R. Rohe, J. Wiggins, L. Grodzins, H. Enge, C. Ordonez, M.K. Salomaa and E. Vulgaris, MIT Conference on Fusion Reactions below the Coulomb Barrier, June 1984.

5. R.M. Diamond and F.S. Stephens, Ann. Rev. Nucl. Sci. 30, 85(1980)

6. H.A. Enge and D. Horn, Nucl.Inst.Meth., 145, 271 (1977)
M. Salomaa and H.A. Enge, Nucl. Inst. Meth., 145, 279 (1977)

7. S. Cohen, F. Plasil and W.J. Swiatecki, Ann. Phys. 82, 577 (1974)

8. MIT Conference on Fusion Reactions Below the Coulomb Barrier, S. Steadman, Editor. June, 1984. To be published.

9. R.G. Stokstad and E.E. Gross, Phys. Rev. C 23, 281 (1982) and R.G. Stokstad et al., Phys. Rev. C 21, 2427 (1980).

10. M. Beckerman, J. Ball, H. Enge, M.K. Salomaa, A. Sperduto, S. Gazes, A. DiRienzo and J.D. Molitoris, Phys. Rev. C 23, 1581 (1981).
M. Beckerman, M.K. Salomaa, A. Sperduto, A. DiRienzo and J.D. Molitoris, Phys.Rev. C 25, 837 (1982).
M. Beckerman, MIT Conference on Fusion of Heavy Ions Below the Coulomb Barrier, June, 1984.

11. R. Vandenbosch, B.B. Back, S. Gil, A. Lazarini and A. Ray, Phys. Rev. C 28, 1161 (1983).
B.B. Back, S. Gill, A. Lazzarini, D.K. Lock, A. Ray, and R. Vandenbosch, MIT Conference on Fusion of Heavy Ions Below the Coulomb Barrier, June, 1984.

12. H. Esbensen, Nucl.Phys. A352, 147 (1981).
H. Esbensen, Jian-Qun Wu and G.F. Bertsch, Nucl. Phys. A411, 275 (1983).

RELATIVISTIC MEAN FIELD THEORY WITH INTERACTING THERMAL PIONS

J.Zimányi*, J. Bondorf, I. Mishustin**
Niels Bohr Institute,DK-2100 Copenhagen Ø, Blegdamsvej 17.
DENMARK
and
J. Theis
Gesellschaft für Schwerionenforschung GmbH.
D-6100 Darmstadt 11, Planckstrasse 1. Postfach 110541
Germany

ABSTRACT

A relativistic mean field theory is presented in which both the nucleons and pions are treated as particles. Their interaction is mediated by the σ mean field. Numerical results for high temperatures are given.

1. INTRODUCTION

It is one of the most discussed question in heavy ion physics, whether a phase transition to quark gluon plasma will occour in high energy heavy ion collisions. However, to obtain reliable signatures for this phase transition, the same observable has to be calculated for the hadronic phase as well as for the quark phase. For this purpose one needs a realistic description of the hadronic phase too. Depending on bombarding energy the high energy density hadronic matter is produced with very different characteris-

---

* Permanent address: Central Research Institute for Physics, H-1525 Budapest 114, POB 49., Hungary

** Permanent address: Khurchatov Institute of Atomic Energy, Moscow, USSR

tics. To deal with this problem we propose an effective mean field model[1].

The relativistic mean field theory introduced by Walecka [2] and Chin and Walecka [3] treats the baryons by quantum fields and the bosons (σ and ω mesons) by mean fields. In the later developments the pions were also included. But they were treated either by mean field in investigations on pion condensates (Banerjee, Glendenning, Gyulassy [4]) predicted by Migdal [5], or were completely neglected on the basis that the expectation value of a pseudoscalar field in a symmetric system vanishes.

In the heavy ion collisions a rather large number of pion particles are produced. In different approaches they were treated e.g. by statistical methods (e.g. ref.[6]) or by cascade calculations (e.g. ref. [7,8]). Quantum mechanical approaches were studied e.g. in refs. [9,10]. In ref.[9] the pions having a realistic dispersion relation depending on the baryonic density were added to a nuclear matter described by the variational method.

An obvious way to include pions in relativistic mean field models is just a simple addition of pions produced in the thermal bath. In this paper we try to solve this problem in a more sophisticated way by taking into account the strong interaction of pions with nucleons.

One has to realize, however, that the interaction of pions with a system of antinucleons is exactly the same as the interaction of pions with a system of nucleons in the same state. Therefore, especially in the $\mu=0$ case one should add these contributions instead of subtracting them, which will happen if the pions are coupled to the baryonic density, ρ_b. This type of coupling was considered in most earlier works.

Once one realizes the importance of the pion antinucleon coupling for the description of heavy ion reaction phenomena, one can proceed in different ways. Here the interaction of pions with nucleons and antinucleons will be simulated by an effective coupling of the pions to the scalar σ field.

In the present paper we propose a consistent model based on an effective Lagrangian. Both the nucleons and pions are described by quantum fields, while the σ and ω mesons are treated by mean fields. Instead of having a direct pion nucleon coupling, in which case it would be very difficult to solve the problem, we approximate the pion nucleon coupling by an interaction mediated by the σ field. This σ coupling corresponds to second order and simplest higher order loop contributions to the pion polarization operator.

The σ field is related to the scalar density, which, on the other hand, is related to the sum of the densities of nucleons and antinucleons. Thus at finite temperatures the pions will have interaction with the nucleon - antinucleon system even in the zero chemical potential case (i.e. when ρ_b is zero) . The present model yields pionic properties similar to those obtained in earlier calculations in the neighborhood of normal nuclear density, but it can be used in a much larger range of thermodynamical parameters.

2. PIONS IN HOT AND DENSE MATTER

Since we want to obtain an effective Lagrangian, which yields a relatively realistic description of the pions, in this section we give a short discussion of the pionic properties of the hadronic matter.

The properties of pion-like excitations in hot and dense nuclear matter have been extensively investigated in the last decade. A systematic study of the pion excitation spectrum in nuclear matter was initiated by a series of papers by Migdal and collaborators [5,11,12] . It was realized, that due to the strong pion nucleon interaction the pion excitation spectrum, $\omega(k)$, in nuclear matter is strongly modified in comparison to that in vacuum. Moreover, instead of a one pion branch, which transforms into the free pion spectrum $\omega(k)=\sqrt{m_\pi^2+k^2}$ at zero density, two additional branches of pionlike excitations appear in nuclear matter. One of them is the spin-isospin sound branch, $\omega \sim k$ at $k \to 0$, which is closely related to the pion condensation phenomena. The other one is the so called Δ -resonance branch corresponding to the collective Δ -particle - nucleon hole excitations.

In general form the spectrum of pion-like excitations is given by the poles of the pion propagator

$$\omega^2 - m_\pi^2 - k^2 - \Pi(\omega,k) = 0. \qquad (1)$$

Here $\Pi(\omega,k)$ is the polarization operator, which depends on the energy, ω , and momentum, k , of pions as well as on the baryon density, the temperature and isotopic composition of the nuclear matter. The main contribution to $\Pi(\omega,k)$ comes from the p-wave pion nucleon interaction with Δ -particle - nucleon hole intermediate states. These interactions are essentially renormalized by the short-range nucleon-nucleon and nucleon-Δ correlations. The s-wave pion-nucleon interaction is small in isotropically symmetric nuclear matter. Detailed calculations of $\Pi(\omega,k)$ at $T=0$ were made in ref. [11]) . The main conclusion of these calculations is the observation, that the pion spectrum is essentially softer in nuclear matter than in vacuum.

At finite temperature the situation is more complicated. Above of the trivial generalization of the baryon occupation numbers for the finite temperature case, there is an important temperature dependence of the pion spectrum arising from the interaction of the pions with the thermally produced pion excitations. The softer the pion spectrum is, the larger number of pion excitations are produced at a given temperature, and, consequently, the stronger is the feedback effect of these excitations on the pion spectrum. It was realized in the last few years, that this effect is especially important for the description of pion condensation phenomena at finite temperatures. In this case the zero energy modes of pion excitations bring a singular contribution to the polarization operator, which prevent the second order phase transition. In the present paper we consider a system well above the critical temperature for the pion condensation. At high temperature the low energy spin-isospin sound branch of pion-like excitations, responsible for the pion-condensation, is either filled up or even disappearing due to the destruction of the Fermi surface. So further on we shall consider only the pion branch.

Instead of using rather complicated expressions for the polarization operator in terms of Lindhard functions (see e.g. refs. [11,13]) , which contains a number of not well known parameters, we shall parametrize the pion spectrum in a suitable way. We are guided by the following arguments. The main contribution to the polarization operator $\Pi(\omega,k)$ at $\omega \simeq m_\pi$, is yielded by the attractive p-wave pion-nucleon interaction leading to a Δ particle as an intermediate state. This contribution is proportional to k^2 as a consequence of the p-wave nature of the Δ -particle. Due to the fact, that the difference, Δm , between the mass of the Δ -particle and that of the nucleon, is large, this contribution is also proportional to the nucleon density, and is almost independent of the temperature at a fixed nucleon density. An essential temperature dependence will appear only near $T \simeq \Delta m \simeq 300$ MeV. Therefore it seems reasonable to drop the explicit temperature dependence of the pion spectrum in the first approximation. So we can write the polarization operator for the pion branch rather generally in the following form:

$$\Pi(\omega,k) = -\frac{a\,k^2\,\Lambda^2(k)\,\rho_b}{1+g''\,a\,\rho_b}. \qquad (2)$$

Here a is a constant, which can be determined from the condition, that at normal nuclear density and in the limit $k \to 0$ the coefficient in front of k^2 should be the same as in the Kisslinger's optical potential for pionic atoms. The denominator in eq. (2) stays for simulating the well known Ericson-Ericson effect [14]. In other words, it appears as a result of summation of ladder diagrams responsible for the local Δ-particle - nucleon hole interaction characterized by the amplitude $g''=g' m_\pi^2/f^{*2}$. The f^* is the $\pi N \Delta$ coupling constant and g' is the conventional Fermi-liquid parameter for spin-isospin channel. The factor $\Lambda(k)$ in eq. (2) is introduced to provide the correct k-behaviour of the pion spectrum at large k. Due to the strong interaction between the different elementary excitations, at large k values the pion branch in fact corresponds to the Δ-particle - nucleon hole excitations. (In the limit of vanishing pion interaction, the "bare" pion branch and the Δ-resonance branch cross at an intermediate k value.) Therefore $\Lambda(k)$ is chosen so, that at large k, ω approaches the Δ-resonance branch instead of the free pion branch. For $\Lambda(k)$ we shall use the exponential form

$$\Lambda(k) = \exp(-k^2/b^2). \qquad (3)$$

The numerical values of the parameters a,b and g'', determined from the conditions discussed above, will be given in the next Section.

3. THE EFFECTIVE LAGRANGIAN

A simple representation of the properties of hadronic matter discussed in the previous chapter can be formulated in the framework of a relativistic mean-field theory based on the effective Lagrangian

$$L = L_0 + L_\pi , \tag{4}$$

where L_0 is the effective Lagrangian proposed by Walecka[2])

$$\begin{aligned} L_0 = {} & \bar{\psi}(i\not\partial - m_n)\psi \\ & -\tfrac{1}{2}(m_\sigma^2 \sigma^2 + m_\omega^2 \omega_0^2) \\ & + g_{n\sigma}\sigma\bar{\psi}\psi - g_{n\omega}\omega_0\bar{\psi}\gamma_0\psi . \end{aligned} \tag{5}$$

Here we dropped the space and time derivatives of the σ and ω fields, which will disappear anyhow in the mean field approximation in a spatially uniform stationary system. The units $\hbar = c = 1$ and the notation of Bjorken and Drell is used. The field ψ corresponds to the nucleon field σ to the scalar field and ω_0 is the time like component of the ω vector field. For L_π we use the following form:

$$\begin{aligned} L_\pi = {} & \sum_a \frac{1}{2}\Big\{ \partial_\mu \pi_a \partial^\mu \pi_a - m_\pi^2 \pi_a^2 \\ & + \frac{g_{\pi\sigma}\sigma/m_\sigma}{1 + g'' g_{\pi\sigma}\sigma/m_\sigma}\Big[(\nabla e^{\Delta/b^2}\pi_a)(\nabla e^{\Delta/b^2}\pi_a) + \bar{g}_{\pi\sigma} m_\sigma^2 \pi_a^2\Big]\Big\} \\ & - \frac{\lambda}{4}(\pi_b\pi_b)(\pi_c\pi_c), \\ & \qquad a, b, c = 1, 2, 3 . \end{aligned} \tag{6}$$

Here $\pi_1(x)$, $\pi_2(x)$ and $\pi_3(x)$ are three real pseudoscalar fields; $\pi_\pm = (\pi_1 \pm i\pi_2)/\sqrt{2}$, $\pi_0 = \pi_3$. In the present paper we consider isosymmetric hadronic matter, where all components of the pion field are equal. With that in mind, from now on we shall describe the pions with one pseudoscalar field, $\pi(x)$, which stands for any of π_1, π_2 or π_3. The summation over the three pion fields will be replaced by the multiplication by the degeneracy factor, $d_\pi = 3$.

The Lagrangian L_π has the form given by eq. (6) in the rest frame of hadronic matter. It can be reformulated in a relativistically invariant way.[1]

The Lagrangian (6) with $g'' = \lambda = 1/b^2 = 0$ leads essentially to Kislinger's optical potential for the elastic scattering of pions in homogeneous nuclear matter. The denominator with $g'' = 0.7$ approximates the Ericson-Ericson correction to the pion optical potential. The factors $\exp(\Delta/b^2)$ cut off the unphysical contribution of very large pion momenta. The results are not sensitive to the cut off parameter, b, because in the thermal integrals there are an exponential cut-off anyway. The last term represents the repulsive pion self interaction, which is important for large pion densities.

From the Lagrangian the field equations are obtained as Euler-Lagrange equations. Then we apply the mean field approximation by replacing σ and ω_0 by their average values, $\langle\sigma\rangle$ and $\langle\omega_0\rangle$, respectively. The treatment of the pion self-interaction in the mean field approximation is more problematic. Here we shall make the following approximation in the pion field equation:

$$\pi(x)^2 \; \pi(x) \longrightarrow \langle \pi^2 \rangle \; \pi(x), \tag{7}$$

and at the same time the contribution of the pion self-interaction to the energy density will be approximated as:

$$\pi(x)^2 \; \pi(x)^2 \longrightarrow \langle \pi^2 \rangle \langle \pi^2 \rangle , \tag{8}$$

where $\langle\pi^2\rangle$ denotes the grand canonical ensemble average of the operator $\pi(x)^2$. This replacement, of course, is only an approximation, which, we hope, will give correct qualitative approximation to the realistic situation. Proceeding in this way one obtains the equation of motions of our model as follows:

$$[\, i\not\partial - g_{n\omega}\, m_\omega\, v\, \gamma_0 \; - \; m_n^* \,]\, \psi(x) = 0, \tag{9a}$$

$$\left[\frac{\partial^2}{\partial t^2} - \left(1 - \frac{g_{\pi\sigma}\, S}{1 + g'' g_{\pi\sigma}\, S} e^{2\Delta/b^2}\right) \Delta + m_\pi^{*2} \right] \pi(x) = 0, \tag{9b}$$

$$S \equiv \frac{\sigma}{m_\sigma} = g_{n\sigma}\, \frac{1}{m_\sigma^3} \langle \bar\psi \psi \rangle + \frac{g_{\pi\sigma}}{(1 + g'' g_{\pi\sigma} S)^2} \, \frac{1}{2} \, \frac{1}{m_\sigma^4} \, *$$

$$* \, d_\pi \langle (\nabla e^{\Delta/b^2} \pi)(\nabla e^{\Delta/b^2} \pi) + \bar g_{\pi\sigma} m_\sigma^2 \pi^2 \rangle , \tag{9c}$$

$$v \equiv \frac{\omega_0}{m_\omega} = g_{n\omega}\, \frac{1}{m_\omega^3} \langle \bar\psi \gamma_0 \psi \rangle , \tag{9d}$$

where the effective masses are given as

$$m_n^* = (1 - g_{n\sigma} \frac{m_\sigma}{m_n} S)\, m_n \tag{10}$$

$$m_\pi^* = \left(1 - \frac{g_{\pi\sigma}\, S}{1 + g'' g_{\pi\sigma}\, S} \left(\frac{m_\sigma}{m_\pi}\right)^2 \bar g_{\pi\sigma} + \lambda\, d_\pi \langle \pi^2 \rangle / m_\pi^2 \right)^{1/2} m_\pi ; \tag{11}$$

These equations of motion have plane wave solutions. This property makes it possible to evaluate the grand canonical averages in an easy way.

Inserting a plane wave form for the pion field in eq. (9b) one obtains the the dispersion relation for the pion field as

$$\omega(k) = \left[\left(1 - \frac{g_{\pi\sigma}\, S}{1 + g'' g_{\pi\sigma}\, S} e^{-2k^2/b^2}\right) k^2 + m_\pi^{*2} \right]^{1/2} , \tag{12}$$

which is consistent with the polarization operator, eq. (2). The corresponding expression for the baryon field has the form

$$E_n(k) = g_{n\omega}\; m_\omega V \pm \left[k^2 + m_n^{*2} \right]^{1/2}. \qquad (13)$$

The π-σ coupling constants $g_{\pi\sigma}$=29.3 and $\bar{g}_{\pi\sigma} = -3.29 \times 10^{-3}$ correspond to b_o =-0.02 and c_o = 0.15 in the Kisslinger's optical potenital. With this choice of parameters we can reproduce the density dependence of the pion spectrum of ref. [9] . For g" 0.7 was used.

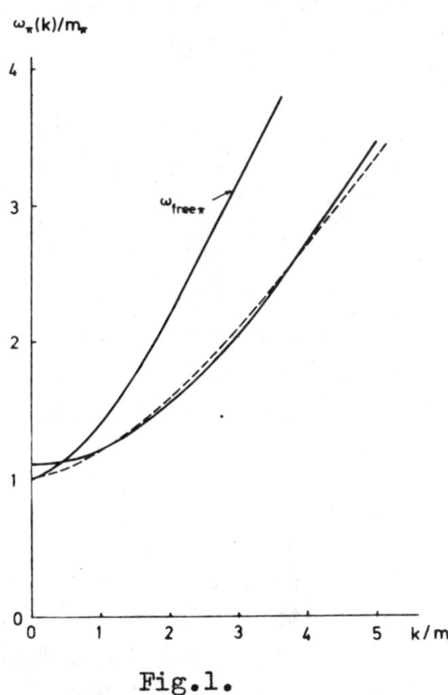

Fig.1.

In Fig.1. the $\omega = \omega(k)$ dispersion relation for T=100 MeV and for $\rho = \rho_o$ as calculated with our model (full line) and given in ref. [9] (dashed line) are displayed. The parameter b appearing in the formafactor corresponding to the π -n interaction will be set equal to b=14m_π . We had to chose this b value to fit the pion dispersion relation of ref. [9]. One can observe, that up to k=5 the two curves are near to each other. On the other hand, the k values higher than that, do not contribute appreciably to the integrals determining the thermodynamical properties of the system. Thus the precise behaviour of $\omega(k)$, for higher k values, if not pathological, is not important for our present investigation.

Since the pionic degress of freedom are not effective in our model at zero temperature, where the coupling constants $g_{n\sigma}$ and $g_{n\omega}$ are determined from the ground state properties of the nuclear matter, we have chosen for them the values given earlier in the literature (see e.g. ref.[16]). These values are

$$g_{n\sigma} = 12.1 \qquad g_{n\omega} = 12.6$$
$$m_{\sigma} = 650 \text{ MeV} \qquad m_{\omega} = 780 \text{ MeV} \qquad m_{n} = 940 \text{ MeV}$$
$$c_s^2 = 307 \qquad c_v^2 = 230.$$

4. RESULTS AND DISCUSSION

We have calculated the simultaneous selfconsistent solutions of eqs. (9a-9d) for T=100 MeV and different chemical potentials, μ, and also for μ=0 and different temperatures.

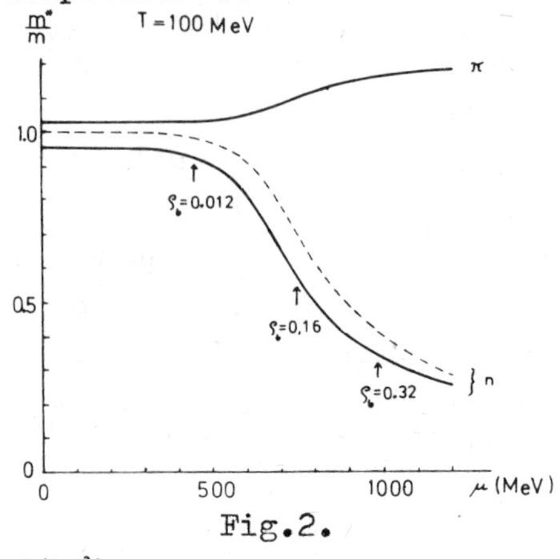

Fig.2.

Fig.2. shows the effective mass of nucleons and pions at T=100 MeV as a function of the chemical potential calculated with the present model (full line) and with free non-interacting pions (dashed line).

Fig.3.

In Fig.3. the pion density is plotted versus the baryon density. Here we can observe a very large increase in pion density (full line) over the free thermal pion gas (dashed line) density. For comparison same values taken from ref.[9] are also plotted (dotted line) . We have to realize, however, that these pions are "soft", off energy shell pions.

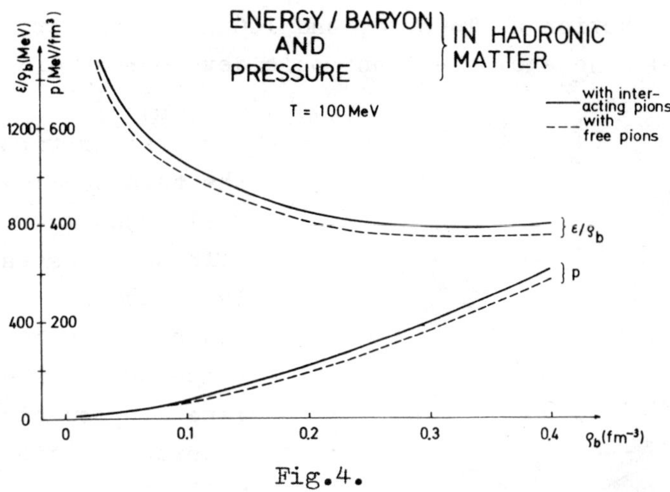

Fig.4.

The energy/baryon and the pressure is not effected strongly at this temperature and at these densities, as can be seen in Fig.4. (Present model: full line, free pion case: dashed line.)

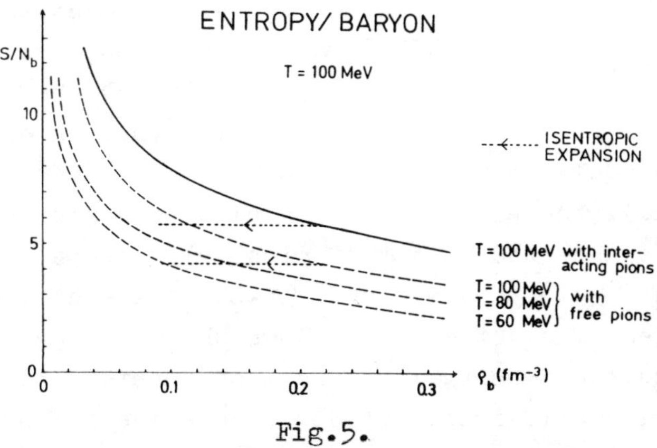

Fig.5.

There is, however, an essential change in the total entropy of the system. (This effect was pointed out also in ref.[17]) . In Fig.5. the entropy/baryon is plotted versus the baryon density.The full line displays the values obtained at T=100 MeV with the present model, while the dashed lines correspond to the values calculated with free thermal pion contribution at T=60, 80 and 100 MeV, respectively. The dotted line show the isentropic development of the system.

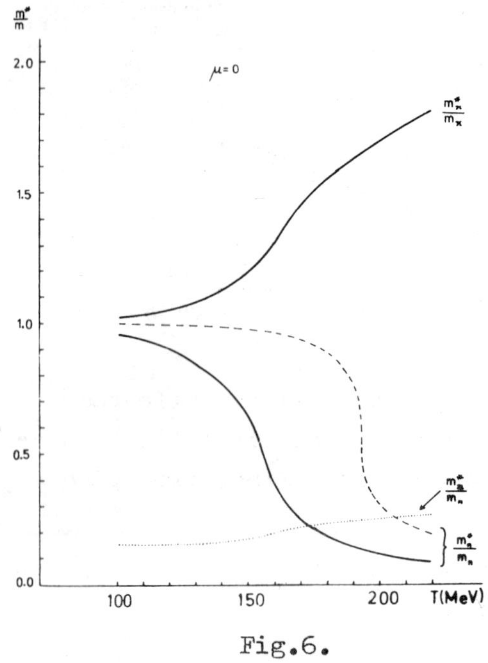

Fig.6.

Let us turn now to the zero chemical potential case. In Fig.6. the effective masses of nucleons and pions are plotted as a function of temperature. The full line corresponds to the interacting pion case, while the dashed line to the free pion case. The dotted line shows the pion effective mass in units of nucleon mass. Here we meet the unusual situation, that at T=175 MeV the pion effective mass is equal to that of the nucleons.

In Fig.7. the pion density is plotted versus the temperature. To show the effect of pion self-interaction, the curves with (λ=1) and without (λ=0) self-interaction are displayed. The dotted line shows the total particle density: nucleon density + antinucleon density + pion density (with self-interaction) . Here one finds the interesting situation, that at T=170 MeV we have one particle/fm^3 . Assuming an R=0.5 fm average radius for the hadrons, we would have at 170 MeV such a dense system,

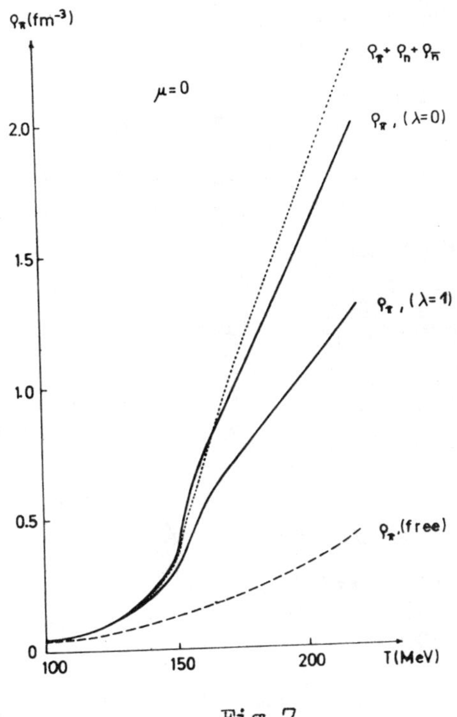

Fig.7.

that the particles would touch each other. That situation may indicate the beginning of the Mott-transition to the quark gluon plasma phase.

REFERENCES

1) J.Zimányi, J.Bondorf, I.Mishustin Nucl.Phys. A 435 /1985/ 810
2) J.D.Walecka, Ann.of Phys. 83 /1974/ 491
3) S.A.Chin, J.D.Walecka, Phys.Lett 52B /1974/ 2
4) B.Bannerjee, N.K.Glendenning, M.Gyulassy, Nucl.Phys. A361 /1981/ 326
5) A.B.Migdal, ZhETF/USSR/, 63 /1972/ 1993
6) V.M.Galitski, I.N.Mishustin, Sov. J. Nucl. Phys. 29 /1979/ 181
I.Montvay, J.Zimányi, Nucl.Phys. A316 /1979/ 490
T.S.Bíró, B.Lukács, J.Zimányi, H.W.Barz, Nucl.Phys. A386 /1982/ 617

7) J.Cugnon, T.Mizutani, J.Vandermeulen, Nucl.Phys. A379 /1982/ 505
J.Cugnon, J.Kinet, J.Vandermeulen, Nucl.Phys. A379 /1982/ 553

8) R.Stock, R.Bock, R.Brockman, J.W.Harris, A.Sandoval, H.Stroebele, K.Wolf, H.G.Pugh, L.S.Schroeder, H.Maier, R.E.Renford, A.Dacol, M.E.Orite, Phys.Rev.Lett. 49 /1982/ 1236

9) B.Friedman, V.R.Pandharipande, Q.N.Usmani, Nucl.Phys. A372 /1981/ 483

10) H.Schulz, D.N. Voskresensky, Phys. Lett /in press/

11) A.B.Migdal, O.A.Markin, I.N.Mishustin, ZhETF/USSSR/, 66 /1974/ 443

12) G.G.Bunatyan, Sov.J.of Nucl.Phys. 31 /1980/ 613

13) L.Celenza, H.J.Pirner, Nucl.Phys. A294 /1978/ 357

14) S.Barshay, G.E. Brown, M.Rho, Phys.Rev.Lett. 32 /1974/ 787

15) S.A.Garpman, N.K.Glendenning, Y.Karant, Nucl.Phys. A322 /1979/ 382

16) J.Theis, G.Graebner, G.Buchwald, J.Maruhn, W.Greiner, H.Stöcker, J.Polonyi, Preprint, GSI-83-7 /1983/

17) I.N.Mishustin, F.Myhrer, P.Siemens, Phys. Lett. 95B /1980/ 361

REACTIONS WITH LITHIUM PROJECTS

M.A. Nagarajan
SERC, Daresbury Laboratory, Daresbury, Warrington WA4 4AD
ENGLAND

ABSTRACT

The scattering and reactions with ^{6}Li and ^{7}Li projectiles are analyzed. It is shown that the reaction cross sections provide information on the structure of the Lithium projectiles and are consistent with a cluster-like description of these nuclei.

1. INTRODUCTION

The elastic scattering of heavy ions are often analyzed in terms of optical potentials whose real parts are generated by a convolution of the densities of the colliding nuclei with an effective nucleon-nucleon interaction. A commonly used double-folded potential[1]), which has been very successful in interpreting the elastic scattering of ^{12}C and ^{16}O ions by various targets at projectile energies ranging from a few MeV to about 20 MeV per nucleon, employs the M3Y interaction[2]) which is a sum of three Yukawa potentials for the effective reaction. The first instance where it was noted that the double folded potential had to be renormalized strongly in order to interpret the elastic scattering occurred in the analyses of the scattering of ^{6}Li ions[3]) by various target nuclei. Since then other projectiles, such as ^{7}Li[4]) and ^{9}Be[5]), have exhibited similar behaviour.

The nuclei ^{6}Li, ^{7}Li and ^{9}Be are known to have very low break-up thresholds for breaking up into (α+d), (α+t) and ($\alpha+\alpha$+n) components respectively. It thus seems reasonable to expect that the ease of breakup of these nuclei in an external field should have a strong influence on their elastic scattering. In this lecture, we show explicitly the effect of breakup on the elastic and inelastic scattering of Lithium projectiles.

2. BREAK-UP EFFECTS ON THE ELASTIC AND INELASTIC SCATTERING OF LITHIUM IONS

The structure of ^{6}Li and ^{7}Li nuclei have been studied by several authors within the framework of the cluster model. The properties of the low-lying states of these nuclei are extremely well described by the resonating group method (RGM)[6] as well as by the generator coordinate method (GCM)[7]. These models predict the binding energy, the charge and magnetic form factors as well as the break-up thresholds of these nuclei. ^{6}Li has only one discrete bound state, namely its ground state, and ^{7}Li has two discrete bound states below the break-up threshold. In order to study the effect of the break-up states on the elastic scattering, a reasonable method has to be considered for the treatment of three body channels. One such method, originally employed in the study of the elastic scattering of deuterons[8], is to discretize the continuum states into narrow momentum bins and perform a large coupled channel calculation. Such a procedure has been used by Kamimura and collaborators[9] for the analyses of the scattering of ^{6}Li and ^{7}Li ions. An alternate method, applicable when the energy of the projectile is large compared to its binding energy, is the adiabatic model originally suggested by Johnson and Soper[10] in their study of (d,p) reactions and since adopted to the elastic scattering of deuterons by Amakawa and Yazaki[11]. In the adiabatic model, the scattering amplitude for the elastic scattering of the projectile is calculated for a given separation of the cluster constituents of the projectile, and the total scattering amplitude is obtained by averaging the above amplitudes over the ground state density of the projectile. (This approximation is similar to the one that has been used to describe the elastic scattering of deformed nuclei[12]).

2
1
10^{-1}
10^{-2}
10^{-3}
σ/σ_R
0
20
40
60
θ_{cm} (deg)

Fig. 1(a)

1
10^{-1}
10^{-2}
10^{-3}
σ/σ_R
0
10
20
30
40
50
θ_{cm} (deg)

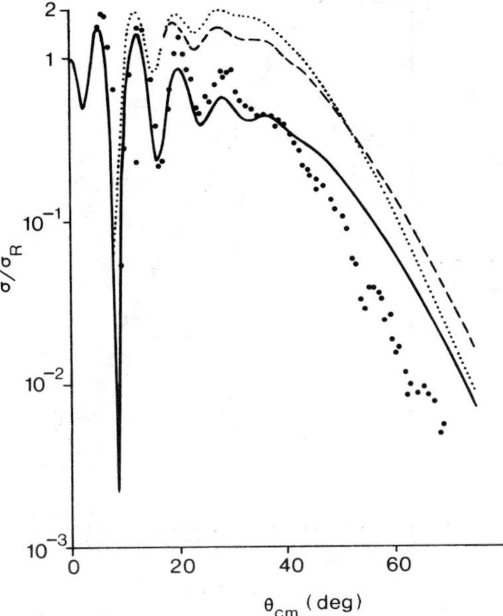

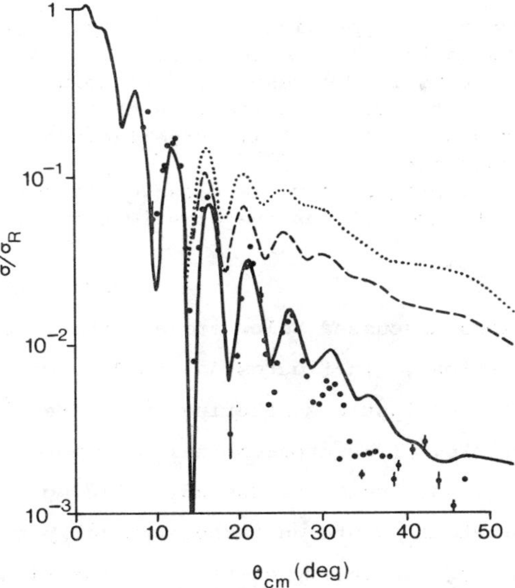

Fig. 1(b)

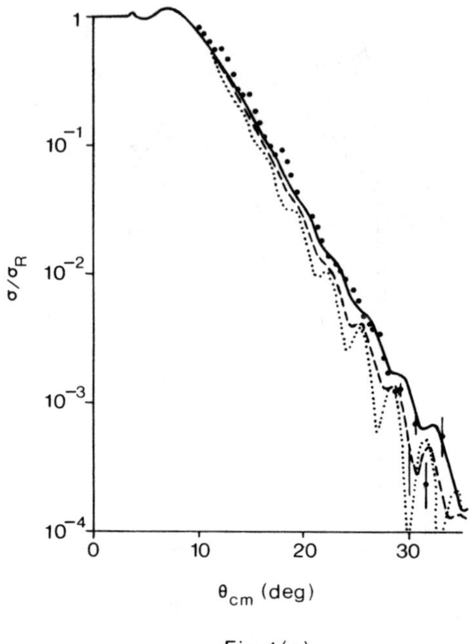

Fig. 1(c)

Fig. 1a) The adiabatic model predictions are compared with the experimental data in the elastic scattering of 156 MeV ^{6}Li by ^{12}C. The dotted curve is the result of including only the ground state of ^{6}Li, the dashed curve is with the inclusion of S-wave break-up channel and the full curve includes the S and D-wave break-up channels.

1b) Same as in figure 1a) for the elastic scattering of ^{6}Li by ^{40}Ca.

1c) Same as in figure 1a) for the elastic scattering of ^{6}Li by ^{208}Pb.

The calculations discussed below are all the results of the adiabatic model calculations. In figures 1a) to 1c), we show the results of the analyses of the elastic scattering of 156 Mev ^{6}Li ions by ^{12}C, ^{40}Ca and ^{208}Pb targets. (The experimental data were kindly provided by Professor H. Rebel). The input for the adiabatic model calculation comprise of alpha-target and deuteron-target optical potentials at approximately 2/3 and 1/3 of the energy of ^{6}Li. In the figures, the dotted curve represents the results of the calculation keeping only the ground

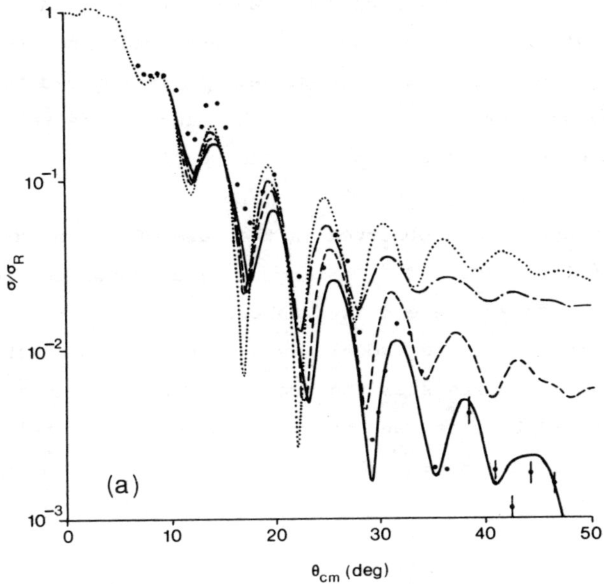

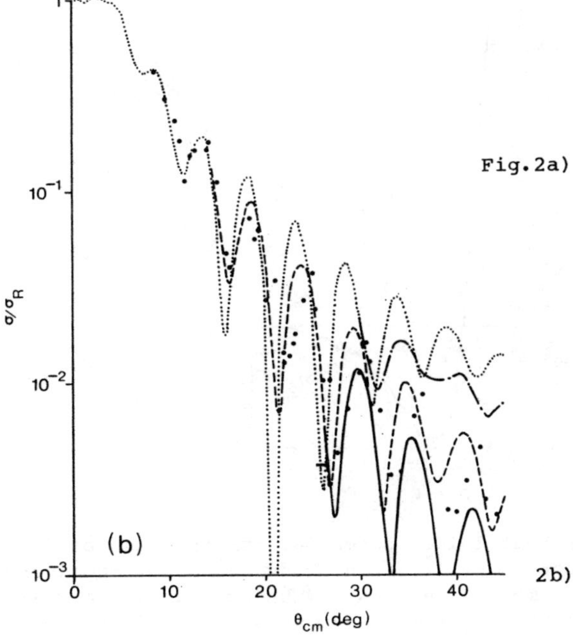

Fig.2a) Comparison of the adiabatic model prediction and the experimental data for the elastic scattering of 89 MeV ^{7}Li by ^{40}Ca. The dotted curve considers only the ground state of ^{7}Li, the dot-dashed curve takes the re-orientation effects into account, the dashed curve includes the p-wave break-up channel and the full curve considers the f-wave break-up channels in addition to the others.

2b) Same as in figure 2a) for the elastic scattering of 89 MeV ^{7}Li by ^{48}Ca.

state of ^{6}Li. Including break-up components with alpha and deuteron in relative L=0 state yields the dashed curve and including the L=2 break-up components yields the full lined curve. It is noticed that the break-up states are strongly coupled to the elastic channel and cause an appreciable change in the elastic cross section[13]).

The same features are observed in the case of the elastic scattering of 89 MeV ^{7}Li by ^{40}Ca and ^{48}Ca targets[14]) as shown in figures 2a) and 2b). In this case, there exists an excited state of ^{7}Li ($1/2^{-}$) at 470 keV, which in the cluster model, is a spin orbit doublet of the ground state. The figures show the successive changes in the elastic scattering as the $1/2^{-}$ state and the break-up states in relative L=1 and L=3 states are introduced. It can be shown that among the break-up

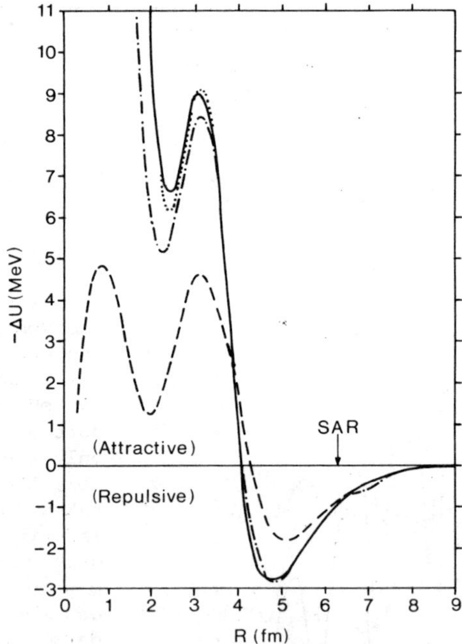

Fig.3 The correction to the real folded potential for the case of 156 MeV ^{6}Li elastic scattering on ^{12}C. The convergence of the iteration of the inversion method is shown. The dashed line corresponds to the first iteration, the dash-dotted one to iteration 2, the short dashed to 3 and the full line to iteration 6.

states, the most important is the resonant $7/2^-$ state of ^7Li which could very well be approximated as a (discrete) excited state in a standard coupled channel calculation[15].

The polarization potential caused by the strong coupling to the continuum states can be calculated. This was done by Mackintosh and Kobos[15] for the elastic scatterong of 156 MeV ^6Li on ^{12}C. The correction to the zeroth order potential is shown in figure 3. The important feature to note is the repulsive nature of the real potential at the strong absorption radius (SAR).

The inelastic scattering of 70 MeV ^7Li by ^{208}Pb to the $1/2^-$ state was also analyzed successfully with the adiabatic model[17]. There were indications that cross sections at small angles ($\theta \lesssim 15°$) are larger than the predictions of Coulomb (E2) excitation. It is known that the M1 Coulomb excitation in ^7Li is a non-negligible effect and one also expects a strong second order effect from E1 excitation. More data at small angles are desirable and a more complete calculation including the E1 and M1 excitations are needed.

Finally, the measured break-up cross section of 70 MeV ^7Li on ^{120}Sn target was analyzed indirectly[18]. In figure 4a) the expected α-energy spectrum for the sequential break-up of ^7Li after its excitation to the $7/2^-$ state at 4.63 MeV is shown. In figures 4b) to 4d), the experimental spectrum is shown as the combined (α-t) angle is changed from 22° to 11.5°. It can be noticed that the 22° data resembles the predicted spectrum whereas at the smaller angles there is evidence of an additional process filling in the alpha energy spectrum between the peaks. To see whether this is a consequence of the direct break-up, we used the known photodisintegration data, i.e.[19]):

$$\sigma_{\gamma_1 {}^7Li}(E_\gamma) = \frac{16\pi^2}{9} \frac{h\omega}{hc} B(\varepsilon_\gamma)$$

where $B(\varepsilon_\gamma)$ is an effective E1 matrix element. The direct Coulomb break-up cross section due to E1 excitation is given by

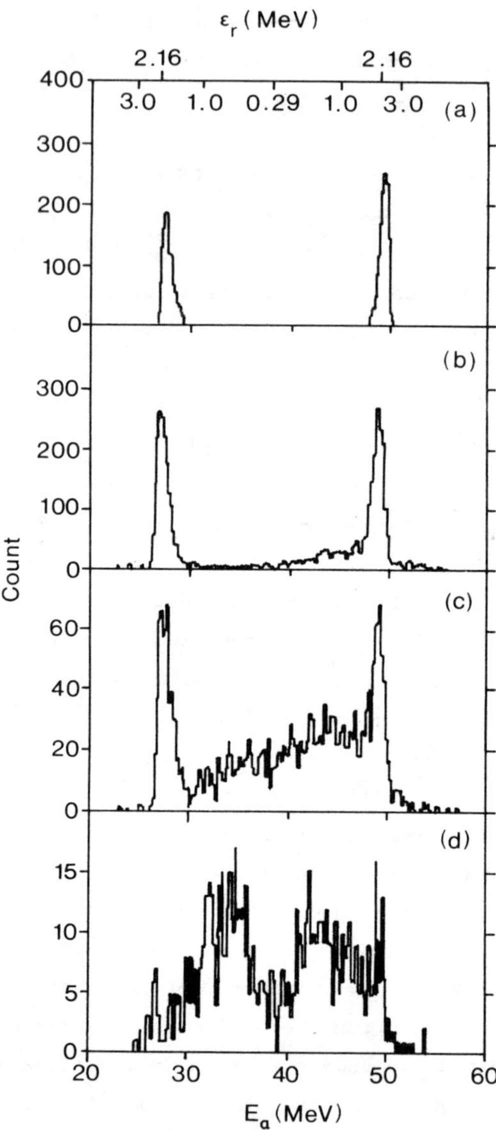

Fig.4a) The Monte-Carlo simulation of the α-energy spectrum for sequential break-up of 70 MeV ^{7}Li by ^{120}Sn at $\theta_{lab} = 22.0°$. The alpha and triton counters were placed at the same angle (close geometry). The α-t relative energy ε_γ, is shown at the top of the figure.

4b)-4d) The experimental α-energy spectra at $\theta_{lab} = 22°$, 15° and 11.5°.

$$d\sigma_{E_1} = \left(\frac{Ze^2}{hv}\right)^2 B(\varepsilon_\gamma)\ df_{E1}(\theta,\varepsilon_\gamma| d\varepsilon_\gamma$$

where $df_{E_1}(\theta,\varepsilon_\gamma)$ is the classical Coulomb excitation cross section, Z is the charge of the target nucleus and V is the relative velocity of ^{7}Li and the target. Using the known photodisintegration cross section, the simulated break-up spectrum is shown in figure 5a) along with the experimental spectrum shown in figure 5b). The agreement between the two is surprisingly good suggesting that a large proportion of direct break-up is concentrated at small angles and is primarily due to Coulomb dipole excitation. This suggests a very useful method for separation of direct and sequential break-up mechanisms.

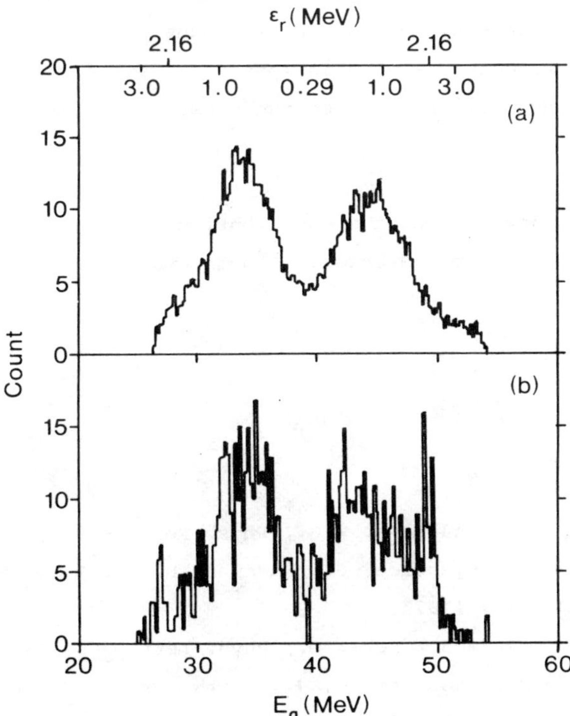

Fig.5a) A Monte-Carlo simulation for Coulomb break-up of 70 MeV ^{7}Li by ^{120}Sn for the experimental set up at 11.5°.
5b) The experimental α-energy spectrum at 11.5°.

ACKNOWLEDGEMENT AND CONCLUSIONS

The elastic and inelastic scattering of Lithium ions as well as their break-up in the field of target nuclei provides a convincing evidence of their cluster structure. The earlier discussion of Professor Johnson on the analyzing powers from the scattering of polarized Lithium ions provides further support. It is of great interest to study the energy dependence of the break-up effects which were found to be so important at high energies. Preliminary analyses seem to indicate that the very low energies (E ≃ 12-15 MeV) in the elastic scattering of Lithium by ^{58}Ni, the rquired real potential is more attractive than the double folded potential[20]). This is an indication that other mechanisms are begining to play an important role at these energies and that break-up effects diminish in imortance. It is of interest to study the effect of nucleon transfer on the elastic scattering at these energies.

I wish to acknowledge the contributions of Ron Johnson, Ian Thompson, Alan Shotter and Joaquin Gomez-Comacho to the work discussed in this talk.

REFERENCES

1) Satchler, G.R. and Love, W.G., Phys. Reports 55 , 183 (1979)

2) Bertsch, G., Borysowicz, J., McManus, H. and Love, W.G., Nucl. Phys. A284, 399 (1977).

3) Satchler, G.R. and Love, W.G., Phys. Lett. 76B, 23 (1978).

4) Steeden, M.F., Coopersmith, J., Cartwright, S.J., Cohler, M.D., Clarke, N.M. and Griffiths, R.J., J. Phys. G6, 501 (1980).

5) Satchler, G.R., Phys. Lett. 83BN, 284 (1979).

6) Kanada, H., Liu, Q.K.K. and Tang, Y.C., Phys. Rev. C22, 813 (1980).

Brockmann R.
Lawrence Berkeley Laboratory
Building 50D
1, Cyclotron Road
Berkeley
Calif 94720
USA

Chakravarti D.R.
Nuclear Physics Division
Bhabha Atomic Research Centre
Bombay 400 085

Chakravarti S.K.
Department of Applied Physics, REC
Kurukshetra 132 119

Chandra H.
Department of Physics
Banaras Hindu University
Banaras

Cheema T.S.
Punjab University
Chandigarh 160 014

Chidambaram R.
Physics Group
Bhabha Atomic Research Centre
Bombay 400 085

Chintalapudi S.N.
Variable Energy Cyclotron Centre
Bhabha Atomic Research Centre
Calcutta 700 064

Chitambar S.A.
Radiochemistry Division
Bhabha Atomic Research Centre
Bombay 400 085

Chaudhury R.K.
Nuclear Physics Division
Bhabha Atomic Research Centre
Bombay 400 085

Datar V.M.
Nuclear Physics Division
Bhabha Atomic Research Centre
Bombay 400 085

Datta B.
Indian Institute of Astrophysics
Bangalore 560 034

Dasgupta S.
Physics Department
Mcgill University 3600
University St., Montreal
PQ H3A2T8,
Canada

Datta Tarun
Radiochemistry Division
Bhabha Atomic Research Centre
Calcutta 700 064

Dey J.
Laboratoire De
Physique Nucleaire
Universite De Monitreal
CP 6128
SUCC 'A', Montreal, PQ
Quebec, Canada

De J.N.
Nuclear Physics Division
Variable Energy Cyclotron Centre
Bhabha Atomic Research Centre
1/AF, Bidhan Nagar
Calcutta 700 064

Devare H.G.
Tata Institute of Fundamental
Research
Bombay 400 076

Devare S.H.
Tata Institute of Fundamental
Research
Bombay 400 005

Divatia A.S.
Variable Energy Cyclotron Centre
Bhabha Atomic Research Centre
Calcutta 700 064

Eswaran M.A.
Nuclear Physics Division
Bhabha Atomic Research Centre
Bombay 400 085

Gambhir Y.K.
Department of Physics
Indian Institute of Technology
Bombay 400 076

Gautam R.P.
Physics Department
Aligarh Muslim University
Aligarh 202 001

Garg U.
Physics Department
University of Notredam
Notre Dame
In 46556, USA

Ghosh B.
TPPED
Bhabha Atomic Research Centre
Bombay 400 085

Ghosh B.
Department of Physics
Indian Institute of Technology
Kanpur 208 016

Ghosh S.K.
Saha Institute of Nuclear Physics
92, Acharya Prafulla Chandra Road
Calcutta 700 009

Goder S.S.
Indian Institute of Technology
Kanpur 208 016

Goswami K.
Physics Department
Gauhati University
Assam 781 014

Grodzins L.
Lab. for Nuclear Science
MIT, Cambridge
MA, USA

Gupta D.K.
Physics Department
S.V. College
Aligarh 202 001

Gupta J.B.
Ramjas College
42/1, Chhatra Marg
Delhi 110 007

Gupta R.K.
Department of Physics
Punjab University
Chandigarh 160 014

Gupta S.K.
Nuclear Physics Division
Bhabha Atomic Research Centre
Bombay 400 085

Hüfner J.
Institute of Physics
Philosophenweg 19
Heidelberg, FRG

Ismail Md.
Variable Energy Cyclotron Centre
Bhabha Atomic Research Centre
1/AF, Bidhan Nagar
Calcutta 700 064

Iyengar K.N.
Nuclear Physics Division
Bhabha Atomic Research Centre
Bombay 400 085

Iyengar P.K.
Bhabha Atomic Research Centre
Bombay 400 085

Iyer R.H.
Radiological Group
Bhabha Atomic Research Centre
BOmbay 400 085

Jacquemin M.
Institut De Physique Nucleaire
Division De Physique Theorique
University De Paris
Centre D 'Orsay
91406, Orsay
France

Jain A.K.
Physics Department
University of Roorkee
Roorkee 247 667

Jain Arun
Nuclear Physics Division
Bhabha Atomic Research Centre
Bombay 400 085

Jain Animesh
Nuclear Physics Division
Bhabha Atomic Reserch Centre
Bombay 400 085

Jain B.K.
Nuclear Physics Division
Bhabha Atomic Research Centre
Bombay 400 085

Jain H.C.
Tata Institute of Fundamental Research
Homi Bhabha Road
Bombay 400 005

Jhingan M.L.
Tata Institute of Fundamental Research
Homi Bhabha Road
Bombay 400 005

Johnson R.C.
Department of Physics
University of Surrey
Guildford, UK

Jyrwa B.B.
Department of Physics
North Eastern Hill University
Shillong 793 003
Assam

Kailas S.
Nuclear Physics Division
Bhabha Atomic Research Centre
Bombay 400 085

Kalekar Aruna
Nuclear Physics Division
Bhabha Atomic Research Centre
Bombay 400 085

Kane P.P.
Department of Physics
Indian Institute of Technology
Powai
Bombay 400 076

Kapoor S.S.
Nuclear Physics Division
Bhabha Atomic Research Centre
Bombay 400 085

Kar K.
Saha Institute of Nuclear Physics
92, Acharya Prafulla Chandra Road
Calcutta 700 009

Kaushal R.S.
Department of Physics
University of Delhi
Delhi 110 007

Kerekatte S.S.
Nuclear Physics Division
Bhabha Atomic Research Centre
Bombay 400 085

Khadkidar S.B.
Department of Physics
Physics Research Laboratory
Ahmedabad 380 009

Krishan K.
Variable Energy Cyclotron Centre
Bhabha Atomic Research Centre
1/AF, Bidhan Nagar
Calcutta 700 064

Kühn B.
Department of Physics
University of Rossendorf
German Democratic Republic

Kulkarni R.J.
Department of Physics
University of Bombay
Vidya Nagari
Bombay 400 098

Kumar A.
2nd Floor, Staff Flats
12, Nalapani
Dehradun 248 001

Kumar Kiran
Radiological Group
Bhabha Atomic Research Centre
Bombay 400 085

Kumar Shyam
Department of Physics
University of Kurukshetra
Kurukshetra 132 119

Mathur J.
Department of Physics
University of Wollongo, NG
Australia

Mishra R.C.
Department of Physics
VSSD College
Kanpur 208 002

Mukherjee G.
Physics Department
Indian Institute of Technology
Kanpur 208 016

Mukhopadhyaya S.
Physics Department
Viswa-Bharati University
Shantiniketan 731 235

Murthy S.R.S.
Nuclear Physics Division
Bhabha Atomic Research Centre
Bombay 400 085

Moretto L.G.
University of California
Lawrence Berkeley Lab.
Berkeley
California 94720
USA

Mythili R.
Nuclear Physics Division
Bhabha Atomic Research Centre
Bombay 400 085

Nadkarni D.M.
Nuclear Physics Division
Bhabha Atomic Research Centre
Bombay 400 085

Nag R.
Department of Physics
Banaras Hindu University
Varanasi 221 005

Nagamiya S.
Department of Physics
Faculty of Science
University of Tokyo, Hongo
Mujkyo-KU, Tokyo
Japan

Nagarajan T.
University of Madras
AC College Buildings
Madras 600 025

Nagarajan M.A.
Daresbury Lab.
Warrington - WA4 4AD
UK

Namboodiri M.N.
Lawrence Livermore National Lab.
L 234, LLNL Livermore
CA 94550
USA

Narayana D.G.S.
Visakhapatnam 530 003

Narasimhan V.L.
Indian Institute of Technology
Bombay 400 076

Natarajan R.K.
Department of Nuclear Physics
University of Madras
Guindy Campus
Madras 600 025

Natowitz J.B.
Texas A & M University
College Station
Texas 77843
USA

Nayak R.C.
Department of Physics
Khalikote College
Behrampur (G.M.)
Orissa 760 001

Nair S.C.K.
Department of Physics
University of Calicut
Calicut

Nikam R.S.
Department of Physics
Indian Institute of Technology
Bombay 400 076

Nix J.R.
Lawrence Livermore National Lab.
Los Alamos, T-9, MS B279
NM 87545
USA

Padmini M.D.
Department of Physics
Quaid & Millat Government
College for Women
Madras 600 002

Pai V.N.
Department of Physics Univ. Bombay
Kalina
Bombay 400 098

Pal D.
Saha Institute of Nuclear Physics
92, Acharya Prafulla Chandra Road
Calcutta 700 009

Pal M.K.
Saha Institute of Nuclear Physics
92, Acharya Prafulla Chandra Road
Calcutta 700 009

Panda K.C.
Sambalpur University
Jyotivihar
Sambalpur 768 107

Pandharipande V.R.
Department of Physics
University of Illinois
1110 W. Green St.
Urbana, Illinois 61801
USA

Pandit V.S.
VECC, BARC
1/AF, Bidhan Nagar
Calcutta 700 064

Pandya S.P.
PHysics Research Laboratory
Ahmedabad 380 009

Pardhasarathi S.K.
VECC, BARC
1/AF, Bidhan Nagar
Calcutta 700 064

Parija B.
Institute of Physics
Sachivalaya Marg
Bhubaneswar 751 005

Patro A.P.
Nuclear Science Centre
Delhi

Phadke S.D.
University of Bombay
Kalina
Bombay 400 098

Phatak S.C.
Institute of Physics
Scahivalaya Marg
Bhubaneswar 751 005

Plasil F.
Oak Ridge National Laboratory
PO Box X, Oak Ridge
Tennessee 37831
USA

Praharaj C.R.
Institute of Physics
Sachivalaya Marg
Bhubaneswar 751 005

Prasad K.G.
Tata Institute of Fundamental Research
Homi Bhabha Road
Bombay 400 005

Puttaswamy N.G.
Department of Physics
Central College
Bangalore University
Bangalore 560 001

Ragoowansi N.L.
Nuclear Physics Division
Bhabha Atomic Research Centre
Bombay 400 085

Raj V.S.
Physics Department
Osmania University
Hyderabad 500 007

Raja Rao M.
Department of Physics
Central College
Bangalore University
Bangalore 560 001

Rajasekaran T.R.
Department of Nuclear Physics
University of Madras
Madras 600 025

Rajeswari V.
Institute of Mathematical Science
Madras 600 113

Ramamurthy S.
Lab. for Nuclear Research
Andhra University
Waltair - 500 003

Ramamurthy V.S.
Nuclear Physics Division
Bhabha Atomic Research Centre
Bombay 400 085

Rama Rao P.N.
Nuclear Physics Division
Bhabha Atomic Research Centre
Bombay 400 085

Ramaswamy C.R.
Department of Physics
Central College
Bangalore University
Bangalore 500 001

Ramanna R.
Anushaktibhavan
Department of Atomic Energy
C.S.M. Marg
Bombay 400 039

Rangarajan G.
107, Ashok Bhavan
B.I.T.S. Pilani 333 001
Rajasthan

Rao A.V.M.
Department of Physics
Nuclear Physics Section
Banaras Hindu University
Varanasi 221 005

Rao K.S.

Institute of Mathematical Sciences
Taramani S.O.
Madras 600 110

Rao M.N.
Institute of Physics
Bhubaneswar 751 005

Rao Y.S.T.
Department of Physics
North Eastern Hill University
Shillong 793 003

Rastogi B.P.
Theoretical Physics Division
Bhabha Atomic Research Centre
Bombay 400 085

Ratna Raju R.D.
Department of Physics
Andhra University
Visakhapatnam 503 003

Rath A.K.
Institute of Physics
Bhubaneshwar 751 005

Ravishankar V.
Department of Physics
Mysore University
Manasagangotri
Mysore 570 006

Ray S.
Physics Department
Banaras Hindu University
Varanasi 221 005

Ray R.S.
Department of Physics
Banaras Hindu University
Varanasi 221 005

Reddy A.V.R.
Radiochemistry Division
Bhabha Atomic Research Centre
Bombay 400 085

Reddy S.B.
Lab. For Nuclear Research
Waltari 530 003

Rodney W.
National Science Foundation
1800 G. Street, NW
Room No. 341
Washington DC 20550, USA

Roy A.
Tata Institute of Fundamental Research
Homi Bhabha Road
Bombay 400 005

Sagina R.
Physics Department
Banaras Hindu University
Varanasi
Varanasi 221 005

Saha A.
Department of Physics
Calcutta University
92, Acharya Prafulla Chandra Road
Calcutta 700 009

Sahu R.
Physics Department
N.C. College
Jajpur, Cuttack
Orissa

Samaddar S.
Saha Institute of Nuclear Physics
92, Acharya Prafulla Chandra Road
Calcutta 700 009

Samanta B.C.
Department of Physics
Burdwan University
Burdwan 713 104 (W.B.)

Samanta C.
Saha Institute of Nuclear Physics
1/AF, Bidhan Nagar
Calcutta 700 064

Santra A.B.
Nuclear Physics Division
Bhabha Atomic Research Centre
Bombay 400 085

Sanyal S.
Department of Physics
Banaras Hindu University
Varanasi 221 005

Sarangi P.
Institute of Physics
Sachivalaya Marg
Bhubaneswar 751 006

Sarkar S.
Department of Theoretical Physics
Indian Association For The Cultivation of Science
Jadavpur
Calcutta 700 032

Sarma N.
Nuclear Physics Division
Bhabha Atomic Research Centre
Bombay 400 085

Saroha P.R.
Punjab University
Chandigarh 160 014

Satpathy L.
Institute of Physics
Sachivalaya Marg
Bhubaneswar 751 005

Saxena Alok
Nuclear Physics Division
Bhabha Atomic Research Centre
Bombay 400 085

Sanmugam G.
Department of Physics
Presidency College
Madras 600 005

Schuck P.
Inst. Science Nucleaire
Avenue Des Martyrs
Frenoble Cedex
France

Sen S.
Saha Institute of Nuclear Physics
92, Acharya Prafulla Chandra Road
Calcutta 700 009

Sehgal M.L.
Physics Department
Aligarh Muslim University
Aligarh 202 001

Sethi B.
Saha Institute of Nuclear Physics
92, Acharya Prafulla Chandra Road
Calcutta 700 009

Sett G.C.
Department of Physics
Visva Bharati
Santiniketan 701 235 (W.B.)

Sharma C.R.
Indian Institute of Technology
Bombay 400 076

Sharma M.N.
Physics Department
Indian Institute of Technology
Kanpur 208 016

Sharma N.R.
M.S.J. College
Bharatpur 321 001

Sharma O.D.
Department of Physics
Meerut College
Meerut

Sharma P.R.
Variable Energy Cyclotron Centre
Bhabha Atomic Research Centre
1/AF, Bidhan Nagar
Calcutta 700 064

Sharma R.P.
Variable Energy Cyclotron Centre
Bhabha Atomic Research Centre
1/AF, Bidhan Nagar
Calcutta 700 064

Sharma S.K.
Indian Institute of Technology
Kanpur 208 016

Sharma R.P.
Tata Institute of Fundamental Research
Homi Bhabha Road
Bombay 400 005

Sharma V.K.
Institute of Advanced Studies
Meerut University
Meerut 250 005

Shivananda A.H.
Department of Physics
University of Mysore
Manasagangothri
Mysore 570 006

Singh V.
Tata Institute of Fundamental Research
Homi Bhabha Road
Bombay 400 005

Sheikh J.A.
Indian Institute of Technology
Bombay 400 076

Singh C.P.
Department of Physics
Banaras Hindu University
Varanasi 221 005

Singh N.L.
Banaras Hindu University
Department of Physics
Varanasi 221 005

Singh Nirmal
Physics Department
Punjab University
Chandigarh 160 014

Singh P.
Department of Physics
St. John's College
Agra 282 002

Singh Pitambar
Nuclear Physics Division
Bhabha Atomic Research Centre
Bombay 400 085

Singh T.M.
Department of Physics
Gauhati University
Gauhati 781 014
Assam

Sinha B.C.
Variable Energy Cyclotron Centre
Bhabha Atomic Research Centre
1/AF, Bidhan Nagar
Calcutta 700 064

Sinha R.P.
Physics Research Laboratory
Ahmedabad 380 009

Scott D.K.
National Superconducting
Cyclotron Laboratory
Michigan State University
Michigan 48824
USA

Sood P.C.
Banaras Hindu University
Varanasi 221 005

Sharma R.C.
Physics Department
Concordia University
Montreal, PQ H3G1M8
Canada

Shyam R.
GSI Darmstadt
Postfach 110 541
D-6100 Darmstadt 11
FRG

Singh P.P.
Indiana University Cyclotron Facility
Bloomington, Indiana
USA

Somayajulu D.R.S.
Department of Physics
Sukhadia University
Udaipur 313 001

Sridhar A.
Mangalore University
Mangalore

Sridhar V.N.
Physical Research Laboratory
Ahmedabad 380 009

Srikantiah R.V.
TPPED
Bhabha Atomic Researh Centre
Bombay 400 085

Srikantan B.V.
Tata Institute of Fundamental
Research
Colaba
Bombay 400 005

Srivastava B.V.
Physics Department
Meerut University
Meerut 250 005

Srivastava M.K
Physics Department
University of Roorkee
Roorkee 247 667

Srinivasan J.
Institute of Mathematical Science
Madras

Specht H.J.
Physikalisches
Institut der Univ.
Philosophenweg 12
D-6900 Heidelberg 1
FRG

Sperber D.
Renssler Polytechnic Institute
Department of Physics, Troy
New York 12181
USA

Stockstad R.G.
Lawrence Berkeley Laboratory
Berkeley
USA

Sud S.P.
Department of Physics
Himachal Pradesh University
Simla 171 005

Symons T.J.
Lawrence Berkeley Laboratory
Berkeley
USA

Talukdar B.
Visva Bharati
Department of Physics
Shantiniketan 731 235

Taras P.
Universaite De Montreal
Laboratoirede Physique
Nucleaire
Univesitede Montreal
Montreal Quebec M3CJ7
Canada

Tandon P.N.
Tata Institute of Fundamental
Research
Bombay 400 005

Thiagasundaram M.
Pachaiyappa's College
Madras 600 030

Thomas A.W.
Physics Department
University of Adelaide
PO 498, GPO Adelaide
SA 5001
Australia

Tripathi P.N.
VSSD College
Kanpur 208 002

Tripathi R.K.
Institute of Physics
Sachivalaya Marg
Bhubaneswar 751 005

Trivedi M.D.
Variable Energy Cyclotron Centre
Bhabha Atomic Research Centre
1/AF, Bidhan Nagar
Calcutta 700 064

Ullah N.
Tata Institute of Fundamental
Research
Homi Bhabha Road
Bombay 400 005

Upadhyaya
Government Girls College
Vikram University
Ujjain 456 006 (M.P.)

Usmani Q.N.
Aligarh Muslim University
Department of Physics
Aligarh 202 001

Varier K.M.
Calicuta
Department of Physics
University of Calicut
Kerala 67335

Varma J.
School of Basic Science &
Humanities
Sukhadia University
Udaipur 313 001

Varma R.
Department of Physics
Indian Institute of Technology
Kanpur 208 016

Varshney A.K.
Physics Section
Z.H. College of Engineering
and Technology
Aligarh Muslim University
Aligarh 202 001

Vinod Kumar P.C.
Physical Research Laboratory
Ahmedabad 380 009

Waghmare Y.R.
Department of Physics
Indian Institute of Technology
Kanpur 208 016

Warke C.S.
Tata Institute of Fundamental
Research
Colaba
Bombay 400 005

Yadav J.S.
Tata Institute of Fundamental
Research
Colaba
Bombay 400 005

Zimanyi J.
Central Research Institute
for Physics
Budapest
Hungary